Advances in Intelligent Systems and Computing

Volume 370

About this Series

The series "Advances in Intelligent Systems and Computing" contains publications on theory, applications, and design methods of Intelligent Systems and Intelligent Computing. Virtually all disciplines such as engineering, natural sciences, computer and information science, ICT, economics, business, e-commerce, environment, healthcare, life science are covered. The list of topics spans all the areas of modern intelligent systems and computing.

The publications within "Advances in Intelligent Systems and Computing" are primarily textbooks and proceedings of important conferences, symposia and congresses. They cover significant recent developments in the field, both of a foundational and applicable character. An important characteristic feature of the series is the short publication time and world-wide distribution. This permits a rapid and broad dissemination of research results.

More information about this series at http://www.springer.com/series/11156

Ajith Abraham · Xin Hua Jiang
Václav Snášel · Jeng-Shyang Pan
Editors

Intelligent Data Analysis and Applications

Proceedings of the Second Euro-China Conference on Intelligent Data Analysis and Applications, ECC 2015

Springer

Editors
Ajith Abraham
Machine Intelligence Research Labs
(MIR Labs)
Auburn, WA
USA

Václav Snášel
Department of Computer Science
VŠB—Technical University of Ostrava
Ostrava-Poruba
Czech Republic

Xin Hua Jiang
Fujian University of Technology
Fujian
China

Jeng-Shyang Pan
Fujian University of Technology
Fujian
China

ISSN 2194-5357 ISSN 2194-5365 (electronic)
Advances in Intelligent Systems and Computing
ISBN 978-3-319-21205-0 ISBN 978-3-319-21206-7 (eBook)
DOI 10.1007/978-3-319-21206-7

Library of Congress Control Number: 2015943052

Springer Cham Heidelberg New York Dordrecht London

Printed on acid-free paper

Springer International Publishing AG Switzerland is part of Springer Science+Business Media
(www.springer.com)

Preface

This volume of *Advances in Intelligent Systems and Computing* contains accepted papers presented in the main track of ECC 2015, the Second Euro-China Conference on Intelligent Data Analysis and Applications. The aim of ECC is to provide an internationally respected forum for scientific research in the broad area of intelligent data analysis, computational intelligence, signal processing, and all associated applications of AIs.

The second edition of ECC was organized jointly by VŠB—Technical University of Ostrava, Czech Republic, and Fujian University of Technology, Fuzhou, China. The conference, organized under the patronage of Mr. Miroslav Novák, President of the Moravian-Silesian Region, took place in late June and early July 2015 in the Campus of the VŠB—Technical University of Ostrava, Czech Republic.

The organization of the ECC 2015 conference was entirely voluntary. The review process required an enormous effort from the members of the International Technical Program Committee, and we would therefore like to thank all its members for their contribution to the success of this conference. We would like to express our sincere thanks to the host of ECC 2015, VŠB—Technical University of Ostrava, and to the publisher, Springer, for their hard work and support in organizing the conference. Finally, we would like to thank all the authors for their high-quality contributions. The friendly and welcoming attitude of conference supporters and contributors made this event a success!

April 2015

Ajith Abraham
Xin Hua Jiang
Václav Snášel
Jeng-Shyang Pan

Organization

Honorary Chairs

Ivo Vondrak, VŠB-TU Ostrava, Czech Republic
Xin Hua Jiang, Fujian University of Technology, China

Advisory Committee Chairs

Tzung-Pei Hong, National University of Kaohsiung, Taiwan
Bin-Yih Liao, Kaohsiung University of Applied Sciences, Taiwan

Conference Chairs

Jeng-Shyang Pan, Fujian University of Technology, China
Václav Snášel, VŠB-TU Ostrava, Czech Republic

Program Committee Chairs

Ajith Abraham, Machine Intelligence Research Labs, USA
Jan Platos, VŠB-TU Ostrava, Czech Republic

Invited Session Chair

Tarek Gaber, VŠB-TU Ostrava, Czech Republic

Electronic Media Chair

Mohamed Mostafa Fouad, VŠB-TU Ostrava, Czech Republic

Conference Organizers

Tarek Gaber, VŠB-TU Ostrava, Czech Republic
Mohamed Mostafa Fouad, VŠB-TU Ostrava, Czech Republic
Jana Nowakova, VŠB-TU Ostrava, Czech Republic
Svatopluk Stolfa, VŠB-TU Ostrava, Czech Republic
Jakub Stolfa, VŠB-TU Ostrava, Czech Republic
Pavel Kromer, VŠB-TU Ostrava, Czech Republic

International Program Committee

Siby Abraham, University of Mumbai, India
Giovanni Acampora, University of Salerno, Italy
Sabrina Ahmad, Universiti Teknikal Malaysia Melaka, Malaysia
Jesus Alcala-Fdez, University of Granada, Spain
Alberto Alvarez, European Centre for Soft Computing, Spain
Michela Antonelli, University of Pisa, Italy
Roberto Armenise, Poste Italiane, Italy
Akira Asano, Kansai University, Japan
Ramiro Barbosa, Instituto Superior de Engenharia do Porto, Portugal
Carlos Barranco, Pablo de Olavide University, Spain
Edurne Barrenechea, Universidad Publica de Navarra, Spain
Anna Bartkowiak, University of Wrocław, Poland
Abd. Samad Hasan Basari, Universiti Teknikal Malaysia Melaka, Malaysia
David Becerra-Alonso, ETEA-INSA, Spain
Punam Bedi, University of Delhi, India
Rafael Bello, Universidad Central de Las Villas, Cuba
Robert Berwick, Massachusetts Institute of Technology, USA
Michael Blumenstein, Griffith University, Australia
Fernando Bobillo, University of Zaragoza, Spain
Gloria Bordogna, CNR IDPA, Italy
Silvio Bortoleto, Federal University of Rio de Janeiro, Brazil

Juan Botia, Universidad de Murcia, Spain
Abdel Hamid, Bouchachia University of Klagenfurt, Austria
Alberto Bugarin, University of Santiago de Compostela, Spain
Alberto Cano, University of Cordoba, Spain
Carlos Cano, University of Granada, Spain
Cristobal J. Carmona, University of Jaen, Spain
Paulo Carrasco, Universidade do Algarve, Portugal
Andre Carvalho, University of Sao Paulo, Brazil
Giovanna Castellano, Universita di Bari, Italy
Gladys Castillo, University of Aveiro, Portugal
Oscar Castillo, Tijuana Institute of Technology, Mexico
Swati V. Chande, International School of Informatics and Management, India
Chin-Chen Chang, Feng Chia University, Taiwan
Lee Chang-Yong, Kongju National University, Korea
Chao-Chun Chen, Southern Taiwan University, Taiwan
Ying-Ping Chen, National Chiao Tung University, Taiwan
Sung-Bae Cho, Yonsei University, Korea
Yun-Huoy Choo, Universiti Teknikal Malaysia Melaka, Malaysia
Jyh-Horng Chou, National Kaohsiung First University of Science and Technology, Taiwan
Mario Giovanni Cimino, University of Pisa, Italy
Martine De Cock, Ghent University, Belgium
Marco Cococcioni, University of Pisa, Italy
Leandro Coelho, Pontificia Universidade Catolica do Parana, Brazil
Valentina Colla, Scuola Superiore Sant'Anna, Italy
Chris Cornelis, Ghent University, Belgium
Jose Luis Perez de la Cruz, University of Malaga, Spain
Radu-Codrut David, Politehnica University of Timisoara, Romania
Fernando Delaprieta, University of Salamanca, Spain
Jitender S. Deogun, University of Nebraska, USA
Vivek Deshpande, MIT College of Engineering, India
Anusuriya Devaraju, Forschungszentrum Julich GmbH, Germany
Norberto Diaz-Diaz, Pablo de Olavide University, Spain
Federico Divina, Pablo de Olavide University, Spain
Victor Hugo Menendez Dominguez, Universidad Autonoma de Yucatan, Mexico
Abraham Duarte, Universidad Rey Juan Carlos, Spain
Pietro Ducange, University of Pisa, Italy
Jiri Dvorsky, VŠB-TU Ostrava, Czech Republic
Kubilay Ecerkale, Turkish Air Force Academy, Turkey
Luka Eciolaza, European Centre for Soft Computing, Spain
Wilfried Elmenreich, University of Klagenfurt, Austria
Nurulakmar Emran, Universiti Teknikal Malaysia Melaka, Malaysia
Macarena Espinilla Estevez, Universidad de Jaen, Spain
Anna Fanelli, Universita di Bari, Italy
Carlos Fernandes, GeNeura Team, Spain

Alberto Fernandez, Universidad de Jaen, Spain
Joaquin Lopez Fernandez, University of Vigo, Spain
Francesc J. Ferri, Universitat de Valencia, Spain
Vladimir Filipovic, University of Belgrade, Serbia
Giacomo Fiumara, University of Messina, Italy
Alexandru Floares, Romania
Juan J. Flores, Universidad Michoacana de San Nicolas de Hidalgo, Mexico
Amparo Fuster-Sabater, Institute of Applied Physics (C.S.I.C.), Spain
Tarke Gaber, VŠB-TU Ostrava, Czech Republic
Mikel Galar, Universidad Publica de Navarra, Spain
Ivan Garcia-Magarino, Universidad a Distancia de Madrid, Spain
Carlos Garcia-Martinez, University of Cordoba, Spain
Eva Gibaja, University of Cordoba, Spain
Elizabeth Goldbarg, Federal University of Rio Grande do Norte, Brazil
Alma Gomez, University of Vigo, Spain
Amelia Zafra Gomez, University of Cordoba, Spain
Juan Gomez-Romero, Universidad Carlos III de Madrid, Spain
Fernando Gomide, University of Campinas, Brazil
Norton Gonzalez, University of Fortaleza, Brazil
Pedro Gonzalez, University of Jaen, Spain
Manuel Grana, University of Basque Country, Spain
Crina Grosan, Norwegian University of Science and Technology, Norway
Steven Guan, Xi'an Jiaotong-Liverpool University, China
Ramin Halavati, Sharif University of Technology, Iran
Thomas Hanne, University of Applied Sciences Northwestern Switzerland, Switzerland
Takashi Hasuike, Osaka University, Japan
Ilkka Havukkala, Intellectual Property Office of New Zealand, New Zealand
Francisco Herrera, University of Granada, Spain
Enrique Herrera-Viedma, University of Granada, Spain
Cesar Hervas-Martinez, University of Cordoba, Spain
Aymeric Histace, Universite Cergy-Pontoise, France
Tzung-Pei Hong, National University of Kaohsiung, Taiwan
Wei-Chiang Hong, Oriental Institute of Technology, Taiwan
Dusan Husek, Institute of Computer Science, Academy of Sciences of the Czech Republic
Liliana Ironi, IMATI-CNR, Italy
Intan Ermahani A. Jalil, Universiti Teknikal Malaysia Melaka, Malaysia
Alex James, Indian Institute of Information Technology and Management—Kerala, India
Richard Jensen, Aberystwyth University, UK
Isabel S. Jesus, Instituto Superior de Engenharia do Porto, Portugal
Maria Jose Del Jesus, Universidad de Jaen, Spain
Fernando Jimenez, University of Murcia, Spain
Chia-Feng Juang, National Chung-Hsing University, Taiwan

Aranzazu Jurio, Universidad Publica de Navarra, Spain
Saurav Karmakar, Georgia State University, USA
Emaliana Kasmuri, Universiti Teknikal Malaysia Melaka, Malaysia
Frank Klawonn, University of Applied Sciences Baunschweig, Germany
Andreas Koenig, Technische Universitat Kaiserslautern, Germany
Mario Koeppen, Kyushu Institute of Technology, Japan
Michal Kratky, VŠB-TU Ostrava, Czech Republic
Bartosz Krawczyk, Politechnika Wrocławska, Poland
Dalia Kriksciuniene, Vilnius University, Lithuania
Pavel Kromer, VŠB-TU Ostrava, Czech Republic
Yasuo Kudo, Muroran Institute of Technology, Japan
Manuel Lama, Universidade de Santiago de Compostela, Spain
Maria Teresa Lamata, University of Granada, Spain
Kelvin Lau, University of York, UK
Chang-Shing Lee, National University of Tainan, Taiwan
Imre Lendak, University of Novi Sad, Serbia
Jorge Nunez Mc Leod, Institute of C.E.D.I.A.C, Argentina
Leida Li, University of Mining and Technology, China
Chu-Hsing Lin, Tunghai University, Taiwan
Wen-Yang Lin, National University of Kaohsiung, Taiwan
Santi Llobet, Universitat Oberta de Catalunya, Spain
Vincenzo Loia, University of Salerno, Italy
Carlos Lopezmolina, Universidad Publica de Navarra, Spain
Ana Lorena, Federal University of ABC, Brazil
Teresa Ludermir, Federal University of Pernambuco, Brazil
Simone Ludwig, North Dakota State University, USA
Jose-Maria Luna, University of Cordoba, Spain
Gabriel Luque, University of Malaga, Spain
Kun Ma, University of Jinan, China
Jose Tenreiro Machado, Instituto Superior de Engenharia do Porto, Portugal
Francesco Marcelloni, University of Pisa, Italy
Pierre-Francois Marteau, Universite de Bretagne Sud, France
Francisco Martinez-Alvarez, Pablo de Olavide University, Spain
Francisco Martinez-Estudillo, University Loyola Andalucia, Spain
Jose M. Merigo, University of Barcelona, Spain
Sadaaki Miyamoto, University of Tsukuba, Japan
Carlos Morell, Universidad Central Marta Abreu de Las Villas, Cuba
Mohamed Mostafa, VŠB-TU Ostrava, Czech Republic
Manuel Mucientes, University of Santiago de Compostela, Spain
Azah Kamilah Muda, Universiti Teknikal Malaysia Melaka, Malaysia
Noor Azilah Muda, Universiti Teknikal Malaysia Melaka, Malaysia
Kazumi Nakamatsu, University of Hyogo, Japan
Francisco Fernandez Navarro, University of Cordoba, Spain
Roman Neruda, Institute of Computer Science, Czech Republic
Maria Nicoletti, Federal University of Sao Carlos, Brazil

Julio Cesar Nievola, Pontificia Universidade Catolica do Parana, Brazil
Yusuke Nojima, Osaka Prefecture University, Japan
Isabel Nunes, UNL/FCT, Portugal
Eliska Ochodkova, VŠB-TU Ostrava, Czech Republic
Jae Oh, Syracuse University, USA
Jose Valente De Oliveira, Universidade do Algarve, Portugal
Paulo Moura Oliveira, University of Tras-os-Montes and Alto Douro, Portugal
Juan-Luis Olmo, University of Cordoba, Spain
Suhail Owais, Applied Science University, Jordan
Nour Oweis, VŠB-TU Ostrava, Czech Republic
Jose Pena, Universidad Politecnica de Madrid, Spain
Antonio Peregrin, University of Huelva, Spain
Javier Perez, University of Salamanca, Spain
Sylvain Piechowiak, Universite de Valenciennes et du Hainaut-Cambresis, France
Vincenzo Piuri, University of Milan, Italy
Jan Platos, VŠB-TU Ostrava, Czech Republic
Silvia Poles, EnginSoft, Italy
Mariantonietta Noemi La Polla, IIT-CNR, Italy
Beatriz Pontes, University of Seville, Spain
Gede Pramudya, Universiti Teknikal Malaysia Melaka, Malaysia
Girijesh Prasad, University of Ulster, UK
Satrya Fajri Pratama, Universiti Teknikal Malaysia Melaka, Malaysia
Kumudha Raimond, Karunya University, India
S. Ramakrishnan, Dr. Mahalingam College of Engineering and Technology, India
Sazalinsyah Razali, Universiti Teknikal Malaysia Melaka, Malaysia
Selva Rivera, Institute of C.E.D.I.A.C, Argentina
Cristobal Romero, University of Cordoba, Spain
Jose Raul Romero, University of Cordoba, Spain
Cristina Rubio-Escudero, University of Sevilla, Spain
Ashraf Saad, Armstrong Atlantic State University, USA
Ramon Sagarna, University of Birmingham, UK
Ozgur Koray Sahingoz, Turkish Air Force Academy, Turkey
Virgilijus Sakalauskas, Vilnius University, Lithuania
Ovidio Salvetti, ISTI-CNR, Italy
Philip Samuel, Cochin University of Science and Technology, India
Javier Sedano, Technological Institute of Castilla y Leon, Spain
Detlef Seese, Karlsruhe Institut of Technology (KIT), Germany
Siti Rahayu Selamat, Universiti Teknikal Malaysia Melaka, Malaysia
Jesus Serrano-Guerrero, University of Castilla-La Mancha, Spain
Patrick Siarry, Universit de Paris, France
Václav Snášel, VŠB-TU Ostrava, Czech Republic
Hussein Soori, VŠB-TU Ostrava, Czech Republic
Luciano Stefanini, University of Urbino "Carlo Bo", Italy
Jakub Stolfa, VŠB-TU Ostrava, Czech Republic
Svatopluk Stolfa, VŠB-TU Ostrava, Czech Republic

Eulalia Szmidt, Systems Research Institute Polish Academy of Sciences, Poland
Kang Tai, Nanyang Technological University, Singapore
Antonio J. Tallon-Ballesteros, University of Seville, Spain
Ayeley Tchangani, University Toulouse III, France
Chuan-Kang Ting, National Chung Cheng University, Taiwan
Maria Torsello, Universita di Bari, Italy
Luigi Troiano, University of Sannio, Italy
Eiji Uchino, Yamaguchi University, Japan
Ashish Umre, University of Sussex, UK
Olgierd Unold, Wrocław University of Technology, Poland
Coral del Val, University of Granada, Spain
Sebastian Ventura, University of Cordoba, Spain
Brijesh Verma, Central Queensland University, Australia
Gregg Vesonder, AT&T Labs—Research, USA
Juan Vidal, Universidade de Santiago de Compostela, Spain
Pablo Villacorta, University of Granada, Spain
Jose Villar, Oviedo University, Spain
Leon Wang, National University of Kaohsiung, Taiwan
Wei Wei, Xi'an University of Technology, China
Michal Wozniak, Wrocław University of Technology, Poland
Min Wu, Oracle, USA
Qinghan Xiao, Defence R&D Canada, Canada
Yunyi Yan, Xidian University, China
Enrique Yeguas, University of Cordoba, Spain
Kaori Yoshida, Kyushu Institute of Technology, Japan
Robiah Yusof, Universiti Teknikal Malaysia Melaka, Malaysia
Anazida Zainal, Universiti Teknologi Malaysia, Malaysia
Ivan Zelinka, VŠB-TU Ostrava, Czech Republic
Qieshi Zhang, Waseda University, Japan
Jun Zhang, Waseda University, Japan
Liang Zhao, University of Sao Paulo, Brazil
Huiyu Zhou, Queen's University Belfast, UK
Shang-Ming Zhou, University of Wales Swansea, UK

Sponsoring Institutions

VŠB—Technical University of Ostrava, Czech Republic

Contents

Part III Security and Privacy

Part IV Signal Processing and Applications

Part V Applications of the Big Data and the Connected Vehicles

Part I
Data Analysis and Applications

Association-Rule-Based Random Walk Method for Personalized Tag Recommendation

Jing Wang and Nianlong Luo

Abstract With the development of social websites during the web2.0, tagging has been playing an important role for users to mark their web resource. By offering personalized tags, recommender systems help users to integrate resource and complete tagging effectively. Graph based methods of tags recommending have been shown to provide high quality results such as RWR, FolkRank, PageRank, etc. However, data sparsity leads to sparseness of graphs limiting the precision in the process of use. In this paper, we propose a new method ARRW to alleviate the sparsity problem. We introduce Association Rules to Random Walk for digging up more relevance among nodes in graphs. We evaluate ARRW on a real-world dataset collected on Delicious. Data tests show that ARRW outperforms other Random Walk methods which not consider intra-relations, and ARRW successfully alleviate the graph sparsity problem.

Keywords Association rule · Tag recommendation · Random walk · Data sparsity

1 Introduction

Social tagging system is an application which provides a function of marking labels for web users. With the prevalence of web2.0, content-sharing websites of which social tagging mechanism is a key part develop rapidly. The mechanism transfers the right of categorizing internet resources from specialists to common persons. Users can freely provide metadata to describe the content of resource on the Internet so as to organize the resource easily and share with their friends conveniently. There are

---

J. Wang (✉) · N. Luo
Information Technology Center, Tsinghua University, 100084 Beijing, China
e-mail: j-wang13@mails.tsinghua.edu.cn

N. Luo
e-mail: lnl@tsinghua.edu.cn

A. Abraham et al. (eds.), *Intelligent Data Analysis and Applications*,
Advances in Intelligent Systems and Computing 370,
DOI 10.1007/978-3-319-21206-7_1

several typical websites with tagging systems, such as Delicious[1] with bookmarks, Flickr[2] with images, Youtube[3] with clips and CiteULike[4] with publications.

Personalized tag recommendation [7] is the main part in a tagging system. The systems predict and offer users a set of tags that they are most likely to use. Users' tagging behaviors can be seen as ternary relations $\langle user, item, tag \rangle$. When a Tag t is recommended for a User u to annotate Item i, t should be highly relevant to both u and i. For the reason that different users have diverse ways of tagging, recommended tags should differ from user to user. At the same time, the tags should be able to describe the content of items.

Our work builds on the personalized tag recommender method by using Random Walk model. Methods based on Random Walk like Random Walk with Restart (RWR) [10], FolkRank (FR) [5] have shown to result in good prediction quality. But the sparsity problem still exists in graph, which limit the prediction quality. Here, we introduce association rules into Random Walk, in order to acquire more semantic information in the graph.

In this paper, our contributions are summarized as follows:

1 Building a new model (Association Rule based Rand Walk) for personalized tag recommendation.
2 Conducting experiments on the real-world dataset and demonstrating that our model outperforms other models based on Random Walk.
3 Our method can alleviate the problem of sparsity.

The rest of this paper is organized as follows. In the next section, we review and discuss the related work. Then, in Sect. 3, we precisely define the problem and notation. In Sect. 4, we completely describe and formulate our ARRW method for tag recommendation. After that, we report result and analysis of our experiments in Sect. 5. Finally, it comes to conclusion and outlines future work in Sect. 6.

2 Related Work

Personalized tag recommendation is a hot topic in recommender system over the years. Different types of tagging systems were detailed by Huberman et al. [6] and Marlow et al. [15]. In general, according to whether the processing procedure involves the resource content, tag recommendation methods fall into two categories: content-based approaches and graph-based approaches.

Content-based methods, which usually collect available information from context of items (e.g., web pages, anchor text, academic papers or other textual resources) to build user models or item models, can predict tags even for cold start. Yin [19]

---

[1] www.delicious.com.

[2] www.flickr.com.

[3] www.youtube.com.

[4] www.citeulike.com.

proposed a Bayesian probabilistic model, the prediction is treated as the reverse of web search, consider a list of words on web pages as a list of tags, then retrieve the potential tags for the given web page.

Liu et al. [13] calculated the similarity between items according textual content, then recommend the tags which are most frequently labelled to similar items. Si [17] put forward a fast tag recommending framework named Feature-Driven Tagging which represents a tag by some features. The feature can be a word, an id or other context information. Content-based approaches usually result in better precision than non-content-based approaches in most cases because of the more information ultilized by the former. However, content-based methods are unavailable where items belong to nonstructural resources or the content can not be obtained directly, such as movies, songs, etc.

Graph-based methods, which focus on the relations between users, items and tags, mostly yield lower computation complexity compared with content-based methods, for they don't parse items content. Adriana et al. [2] presented an efficient top-*k* tag selection algorithm HAMLET that predicts tags by inspecting neighbors in the graph. The model, that takes items as nodes and relations as edges, can be applied for a directed graph (edges for webpage links, academic references, etc.). FolkRank, an adaption of PageRank which has been widely used in search engines, was introduced by Hotho [5]. Wei [3] proposed an optimization framework called OptRank. OptRank incorporates heterogeneous information to represent edges and nodes with features. OptRank can get better prediction than common Rand Walk methods, but it's too time-consuming for a large number of parameters. Krestel [11] extended Latent Dirichlet Allocation for recommending tags, which is on the hypothesis that item content consists of some latent topics. When a topic is represented by a tag probability distribution, tags can be predicted according to the posterior probability of latent topics.

There are many methods based on Factorization models. Higher-Order-Singular-Value-Decomposition (HOSVD) was introduced into tag recommendation to reduce tensor dimensionality in [18]. Steffen [16] put forward a special Tucker Decomposition model called Pairwise Interaction Tensor Factorization (PITF) with linear runtime for learning and prediction. On their experiment dataset, the run time of PITF is shorter than TD, and the former can get better prediction as well. Since methods based on Factorization are quite complex, they are hardly applied to real systems.

In the fields of item recommendation like e-commerce, movie, music or photo websites, collaborative filtering [1] methods are widely used successfully. So, many researches [9] introduced collaborative filtering into tag recommendation.

There were some models considering tag recommendation from other perspectives. Jin [8] proposed an orientation of motivation discrimination model (OMDM), they used five features to measure the motivation of users and items. Liu [12] took location information into consideration to recommend tags for photos. Temporal information was also introduced to build models [14, 20].

3 Definition

When users are marking labels to items, tag recommender systems will predict and offer some tags the users might use. The systems calculate target tags from users' behavior history and item context.

Given a set of items I, tags T, and users U, we represent the user's marking to items with the ternary relations $S \subseteq R \times T \times U$. For tag recommender, the task is to offer a specific pair (user,item) a list of tags. Here, we define distinct user-item pairs Ps: $P_s = \{(u, i) | u \in U, i \in I\}$. All methods presented in this paper, give a scoring function $\hat{Y} : U \times I \times T \rightarrow \mathbb{R}$ to derive a tag ranking list. The number of recommended tags is often restricted, and we define the Top-N tags as:

$$Top(u, i, N) = \underset{t \in T}{\overset{N}{\operatorname{argmax}}}\, \hat{y}_{u,i,t}$$

Here, N is the number of tags in the ranking list.

4 Methodology

In this section, we describe our algorithm for tag prediction and recommendation.

4.1 Random Walk

A tagging recommender system, containing heterogeneous information, can be modeled as a graph, such as Fig. 1. Edges in the graph are derived from annotation behaviors S. We can see the followings from the graph:

- There are three types of nodes in the graph: users, items and tags.
- Inter-relation, the edges between different types of nodes. The weight of edge means the relevance between the two nodes. As a result, for a $i \in I$ and $t \in T$, the weight of $< i, t >$ can be the times of i being annotated by t. It's the same rule to $< u, t >$, $< u, i >$, etc.
- Intra-relation. Users' social network, items' similarity network and tags semantic network.

Inter-relation network has been well studied in many researches, but few models combine inter-relation and intra-relation at the same time. Adding intra-relation into model might alleviate sparsity problem.

When tag t is recommended for user u to annotate item i, t should be highly relevant to both u and i. So, when a random walk with restart at u and i is performed, t should get a high visiting probability. We can format the random walk with restart as follows:

Fig. 1 Social tagging system

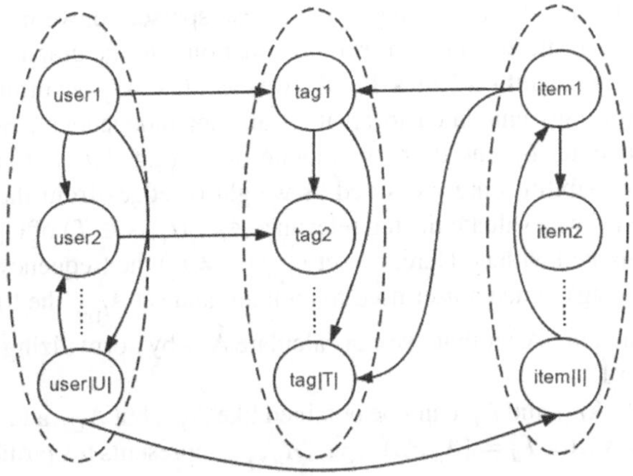

$$\begin{pmatrix} \mathbf{p}_U \\ \mathbf{p}_I \\ \mathbf{p}_T \end{pmatrix}^{t+1} = \alpha\overline{\mathbf{A}} \begin{pmatrix} \mathbf{p}_U \\ \mathbf{p}_I \\ \mathbf{p}_T \end{pmatrix}^{t} + (1-\alpha-\beta) \begin{pmatrix} \mathbf{p}_U \\ \mathbf{p}_I \\ \mathbf{p}_T \end{pmatrix}^{t} + \beta \begin{pmatrix} \mathbf{q}_U \\ \mathbf{q}_I \\ \mathbf{0} \end{pmatrix} \tag{1}$$

Here,

- α is the transfer rate, which means the random walker has the probability of α to jump to the next node.
- β is the restart probability.
- $\mathbf{p}^T = (\mathbf{p}_U^T, \mathbf{p}_I^T, \mathbf{p}_T^T)$ is the visiting probability of all nodes.
- $\mathbf{q}^T = (\mathbf{q}_U^T, \mathbf{q}_I^T, \mathbf{0}^T)$ is the normalized beginning probability, of which sum is 1.
- $\overline{\mathbf{A}}$ is the normalized transfer probability matrix.

4.2 Transition Matrix

The transition matrix contains the whole relation information in the graph, and stores the weight of each edge. We let $G_{MN}(M, N \in \{U, I, T\})$ denote sub-graph of nodes of two types and A_{MN} is its adjacency matrix. So, we can denote A by its sub-matrices:

$$\mathbf{A} = \begin{bmatrix} \mathbf{A}_{UU} & \mathbf{A}_{UI} & \mathbf{A}_{UT} \\ \mathbf{A}_{IU} & \mathbf{A}_{II} & \mathbf{A}_{IT} \\ \mathbf{A}_{TU} & \mathbf{A}_{TI} & \mathbf{A}_{TT} \end{bmatrix} \tag{2}$$

Here, $\mathbf{A}_{MN}(M \neq N)$ describes the edges of two different types of nodes. We can measure the weight of edges from behavior history S, so we set $A_{mn}(m \in M, n \in N, M \neq N)$ the times of m and n appearing together in S. But we can't measure the weight of edges in intra-relation directly. Meantime, because of the sparseness of data, the corresponding graph is also sparse. So, we introduce the association rules which succeed in item recommendation into the graph.

Association Rules. Firstly, we calculate $\mathbf{A}_{TT}$ according to the tag co-occurrence. The tags annotated to an item are not independent. Some tags are often chosen together by one user to annotate items (e.g. *Internet* and *web*). Thus, conditional probabilities are measured as weight of edges from the tag co-occurrence. In this way, we evaluate the tag relevance $A_{t_1,t_2}(t_1, t_2 \in T)$ of tag $\mathbf{t}_1$ for an item having been marked with $\mathbf{t}_2$. Here, we set $A_{t_1,t_2}(t_1 \neq t_2)$ the frequency of both t_1 and t_2 appearing in tags which a user used to an item, and set A_{t_1,t_1} the times of t_1 used by a user for an item. After that, we can calculate $\widehat{\mathbf{A}}_{TT}$ by normalizing each column of $\mathbf{A}_{TT}$ to sum to 1.

$\mathbf{A}_{UU}$ and $\mathbf{A}_{II}$ can't be obtained like $\mathbf{A}_{TT}$, but $\mathbf{A}_{UT}$ and $\mathbf{A}_{IT}$ can be helpful. For user u, vector $T_u = \{A_{u,t_1}, A_{u,t_2}, ..., A_{u,t_{|T|}}\}$ represents u's profile. We could obtain similarities of different users through cosine similarity. But $\mathbf{U}_{TT}$ is sparse, some latent relations may not be obtained. For example, u_1's profile is <Internet(4), data-mining(3), web(0), machine learning(0)>, u_2's profile is <Internet(0), data-mining(0), web(4), machine learning(2)>, though the cosine similarity between u_1 and u_2 is 0, we still think the two users are highly relevant. Here, we use association rules to update user profile for obtaining more relevance. For each user u, before calculating of cosine similarity, u's profile is renewed as follows.

$$\widehat{T}_u^T = \widehat{\mathbf{A}}_{TT} T_u^T \tag{3}$$

Then, obtain the user-user matrix by

$$A_{u_1,u_2} = \frac{\widehat{T}_{u_1} \cdot \widehat{T}_{u_2}}{|\widehat{T}_{u_1}| \cdot |\widehat{T}_{u_2}|} \tag{4}$$

After getting A_{UU}, then to calculate $\widehat{A}_{UU}$ by normalizing each column of A_{UU} to sum to 1.

Next, obtain $\widehat{A}_{II}$ with the same method of $\widehat{A}_{UU}$. Then we normalize sub-matrices of two different types of nodes, $A_{MN}(M \neq N)$ to $\widehat{A}_{MN}(M \neq N)$ by normalizing sum of each column to 1. Finally, we can get our transfer probability matrix.

$$\overline{\mathbf{A}} = \frac{1}{3}\begin{bmatrix} \widehat{\mathbf{A}}_{UU} & \widehat{\mathbf{A}}_{UI} & \widehat{\mathbf{A}}_{UT} \\ \widehat{\mathbf{A}}_{IU} & \widehat{\mathbf{A}}_{II} & \widehat{\mathbf{A}}_{IT} \\ \widehat{\mathbf{A}}_{TU} & \widehat{\mathbf{A}}_{TI} & \widehat{\mathbf{A}}_{TT} \end{bmatrix} \tag{5}$$

Here, the coefficient $\frac{1}{3}$ is to ensure $\overline{\mathbf{A}}$ normalized.

4.3 Derivatives

Usually, to recommend tags for user u to annotate item i, the beginning probability $\mathbf{q}$ is highly relevant to u and i. In our method, we assign q_u and q_i 0.5.

Finally, by the Eq. 1, converged visiting probability vector $\mathbf{p}$ will be obtained.

5 Experiment

In our experiment, we test our model on publicly available bookmark datasets on Delicious. In the raw data, there are 437590 marking records of 1867 users, 69223 items and 40897 tags. The raw data is too sparse so that we select a subset of it as our dataset. First, we select the top 45 users in terms of annotating frequency as user set, in the same way, we select top 50 items being annotated by users in user set as our item set, and select top 50 tags used by users in user set for items in item set. Finally, we get 999 posts.

For validation, we randomly split the record set into meaningful training and test sets (training S 90 %, test S 10 %) five times. We use mean precision and mean recall of five times to measure the performance.

We compare ARRW with two random walk based methods: Random Walk with Restart(RWR) and FolkRank(FR). RWR is a random walk based algorithm which doesn't consider intra-relation, introduced into social network in [10], different types of edges are normalized empirically. FolkRank [4] is a state-of-the-art graph-based algorithm, the graph is defined the same as RWR. We can summarize FolkRank as three steps: (1) calculate a global pagerank score p_{global}. (2) calculate personalized pagerank score p_{pref}. (3) calculate folkrank score $p_{fr} = p_{pref} - p_{global}$.

We set restart start probablity β (or called user preference) of RWR and FR the same as that of ARRW. For RWR, weight of edges is the times of two nodes co-occurred in ternary relations S. Given RWR gets the best result at restart probability of 0.3, we assign its restart probability 0.7. For FR, edges are the same as RWR. When we set its restart probability β 0.2, transfer rate α 0.5, and in this way, it performs the best. For ARWR, we assign its restart probability β 0.3, transfer rate α 0.5.

According to the results of the three methods in Fig. 2 and Table 1, we can see that when N varies from 1 to 10, ARRW has an obvious performance than RWR and FR, while RWR performs worst. Notably, when N is 3, ARRW improves the top-3 precision by 5.1 % compared with FR. Compared with FR, more information are used in ARRW. By using intra-relation calculated by association rules, graph of ARRW is denser than that of FR. Hence, considering the contribution of intra-relation to improve performance, we conclude that ARRW outperforms FR and RWR which only use inter-relations. Furthermore, association rules can also obtain more useful latent relation for tag recommendation.

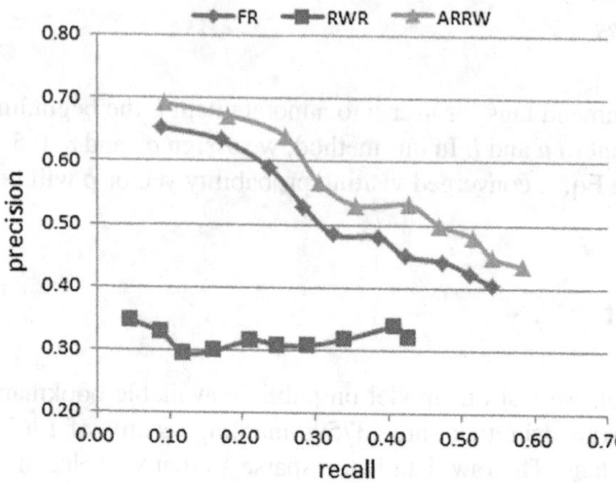

Fig. 2 Performance comparison

Table 1 Precision @N

N	RWR	FolkRank	ARRW
1	0.34615	0.65385	0.69231
2	0.32692	0.63462	0.67308
3	0.29487	0.58974	0.64103
4	0.29808	0.52885	0.56731
5	0.31538	0.48462	0.53077
6	0.30769	0.48077	0.53205
7	0.30769	0.45055	0.50000
8	0.31731	0.44231	0.48077
9	0.33761	0.42308	0.44872
10	0.31923	0.40385	0.43462

6 Conclusion and Future Work

In this paper, we proposed a new method ARRW for personalized tag recommending. By bringing association rules into Random Walk, ARRW alleviate the sparsity problem of graph, and improve the precision of the tag recommendation. We empirically showed that our model outperforms some other random-walk-based methods. In future work, we will try to reduce the graph size, since it takes up too much space and it's also time-consuming, those may block the model's capacity in actual systems. Additionally, we would also bring more information into our methods, such as temporal information.

References

1. Breese JS, Heckerman D, Kadie C (1998) Empirical analysis of predictive algorithms for collaborative filtering. In: Proceedings of the fourteenth conference on uncertainty in artificial intelligence. Morgan Kaufmann Publishers Inc
2. Budura A et al (2009) Neighborhood-based tag prediction. The semantic web: research and applications. Springer, Berlin, pp 608–622
3. Feng W, Wang J (2012) Incorporating heterogeneous information for personalized tag recommendation in social tagging systems. In: Proceedings of the 18th ACM SIGKDD international conference on Knowledge discovery and data mining. ACM
4. Hotho A et al (2006) Folkrank: a ranking algorithm for folksonomies. LWA. Ed. Klaus-Dieter Althoff, vol 1
5. Hotho A et al (2006) Information retrieval in folksonomies: search and ranking. Springer, Berlin
6. Huberman, SGBA (2005) The structure of collaborative tagging systems. No. cs. DL/0508082. cs/0508082
7. Jäschke R, Marinho L, Hotho A, Schmidt-Thieme L, Stumme G (2007) Tag recommendations in folksonomies. In: Kok JN, Jacek, J, Lopez de Mantaras R, Matwin S, Mladenič D, Skowron A (eds) PKDD 2007, vol 4702. LNCS (LNAI) Springer, Heidelberg, pp 506–514
8. Jin Y-A (2013) Approach for tag recommendation based on orientation of motivation. Jisuanji Yingyong Yanjiu 30(1):72–77
9. Jschke R et al (2008) Tag recommendations in social bookmarking systems. Ai Commun 21.4:231–247
10. Konstas I, Stathopoulos V, Jose JM (2009) On social networks and collaborative recommendation. In: Proceedings of the 32nd international ACM SIGIR conference on Research and development in information retrieval. ACM
11. Krestel R, Fankhauser P, Nejdl W (2009) Latent dirichlet allocation for tag recommendation. In: Proceedings of the third ACM conference on Recommender systems. ACM
12. Liu J et al (2014) Personalized geo-specific tag recommendation for photos on social websites. IEEE Trans Multimed 16.3:588–600
13. Lu Y-T et al (2009) A content-based method to enhance tag recommendation. IJCAI, Vol 9
14. Luo J, Pan X, Zhu X (2012) Time-aware user profiling for tag recommendation. Adv Inf Sci Serv Sci 4.19
15. Marlow C et al (2006) HT06, tagging paper, taxonomy, Flickr, academic article, to read. In: Proceedings of the seventeenth conference on Hypertext and hypermedia. ACM
16. Rendle S, Schmidt-Thieme L (2010) Pairwise interaction tensor factorization for personalized tag recommendation. In: Proceedings of the third ACM international conference on Web search and data mining. ACM
17. Si X et al (2009) Content-based and graph-based tag suggestion. ECML PKDD Discov Chall 2009:243–260
18. Symeonidis P, Nanopoulos A, Manolopoulos Y (2008) Tag recommendations based on tensor dimensionality reduction. In: Proceedings of the 2008 ACM conference on recommender systems. ACM
19. Yin D et al (2010) A probabilistic model for personalized tag prediction. In: Proceedings of the 16th ACM SIGKDD international conference on Knowledge discovery and data mining. ACM
20. Yin D et al (2011) Temporal dynamics of user interests in tagging systems

Motion Sequence Recognition with Multi-sensors Using Deep Convolutional Neural Network

Runfeng Zhang and Chunping Li

Abstract With the rapid development of intelligent devices, motion recognition methods are broadly used in many different occasions. Most of them are based on several traditional machine learning models or their variants, such as Dynamic Time Warping, Hidden Markov Model or Support Vector Machine. Some of them could achieve a relatively high classification accuracy but with a time-consuming training process. Some other models are just the opposite. In this paper, we propose a novel designed deep Convolutional Neural Network (DBCNN) model using "Data-Bands" as input to solve the motion sequence recognition task with a higher accuracy in less training time. Contrast experiments were conducted between DBCNN and several baseline methods and the results demonstrated that our model could outperform these state-of-art models.

1 Introduction

Nowadays, intelligent devices and modern life are more and more inseparable. These devices could record humans' motion sequences through embedded sensors precisely. Lots of motion recognition models have been proposed, such as Hidden Markov Models, Dynamic Time Warping, and Feature-Based Support Vector Machine. There are drawbacks more or less. Some of them can not attain a high accuracy, others might be time-consuming. In this paper, we proposed a new model that pre-processes motion sequences to form "data-bands" and then trains special designed deep convolutional neural network using "data-bands" as inputs. We call this model as DBCNN, in which "DB" means "Data-Bands" and "CNN" means "Convolutional Neural Network". Experiments show that our novel model could attain a better recognition accuracy compared with other state-of-art methods. Due to

R. Zhang (✉) · C. Li
School of Software, Tsinghua University, Beijing 100084, BJ, China
e-mail: rf-zhang12@mails.tsinghua.edu.cn

C. Li
e-mail: cli@tsinghua.edu.cn

A. Abraham et al. (eds.), *Intelligent Data Analysis and Applications*,
Advances in Intelligent Systems and Computing 370,
DOI 10.1007/978-3-319-21206-7_2

characteristics of deep convolutional neural network, training process can be accelerated on GPU in extremely short time, which makes DBCNN win more scores.

This paper is organized as follows. Section 2 introduces some backgrounds, including problem description, related works and some basic knowledge about Convolutional Neural Network (CNN) in brief. Section 3 presents DBCNN for motion recognition task, including data process and deep Convolutional Neural Network design. Section 4 takes a glance at some details of parameters learning. Section 5 reports experiment results of our model comparing with two other state-of-art methods. And Sect. 6 concludes this paper.

2 Background

2.1 *Problem Description*

Now we describe motion recognition task more formally : Motion sequence recognition is a classification task with $\mathcal{N}$ categories, and each of these categories has a corresponding training set $\mathcal{D}_i$. The total training set is $\mathcal{D} = \{\mathcal{D}_1, \mathcal{D}_2, \dots, \mathcal{D}_{\mathcal{N}}\}$. Training set $\mathcal{D}_i$ has $\mathcal{N}_i$ action instances: $\mathcal{D}_i = \{\mathcal{M}_1, \mathcal{M}_2, \dots, \mathcal{M}_{\mathcal{N}_i}\}$. As mentioned before, each motion instance is constituted by several sequences gathered by different sensors, so we have $\mathcal{M}_i = \{\mathcal{S}_1, \mathcal{S}_2, \dots, \mathcal{S}_{channel\_num}\}$. Assuming that processed sequences have t dimension, sequence $\mathcal{S}_i$ would be a vector with length t. The motion recognition task is to build a classifier given training dataset $\mathcal{D}$.

2.2 *Related Works*

A wide variety of motion sequence recognition models have been proposed, such as Hidden Markov Models (HMMs) [1–4], Dynamic Time Warping (DTW) [5], and Feature-Based Support Vector Machine (Feature-SVM) [6] etc. Among models of HMMs, continuous-HMM proposed in [3] claimed to gain the highest accuracy. In a task with 10 pre-setting motions, it got a classification accuracy of 96.76 %. This result has beaten the discrete-HMM [4], which could achieve an accuracy of 96.1000 % in the same dataset. Liu et al. has given a better result than both discrete-HMM and continuous-HMM in [5] using Dynamic Time Warping. SVM could not be used in our task directly due to the large scale and high dimensionality of sequences. Wu et al. [6] proposed a feature-based SVM algorithm, which extracts new features with a lower dimension from the original sequences in a manual procedure. Wang et al. [7] proposed a similar method in different feature extraction rules. Experiment results in [7] show that Wang's method could achieve higher accuracy comparing with HMMs [1–4], DTW [5] and Feature-SVM designed by Wu et al. [6], we adopt this Feature-SVM model as one of our baseline methods in our experiment.

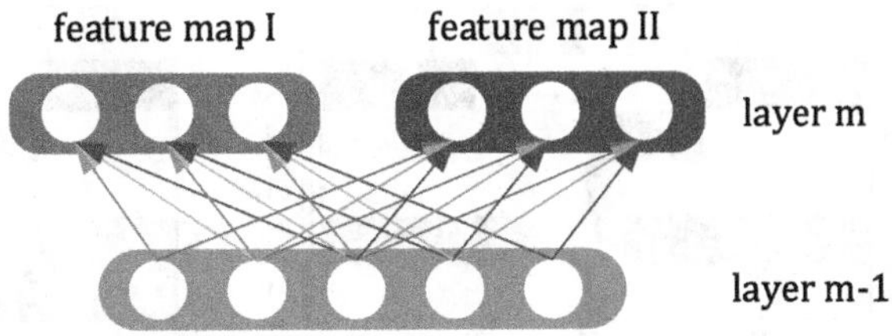

Fig. 1 Shared weights of convolutional layers in CNN

2.3 Convolutional Neural Network

Convolutional Layer in CNN Although full connection structure in multi-layer perception could make the networks to have more ability in expression, model will be more possible to overfit the data. CNN solves these problems via sparse connection structure, as shown in Fig. 1. First, nodes in high layer only connect to parts of nodes in low layer, which could reduce the number of parameters efficiently. Moreover, nodes in high layer share identical weights. Just like Fig. 1, arrows with the same weight value have identical colors. This structure make the model to detect position independent features, from which we benefit a lot.

Sub-sampling Layer in CNN Pooling is a form of non-linear down-sampling in CNN. Among several different pooling methods, max-pooling and average-pooling are the most common. LeCun et al. [9] used this method in their LeNet-5 for two reasons: reducing the computational complexity for upper layers and providing a form of translation invariance.

3 The DBCNN Model for Motion Sequence Recognition

3.1 Data Preprocess

Different people might do the same motion in different duration time. Sequences of motion gathered by intelligent devices are usually sampled in a constant time interval. So our obtained sequences of motion instances have different dimensionality.

To overcome this, our model introduces an approximation, linear interpolation. We get values of a sequence in specific proportional position to form new feature vectors, which have a fixed dimensionality. Actually, the value in specific proportional position is a linear combination of two real sample values near the position.

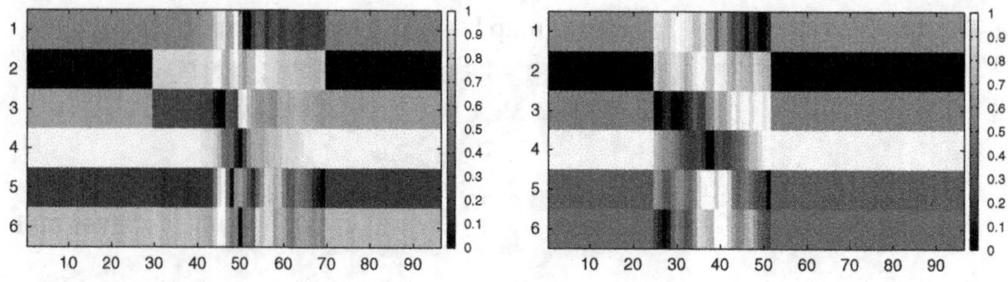

Fig. 2 Data band of the motion swipe right

3.2 Data-Band

Now, the new sequences of motion instances are in a same dimension. In our experiment, each instance of motion data consists of six sequences. Instead of arranging the six sequences into a long vector, we organize them into a "data-band", which is a narrow matrix with a small row number and a large column number, just like a band.

Figure 2 shows "data-bands" of two motion instances belong to a same motion category, "Swipe Right". It is obvious that they have position invariance: there are similar parts appearing in distinct positions. This invariance in a same motion class is an inherent attribute of motion itself and irrelevant to the motion's duration or executors. Our model use "data-bands" as input could grasp the position invariant successfully, which could not be achieved if we organize the data into a long vector.

3.3 Specially Designed CNN for Recognition of Motion Sequence

Input of CNN In our proposed CNN for sequences of motion, the input of network are actually "data-bands" as mentioned before. In Fig. 3, the input of the network is showed in the left side clearly. These "data-bands" are matrices with dimensionality of 6×96. The rows of matrices correspond to 6 axes (e.g. x, y, z axes in both accelerator and angular speed), and the columns correspond to 96 sampled points.

Design of Convolutional Layer LeCun et al. [9] have shown that CNN model is extremely suitable for data with high dimensionality and invariant attributes. Sequences of motion instance just fit the properties mentioned above. However, in LeNet-5, receptive fields are little squares to capture the invariance between images pixels. In our model, we treat our pre-processed sequences, "data-bands", as images. The difference is that the invariance of "data-bands" exhibits in a narrow form spreading in the lengthways dimension in "data-bands" matrices. If we use small squares as receptive fields, more receptive fields are necessary to catch the invariance. At the same time, the CNN model has more parameters and is more possible to be overfitting. Our model solves the invariance extraction problem in a very elegant

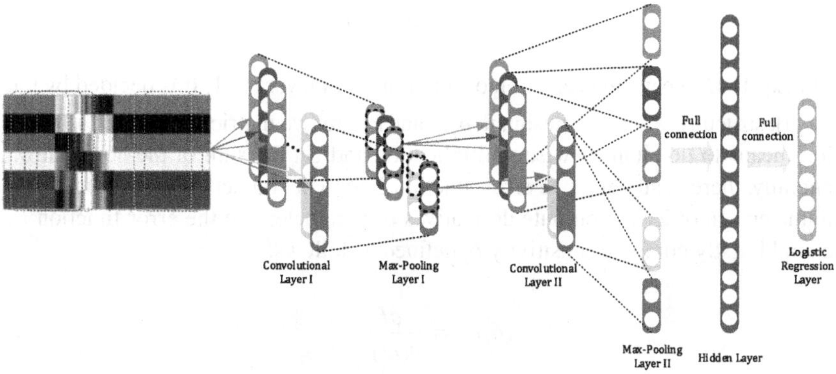

Fig. 3 The DBCNN model for sequences of motion data

way: we set up several convolutional windows to cover the lengthways of "data-bands" matrices, just like dashed rectangles with distinct colors in left of Fig. 3. This design could catch the invariance in lengthways using relatively small number of receptive fields. Moreover, each convolutional window scans a "data-band" and could get a 1-D feature vector, not a matrix in general CNN for image recognition.

Pooling Layer A pooling layer provides a further dimensionality reduction. We concatenate a max-pooling layer after a convolutional layer to shrink the feature space. At the same time, pooling layer provides an auxiliary way to grasp position invariant feature.

Classifier After two convolution-pooling structures, we connect a hidden layer and a logistic regression layer. They form a general multi-layer perception. In fact, two convolution-pooling layers mentioned before could be regarded as a function of feature transformation.

4 Details of Learning

4.1 Gradients in Convolutional Layers

In convolutional layers, a small patch $(\boldsymbol{p}_i^{l-1})_{uv}$ in previous layer's feature maps $\boldsymbol{x}_i^{l-1}$ are convolved with kernels $\boldsymbol{k}_{ij}^l$ and added a bias b_i^l to get a value $(\boldsymbol{u}_j^l)_{uv}$, then activation function $f(\cdot)$ would act on this value to form the corresponding element at (u, v) in the output feature maps $\boldsymbol{x}_i^l$, which is noted by $(\boldsymbol{x}_j^l)_{uv}$:

$$(\boldsymbol{u}_j^l)_{uv} = \sum_{i \in M_j} \boldsymbol{p}_i^{l-1} * \boldsymbol{k}_{ij}^l + b_j^l, \tag{1}$$

$$(\boldsymbol{x}_j^l)_{uv} = f((\boldsymbol{u}_j^l)_{uv}). \tag{2}$$

In Eq. 1, M_j means the selection of input maps in layer $l-1$. It is decided by the network structure when you design it. $\boldsymbol{p}_i^{l-1}$ and $\boldsymbol{k}_{ij}^l$ are two matrices, and the "*" operation means to do element-wise multiplication and get the sum of the new matrix. Generally, there is an objective error function E for a neural network to optimize the parameters. In order to compute derivatives of parameters on the error function E, we could firstly compute sensitivity $\boldsymbol{\delta}_j^l$ defined as follows:

$$(\boldsymbol{\delta}_j^l)_{uv} = \frac{\partial E}{\partial(\boldsymbol{u}_j^l)_{uv}}. \tag{3}$$

Sensitivity computation is a top-down process similar to the situation in general neural network but with a little difference. We compute sensitivity $(\boldsymbol{\delta}_j^l)_{uv}$ in layer l via the sensitivity $(\boldsymbol{\delta}_j^{l+1})_{uv}$ in layer $l+1$:

$$\boldsymbol{\delta}_j^l = \beta_j^{l+1}(f'(\boldsymbol{u}_j^l) \circ up(\boldsymbol{\delta}_j^{l+1})), \tag{4}$$

where $up(\cdot)$ function is exactly an inverse process of the down-sampling and β_j^{l+1} is a parameter in next sub-sampling layer (please refer to sect. 4.2). In Eq. 4, we use the notation "∘" to indicate the element-wise multiplication of two matrices. After getting sensitivities for feature maps in convolutional layer, we could compute gradients about parameters in convolutional layers as follows:

$$\frac{\partial E}{\partial \boldsymbol{k}_{ij}^l} = \sum_{u,v} (\boldsymbol{\delta}_j^l)(\boldsymbol{p}_i^{l-1})_{uv}, \tag{5}$$

$$\frac{\partial E}{\partial b_j^l} = \sum_{u,v} (\boldsymbol{\delta}_j^l)_{uv}. \tag{6}$$

4.2 Gradients in Sub-sampling Layers

Sub-sampling layers don't change the number of input maps, and they are really down-sampled versions which would downsize the output maps as follows:

$$\boldsymbol{x}_j^l = f(\beta_j^l down(\boldsymbol{x}_j^{l-1}) + b_j^l). \tag{7}$$

Here, $down(\cdot)$ means the sub-sampling function, such as average-pooling or max-pooling. The parameter β_j^l is a scalar to control the efficiency of sub-sampling, and b_j^l is still a bias item just like in convolutional layers. To compute the gradients about

β_j^l and b_j^l, we adopt similar steps to compute the sensitivities and gradients:

$$\boldsymbol{\delta}_j^l = f'(\boldsymbol{u}_j^l) \circ \boldsymbol{q}_j^l. \tag{8}$$

In Eq. 8, $\boldsymbol{q}_j^l$ is a matrix which has the same size as $\boldsymbol{\delta}_j^l$, and its element at (u, v) is noted by $(\boldsymbol{q}_j^l)_{uv}$, which has a complex definition form as follows:

$$(\boldsymbol{q}_j^l)_{uv} = \sum_{i \in \Phi(j)} \sum_{s,t} (\boldsymbol{k}_{i,j}^{l+1})_{s,t} (\boldsymbol{\delta}_i^{l+1})_{u'(s,t),v'(s,t)}, \tag{9}$$

where $\Phi(j)$ is a layer $l+1$ feature maps' set for which feature map j in layer l, $\boldsymbol{x}_j^l$, is participated the convolutional process. $(\boldsymbol{\delta}_i^{l+1})_{u'(s,t),v'(s,t)}$ is a tricky representation that means an element in feature map $\boldsymbol{x}_i^{l+1}$ which use the element $\boldsymbol{x}_j^l)_{uv}$ during its convolutional process with weight $(\boldsymbol{k}_{i,j}^{l+1})_{s,t}$. It is worth to notice that if the element corresponding to weight $(\boldsymbol{k}_{i,j}^{l+1})_{s,t}$ does not exist, the value of $(\boldsymbol{\delta}_i^{l+1})_{u'(s,t),v'(s,t)}$ would be zero.

Now, we could compute the gradients for b_i^l and β_j^l as the following equations:

$$\frac{\partial E}{\partial b_j^l} = \sum_{u,v} (\boldsymbol{\delta}_j^l)_{uv}, \tag{10}$$

$$\boldsymbol{d}_j^l = down(\boldsymbol{x}_j^{l-1}), \tag{11}$$

$$\frac{\partial E}{\partial \beta_j^l} = \sum_{u,v} (\boldsymbol{\delta}_j^l \circ \boldsymbol{d}_j^l)_{uv}. \tag{12}$$

After combining all the details mentioned before, we have our parameters optimization algorithm for CNN in **Algorithm 1**.

5 Experiment Results

5.1 Introduction to 6DMG

Our experiments are based on the dataset 6DMG [8], which is a 6-D Motion gesture database developed by Chen et al. in Georgia Institute of Technology. In our experiments, we use parts of original sequences: acceleration sequences (including $x, y, z-$axies) and angular speed sequences (including $x, y, z-$axies). Figure 4 shows 20 kinds of motion in 6DMG. The size of training set is 32300. The size of validation set is 4000 and the one of test set is 5100.

Algorithm 1 Parameters Optimization for Convolutional Neural Network

Require: processed dataset $\mathcal{D}$; learning rate α; epoch number $\mathcal{N}$; batch size n_b; Network Structure $\mathcal{M}$

Ensure: CNN's parameters $\boldsymbol{W}, \boldsymbol{k}, \boldsymbol{b}, \boldsymbol{\beta}$

initialize network parameters $\boldsymbol{W}, \boldsymbol{k}, \boldsymbol{b}, \boldsymbol{\beta}$ randomly
for each $i \in [1, \mathcal{N}]$ **do**
  for each batch data $\mathcal{D}_i$ with size n_b **do**
    for layer $l := L$ to 1 **do**
      1. compute **sensitivities** of layer l according to back-propagation
      2. compute **gradient** of each parameters in layer l
      3. **update parameters** in layer l in gradient descent method with learning rate α
    end for
  end for
end for

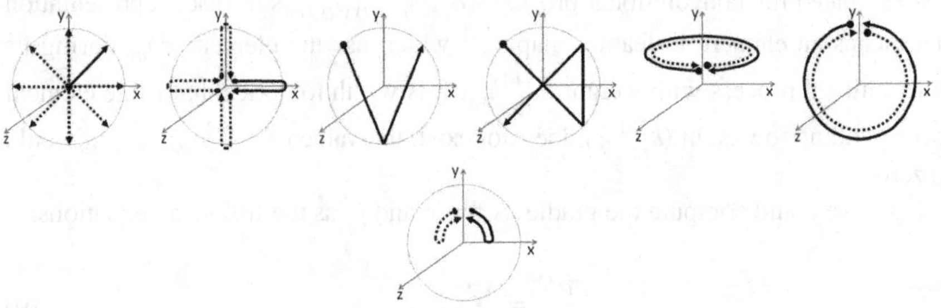

Fig. 4 Twenty kinds of motion in 6DMG

5.2 Feature-SVM

Feature-SVM proposed by Wang et al. [7] is claimed to outperform HMMs [1–4], DTW [5] and another featured-SVM given by Wu et al. [6]. We adopt this model as one of baseline experiments. After feature transformation, new features have the length of 48. We list the feature design rules in Table 1.

New 48-d features are trained using Gaussian Kernel SVM and Linear Kernel SVM with a 5-fold cross validation. For Linear Kernel SVM, the best accuracy 96.9125 % is attained when the cost parameter $log_2 C$ is 2.3206. And for Gaussian Kernel SVM, the best result appears when $log_2 C$ is 1.9459 and $log_2 g$ is -2.8332, and the best classification accuracy is 93.7011 %.

5.3 Deep Belief Network

Hinton et al. [14] proposed a fast learning algorithm for deep belief network, which use a greedy layer-wise training method. Deep architecture could be built with blocks (e.g. restricted boltzmann machine). Parameters in DBN would be fine-tuned finally.

Table 1 Feature extraction map in feature-SVM

Stochastic characteristic	Dimension
Acceleration (x, y, z) and angular speed (x, y, z) mean	6
Acceleration and angular speed standard deviation	6
Acceleration and angular speed variance	6
Acceleration and angular speed interquartile range	6
Acceleration correlation coefficient	3
Angular speed correlation coefficient	3
Acceleration and angular speed MAD	6
Acceleration and angular speed root mean square	6
Acceleration and angular speed energy (mean square of DFT items)	6
Total	48

Table 2 Deep belief network's results for motion sequence recognition

Learning rate	DBN structure	1000 epoch accuracy (%)	2000 epoch accuracy (%)
0.10	[600, 600]	97.5736	97.5736
	[1000, 1000]	97.5713	97.5713
	[600, 600, 600]	97.7295	97.7736
	[600, 600, 600, 600]	97.7104	97.7210
0.03	[600, 600]	97.3512	97.6059
	[1000, 1000]	97.4776	97.7585
	[600, 600, 600]	97.8736	**97.9571**
	[600, 600, 600, 600]	97.7753	97.8560
0.01	[600, 600]	97.3340	97.5913
	[1000, 1000]	97.3933	97.5407
	[600, 600, 600]	97.5852	97.7107
	[600, 600, 600, 600]	97.5115	97.6781

We get the best result with accuracy of 97.9571 % using the DBN model. The structure with the highest classification accuracy has 3 hidden layers, and each layer has 600 nodes. Learning rate in the process is 0.03 with 20 % attenuation every 500 epochs. Table 2 records experiment results in more details.

5.4 The DBCNN Model

Table 3 gives the classification accuracy under distinct parameters configuration. As we could see, our DBCNN model has achieved a recognition accuracy of 98.66 %, which outperforms the other two methods. Left part of Fig. 5 gives a results comparison among these three models.

Table 3 DBCNN's results for motion sequence recognition

#Kernels	1st conv filter	1st pooling	2nd conv filter	2nd pooling	Accuracy (%)
[20,40]	25 × 6	3 × 1	5 × 1	2 × 1	98.06
[30,40]	25 × 6	3 × 1	5 × 1	2 × 1	98.22
[30,50]	25 × 6	3 × 1	5 × 1	2 × 1	98.40
[40,40]	25 × 6	3 × 1	5 × 1	2 × 1	**98.66**
	33 × 6	4 × 1	5 × 1	2 × 1	98.57
[40,50]	25 × 6	3 × 1	5 × 1	2 × 1	98.58
[45,35]	25 × 6	3 × 1	5 × 1	2 × 1	98.42
[45,45]	25 × 6	3 × 1	5 × 1	2 × 1	98.52

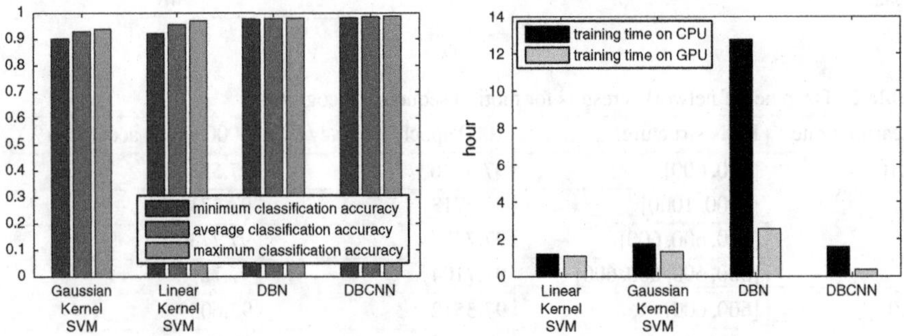

Fig. 5 Recognition accuracy comparison and training time comparison

In our experiments, a GTX-660 graphic unit with 2G GPU RAM is used. Training time using DBCNN model is about 22 min on GPU. As a contrast, it takes about 3 times longer to train the Feature-SVM model on GPU, and about 7 times longer to train the DBN model on GPU. Training Feature-SVM model and DBN model on CPU might take much longer time. Right part of Fig. 5 shows an approximate training time of these models on CPU and GPU.

6 Conclusion

In this paper, we proposed a novel motion sequence recognition model, DBCNN. First of all, we pre-process motion sequences to form "data-bands". Then we train a specially designed convolutional neural network using these "data-bands" as input. Our experiments on 6DMG demonstrate that our model can get a higher recognition accuracy than state-of-art algorithms in less training time.

References

1. Amma C, Gehrig D, Schultz T (2010) Airwriting recognition using wearable motion sensors. In: Proceedings of the 1st augmented human international conference. ACM
2. Schlömer T et al (2008) Gesture recognition with a Wii controller. In: Proceedings of the 2nd international conference on Tangible and embedded interaction. ACM
3. Timo P (2005) Accelerometer based gesture recognition using continuous HMMs. Springer, Berlin, pp 639–646 (Pattern Recognit Image Anal)
4. Kallio S, Juha K, Jani M (2003) Online gesture recognition system for mobile interaction. In: IEEE International conference on systems, man and cybernetics
5. Liu J et al (2009) uWave: accelerometer-based personalized gesture recognition and its applications. Pervasive Mobile Comput 5:657–675
6. Wu J et al (2009) Gesture recognition with a 3-d accelerometer. Ubiquitous intelligence and computing. Springer, Berlin, pp 25–38
7. Wang J-S, Chuang F-C (2012) An accelerometer-based digital pen with a trajectory recognition algorithm for handwritten digit and gesture recognition. IEEE Trans Ind Electron 2998–3007
8. Chen M, AlRegib G, Juang B-H (2012) A new 6d motion gesture database and the benchmark results of feature-based statistical recognition. In: IEEE international conference on emerging signal processing applications
9. LeCun Y et al (1998) Gradient-based learning applied to document recognition. Proc IEEE 2278–2324
10. Cho S-J et al (2006) Two-stage recognition of raw acceleration signals for 3-D gesture-understanding cell phones. In: Tenth international workshop on frontiers in handwriting recognition
11. Bishop CM (1995) Neural networks for pattern recognition
12. Bengio Y (2009) Learning deep architectures for AI. In: Machine learning foundations trends, pp 1–127
13. Hinton GE (2007) Learning multiple layers of representation. Trends Cognit Sci 428–434
14. Hinton G, Osindero S, Teh Y-W (2006) A fast learning algorithm for deep belief nets. Neural Comput 1527–1554
15. Hinton GE, Salakhutdinov RR (2006) Reducing the dimensionality of data with neural networks. Science 504–507
16. Hinton GE (2012) A practical guide to training restricted boltzmann machines. In: Montavon G, Orr GB, Müller K-R (eds) Neural networks: tricks of the trade, vol 7700, 2nd edn. LNCS. Springer, Heidelberg, pp 599–619
17. Bergstra J et al (2011) Theano: Deep learning on gpus with python. In: BigLearning workshop, NIPS
18. Bengio Y et al (2007) Greedy layer-wise training of deep networks. In: Advances in neural information processing systems
19. Chang C-C, Lin C-J (2011) LIBSVM: a library for support vector machines. ACM Trans Intell Syst Technol

Vibration Attenuation of the Electromechanical System by a Double Impact Element

Marek Lampart and Jaroslav Zapoměl

Abstract The goal of this paper is to concentrate on dynamics and vibration attenuation of the electromechanical system flexibly coupled with a baseplate and damped by a double impact element. The model is constructed with four degrees of freedom in the mechanical oscillating part, three translational and one rotational. The system movement is reported by five mutually coupled second-order ordinary differential equations. There are the most important nonlinearities: stiffness of the support spring elements and internal impacts. As the main results it is shown that the double impact damping device massively attenuates vibrations of the rotor frame for a suitable choice of mass of the impact elements in dependence of the excitation amplitude of the baseplate.

Keywords Electromechanical system · Impact damper · Nonlinear stiffness · Impacts · Vibration attenuation

1 Introduction

The impacts of solid bodies are an important mechanical phenomena. Their presence can be observed during a large number of natural and technological processes. The impacts are characterized by short duration body collisions, very large impact forces and by near sudden changes of the system state parameters. The experience and

M. Lampart (✉)
Department of Applied Mathematics & IT4Innovations,
VŠB—Technical University of Ostrava, Ostrava, Czech Republic
e-mail: marek.lampart@vsb.cz; http://www.vsb.cz

J. Zapoměl
Department of Applied Mechanics, VŠB—Technical University
of Ostrava, Ostrava, Czech Republic
e-mail: jaroslav.zapomel@vsb.cz; http://www.vsb.cz

J. Zapoměl
Department of Dynamics and Vibrations,
Institute of Thermomechanics, Ostrava, Czech Republic

A. Abraham et al. (eds.), *Intelligent Data Analysis and Applications*,
Advances in Intelligent Systems and Computing 370,
DOI 10.1007/978-3-319-21206-7_3

theoretical analyses show that the behavior of the impact systems is highly nonlinear, highly sensitive to initial conditions and instantaneous excitation effects, leading frequently to irregular vibrations and nearly unpredictable movements. The behavior of each system where the body collisions take place is different, and therefore, each of them must be investigated individually.

Because of the practical importance, a good deal of attention is focused on analysis of vibro-impact systems, where the vibrations are governed by the momentum transfer and mechanical energy dissipation through the body collisions. This is utilized for impact dampers applied to attenuate high-amplitude oscillations, such as those appearing in subharmonic, self-excited and chaotic vibrations.

Even though the problem of impacts is very old, the new possibilities of its investigation enabled by efficient computational simulations appeared at the end of the 20th century. [1] studied the dynamic behavior of an impact damper for vibration attenuation of an externally loaded and self-excited cart moving in one direction. The damping was produced by impacts of a point body colliding with the cart walls. A number of authors have dealt with the so-called non-ideal problem, which means that the power of the source exciting the oscillator is limited.

The article of [12], who extended Chaterjee's model with the cart by attaching a rotor driven by a motor, belongs in this category. Application of a non-ideal model to the gear rattling dynamics was done by [2]. A new mechanical model with clearances for a gear transmission was reported by [7]. Their model has time varying boundaries and impacts between two gears occur at different locations. The horizontal movement of a cart excited by a rotating particle and damped by an impact damper formed by a point body bouncing on the cart walls was investigated by [3].

Wagg [11] studied periodic sticking motion of a two degree-of-freedom impact oscillator for several parameter ranges. He concentrated on influence of the forcing frequency and the coefficient of restitution on the system vibrations and described three types of post-bifurcation behavior which occur if the coefficient of restitution takes a zero value. Metallidis and Natsiavas [9] generalized their previous works on the dynamics of discrete oscillators and investigated the response of a continuous system with clearance and motion-limiting constraints. The approach used by the authors is based on the exact solution form obtained within time intervals where the system parameters remain constant. Results of the parametric studies demonstrated a strong influence of super- and subharmonic modes on the system vibrations.

Luo et al. [8] analyzed oscillation of a two degree-of-freedom system with a clearance subjected to harmonic oscillations concentrating on influence of the exciting frequency and the clearance width. Based on the sampling ranges of dynamic parameters, investigation of their influence on impact velocities, existence regions and correlative distribution of different types of periodic-impact motions of the system was the main objective of the analysis. The computational simulations confirmed a chaotic character of the oscillations. Consequently, the influence of the impact body to the cart mass ratio on its suppression was examined.

The non-ideal impact problem completed by flexible stops was analyzed by [14]. The main objection was to study the character of the vibration of the oscillator excited by an unbalanced rotor driven by a motor of limited power. The mutual interaction of a mechanical and electrical system was analyzed by [10]. The investigated unbalanced rotor of an electric motor was attached to a cantilever beam. The amplitude of its bending vibration was limited by a flexible stop. The performed simulations were aimed at studying the character of the induced vibration and at fluctuations of the current in the electric circuit. Vibrations of a rotor supported by bearings with nonlinear stiffness and damping characteristics considering its impacts against the stationary part were investigated by [13]. The shaft was represented by a beam like-body and rotation of the disc was taken into account. The impacts were described both by collisions of rigid bodies utilizing the Newton theory and by impacts with soft stops. Vibration reduction of an electromechanical system by an impact damper having rigid stops was investigated by [5]. Emphasis was put on observing the influence of the inner impacts on the character and reduction of the system vibration dependent on the geometric parameters.

In this paper, a system formed by a rotor and its casing flexibly coupled with a baseplate and of an impact body, which is separated from the casing by two, lower and upper, gaps is analyzed. The rotor is driven by a motor of limited power and from this point of view the investigated model system can be classified as non-ideal. A new contribution of the presented work consists of investigating the system oscillations as a result of a combined time variable loading caused by two sources, the rotor unbalance and the baseplate vibrations, and in investigating the interaction between the motor and its feeding electric circuit. Emphasis is put on observing the influence of the inner impacts on the character and reduction of the system vibration dependent on the width of the upper and lower clearances between the rotor frame and the impact body.

The investigated system is of great practical importance as it represents a simplified model of a rotating machine, which is excited by a ground vibration and unbalance of the rotating parts and damped by an impact damper. Results of the performed simulations contribute to better understanding of the dynamic behavior of such technological devices and of impact systems with complicated loading, in general.

2 The Vibrating System

The investigated system consists of a rotor (body 1, Fig. 1), of its cas ing (body 2, Fig. 1) and of a baseplate (body 3, Fig. 1), with which the rotor casing is coupled by a spring and a damping element. The casing and the baseplate can move in a vertical direction and the rotor can rotate and slide together with its casing. Vibration of the baseplate and unbalance of the rotor are the main sources of the casing excitation. To attenuate its oscillation a new impact damper was proposed. It consists of a housing fixed to the rotor casing and of two impact elements (body 4 and 5, Fig. 1), which are coupled with the housing by a non-linear spring. The impact bodies can move only in a vertical direction and are separated mutually and from the housing by the middle, upper and lower clearances that limits their vibration amplitude. The rotor is loaded by an external moment produced by a DC motor. Its behavior is described

Fig. 1 Model of vibrating system

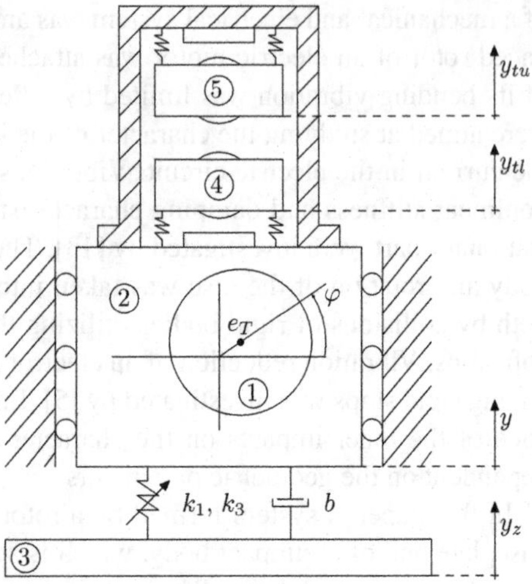

by a moment characteristic, which enables implementation of the influence of the electric parameters of the motor feeding circuit into the mathematical model of the investigated impact system.

The task was to analyze the influence of the upper and lower clearances and the mass of the impact body respectively on attenuation of the rotor frame oscillation and character of its motion.

The main goal was also to investigate the damping device where synchronization of movements can be observable, that is movements are regular, non-chaotic.

In the computational model all bodies are considered as absolutely rigid except the contact areas between the impact elements and the rotor frame. The contact damping is assumed to be proportional to the contact stiffness. The disc springs coupling the rotor casing and the baseplate have a nonlinear cubic characteristic

$$F_K = k_1\Delta + k_3\Delta^3 \tag{1}$$

where F_K is the spring force, k_1, k_3 are the stiffness parameters and Δ is the spring deformation (compression or extension). The damper between the rotor frame and the baseplate and the spring coupling the impact body with the damper housing are linear. The Hertz theory has been accepted to describe the impacts.

The investigated system has four mechanical degrees of freedom. The equation of motion have been derived by means of the Lagrange equations of the second kind. An alternative approach utilizing the liberation method is discussed in [6]. Its instantaneous position is defined by three generalized coordinates: y—vertical displacement of the rotor casing, y_{tu}—vertical displacement of the upper impact body, y_{tl}—vertical displacement of the lower impact body and Φ—angular rotation of the rotor:

$$
\begin{aligned}
(m+m_R)\ddot{y} + m_R e_T \cos(\Phi)\ddot{\Phi} &= m_R e_T \dot{\Phi}^2 \sin(\Phi) - (k_{tu}+k_{tl})y \\
&\quad + k_{tu}y_{tu} + k_{tl}y_{tl} \\
&\quad + F_{ZK} + F_{ZB} + F_{Ru} - F_{Rl} - (m+m_R)g, \\
m_{tl}\ddot{y}_{tl} &= -F_{Rs} - F_{Rl} - k_{tl}(y_{tl}-y) - m_{tl}g, \\
m_{tu}\ddot{y}_{tu} &= F_{Rs} - F_{Ru} - k_{tu}(y_{tu}-y) - m_{tu}g, \\
(J_{RT} + m_R e_T^2)\ddot{\Phi} &= -m_R g e_T \cos(\Phi) - m_R e_T \cos(\Phi)\ddot{y} + k_p \dot{\mathrm{I}}, \\
k_p \tau \dot{\mathrm{I}} &= -k_p \mathrm{I} + M_z - k_M \dot{\Phi}.
\end{aligned}
\tag{2}
$$

where ($\dot{}$), ($\ddot{}$) denote the first and second derivative with respect to time respectively.

Here, $F_\bullet$ stand for the forces:

$$
\begin{aligned}
F_{ZK} &= -k_3(y-y_z)^3 - k_1(y-y_z), \\
F_{ZB} &= -b(\dot{y}-\dot{y}_z), \\
F_{Ru} &= \begin{cases} k(y_{tu}-c_u-y)^{\frac{1}{2}}(y_{tu}-c_u-y+\frac{3}{2}\beta_u(\dot{y}_{tu}-\dot{y})) & \text{if } y_{tu}-c_u-y>0, \\ 0 & \text{if } y_{tu}-c_u-y\le 0, \end{cases} \\
F_{Rs} &= \begin{cases} k(y_{tl}-c_s-y_{tu})^{\frac{1}{2}}(y_{tl}-c_s-y_{tu}+\frac{3}{2}\beta_s(\dot{y}_{tl}-\dot{y}_{tu})) & \text{if } y_{tl}-c_s-y_{tu}>0, \\ 0 & \text{if } y_{tl}-c_s-y_{tu}\le 0, \end{cases} \\
F_{Rl} &= \begin{cases} k(y-c_l-y_{tl})^{\frac{1}{2}}(y-c_l-y_{tl}+\frac{3}{2}\beta_l(\dot{y}-\dot{y}_{tl})) & \text{if } y-c_l-y_{tl}>0, \\ 0 & \text{if } y-c_l-y_{tl}\le 0, \end{cases}
\end{aligned}
$$

where c_u, c_s and c_l stand for clearances between the frame and the upper impact element, upper and lower impact elements and the frame and the lower impact element, respectively, referred to the case when the system takes the equilibrium position.

It can be assumed, without loss of generality, that in the beginning, the system is at rest and takes the equilibrium position with no contacts between the impact bodies and the rotor frame. Then the initial conditions are given as follows

$$
\begin{aligned}
&\dot{y}(0)=0,\ \dot{y}_{tu}(0)=0,\ \dot{y}_{tl}(0)=0,\ \dot{\Phi}(0)=0,\ \Phi(0)=3/2\,\pi,\ i(0)=0 \\
&k_3 y(0)^3 + k_1 y(0) + (m+m_{tu}+m_{tl}+m_R)g = 0, \\
&k_{tu}(y_{tu}(0)-y(0)) + m_{tu}g = 0, \\
&k_{tl}(y_{tl}(0)-y(0)) + m_{tl}g = 0.
\end{aligned}
\tag{3}
$$

The last three conditions of (3) can be simplified and solved in algebraic form using Cardano's formulas, analogous proof was provided in [5].

Let us simulate the vibration of the baseplate with the map

$$
y_z(t) = A\,(1-\mathrm{e}^{-\alpha t})\sin(\omega t) \tag{4}
$$

where A is the amplitude, α is the constant determining how fast the vibration of the baseplate becomes a steady state and ω stands for the excitation frequency, so $y_z(0)=0$ and $\dot{y}_z(0)=0$.

Table 1 Parameters of the system (2)

Value	Quantity	Format	Description
m	100	kg	Mass of the damping body
m_R	40	kg	Mass of the rotor
m_{tu}		kg	Mass of the upper impact element
m_{tl}		kg	Mass of the lower impact element
k_1	1.5×10^5	N m^{-1}	Linear stiffness coefficient
k_3	6×10^{10}	N m^{-3}	Cubic stiffness coefficient
J_{RT}	5	kg m^2	Moment of inertia of the rotor
b	1.5×10^3	N s m^{-1}	Damping coefficient of the suspension
k	4×10^7	N m$^{-3/2}$	Contact stiffness parameter
β_u	3000	s	Contact damping parameter
β_s	3000	s	Contact damping parameter
β_l	3000	s	Contact damping parameter
k_{tu}	8×10^4	N m^{-1}	Coupling stiffness of the upper impact element
k_{tl}	8×10^4	N m^{-1}	Coupling stiffness of the lower impact element
e_T	2	mm	Eccentricity of the rotor center of gravity
Φ		rad	Rotation angle of the rotor
M_Z	100	N m	Starting moment
k_M	8	N m s rad^{-1}	Negative of the motor characteristic slope
α	1	s^{-1}	Parameter of the baseplate excitation
ω		rad s^{-1}	Baseplate excitation frequency
A		mm	amplitude of y_z
k_c	4×10^7	N m^{-1}	Contact stiffness
b_c	3×10^3	N s m^{-1}	Coefficient of contact damping
c_u	5	mm	Upper clearance
c_s	5	mm	Middle clearance
c_l	5	mm	Lower clearance

3 Main Results

The following simulations are performed for the baseplate excitation frequencies of $\omega = 101$ rad s^{-1}and $\omega = 141$ rad s^{-1} for the system parameters summarized in Table 1. Next, parameters c_u, c_s and c_l are assumed to be equal for simplicity. The mass of the impact elements m_{tu}, m_{tl} and A the amplitude of y_z were taken as variables in the performed analysis. In the following, behavior of the system in dependence on the impact elements mass and amplitude of the baseplate excitation.

Figure 2 shows dependence of the peak-to-peak amplitude of the frame vibration on the excitation amplitude for two resonance frequencies of 101 rad s^{-1} and 141 rad s^{-1}. It is evident that it has an increasing character in both cases. In the case

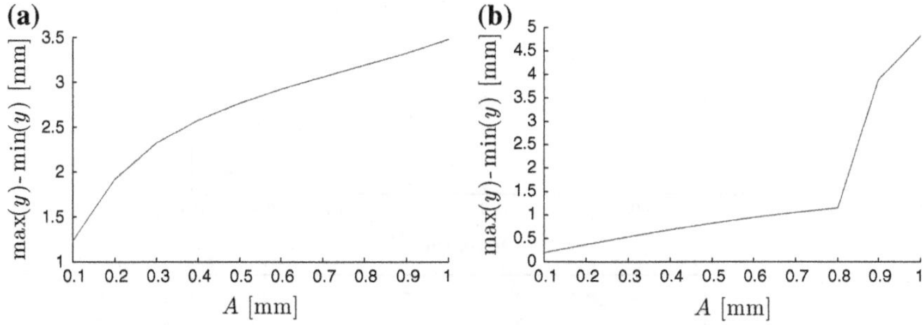

Fig. 2 Dependence of $\max y - \min y$ on A the amplitude of y_z, without damping device, for: **a** $\omega = 101$ rad s^{-1}, and **b** $\omega = 141$ rad s^{-1}

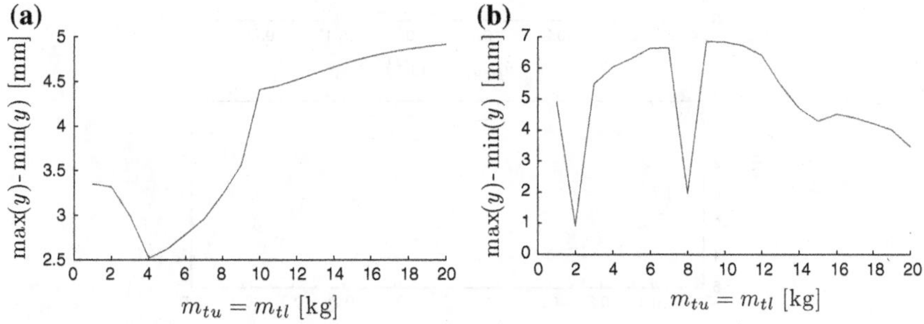

Fig. 3 Dependence of $\max y - \min y$ on $m_{tu} = m_{tl}$ the mass of the upper and lower impact element for $A = 1$ mm and: **a** $\omega = 101$ rad s^{-1}, and **b** $\omega = 141$ rad s^{-1}

of the resonance frequency of 141 rad s^{-1} there is a sharp increase after exceeding the excitation amplitude of 0.8 mm.

The maximum damping effect for the excitation frequency 101 rad s^{-1} is achieved if the impact elements have the mass of 4 kg, Fig. 3a. As evident from Fig. 4 the vibrations are regular and the damping elements collide both mutually and with the frame.

For the excitation frequency 141 rad s^{-1} the maximum vibrations attenuations is observed for the masses of the impact elements 2 and 8 kg, Fig. 3b. In this case the damping is dominantly produced by inertial effects of the impact elements and the collisions almost do not occur, Fig. 5.

The bifurcation diagram drawn in Fig. 6a shows that vibration of the frame induced by the excitation frequency 101 rad s^{-1} is regular or very close to regular in the whole extent of the investigated masses of the impact bodies. It is also evident that the frame performs complicated oscillations which except the principal contain the subharmonic components and that their character is independent on the damping element masses.

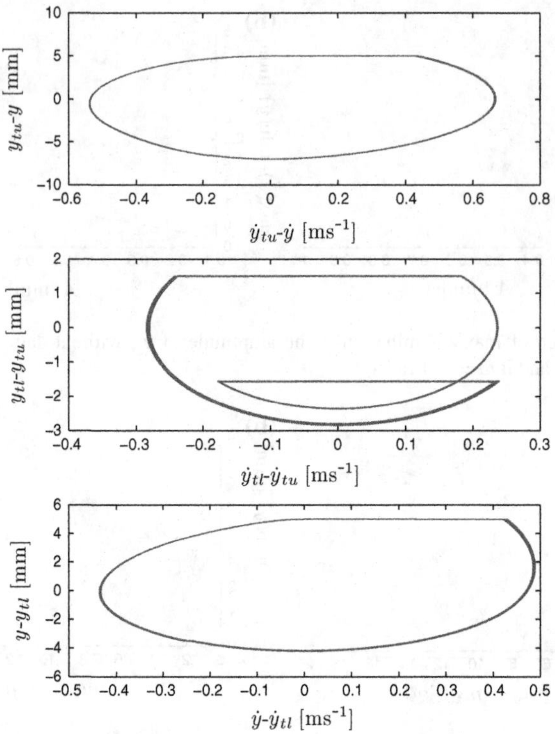

Fig. 4 Phase portraits with respect to the change of the angular frequency and mass of the impact elements, parameters of the system are given in Table 1, here $A = 1$ mm, $\omega = 101$ rad s^{-1} and $m_{tu} = m_{tl} = 4$ kg

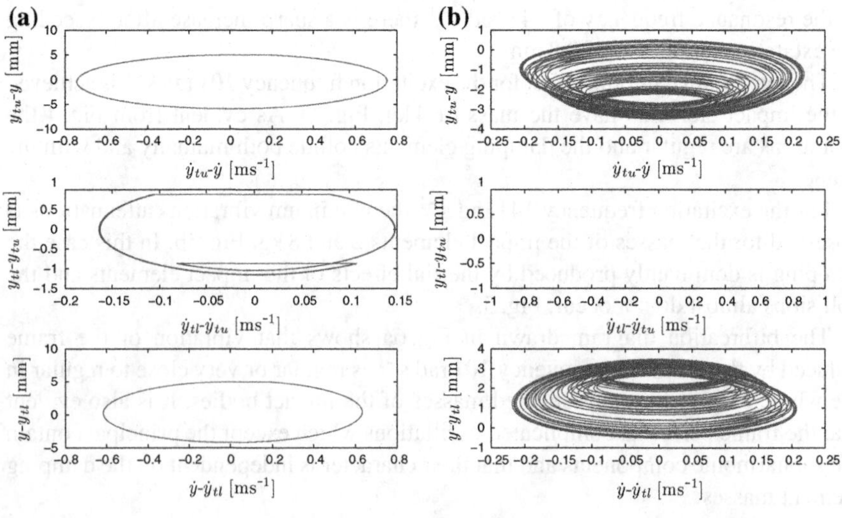

Fig. 5 Phase portraits with respect to the change of the angular frequency and mass of the impact elements, parameters of the system are given in Table 1, here $A = 1$ mm and: **a** $\omega = 141$ rad s^{-1} and $m_{tu} = m_{tl} = 2$ kg, **b** $\omega = 141$ rad s^{-1} and $m_{tu} = m_{tl} = 8$ kg

For the excitation frequency 141 rad s^{-1} motion of the frame is two-periodic (or close to it) for the weight of the impact bodies less than 14 kg with two exceptions referred to masses of 2 and 8 kg, see Fig. 6b. They correspond to the case when amplitude of the frame vibration is minimum (compare with Fig. 3). Change of the vibration character occurrs after the mass of the impact elements exceeds the value of 14 kg. The regular motion converts into quasi-periodic or even chaotic. The exception is a narrow band around the mass of 15 kg for which the motion starts to be regular.

The Fourier spectrum of the time history of the frame vertical displacement in Fig. 7a which is referred to the excitation frequency 141 rad s^{-1} and mass of the impact elements 8 kg has a line character with several dominant frequencies. This corresponds to a regular motion consisting mostly of the principal and two subharmonic components. If mass of the impact elements is increased to 14 kg, character of the Fourier spectrum changes Fig. 7b. Side bands appear around the dominant frequencies and in some areas the spectrum takes a broad-band character.

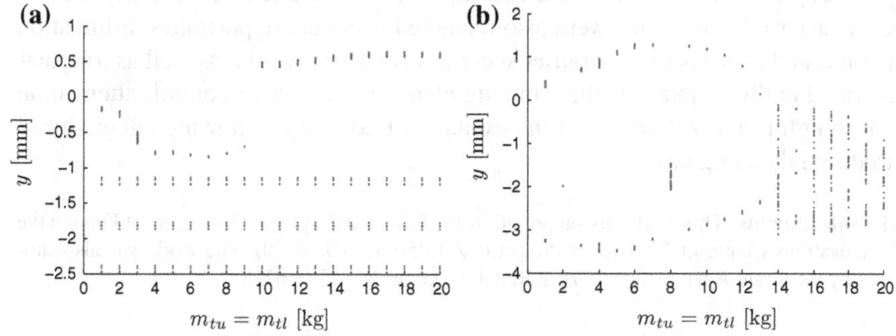

Fig. 6 Bifurcation diagrams of y with respect to mass of the impact elements $m_{tu} = m_{tl}$ for the angular frequency: **a** $\omega = 101$ rad s^{-1}, **b** $\omega = 141$ rad s^{-1}

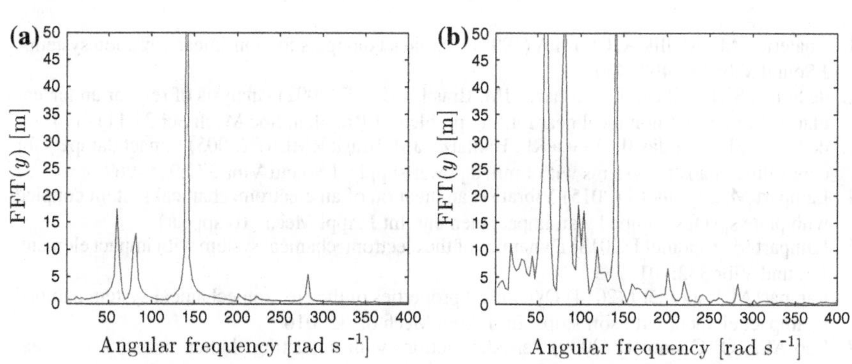

Fig. 7 Fourier spectra for parameters given in Table 1 for the angular frequency $\omega = 141$ rad s^{-1} and mass of the impact elements: **a** $m_{tu} = m_{tl} = 8$ kg, **b** $m_{tu} = m_{tl} = 14$ kg

4 Conclusions

In this paper, it was initiated and analyzed a new electromechanical system damped by a newly designed impact element with soft stops dependent on several parameters: the excitation amplitude, the excitation frequency and namely mass of the impact element. This model was inspired by real frequently occurring technological problems when electromechanical rotating machines are excited by a combined loading produced by the rotor unbalance and ground vibrations. The equations of motions were solved numerically by the explicit Runge-Kutta method using Matlab. The computational simulations showed that the excitation amplitude of vibrations of the baseplate played a key role here. These simulations also proved that application of the impact body arrived at a significant decrease of vibration amplitude of the rotor frame.

It was observed that for given parameters of the model, the damping element can work as an inertia (there are no impacts: $A = 1\,\mathrm{mm}$, $\omega = 141\,\mathrm{rad\,s^{-1}}$, $m_{tu} = m_{tl} = 8\,\mathrm{kg}$) or as an impact damper (impacts occur: $A = 1\,\mathrm{mm}$, $\omega = 141\,\mathrm{rad\,s^{-1}}$, $m_{tu} = m_{tl} = 2\,\mathrm{kg}$). Both situations were investigated and commented on. Furthermore, the character of movements were also remarked using phase portraties, bifurcation diagrams and analysis of the Fourier spectra. They show regular as well as irregular patterns. Finally, if mass of the damping element is in active control, attenuation is meaningful in a wide range of the excitation frequencies, showing effectivity of introduced damping device.

Acknowledgments This work was supported by the European Regional Development Fund in the IT4Innovations Centre of Excellence Project (CZ.1.05/1.1.00/02.0070). The work was also supported by the Grant Agency of the Czech Republic, Grant No. 15-06621S.

References

1. Chaterjee AK, Mallik A, Ghosh A (1995) On impact dampers for non-linear vibration systems. J Sound Vibr 187:403–420
2. de Souza SLT, Caldas IL, Balthazar JM, Brasil RMLRF (2002) Analysis of regular and irregular dynamics of a non ideal gear rattling problem. J Brayilian Soc Mech Sci 24:111–114
3. de Souza SLT, Caldas IL, Viana RL, Balthazar JM, Brasil RMLRF (2005) Impact dampers for controlling chaos in systems with limited power supply. J Sound Vibr 279:955–967
4. Lampart M, Zapoměl J (2015) Vibration attenuation of an electromechanical system coupled with plate springs damped y an impact element. Int J Appl Mech (To appear)
5. Lampart M, Zapoměl J (2013) Dynamics of the electromechanical system with impact element. J Sound Vibr 332:701–713
6. Lampart M, Zapoměl J (2014) Dynamical properties of the electromechanical system damped by impact element with soft stops. Int J Appl Mech 6:1450016
7. Luo ACJ, O' Connor D (2009) Periodic motions with impacting chatter and stick in a gear transmission system. ASME J Vibr Acoust 131
8. Luo GW, Lv XH, Shi YQ (2014) Vibro-impact dynamics of a two-degree-of freedom periodically-forced system with a clearance: diversity and parameter matching of periodic-impact motions. Int J Non-Linear Mech 65:1–286
9. Metallidis P, Natsiavas S (2000) Vibration of a continuous system with clearance and motion constraints. Int J Non-Linear Mech 35:675–690

10. Půst L (2008) Electro-mechanical impact system excited by a source of limited power. Eng Mech 6:391–400
11. Wagg DJ (2005) Periodic sticking motion in a two-degree-of-freedom impact oscillator. Int J Non-Linear Mech 40:1076–1087
12. Warminski J, Balthazar JM, Brasil RMLRF (2001) Vibrations of a non-ideal parametrically and self-excited model. J Sound Vibr 245:363–374
13. Zapoměl J, Fox CHJ, Malenovský E (2001) Numerical investigation of a rotor system with disc-housing impact. J Sound Vibr 243:215–240
14. Zukovic M, Cveticanim L (2009) Chaos in non-ideal mechanical system with clearance. J Vibr Control 8:1229–1246

Ear Feature Extraction Using a DWT-SIFT Hybrid

Lamis Ghoualmi, Amer Draa and Salim Chikhi

Abstract Human ear recognition is a new biometric technology which competes with other powerful biometrics modalities such as fingerprint, face and iris. In this paper we present a hypridised approach for ear biometric feature extraction named DWT-SIFT based on the combination of global and local approach named Wavelets and SIFT respectively. The proposed approach has been evaluated on two ear biometric databases, namely IIT Delhi and USTB 2. For performance evaluation of the proposed method we compute the false rejection rate (FRR), the false acceptance rate (FAR), accuracy and the needed time for ear authentication. Experimental results show that the proposed approach allows getting a higher accuracy and less time consumption compared to basic SIFT and wavelets based ear authentication systems taken individually.

Keywords Ear biometric authentication · Discrete wavelet transform · Scale invariant feature transform.

1 Introduction

Biometric authentication [1] can be defined as the automatic recognition of a person based on his physiological or behavioral characteristics. Biometrics has been widely used in many official and commercial identification systems, especially those involving automatic access control. The most common modalities used in biometrics are face, iris and fingerprint. However, these modalities suffer from some drawbacks.

L. Ghoualmi (✉) · A. Draa · S. Chikhi
MISC Laboratory, Constantine2-Abdelhamid Mehri University,
25000 Constantine, Algeria
e-mail: lamis.ghoualmi@univ-constantine2.dz

A. Draa
e-mail: draa.amer@gmail.com

S. Chikhi
e-mail: salim.chikhi@univ-constantine2.dz

A. Abraham et al. (eds.), *Intelligent Data Analysis and Applications*,
Advances in Intelligent Systems and Computing 370,
DOI 10.1007/978-3-319-21206-7_4

For instance the face can be affected by age, health and facial expressions, the iris is an intrusive modality in addition to its high cost, and the fingerprint can be also affected by some medicine, burns or ink on fingers [2]. Human ear recognition is a new biometric technology, the french criminologist Bertillon [3] was the first to suggest that people can be identified by the shape of their outer ear. A bit later, the American police officer Iannarelli [4] proposed the first ear recognition system based on 12 features . Iannareli experimentally found that ten thousand ears are different even in identical twins. In fact, the ear has attracted the interest of biometrics community because of its advantages; it has small size which allows speeding up the system and increasing its efficiency. The ear has also a uniform distribution color which insures that all information will be conserved when converting it into gray scale [5]. The ear does not need much collaboration from the subject and it can even be captured without his knowledge from far distances [6]. Thus, ear biometrics is a good choice because it offer a very good compromise between accuracy and cost.

In this paper, we present a new approach for ear biometrics that combine global and geometrical approaches for ear feature extraction which are: the Discrete Wavelet Transform and the Scale Invariant Feature Transform. The aim of the proposed approach is selecting the most discriminative features, increasing the accuracy of the system and reducing the authentication time by extracting the SIFT features from the low frequency sub-band (LL) only. Some common work applied this combination for biometrics recognition. Kumar and Sathidevi [7] proposed an approach for face recognition that extract the SIFT features from low frequency (LL) and high frequency sub-band (HH). Karanwal et al. [8] design a method that decompose face and fingerprint images in low frequency sub-band (LL) and three high frequency sub-bands (LH/HL/HH).The latter are fused together to produce a unique image then the SIFT features are extracted. The rest of the paper is organised as follows. Literature review related to ear biometrics is provided in Sect. 2. In Sect. 3, we present the proposed DWT-SIFT based ear biometrics feature extraction approach. In Sect. 4, experimental evaluation is carried out on two biometric databases. Finally, some conclusions and future research directions are drawn up in Sect. 5.

2 Related Work

The most popular approaches to ear recognition can be divided into two categories [9] including geometrical and global approaches. In geometrical approaches category, we cite the work of Burge and Burge [10], this approach is based on building neighborhood graphs and Voronoi diagrams of the detected edges. Choras [11] proposed a geometric feature extraction method inspired by the work of Iannarelli called the Concentric Circles approach which uses a reference points for a number of concentric circles with predefined radii. The intersection points of the circles and the ear contours are used as feature points [12]. Yuan and Tian [13] applied a method that allows localising human ears form side face images named Ear Contour Detection, it is based on Edge Tracking. The Scale Invariant Feature Transform (SIFT) has been used on ear feature extraction in [6]. SIFT allows extracting the landmark even in images with small pose variations and varying brightness conditions [14].

The global approaches represents the second category of feature extraction methods. A Principal Component Analysis (PCA)-based technique was applied for ear biometrics by Victor et al. [15] and Chang et al. [16]. The PCA-based technique uses the computed eigenvalues and eigenvectors from the set of training images to select an ear space within the largest eigenvalues. Active Shape Models (ASMs) were applied by Lu [17] to model the shape and local appearance of the ear in a statistical manner. The ASMs were used to is to extract ear shape features for ear biometric recognition. The Force Field Transform (FFT) was introduced by Hurley et al. [18, 19], where each pixel in the image is treated as a force source directed to all the other pixels. Sana and Gupta [20] developed an approach for ear biometrics based on Haar Wavelets who decompose the detected image and compute coefficient matrices of the wavelet transforms which are clustered in its feature template. HaiLong and Mu [21] used the low-frequency sub-images, obtained by utilising a two-dimensional wavelet transform, and then extracted features by applying an orthogonal centroid algorithm.

3 The Proposed Approach

As described in the chart of Fig. 1, the proposed system is composed of three steps: pre-processing, feature extraction and matching steps.

3.1 Preprocessing Stage

Image enhancement aims at improving the quality of an image by increasing the contrast and removing noise and blur from it in order to improving the recognition rate. In the literature, many techniques for improving image contrast have been proposed [22]. Among these techniques, Histogram equalisation (HE) which is a technique for adjusting image intensities in order to enhance its contrast. It operates by dividing the histogram into classes containing equal numbers of pixels. However, HE might produce an over-enhancement resulting in an unnatural output image [23]. To overcome this limit, Contrast Limited Adaptive Histogram Equalisation (CLAHE) [24] was proposed. CLAHE is an extension to traditional Histogram Equalisation (HE) technique. CLAHE enhances the contrast by dividing an image into multiple non-overlapping regions and performs histogram equalisation for each region separately. Then these regions are combined to get the entire enhanced image. In our work, we use the Contrast Limited Adaptive Histogram Equalisation for improving the contrast of ear images. Figure 2 shows a sample ear image before and after enhancement.

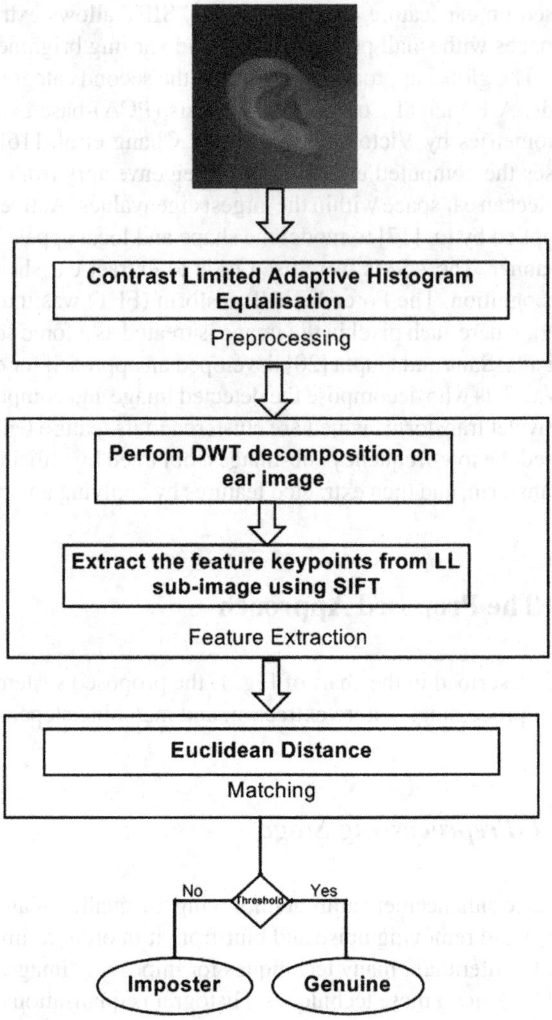

Fig. 1 The proposed approach

3.2 Feature Extraction

In this subsection, we present the used techniques for the proposed approach which are: the Discrete Wavelets Transform and the Scale Invariant Feature Transform.

3.2.1 Wavelets Transform

In recent years, the wavelet transform emerged in the field of Biometrics. Wavelets Transform can be thought of as an extension of the classic Fourier transform.

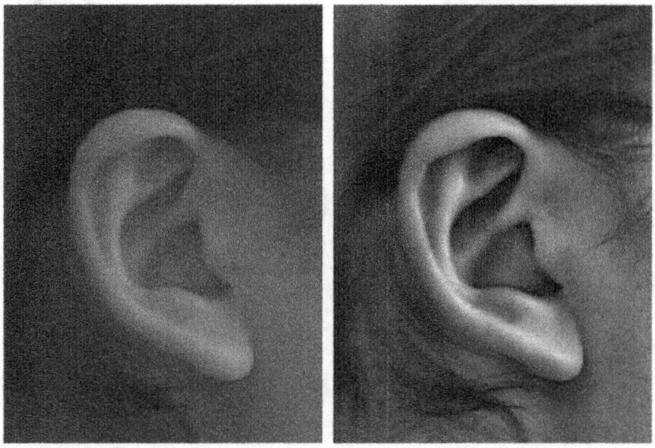

Fig. 2 Ear biometrics before and after enhancement

It has better space-frequency localization compared with the traditional Fourier analysis. Wavelets have a varying window size, being wide for slow frequencies and narrow for the fast ones, thus leading to an optimal time-frequency resolution in all frequency ranges [25]. In mathematics, the wavelet transform function can be described in formula 1:

$$X_w(a,b) = \frac{1}{\sqrt{|a|}} \int_{-\infty}^{+\infty} x(t)\psi(\frac{t-b}{a})dt \tag{1}$$

where $\psi(t)$ is called mother wavelet and b is the translation factor and a is the scale factor. There are two types of wavelet transform that are the Continuous Wavelet Transform(CWT) and the Discrete Wavelet Transform(DWT). The discrete wavelet transform is a special case of the continuous wavelet transform that provides a compact representation of a signal in time and frequency that can be computed efficiently. The 2D DWT is a very modern mathematical tool. It is used in compression, denoising and watermarking and biometrics applications.

In this paper, we use the Haar transform wavelets which is conceptually simple and fast and allows increasing detail in a recursive manner. Haar transform decomposes an ear image into one low frequency sub-band (LL) and three high frequency sub-bands (LH/HL/HH). In the proposed approach, only LL sub-bands of the first level is kept. The latter represents the original ear image with a smallest dimension and preserves the ear image main information. Figure 3 illustrates wavelet decomposition.

3.2.2 Scale Invariant Feature Transform

The Scale Invariant Feature Transform (SIFT) was originally developed for general purpose of object recognition. Recently, SIFT has been applied in biometric recognition systems and has proven its efficiency [26, 27]. It transform image data into scale

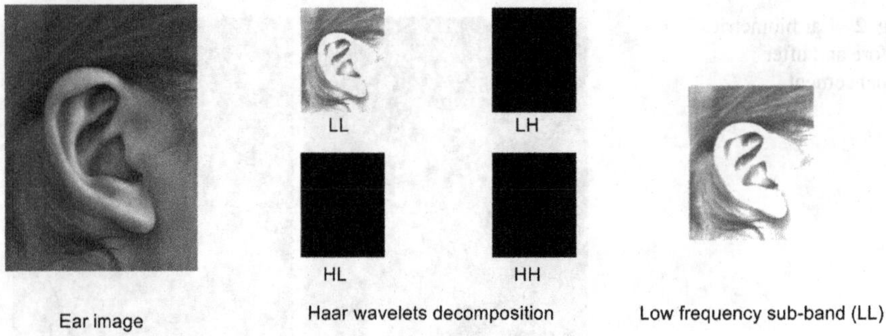

Fig. 3 Wavelet decomposition

invariant coordinates relative to local features that are invariant to rotation, scaling and partially invariant to changes to illumination conditions, as well as affine transformation of images. The generation of image feature sets involves the following stages [14]:

1. **Scale-space extrema detection**: This step identifies all scales and image locations. It is efficiently implemented using a Difference-of-Gaussian function to detect potential interest points that are invariant to scale. The scale space of an image is defined as a function $L(x, y, \sigma)$ that is produced from the convolution of a variable scale Gaussian function $G(x, y, \sigma)$ with an input image $I(x, y)$ as described in formula (2).

$$L(x, y, \sigma) = G(x, y, \sigma) * I(x, y) \tag{2}$$

Where, σ is the scale of blurring. The $2D$ Gaussian kernel is given by equation (3).

$$G(x, y, \sigma) = \frac{1}{2\Pi\sigma^2} e^{\frac{-(x^2+y^2)}{2\sigma^2}} \tag{3}$$

To efficiently detect the stable keypoint locations in scale space, Lowe [14] used scale space extrema in the difference of gaussian function convolved with the image $D(x, y, \sigma)$, which is computed from the difference of two nearby scales separated by a constant multiplicative factor k as described in Formulas (4) and (5).

$$D(x, y, \sigma) = (G(x, y, k\sigma) - G(x, y, \sigma)) * I(x, y) \tag{4}$$

$$D(x, y, \sigma) = L(x, y, K\sigma) - L(x, y, \sigma)) \tag{5}$$

To find the local maxima and minima of $D(x, y, \sigma)$, each sample point is compared to its eight neighbors in the current image and nine neighbours in the scale above and below the image. A local extrema is selected only if the current pixel is larger or smaller than the remaining pixels and is discarded otherwise.

2. **Keypoint localization**: At each candidate location, a detailed model is fit to determine location and scale. The keypoint selection is based on the measures of their stability.
3. **Orientation assignment**: Consistent orientation is assigned to each keypoint location, based on local image gradient directions. All future operations are performed on image data that are transformed relatively to the assigned orientation, scale and location for each feature, there by providing invariance to these transformations.
4. **Keypoint descriptor**: The local image gradients are measured at the selected scale in the region around each keypoint. These are transformed into a representation that allows for significant levels of local shape distortion and change in illumination.

In our approach, we use the low frequency sub-band (LL) to represent the ear image, then the most discriminant feature are extracted from the LL sub-image using SIFT.

3.3 Matching Stage

In the matching phase, the detected key-points of the input image is compared with the template collected during the enrollment phase. The matching of two feature key-points is achieved by finding candidate features to be matched on the basis of Euclidean distance. The decision of acceptance or reject of a person depends on a threshold which represents the number of matched key-points between the two templates.

4 Experimental Results

The performance of a biometric system can be evaluated in terms of recognition rate (accuracy), which is given by the formula of Eq. (6).

$$Accuracy = 100 - \frac{FRR + FAR}{2} \tag{6}$$

where: FAR is the false acceptance rate which indicates the rate at which an imposter is incorrectly accepted as genuine person and FRR is the false reject rate at which a genuine person is incorrectly rejected as an imposter. The Receiver Operator Characteristic (ROC) curve can also be plotted. The curve presents the changes of the Genuine Acceptance Rate (GAR) with changes in FAR where the Genuine Acceptance Rate is defined as: $GAR = (100 - FRR)$.

4.1 Benchmarks

In this work, we have used two databases which are IIT Delhi Database and USTB database collection 2.

4.1.1 IIT Delhi Database

The IIT Delhi Database [28] is provided by the Hong Kong Polytechnic University. It contains ear images that were collected at the Indian Institute of Technology Delhi in New Delhi. The database contains 125 subjects, and at least 3 images were taken per subject in an indoor environment, which means that the database consists of 421 images on the whole.

4.1.2 USTB 2 Database

The University of Science and Technology in Beijing (USTB) databases collection 2 [29] contains right ear images from students and teachers from USTB. This time, the number of subjects is 77. Hence the database contains 308 images on the whole, which were taken under different lighting conditions.

The IIT delhi and USTB 2 databases are divided in our experiments into two groups, the first group presents training data. We have taken 3 images per subject as training data. The remaining images are considered as test images. The number of subjects enrolled in the system is 100 and 77 subjects for IIT Delhi and USTB 2 databases respectively.

The experiments have been performed on three biometric architectures which are SIFT based verification system, Haar wavelet based verification system and our approach for ear biometrics. Tables 1 and 2 show the obtained numerical results in terms of: FRR, FAR, accuracy and the authentication mean elapsed time for ear images of each database.

In Table 1 we see that the accuracy of our approach is better than the basic SIFT and Haar wavelet based ear verification systems. Also, it takes less time in ear authentication. In Table 2 we see that Haar wavelet based verification system takes less time compared to the traditional SIFT and our approach, but its accuracy is very less compared to the other. We can observe too that the proposed DWT-SIFT based ear biometrics features extraction and the SIFT based verification system have the same accuracy but the proposed approach takes less time then the traditional SIFT-based verification system. We can say that our approach is faster than the basic SIFT because DWT-SIFT compute the SIFT descriptors on the LL ear sub-image which represents the original image without redundant information and with a smallest resolution. So, the detected key-points resulting from DWT-SIFT are more discriminative then the key-points resulting from the basic SIFT.

Table 1 Performance evaluation of the proposed approach using New Delhi database

Algorithm	FRR (%)	FAR(%)	Accuracy (%)	Time (s)
Haar wavelets	0	68.5393	65.7303	0.5587
SIFT	3.9216	0	98.0392	2.6247
Proposed approach	0	0	100	0.52268

Table 2 Performance evaluation of the proposed approach using USTB 2 database

Algorithm	FRR (%)	FAR(%)	Accuracy (%)	Time (s)
Haar wavelets	27.1605	13.6986	79.5704	0.38191
SIFT	7.4074	4.1096	94.2415	0.73353
Proposed approach	7.4074	4.1096	94.2415	0.46709

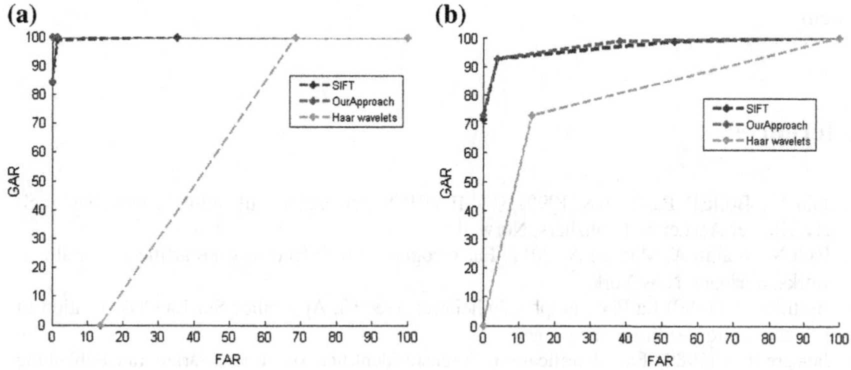

Fig. 4 Roc curves. **a** New Delhi ROC curve. **b** USTB2 ROC curve

The Receiver Operating Characteristic (ROC) curves of SIFT, Haar Wavelets and the proposed approach based ear authentication system for the two databases, IIT Delhi and USTB 2, are shown in Fig. 4. From curve a, we clearly see that the proposed approach outperforms the classical approach of SIFT and Haar wavelets based ear verification. In curve b, we see that the proposed approach and the basic approach of SIFT ear authentication perform identically and outperform the Haar Wavelets based verification system.

The obtained experimental results prove that the combination of Discrete Wavelets Transform and the Scale Invariant Feature Transform offers higher accuracy and less time consumption compared to SIFT based ear authentication system because the use haar wavelets decomposition allows eliminating redundant information and reducing the dimension of the ear image. So, the extract feature will be more discriminative.

5 Conclusion

In this paper an hypridised approach named DWT-SIFT for ear biometrics has been presented. The proposed approach uses the the Discrete Wavelets Transform (DWT) and SIFT descriptors for ear biometric feature extraction. The DWT-SIFT approach has been evaluated on two ear databases which are: IIT Delhi and USTB 2 databases. Experimental results show that the proposed approach outperforms the traditional SIFT and Haar wavelet based ear authentication systems. It allows eliminating the redundant information, reducing the resolution of ear images, extracting the most discriminative features and so offering a higher accuracy and less time consumption. Our prospect for the near future is to be guided by these results and perform a feature level fusion of ear biometrics and an other complementary modality using the proposed approach and trying to add other concepts in order to reduce the time needed for authentication and at the same time increase the accuracy of biometrics system.

References

1. Jain AK, Bolle R, Pankanti S (1999) BIOMETRICS: personal identification in networked society. Kluwer Academic Publishers, Norwell
2. Fadi N, Nuaimi A, Maamri A (2012) Ear recognition with feed-forward artificial neural networks. Springer, New York
3. Bertillon A (1890) La Photographie Judiciaire: Avec Un Appendice Sur La Classification Et L'Identification Anthropometriques
4. Iannarelli A (1989) Ear identification. Forensic identification series. Paramount Publishing company, Fremont
5. Abate AF, Nappi M, Riccio D (2006) Face and ear: a bimodal identification system, in image analysis and recognition. Lecture notes in computer science. Springer, New York, pp 297–304
6. Prakash S, Gupta P (2013) An efficient ear recognition technique invariant to illumination and pose. Telecommun Syst 52:1435–1448 (manuscript, Springer, US)
7. Kumar NAM, Sathidevi PS (2013) Wavelet SIFT feature descriptors for robust face recognition. Adv Comput Inf Technol 177:851–859
8. Karanwal S, Kumar D, Maurya R (2010) Fusion of fingerprint and face by using DWT and SIFT. Int J Comput Appl 0975 8887
9. Kumar P, Rao KN (2009) Pattern extraction methods for ear biometrics—a survey world congress on nature and biologically inspired computing (NaBIC 2009), pp 1657–1660
10. Burge M, Burge W (1998) Ear biometrics. Personal identification in networked society. Springer, New York, pp 273-286
11. Choras M (2005) Ear biometrics based on geometrical methods of feature extraction. In: Perales FJ, Draper BA (eds) Articulated motion and deformable objects, LNCS, vol 3179. Springer, New York, pp 51–61
12. Pflug A, Busch C (2012) Ear biometrics: a survey of detection, feature extraction and recognition methods. The Institution of Engineering and Technology
13. Yuan W, Tian Y (2006) Ear contour detection based on edge tracking. In: Proceedings of intelligent control and automation. IEEE Press, Dalian, China, pp 10450–10453
14. Lowe GD (2004) Distinctive image features from scale-invariant keypoints. Int J Comput Vis
15. Victor B, Bowyer K, Sarkar A (2002) An evaluation of face and ear biometrics. In: 16th international conference on pattern recognition (ICPR), vol 1, pp 429–432

16. Chang K, Bowyer KW, Sarkar A, Victor B (2003) Comparison and combination of ear and face images in appearance based biometrics. IEEE Trans Pattern Anal Mach Intell 1160–1165
17. Lu L, Zhang X, Zhao Y, Jia Y (2006) Ear recognition based on statistical shape model. In: Proceedings of international conference on innovative computing information and control, vol 3, pp 353–356
18. Hurley DJ, Nixon MS, Carter JN (2002) Force field energy functions for image feature extraction. Image Vis Comput J 6:311–318
19. Hurley DJ, Nixon MS, Carter JN (2005) Force field energy functions for ear biometrics. Comput Vis Image Underst 3:491–512
20. Sana A, Gupta P (2006) Ear biometrics: new approach. In: Proceeding ICAPR
21. Haolong Z, Mu Z (2009) Combining wavelet transform and orthogonal centroid algorithm for ear recognition. In: Proceedings of the 2nd IEEE international conference on computer science and information technology
22. Preethi SJ, Rajeswari K (2010) Image enhancement techniques for improving the quality of colour and grayscale medical images. Int J Comput Sci Eng 18–23
23. Kim DH, Cha E (2009) Intensity surface stretching technique for contrast enhancement of digital photography. Multidimens Syst Signal Process 20:81–95
24. Zuiderveld K (1994) Graphics gems IV, chap. Contrast limited adaptive histogram equalization. Academic Press Professional Inc., San Diego, pp 474–485
25. Nanni L, Alessandra L (2008) Wavelet decomposition tree selection for palm and face authentication. Pattern Recogn Lett 29:343–353
26. Badrinath G, Gupta P (2009) Feature level fused ear biometric system. In: Seventh international conference on advances in pattern recognition (ICAPR), pp 197–200
27. Ghoualmi L, Chikhi S, Draa A (2014) A SIFT-based feature level fusion of iris and ear biometrics. Multimodal pattern recognition of social signals in human computer interaction (MPRSS)
28. IIT Delhi Database. http://www4.comp.polyu.edu.hk/csajaykr/IITD/Database-Ear.htm
29. The University of Science and Technology in Beijing Database. http://www1.ustb.edu.cn/resb/en/news/news3.htm

Automatic Fuzzy Classification System for Metabolic Types Detection

Michal Prauzek, Jakub Hlavica and Marketa Michalikova

Abstract Patients suffering from obesity have different demands for medical treatment regarding the causes of their metabolic disorders. To propose new medical solutions to weight reduction, it is desirable to group patients exhibiting similar characteristics. This contribution describes an automatic fuzzy classification system capable of dividing obese patients into groups of diverse metabolic types. Metabolic data were acquired through energometry tests and bioimpedance measurements. Methods considered in this paper are particularly Principal Component Analysis used for data set's reduction and fuzzy clustering method dividing patients into groups called clusters. Newly tested patients are then classified into designed clusters. A set of statistical hypothesis testing methods is eventually applied to verify the performed classification. The designed classification system could be applied in hospitals to help the doctors with design of an individual treatment for obese patients' groups.

Keywords Fuzzy classification · Cluster analysis · PCA · Metabolic data · Obesity · Statistical hypothesis testing

1 Introduction

Obesity is a typical representative of disorders affecting particularly people in developed countries. However, it is not necessarily a consequence of an unhealthy lifestyle. Patients having complex metabolic disorders or serious diseases may suffer from obesity as well. It is important to detect whether the cause is an excessive income of food associated with a limited physical activity or an effect of metabolic problems, such as low fat-burning rate or an increased level of insulin in the blood.

M. Prauzek (✉) · J. Hlavica · M. Michalikova
Faculty of Electrical Engineering and Computer Science, Technical University of Ostrava, 17.listopadu 15/2172, 708 33 Ostrava, Czech Republic
e-mail: michal.prauzek@vsb.cz

A. Abraham et al. (eds.), *Intelligent Data Analysis and Applications*,
Advances in Intelligent Systems and Computing 370,
DOI 10.1007/978-3-319-21206-7_5

Therefore, it is convenient to group patients suffering from obesity and then, on the basis of medical examinations, describe a typology of people depending on the type of metabolism [3, 13], eventually recommending an individual treatment [11]. Such examinations could be based on metabolic energometry tests, bioimpedance measurements [9], blood analyzes, and other methods that are assessed subjectively without standardization [16, 17].

While our previous study [12] introduced fuzzy clustering method dividing obese patients into groups with similar characteristics (using data from metabolic tests mentioned above), this paper presents an automated classification system that classifies new patients into the designed groups, based on experience of medical specialists. Three system's configurations, varying in the number of metabolic parameters, are compared and assessed, and the most accurate one is suggested to use in practical applications.

Similar solution is presented in [4], where cluster analysis along with PCA are also applied. However, it does cluster metabolic parameters instead of directly clustering individual patients, as it is proposed in our contribution. Direct clusterization allows us to individualize medical treatment for obese patients.

Other possible alternatives to use in the field metabolic data analysis are adaptive neuro-fuzzy inference system [14], artificial neural networks (with respect to childhood obesity) [10], fuzzy genetic algorithms [5] or fuzzy unsupervised clustering algorithm [15].

The rest of this paper is organized into five sections as follows. Section 2 briefly introduces a metabolic data set applied in this study and methods used for data conditioning, fuzzy clusterization and system design. Section 3 describes new patients' classification into designed groups and verification methods. Results are evaluated in Sect. 4 and major conclusions are presented in Sect. 5.

2 Background and Methods

The design of the fuzzy classification system is divided into two parts: analytic part and testing part. Figure 1 shows the system's overview. The upper blocks represent the *analytic part* of system's design (described in detail in this section) while the bottom blocks, using parameters calculated in analysis, represent the *testing part* described in Sect. 3.

2.1 Metabolic Data Acquisition

Metabolic data were obtained in collaboration with patients suffering from obesity. Almost 4000 persons underwent a series of energometry tests during eight years of this research in Hradec Kralove hospital, Czech Republic. Eventually, 254 patients with a complete diagnosis in the year 2008 were chosen for this study. Energometry

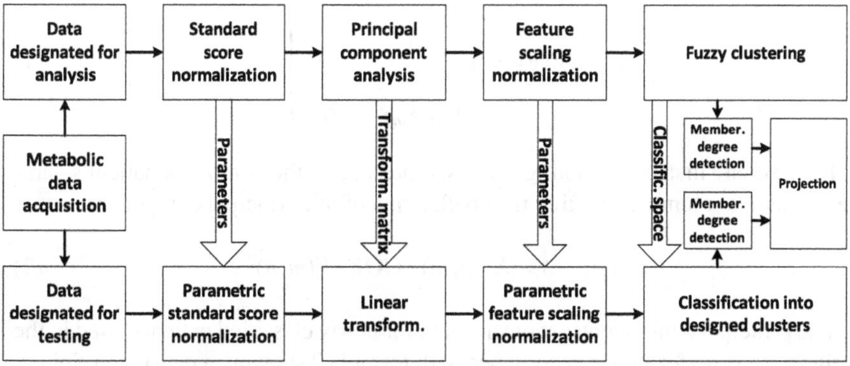

Fig. 1 An overview of the classification system

tests were based on measurements of body composition and reaction to food utilization. Additional metabolic data were obtained by bioimpedance measurements of a human body and biochemical blood samples analyses. Detailed description of the metabolic data set is provided in [12].

2.2 Data Normalization

Data normalization equalizes the metabolic data with different statistical characteristics. It ensures that each parameter of multidimensional data is treated equally in classification space, with the same weight [12]. As the normalization methods applied in the analytic part are very similar to their parametric counterparts used in the testing part, they are all described in Sect. 3.1.

2.3 Principal Component Analysis

Principal Component Analysis (PCA) is a data decorrelation method. Applying PCA, the original data set's quantity of parameters is reduced whereas an excessive information loss is prevented. The new reduced data set is composed of vectors, so-called principal components, and is obtained using a linear transformation based on variance characteristics of the original data set.

PCA was performed on the data set designated for analysis, creating new variables that represent an information entropy of the original data set.

As it is a linear method, the transformation was executed using a linear transformation matrix, described by formula 1, which is composed of PCA coefficients.

$$\mathbf{COEFF} = \begin{pmatrix} k_{11} & k_{12} & \dots & k_{1n} \\ k_{21} & k_{22} & \dots & k_{2n} \\ \vdots & \vdots & \ddots & \vdots \\ k_{m1} & k_{m2} & \dots & k_{mn} \end{pmatrix} \tag{1}$$

It is a square matrix containing m rows, representing the number of patient's parameters, and n columns, denoting the coefficients of nth principal component.

$$\mathbf{T}_2(p,n) = \mathbf{X}_2(p,m) \cdot \mathbf{COEFF}(m,n) \tag{2}$$

The principal components reconstruction for newly classified patients, used in the testing part, is performed in accordance with formula 2 showing a matrix multiplication. The original data $\mathbf{X}_2(p,m)$ are multiplied by the coefficient matrix $\mathbf{COEFF}(m,n)$ where the number of reconstructed principal components is determined by the number of matrix columns. The product is matrix $\mathbf{T}_2(p,n)$ where p is a number of patients and n is a number of reconstructed principal components [7]. Naturally, newly classified testing patients need to have the same quantity and the equal structure of parameters entering the transformation.

2.4 Cluster Estimation and Clusterization

Fuzzy clustering method is used in this study to divide patients into several groups denoted clusters. Patients belonging to particular clusters exhibit similar metabolic characteristics and can undergo a further medical analysis. One patient be classified to more clusters with respect to fuzzy membership degrees that determine the rate of affiliation to clusters.

Estimation of the number of clusters is crucial for the clustering algorithm. In multi-parameter analysis, such as metabolic types detection, two criteria called Partition Coefficient and Classification Entropy are used to estimate the optimal number of clusters. Fuzzy clusterization can be performed using fuzzy c-mean clusters [1, 2].

This clustering algorithm can calculate degree of membership to individual clusters and get the individual cluster parameters [6].

Cluster estimation and clusterization applied in this study are described in detail in [12].

3 Testing

System's testing can be executed after the patients clusterization performed in the analytic part. For testing purposes, three models were applied. Each model uses a

Table 1 Testing models description

Model	Metabolic data source	Parameters
Model A	Energometry and blood analyses tests	10
Model B	Bioimpedance and blood analyses tests	16
Model C	Energometry, bioimpedance and blood analyses tests	28

different part of the original data set, thus varying in a number of metabolic parameters. An overview of these models is presented in Table 1.

Metabolic parameters in these models are then parametrically normalized, classified into the designed clusters and eventually verified, as described in following subsections.

3.1 Parametric Normalization

The first step of the system's testing part is a normalization of new patients' metabolic parameters. It is very similar to the one performed in the analytic part where coefficients' indexes '1' and '2' were not used.

$$X_{2_{norm}} = \frac{X_2 - X_{1_{min}}}{X_{1_{max}} - X_{1_{min}}} \tag{3}$$

The parametric *feature scaling* normalization is shown in formula 3. The new patient's metabolic parameter value $X_{2_{norm}}$ is normalized using parameters $X_{1_{min}}$ and $X_{1_{max}}$ calculated in the analytic part of the design, using the original analytic data set. X_2 is a metabolic parameter to be normalized. This normalization is performed before the classification procedure. Its drawback is the fact that a resulting value might be mapped out of the classification space delimited by values $X_{1_{min}}$ and $X_{1_{max}}$.

$$X_{2_{norm}} = \frac{X_2 - \overline{X_1}}{\sigma_{1_x}} \tag{4}$$

The parametric *standard score* normalization is presented in formula 4. The new patient's metabolic value $X_{2_{norm}}$ is normalized using the statistical parameters of the original analytic data set, particularly the mean value $\overline{X_1}$ and variance σ_{1_x}. This normalization is necessary to ensure data objectivity after applying the linear transformation using the PCA coefficients matrix (see Fig. 1).

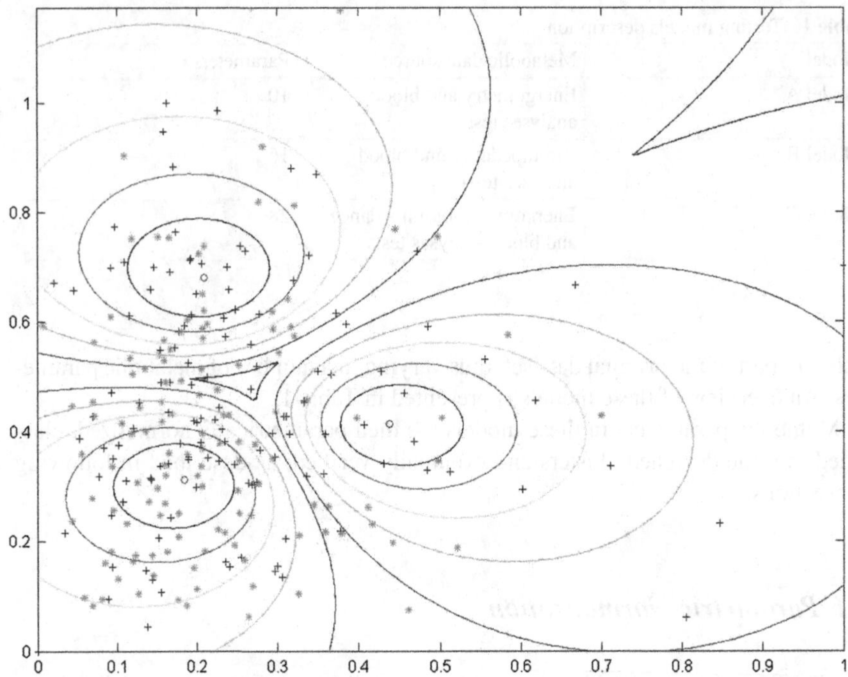

Fig. 2 Classification into classification space

3.2 Classification of New Patients

The second step of system's testing part is a classification of new patients into the classification space that was designed in analytic part in form of clusters.

In the beginning, the new patient's data, conditioned by the parametric normalization algorithms and transformed by the PCA matrix, are presented to the system. Then, the classification into clusters, which were designed in the analytic part, is performed. Afterwards, the new patient's graphic representation is projected in the classification space figure (Fig. 2) and the classification is eventually verified using statistical hypothesis testing.

Figure 2 displays the classification into classification space. Axes represent two principal components normalized into the range of 0–1 using feature scaling method. There are three clusters including particular layers and their centroids represented by tiny red circles. As for the colored marks in the figure, metabolic parameters of patients employed for analysis are represented by blue crosses. Metabolic parameters of newly classified patients (used for testing) are represented by green asterisks.

It is evident that there are also two remote values at the top edge of the figure that were incorrectly calculated by the parametric standard score method and are out of the 0–1 range. It might have been caused by measurement errors or a distinctive patient's pathology.

3.3 Verification Concept

The concept of classification's verification is based on an assessment whether the newly classified patients' exhibit similar characteristics as their counterparts clustered in analysis.

Verification process is divided into three stages:

1. Membership degree limits definition,
2. Statistical hypothesis testing,
3. Results assessment.

3.4 Membership Degree Limits Definition

As the fuzzy classification is applied in this algorithm, the classified patient is characterized by membership degrees to clusters and also by distances to cluster centroids. Therefore, there are two ways of a membership assessment:

1. Definition of limits according to distance from cluster centroids,
2. Definition of limits according to the membership to a cluster.

The first procedure is applicable in case of a uniform distribution of clusters. Thus, it can also be used in classification methods that are not based on fuzzy logic.

The second procedure is more appropriate for the algorithm presented in this study, as it reflects the calculated structure of the clusters. Each patient is assigned as many membership degrees as the specified number of clusters. It is then important to define the membership degree limits, in order to assess whether the classified patient can be considered a significant representative of the cluster.

$$\varphi = \frac{200}{n+1} \tag{5}$$

To correctly define the membership degree limit, a formula 5 was applied. The membership degree is represented by the coefficient φ, in percents. Value of φ is nonlinearly dependent on the number of clusters n. Having one cluster, the membership degree limit is, naturally, 100 %. The limits for a number of two and three clusters are 66.6 and 50 % respectively. However, this algorithm is only applicable for a limited number of clusters. Exceeding the number of approximately five clusters, the membership degree limit is too low and the differences between clusters are diminished.

In all tests, the classification into three patients' clusters was performed.

3.5 Statistical Hypothesis Testing

In this stage, clusters of classified patients are compared using statistical hypothesis testing.

Two following statistical data sets, with respect to each cluster, are compared:

1. Analytic data set with parameters exceeding membership degree limits,
2. Testing data set with parameters exceeding membership degree limits.

In order to compare these two data sets, two statistical hypothesis testing methods were applied: T-test and F-test [8].

T-test method is used to test hypothesis that two normal probability distributions having equal variance originate from the same statistical data set, thus having equal statistical characteristics. To verify the fact that both data sets are characterized by the same variance, the F-test is applied.

Statistical hypothesis testing is split into two tests:

1. Congruent test—evaluation of a correct classification,
2. Cross test—evaluation of a correct separation.

Both congruent and cross testing procedures were designed within the scope of this research. The congruent test verifies whether the patients having alike statistical characteristics were classified into the same cluster. Cross test, on the other hand, verifies whether the designed classification system distinctively separates patients' clusters, considering a particular observed parameter.

As for the results evaluation, a satisfactory result is achieved when a maximum number of accepted null hypothesis in congruent test is reached along with a maximum number of rejected null hypothesis in cross test. The overall evaluation criterion is then the result of T-test, as the F-test only acts as a support whether the T-test was performed correctly.

$$\epsilon_s = 1 - \frac{\sum_{j=1}^{c} \sum_{i=1}^{n} H_{ij}}{n.c} \tag{6}$$

The overall evaluation of the congruent T-test is performed in compliance with formula 6, where coefficient ϵ_s represents a measure of correct classification, with value 1 meaning the most accurate classification rate. ϵ_s is calculated using all test parameters n and quantity c of clusters. H value represents the result of hypothesis testing where value 0 means the acceptance of the null hypothesis and value 1 its rejection. The significance level was set to 0.05.

$$\epsilon_k = \frac{\sum_{j=1}^{c} \sum_{i=1}^{n} H_{ij}}{n.c} \tag{7}$$

The cross T-test evaluation is performed using formula 7, where coefficient ϵ_k determines the separation value with respect to all parameters, with value 1 meaning that a patient is uniquely classified regarding to every parameter into a particular

Table 2 Test evaluation

Model	Congruent F-test (%)	Congruent T-test (%)	Cross F-test (%)	Cross T-test (%)
Model A	76.6	100	60.0	70.0
Model B	70.8	89.6	68.8	66.7
Model C	87.0	97.8	65.2	60.9

cluster. Formula inputs *n*, *c* and *H* have the same meanings as their counterparts in formula 6.

The F-test evaluation is performed for both congruent and cross test using formula 6.

4 Results

The results of statistical hypothesis testing are presented in Table 2. The best result in the congruent F-test was achieved by model C, reaching the value of 87.0 % and the worst result of 70.8 % by model B.

Model A performed the best result in the congruent T-test, reaching 100 %, followed by the value of 97.8 % achieved by model C.

Cross F-test results within the range of 60.0–68.8 %. The best parameter separation assessed by cross T-test was achieved by model A (70.0 %) and the worst by model C (60.9 %).

The overall evaluation shows that the best results were achieved by model C. In succeeded in both congruent tests. The separation value is the worst one, 60.9 %. However, model C is the most complex one and uses 28 parameters.

Model A reached the second place. It performed the best result in congruent T-test. The separation value is 70 % but it only uses 10 parameters, assuming that with an increasing number of parameters the accuracy would decrease.

The worst results were achieved by model B, having the worst accuracy in both congruent tests. The separation values reach standard ranges as other models. It might have been caused by a wrong cluster approximation, not having been possible to definitely determine the correct quantity of clusters.

For practical applications the algorithm designed in model C can be suggested as it utilizes the most parameters. Also, the model A can be applied, however, for a limited quantity of parameters.

The overall assessment, performed by medical specialists, implies that in particular clusters were classified the following types of patients:

- Cluster 1—patients suffering from obesity not capable of fat-burning. Half of his body mass is composed of body fat, also having high values of insulin.

- Cluster 2—patients suffering from obesity that is able to burn fats. One third of his body mass comprises body fat.
- Cluster 3—patients being overweight having problems with fat-burning. However, their body mass is not as high, one third of body mass is body fat.

5 Conclusions and Future Work

This contribution describes a fuzzy classification system designed for metabolic types detection. Using this system, patients suffering from obesity were divided into groups denoted clusters according to similarity of their metabolic characteristics. Grouping was performed on the basis of Principal component analysis and fuzzy clusterization methods.

In the testing part of the system, new patients were classified into the designed clusters. The most effective classification algorithm is represented by testing model C. It's classification accuracy was 97.8 % and the separation value, indicating the patients groups' distinction, was 60.9 %. This model uses 28 input metabolic parameters acquired through energometry, bioimpedance and blood analyses tests.

Implementation of multi-cluster layout allows to diagnose specific patient's metabolic symptoms and to propose an individual medical treatment for obese patients. Limiting the classification clusters' quantity results in a robust patients classification tool.

To further improve the accuracy of the system, it is now important to determine a set of reference groups of patients to be able to specify diagnostic patterns. This would significantly increase the system's practical applicability.

Acknowledgments This work was supported by project SP2015/154, 'Development of algorithms and systems for control, measurement and safety applications.' of Student Grant System, VSB-TU Ostrava.

References

1. Abonyi J, Feil B (2007) Cluster analysis for data mining and system identification. Birkhauser, Basel
2. Almeida R, Sousa J (2006) Comparison of fuzzy clustering algorithms for classification. In: Proceedings of the 2006 International symposium on evolving fuzzy systems, EFS'06
3. Deng J, Hu J, Chi H, Wu J (2010) An improved fuzzy clustering method for text mining. In: NSWCTC 2010—The 2nd international conference on networks security, wireless communications and trusted computing, vol 1, pp 65–69
4. Ferenci T, Almassy Z, Kovacs A, Kovacs L (2011) Effects of obesity: a multivariate analysis of laboratory parameters. In: 2011 6th IEEE International symposium on applied computational intelligence and informatics (SACI), pp 629–634
5. Hong TP (2007) On genetic-fuzzy data mining techniques. In: IEEE International conference on granular computing, GRC 2007, p 3

6. Jing H (2009) Application of fuzzy data mining algorithm in performance evaluation of human resource. In: IFCSTA 2009 Proceedings—2009 International forum on computer science-technology and applications, vol 1, pp 343–346
7. Jolliffe IT (2002) Principal component analysis, 2nd edn. Springer, New York
8. Kassam S (1988) Elements of statistical hypothesis testing. In: Signal detection in non-gaussian noise, pp 1–23. Springer Texts in electrical engineering. Springer, New York
9. Michalikova M, Prauzek M (2014) A hybrid device for electrical impedance tomography and bioelectrical impedance spectroscopy measurement. In: Canadian conference on electrical and computer engineering
10. Novak B, Bigec M (1996) Childhood obesity prediction with artificial neural networks. In: Proceedings ninth IEEE symposium on computer-based medical systems, pp 77–82
11. Paul R, Hoque A (2010) Clustering medical data to predict the likelihood of diseases. In: 2010 5th International conference on digital information management, ICDIM 2010, pp 44–49
12. Prauzek M, Hlavica J, Michalikova M, Jirka J (2015) Fuzzy clustering method for large metabolic data set by statistical approach. In: Proceedings of the 7th Cairo international biomedical engineering conference, CIBEC 2014 pp 87–90
13. Saha I, Maulik U (2009) Nilanjan: Differential fuzzy clustering for categorical data. In: Proceedings of international conference on methods and models in computer science, ICM2CS09
14. Tafti A, Sadati N (2009) Fuzzy clustering means data association algorithm using an adaptive neuro-fuzzy network. In: Aerospace conference. IEEE, pp 1–5
15. Thomas B, Raju G (2010) A fuzzy threshold based unsupervised clustering algorithm for natural data exploration. In: ICNIT 2010–2010 International conference on networking and information technology, pp 473–477
16. Vanisri, D., Loganathan, C.: Fuzzy pattern cluster scheme for breast cancer datasets. In: Proceedings of 2010 International Conference on Communication and Computational Intelligence, INCOCCI-2010. pp. 410–414 (2010)
17. Zhang Y, Ren Y, Liu X (2008) A new method for fuzzy clustering analysis based on afs fuzzy logic. In: 2008 International conference on wireless communications, networking and mobile computing, WiCOM 2008

A Hybrid Approach for Predicting River Runoff

Hieu N. Duong, Hien T. Nguyen and Vaclav Snasel

Abstract Time series prediction has attracted attention of many researchers as well as practitioners from different fields and many approaches have been proposed. Traditionally, sliding window technique was employed to transform data first and then some learning models such as fuzzy neural networks were exploited for prediction. In order to improve the prediction performance, we propose an approach that combines chaotic theory, recurrent fuzzy neural network (RFNN), and K-means. In the past few decades, fuzzy neural networks have been proven to be a great method for modeling, characterizing and predicting many kinds of nonlinear hydrology time series data such as rainfall, water quality, and river runoff. Chaotic theory is a field of physics and mathematics, and having been used to solve many practical problems emerging from industrial practices. In our proposed approach, chaotic theory is firstly exploited to transform original data to a new kind of data called phase space. Then, a novel hybrid model namely RFNN-KM including several RFNNs that are mixed together by K-means algorithm is used to perform prediction. We conduct experiments to evaluate our approach using runoff data of Srepok River in the Central Highland of Vietnam. The experiment results show that the proposed approach outperforms the one combining RFNN and sliding window technique on the same experiment data.

Keywords RFNN · K-means · Chaotic theory · Time series · River runoff

H.N. Duong
Faculty of Computer Science and Engineering, HCM City University of Technology, Ho Chi Minh City, Vietnam

H.T. Nguyen
Faculty of Information Technology, Ton Duc Thang University, Ho Chi Minh City, Vietnam

V. Snasel (✉)
Faculty of Electrical Engineering and Computer Science, Department of Computer Science and IT4Innovations, VŠB-Technical University of Ostrava, Ostrava, Czech Republic
e-mail: vaclav.snasel@vsb.cz

A. Abraham et al. (eds.), *Intelligent Data Analysis and Applications*,
Advances in Intelligent Systems and Computing 370,
DOI 10.1007/978-3-319-21206-7_6

1 Introduction

The Srepok River is located in the Central Highland of Vietnam. Its watershed has a plentiful lake system, districts evenly. Due to slope terrain, the water maintaining ability is not good and most of small streams of the river almost run out of water in the dry season and the water of several big lakes drop into a very low level. Together with the impact of climate change, unusually changes of the watershed of Srepok River as observed recently has posed many challenging problems to water resource security of Vietnam. Naturally, Srepok runoff varies with seasonal rule by year; it is low in dry seasons and high in rainy seasons. But in a few years, it decreases suddenly in dry season or increases suddenly in rainy seasons and it is worth to warn because it impacts directly to people life in Srepok basin. That raises a challenge is how to simulate and predict Srepok runoff in order to help managers and civilian adapt with its anomaly.

In the past few decades, fuzzy neural networks have been applied widely in various fields [4, 6, 9]. In particularly, fuzzy neural networks have been proven to be a great method for modeling, characterizing and predicting many kinds of non-linear hydrology time series data such as rainfall, water quality, and river runoff [9]. In order to improve performance of time series prediction, several researchers have proposed many hybrid methods which are composed by fuzzy neural networks and other theories such as hidden markov model [2], particle swarm optimization (PSO) algorithm [5], chaotic theory [1, 13, 14] and genetic algorithm (GA) [12]. In [2, 5, 15], the authors also pointed out that artificial neural networks are specifically suitable when applied to predict chaotic time series. Moreover, for those time series data resembling Srepok runoff, in [1, 13, 14], the authors applied chaotic theory to reconstruct a kind of new chaotic data namely phase space from the original time series. In [7], we employed RFNN to find out correlations between climate data and Srepok runoff and then to model and predict monthly Srepok runoff in long-term in the future. The proposed method gave good performance for predicting monthly Srepok runoff in long-term, but, unfortunately, for predicting daily Srepok runoff in short-term, the efficiency of this method is not good as expectation.

In the paper, we employ chaotic theory to reconstruct a new kind of data called phase space from the original Srepok runoff in order to enrich its temporal information before applying RFNN. Due to non-chaotic characteristic of the phase space, to improve the performance of prediction, we propose a new hybrid model namely RFNN-KM including several RFNNs that are mixed together by K-means algorithm. By experiments, we also prove that our method outperforms a popular method that combines RFNN with sliding window technique.

The rest of this paper is organized as follows. Section 2 presents how to employ chaotic theory to construct phase space from the original Srepok runoff. Section 3 presents a hybrid model of RFNNs and K-means algorithm namely RFNN-KM. Section 4 introduces shortly about recurrent fuzzy neural network. In Sect. 5, we present experiments. Some finding also will be pointed out in Sect. 5. Finally, we draw conclusion and perspectives for future work.

2 Background on Chaotic Theory

The Srepok runoff gathered daily automatically at several hydrology stations is a kind of time series data. In order to employ RFNN for predicting Srepok runoff, firstly the data must be preprocessed. One main task of data preprocessing is how to transform the original time series data into a new kind of data that can be applied by RFNN. The most popular transforming method is sliding window technique. When applying sliding window technique, the original time series, denoted $X = \{x_1, x_2, \dots, x_n\}$, will be converted to Y including several $Y_i = \{x_i, x_{i+\tau}, \dots, x_{i+(m-1)*\tau}\}$, $i = 1, 2, \dots, n - (m-1)\tau$. In which, τ and m are parameters used for configuration settings, normally $\tau = 1$ and m is set by experience of analyst. The various values of τ and m will influence overall performance of RFNN. Unfortunately, it is quite hard to choose the most suitable m if only based on the experience of analyst. Besides, the main disadvantage of this technique is that we just are able to predict in short-term because the accumulative error becomes larger while predicting in long-term.

Algorithm 1: Pseudo-code of false nearest neighbor method

input : The original time series data X, the minimum delay time τ, neighbor threshold R and false neighbor threshold R_{tol}
output: The optimal embedding dimension m

$m \leftarrow 1$
Build the phase space Z based on X, τ and the current value of m
foreach Y_i *in the new phase space Z* **do**
    Identify all nearest neighbors of Y_i called Y_i^j if their square of distance in m dimension phase space $D_m^2(ij) = \Sigma_{k=0}^{m}(x_{i+k\tau} - x_{i+k\tau}^j)^2$ is less than neighbor threshold R
end
repeat
    $m \leftarrow m + 1$
    Rebuild the phase base Z based on X, τ and the new value of m
    foreach Y_i *in the phase space Z* **do**
        foreach *neighbor* Y_i^j *of* Y_i **do**
            if $\frac{|x_{i+m\tau} - x_{i+m\tau}^j|}{D_m(ij)} > R_{tol}$ **then**
                A false nearest neighbor is found and break to Repeat Loop
            end
        end
    end
until *no false nearest neighbor is found*;

We employ chaotic theory to convert the time series data into a phase space namely Z including several X_i and $X_i = (x_i, x_{i+\tau}, \dots, x_{i+(m-1)*\tau})$, $i = 1, 2, \dots, n - (m-1)\tau$. In that τ is delay time and m is embedding dimension. The phase space Z is able to enrich the temporal information of original time series data that will help RFNN learn several behind natural rules of the data faster and more accuracy [14].

2.1 Minimum Delay Time

One of the most popular methods employed to find out the minimum delay time is applying auto-correlation function:

$$R = \frac{\Sigma_{i=1}^{n}(x_i - \overline{x})(x_{i+\tau} - \overline{x})}{\Sigma_{i=1}^{n}(x_i - \overline{x})^2} \text{ where } \overline{x} \text{ is the average value of } X \quad (1)$$

In this method, τ is identified iteratively and firstly it is set to zero. Then, τ is increased by one in each iteration until R is the first time less than $(1 - 1/e)$. The final value of τ is the delay time of reconstructed phase space.

2.2 False Nearest Neighbor Method

In [3, 14], the authors introduced many methods finding the embedding dimension effectively such as method of Schuster, Grassberger-Procaccia algorithm (G-P), false nearest neighbor method, etc. The authors also concluded that G-P is the most popular but false nearest neighbor is the most effective method.

We define one point as a neighbor of another if their Euclid distance is smaller than a threshold R. Let $y(r)(n)$ be the r_th nearest neighbor of $y(n)$ and m be the current dimension of $y(n)$ and $y(r)(n)$. When increasing m by one, the distance of $y(n)$ and $y(r)(n)$ will be changed (larger) and then $y(r)(n)$ may not be a neighbor of $y(n)$ anymore. In this case, we call $y(r)(n)$ is a false nearest neighbor of $y(n)$. Based on this principle, Algorithm 1 will find out the optimal embedding dimension.

3 Hybrid Model of K-means and RFNN

Not all of time series are able to generate chaos, so it is necessary to check chaotic characteristic of the time series after converted to the phase space. Until now, the largest Lyapunov index has been a main indicator to judge if a time series data has chaotic characteristic. In this method, if the largest Lyapunov index is greater than zero, the chaotic characteristic of data can be determined [11].

According to m and τ value that are calculated from the Srepok runoff data ranging from 2006 to 2010, the largest Lyapunov index calculated by TISEAN software [8] is -1.8. The value of Lyapunov index points out that our data do not have chaotic characteristic. That also means the received phase space is not able to be learned by only one RFNN, even the RFNN is learned in very long time [13, 14]. Meanwhile we observe a few smaller subphase spaces, we realize the subphase spaces have chaotic characteristic. From the observation, we propose a novel hybrid model

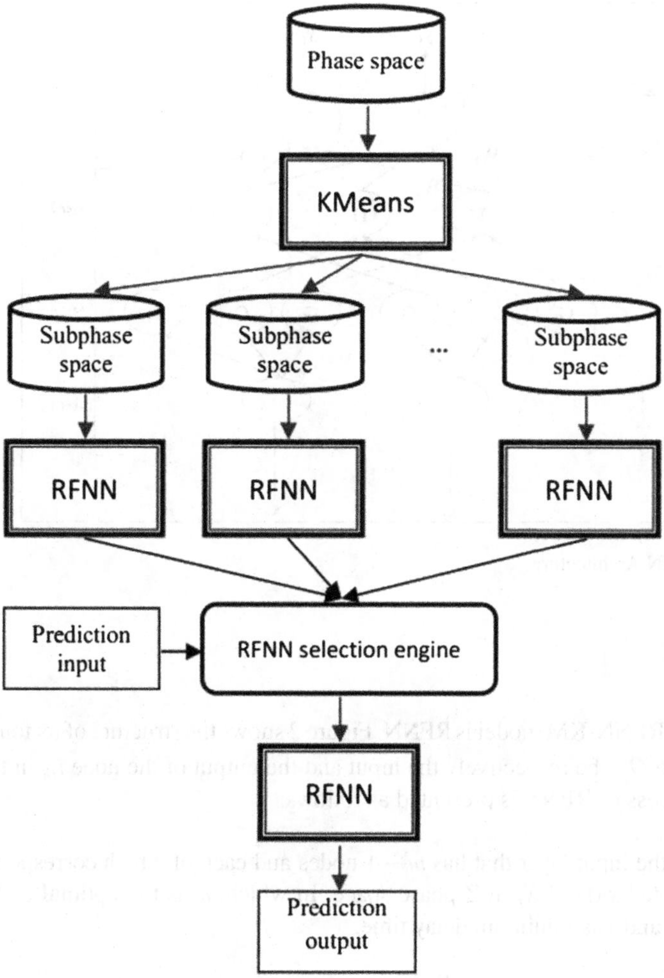

Fig. 1 RFNN-KM Architecture

namely RFNN-KM including several RFNNs that are mixed together by the K-means algorithm as Fig. 1.

In the hybrid model, the phase space is divided to k subphase spaces by K-means algorithm. Then, every subphase space will be learned by a single RFNN. Because all data in every subphase space have similar characteristics, so the training phase of every RFNN will be faster and more accuracy. In prediction phase, we employ a traditional strategy that applies predicted value in a certain step as a prediction input for the following step. Then, the prediction input will be used to find out what subphase space is the nearest with it. Subsequently, the corresponding RFNN of the nearest subphase space will be used to predict.

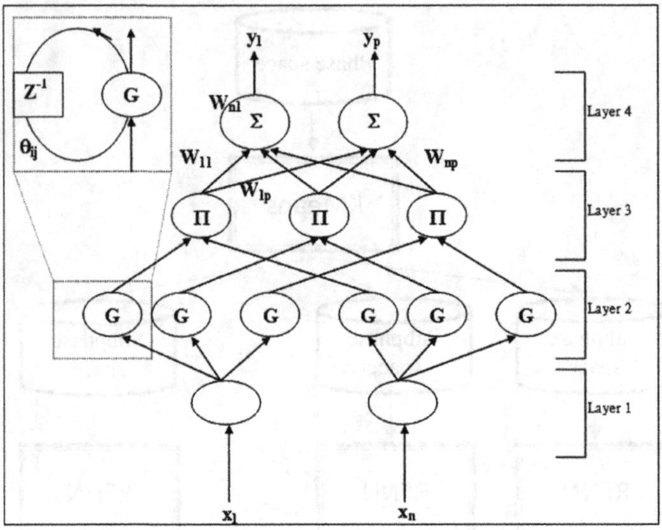

Fig. 2 RFNN Architecture

4 RFNN

The key of RFNN-KM model is RFNN. Figure 2 shows the structure of its four layers. Let $u_i^{(k)}$ and $O_i^{(k)}$ be respectively the input and the output of the node i_{th} in the layer k. The process of RFNN is presented as follows:

Layer 1

This is the input layer that has $m - 1$ nodes and each of which corresponds with $x_{t+i*\tau}$ denoted $x_i(t)$ of X_t in Z phase space. In which, m is the optimal embedding dimension and τ is minimum delay time.

$$O_i^{(1)} = u_i^{(1)} = x_i(t) \; where \; i = 0 \div m - 2 \tag{2}$$

Layer 2

This is called membership layer. Nodes in this layers are responsible for converting crisp data into fuzzy data by applying membership functions such as Gauss function. The number of neural nodes in this layer is $(m - 1) * N$ where N is the number of fuzzy rules. Every node has three parameters, namely m_{ij}, σ_{ij} and θ_{ij} respectively.

$$O_{ij}^{(2)} = \exp\left[-\frac{(u_{ij}^{(2)} - m_{ij})^2}{(\sigma_{ij})}\right] \; where \; i = 0 \div m - 2, \; j = 1 \div N \tag{3}$$

In (3) m_{ij} and σ_{ij} are the mean and the variance of Gauss distribution function.

$$u_{ij}^{(2)}(t) = O_i^{(1)} + \theta_{ij} O_{ij}^{(2)}(t-1) \text{ where } i = 0 \div m-2,\ j = 1 \div N \quad (4)$$

In (4), θ_{ij} denotes the weight of a recurrent node.

We easily realize that the input of nodes in this layer has the factor $O_{ij}^{(2)}(t-1)$. This factor denotes a remaining information of the previous learning step. Therefore, after replacing $u_{ij}^{(2)}$ in (2) by (3), we get eq. (4) as follows:

$$O_{ij}^{(2)} = \exp\left[-\frac{\left[O_i^{(1)} + \theta_{ij} O_{ij}^{(2)}(t-1) - m_{ij}\right]^2}{(\sigma_{ij})}\right] \text{ where } i = 0 \div m-2,\ j = 1 \div N \quad (5)$$

Layer 3

This is the layer of fuzzy rules and has N nodes. Each node in this layer plays the role of a fuzzy rule. Connecting between Layer 3 and Layer 4 presents for fuzzy conclusion. Each node in this layer corresponds with an *AND* expression. Each *AND* expression is defined as follows:

$$O_j^{(3)} = \prod_{i=1}^{N} O_{ij}^{(2)} \text{ where } i = 0 \div m-2,\ j = 1 \div N \quad (6)$$

Layer 4

This is the output layer including P nodes. In our model, P will be set to one and this is $x_{t+(m-1)\tau}$ of X_t in Z phase space. Nodes of this layer are responsible for converting fuzzy to crisp.

$$y_k = O_k^{(4)} = \sum_{j=1}^{M} u_{jk}^{(4)} w_{jk} = \sum_{j=1}^{M} O_j^{(3)} w_{jk} \text{ where } k = 1 \div P \quad (7)$$

After defining process of RFNN and detail operation of every layer, we employ back propagation (BP) algorithm to learning RFNN. BP algorithm has a big disadvantage that learning phase usually falls into local minima. In our study, we improve BP by applying momentum technique to speed up the learning phase and overcome local minima.

5 Experiment

The performance of the models developed in this study are assessed by using various standard statistical performance evaluation criteria. The statistical measures considered are coefficient of correlation (R), root mean squared error (RMSE), and mean absolute relative error (MARE).

$$R = \frac{\sum_{i=1}^{n}((O_i - \overline{O})(P_i - \overline{P}))}{\sqrt{\sum_{i=1}^{n}(O_i - \overline{O})^2}\sqrt{\sum_{i=1}^{n}(P_i - \overline{P})^2}} \tag{8}$$

$$RMSE = \sqrt{\frac{1}{n}\sum_{i=1}^{n}(O_i - P_i)^2} \tag{9}$$

$$MARE = \sum_{i=1}^{n}\frac{|O_i - P_i|}{O_i} \tag{10}$$

where O_i is the observed runoff at time i; $\overline{O}$ is the average observed runoff; P_i is the simulated runoff at time i; $\overline{P}$ is the average simulated runoff and n is the number of observed runoff data.

For short-term prediction objective, we collected Srepok runoff data ranging from 2006 to 2011 at a hydrology station called BUON DON for evaluating RFNN-KM and RFNN combining with sliding window technique. We apply data from 2006 to 2009 for training, 2010 for validation and 2011 for prediction. After calculating τ and m with the data ranging from 2006 to 2010, we have $\tau = 11$ and $m = 23$. For sliding window technique, we choose $m = 15$ because 15 is haft-monthly tidal cycle of most rivers in Vietnam [10]. We also set number of cluster $k = 15$ for K-means algorithm. In prediction phase, we employ a traditional strategy that applies predicted value in a certain step as a prediction input for the following step. In Table 1, Figs. 3 and 5, we can realize that RFNN-KM outperforms RFNN in training, validation and prediction phases. Specifically, RFNN-KM shows its obvious advantage if comparing with RFNN in prediction phase. This advantage can be explained that RFNN-KM captures several temporal natural rules contained in several RFNNs. In Fig. 4, the performance of two methods are quite equal when predicting in one month. In contrast, for predicting in one year, Fig. 5 points out that the predicted runoff by RFNN-KM and the observed runoff are similar while RFNN combining with sliding window technique gives worse results. The prediction results of RFNN gradually become a straight line by time because of the increasing accumulative error. Because the observed runoff and the predicted runoff by RFNN-KM are similar but locally

Table 1 The performance of RFNN combining with sliding window and RFNN-KM combining with chaotic during training and validation phases

	RFNN-KM and chaotic			RFNN and sliding window		
	MARE	R	RMSE	MARE	R	RMSE
Training	4.921	0.983	34.010	8.814	0.986	50.209
Validation	16.261	0.952	76.139	22.396	0.913	85.541
Predicting for 1 month	29.834	0.656	56.99	30.3	−0.089	62.135
Predicting for 1 year	44.947	0.738	128.891	56.603	−0.535	296.007

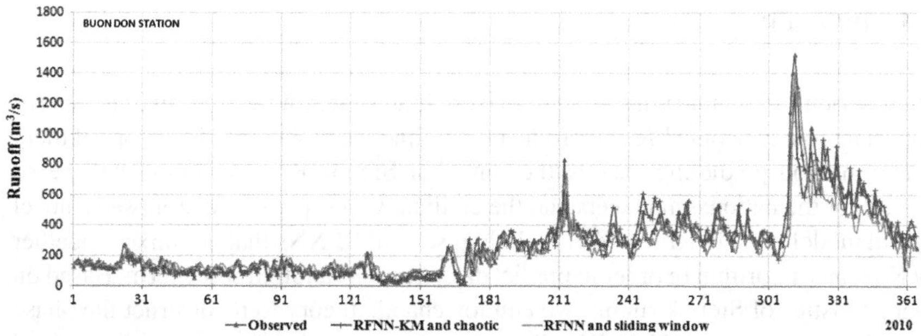

Fig. 3 The observed and simulated runoff in validation phase

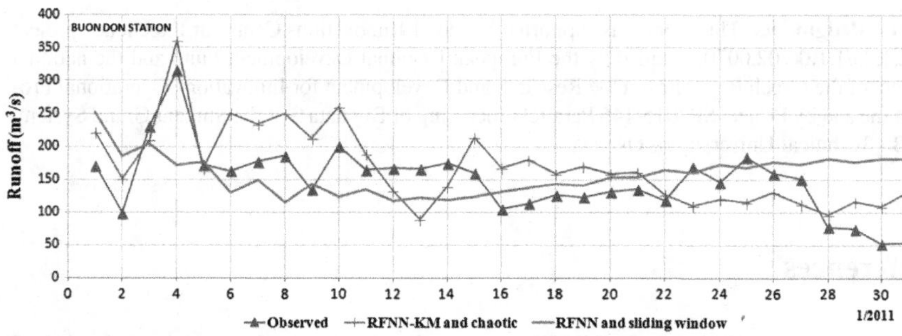

Fig. 4 The observed and simulated runoff in one month period prediction

out of phase, the values of evaluation criteria are not good. The main reason is that the rainy season in 2011 came earlier than 2006–2010 and this caused the Srepok runoff rule in 2011 was different from 2006–2010. Therefore, to improve the performance of the method in this case, we have to change supervised learning method to reinforcement learning method so that RFNNs can be adjusted immediately when appearing any extraordinary events in real time [16]. Then, our problem will become a kind of real time forecasting and we will take it in consideration of the further research.

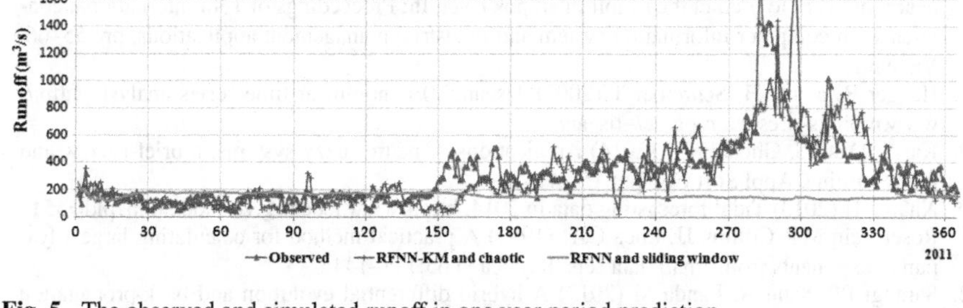

Fig. 5 The observed and simulated runoff in one year period prediction

6 Conclusion

In an agriculture country as Vietnam, rivers like Srepok have been playing a very important role in people life and production around the basin area. Therefore, if there is any method predicting exactly the runoff of Srepok River in future will be so helpful for the national managers and the civilian. In this paper, we proposed a novel hybrid model namely RFNN-KM including several RFNNs that are mixed together by K-means algorithm in order to predict daily Srepok runoff in short-term. Based on characteristics of Srepok runoff, we employ chaotic theory to reconstruct the phase space before applying RFNN-KM. By experiments, we also show that the proposed method outperforms RFNN combining with sliding window technique.

Acknowledgments This work was supported by the IT4Innovations Centre of Excellence project (CZ.1.05/1.1.00/02.0070), funded by the European Regional Development Fund and the national budget of the Czech Republic via the Research and Development for Innovations Operational Programme and by Project SP2015/146 Parallel processing of Big data 2 of the Student Grand System, VŠB - Technical University of Ostrava.

References

1. Benmouiza K, Cheknane A (2013) Forecasting hourly global solar radiation using hybrid k-means and nonlinear autoregressive neural network models. Energy Convers Manage 75:561–569
2. Bhardwaj S et al (2010) Chaotic time series prediction using combination of hidden markov model and neural nets. In: Proceedings of the 9th international conference on computer information systems and industrial management applications, pp. 585–589
3. Cellucci CJ, Albano AM, Rapp PE (2003) Comparative study of embedding methods. Phys Rev 67
4. Chang FJ, Chiang YM, Tsai MJ, Shieh MC, Hsu KL, Sorooshian S (2014) Watershed rainfall forecasting using neuro-fuzzy networks with the assimilation of multi-sensor information. J Hydrol 508:374–384
5. Du B, Xu W, Song B, Ding Q, Chu SC (2014) Prediction of chaotic time series of rbf neural network based on particle swarm optimization. Intell Data Anal Appl 298:489–497
6. Duong HN (2006) Using rfnn to predict price of products in market. In: Proceedings of 9th Conference on Science and Technology, Ho Chi Minh City University of Technology, pp. 34–43
7. Duong HN, Nguyen QNT, Bui LT, Nguyen HT, Snasel V (2014) Applying recurrent fuzzy neural network to predict the runoff of srepok river. In: Proceedings of 13th international conference on computer information system and industrial management applications, pp. 55–66, Vietnam
8. Hegger R, Kantz H, Schreiber T (2007) Tisean 3.0.1, nonlinear time series analysis. http://www.mpipks-dresden.mpg.de/~tisean/
9. Kar S, Dasb S, Ghoshb PK (2014) Applications of neuro fuzzy systems: a brief review and future outline. Appl Soft Comput 15:243–259
10. Nhan NH (2013) Tidal forecasting data in 2014. http://www.icoe.org.vn/index.php?pid=551
11. Rosenstein MT, Collins JJ, Luca CJD (1993) A practical method for calculating largest lyapunov exponents from small data sets. Physica D 65:117–144
12. Sarangi PP, Sahu A, Panda M (2013) A hybrid differential evolution and back-propagation algorithm for feedforward neural network training. Int J Comput Appl 84

13. Shang P, Li X, Kamae S (2005) Chaotic analysis of traffic time series. Chaos Solitons Fractals 25:121–128
14. Shudong Z, Weiguang L, Jun N, Lina WGZ (2010) Combined method of chaotic theory and neural networks for water quality prediction. J Northeast Agric Univ 17:71–76
15. Todorov Y, Terziyska M (2014) Modeling of chaotic time series by interval type-2 neo-fuzzy neural network. Artif Neural Netw Mach Learn 8681:643–650
16. Yadav AK, Sachan AK (2012) Research and application of dynamic neural network based on reinforcement learning. In: Proceedings of the international conference on information systems design and intelligent applications, pp. 931–942

Critical Evaluation of Seven Lactation Curve Estimation Models

Jaroslav Marek, Radko Rajmon and Tomas Haloun

Abstract A mathematical model of the lactation curve approximation provides summary information about dairy cattle production, which is useful in making management and breeding decisions and in simulating a dairy enterprise. Several nonlinear regression models have been developed during the past decades. Unfortunately, there is no unique algorithm in literature. The question of choosing the best function is brought up. We tested seven such models (Gaines, Nelder, Ning-Yang, Marek-Zelinkova, McMillan, Papajesic-Bodero, Wood) using milk yield data of more than 5 000 lactation cycles. Non-linear approximations showed high level of confidence in all models. Critical limitations of individual approaches are discussed using data of several individual cows with problematic approximations. We stress the limitations of individual—generally well-fitting—mathematical models at the level of individual animals. The best results were obtained with Wood, Nelder or Marek-Zelinkova models.

Keywords 305-day yield estimate · Fitting of lactation curve · Approximation of nonlinear function · Nonlinear regression

J. Marek (✉)
Department of Mathematics and Physics, Faculty of Electrical Engineering and Informatics, University of Pardubice, Studentska 95, 53210 Pardubice, Czech Republic
e-mail: jaroslav.marek@upce.cz

R. Rajmon · T. Haloun
Department of Veterinary Science, Faculty of Agrobiology, Food and Natural Resources, Czech University of Life Sciences Prague, Kamycka 129, 165 21 Praha 6 - Suchdol, Czech Republic
e-mail: rajmon@af.czu.cz

T. Haloun
e-mail: tomas.haloun@email.cz

A. Abraham et al. (eds.), *Intelligent Data Analysis and Applications*,
Advances in Intelligent Systems and Computing 370,
DOI 10.1007/978-3-319-21206-7_7

1 Introduction

A mathematical model of the lactation curve provides summary information about dairy cattle production, which is useful in making management and breeding decisions and in simulating a dairy enterprise. Cf. [16, 17]. Generally, the objective in modelling the lactation curve is to predict the yield on each day of lactation with minimum error so as to elucidate the underlying pattern of milk production in the presence of high local variation due to the effect of the environment, see [9]. Various models have been developed and tested in the past, often with a high parameters of fitting (small residual sum of squares, large index of determination), cf. [1, 2, 6, 7, 9, 11, 13, 15, 18]. A different approach based on neural networks and auto-regressive model is studied in [8].

The aim of the paper is to explore the quality of the mathematical models proposed by the various authors for an approximation of the lactation curve based on the monthly performed measurements of daily yields, not only on the general evaluation of criterion of approximation but also with the respect to individual animal data.

2 Measurement

The milk yield of cows during lactation is not always the same and measurements of daily yield are often performed only once a month. Therefore the necessity to describe the dependence of milk yield and days of lactation has become apparent. Test-day data for milk production of the Holstein breed were extracted from the database of the cowshed in Záhoří.

Milk yields from monthly measurements of Holsteins cows were used to estimate model parameters, where Y_t is daily milk production on day t. The data included 76724 measurement—2589, 2080, 1185 cycles for first, second and third and larger lactation, respectively, sampled from 2589 cows between the years 2001 and 2013.

A numerical study was performed to calculate 305-day yield estimates using models defined in the Sect. 3. Samples of the data on the daily milk yield and the day in lactation of monthly measurements during particular lactation of a particular cow can be seen in Tables 1 and 2.

Estimates of unknown parameters were computed by method of nonlinear regression. Then the estimates of milk yield for the entire lactation cycle can be found.

3 The Approximating Function

Let x be the day of lactation and Y is the daily yield. Let $\beta = (\beta_1, \ldots, \beta_k)'$. Let's consider that function f model is how variable x explain variable Y.

$$Y = f(\beta, x) + \varepsilon. \tag{1}$$

Table 1 Samples of the data on monthly measurements of the daily milk yield during particular lactation of several cows (kg)

Cow No/lactation	Measurement of daily yields									
43539/7	41.2	42.0	39.6	35.2	26.2	30.8	30.6	28.4	23.2	20.6	23.0
411503/1	35.6	38.9	41.5	41.3	38.0	38.4	39.6	41.4	39.5	37.5	32.0
411896/1	37.2	45.2	44.2	37.0	39.4	37.6	38.2	34.2	34.9	24.6	–
411905/1	25.0	42.7	31.0	35.5	35.3	34.2	34.9	33.3	30.0	26.6	23.1
411921/1	31.3	35.7	27.1	26.6	26.9	30.6	28.0	27.4	22.9	20.0	13.7
411572/1	52.4	50.0	43.8	51.1	49.3	45.6	50.9	45.5	44.7	39.8	42.0
411578/1	46.0	39.1	37.1	39.6	44.1	41.6	37.1	39.0	34.2	30.1	–
411583/1	43.6	36.1	38.7	38.5	39.5	42.1	35.1	36.0	32.3	27.4	–
411587/1	47.6	52.2	52.4	50.0	43.4	41.0	–	–	–	–	–
411605/1	43.8	39.4	37.3	39.6	39.9	38.8	35.8	35.2	26.9	31.2	26.0
411898/1	40.2	33.6	39.2	37.8	36.7	32.2	27.4	29.6	25.6	–	–

Table 2 Samples of the data on the day in lactation of monthly measurements during particular lactation of several cows (day in lactation)

Cow No/lactation	The order of the day in lactation										
43539/7	19	51	81	114	143	172	205	214	244	271	325
411503/1	16	40	72	102	130	166	194	226	254	283	313
411896/1	29	60	91	119	150	180	211	241	272	303	–
411905/1	17	52	76	108	138	166	202	230	262	290	319
411921/1	43	71	103	131	160	190	222	257	285	314	343
411572/1	43	67	99	129	157	193	221	253	281	310	340
411578/1	21	53	88	112	144	174	202	238	266	298	–
411583/1	28	60	95	119	151	181	209	245	273	305	–
411587/1	32	63	94	124	155	185	–	–	–	–	–
411605/1	41	76	100	132	162	190	226	254	286	314	343
411898/1	35	70	94	126	156	184	220	248	280	–	–

Our problem actually lies in estimating the values of the parameter vector based on nonlinear regression. The procedure is described in the following section.

In literature, see [6, 15, 18], we can find models of for approximation of lactation curve:

$$\text{Gaines (2010)}\quad f(\beta,x) = \beta_1 e^{-\beta_2 x}, \tag{2}$$

$$\text{Nelder (1966)}\quad f(\beta,x) = \frac{x}{\beta_1 + \beta_2 x + \beta_3 x^2}, \tag{3}$$

$$\text{Wood (1967)}\quad f(\beta,x) = \beta_1 x^{\beta_2} e^{-\beta_3 x}, \tag{4}$$

$$\text{Papajesic-Bodero (1988)}\quad f(\beta,x) = \frac{\beta_1 x^{\beta_2}}{\cosh(\beta_3 x)}, \tag{5}$$

$$\text{McMillan (1970)}\ f(\beta,x) = \beta_1 e^{-\beta_2 x}(1 - e^{-\beta_3(x-\beta_4)}), \tag{6}$$

$$\text{Ning-Yang (1983)}\ f(\beta,x) = \beta_1 e^{\frac{-\beta_2 x}{1-e^{-\beta_3(x-\beta_4)}}}, \tag{7}$$

$$\text{Marek-Zelinkova (2010)}\ f(\beta,x) = \beta_1 + \frac{2\beta_2\beta_3}{(x-\beta_3)^2+\beta_4^2}. \tag{8}$$

4 Numerical Study

Estimates of unknown parameters can be computed by method of nonlinear regression. We can estimate the values of unknown parameters occurring in nonlinear function by linearization of the model and by ordinary method of least squares. For readers who want an extensive coverage of linearization method and parameter estimation models, we refer to the books of [4, 10, 12].

We need an appropriate initial solution, which is near to the true value of unknown parameters. In this context, linearization domains can be constructed, see [5]. However, we will present for each of the studied function a suitable initial solution and an estimate. Values in Table 3 can serve the reader for approximation of another data as initial parameters.

We counted initial estimates of unknown parameters for individual models (Table 3) on the basis of 5187 cow data. Than, we estimated the 305-d yields of individual cows using all seven models. The validity of estimates was generally high – for Nelder, and Wood function index of determination varied among 0.83. Estimates obtained for all cows from Table 1 are presented in Table 4.

Table 3 Initial estimates of unknown parameters of regression function and estimates for cow No. 43539, lactation No. 7 (3*rd* column)

Function	Initial solution	Estimate for cow no. 43539
Gaines	$\widehat{\beta}_0 = (45, 0.002)'$	$\widehat{\beta} = (45.19, 0.0024)'$
Nelder	$\widehat{\beta}_0 = (0.09, 0.02, 0.005)'$	$\widehat{\beta} = (0.0949, 0.0174, 0.0001)'$
Wood	$\widehat{\beta}_0 = (5, 0.8, 0.02)'$	$\widehat{\beta} = (39.4292, 0.0408, 0.0028)'$
Papajesic-Bodero	$\widehat{\beta}_0 = (35, 0.06, -0.007)'$	$\widehat{\beta} = (52.96, -0.0714, -0.0036)'$
McMillan	$\widehat{\beta}_0 =$ $(40, 0.05, -8, 0.005)'$	$\widehat{\beta} = (47.02, 0.37, 12.10, 0.0026)'$
Ning-Yang	$\widehat{\beta}_0 =$ $(20, -0.3, -0.1, 0.7)'$	$\widehat{\beta} = (24.51, -17.13, -0.02, -266.76)'$
Marek-Zelinkova	$\widehat{\beta}_0 =$ $(14; -1241; -333; 82)'$	$\widehat{\beta} = (19.6, -7992.1, -226.9, 34.4)'$

For 7th lactation cycle of cow No. 53539 are approximation by models (1–2) drawn in Fig. 1. Approximation by models (1–7) are drawn in Figs. 1, 2, 3, 4, 5 and 6.

Table 4 Estimators of 305-d lactation yield

Function/cow no.	Gaines	Nelder	Wood	Papajesic-Bodero	McMillan	Ning-Yang	Marek-Zelinkova
43539/7	9760	9605	9741	9795	956	9625	9725
411503/1	11773	11707	11792	11820	11845	11748	11863
411896/1	11537	12258	11305	11393	8183	11361	11503
411905/1	9856	10234	9872	9911	−47992	9749	10008
411921/1	8607	8403	8391	8527	−INF	8235	8609
411572/1	14619	14550	14551	14626	14619	14390	14617
411578/1	11902	14084	11899	11913	11894	11958	11895
411583/1	11421	12705	11351	11394	11421	INF	11402
411587/1	13897	16104	12475	12404	12600	13382	13312
411605/1	11520	13113	11353	11483	11468	10922	11515
411898/1	10334	10392	10131	10207	10209	10070	10209

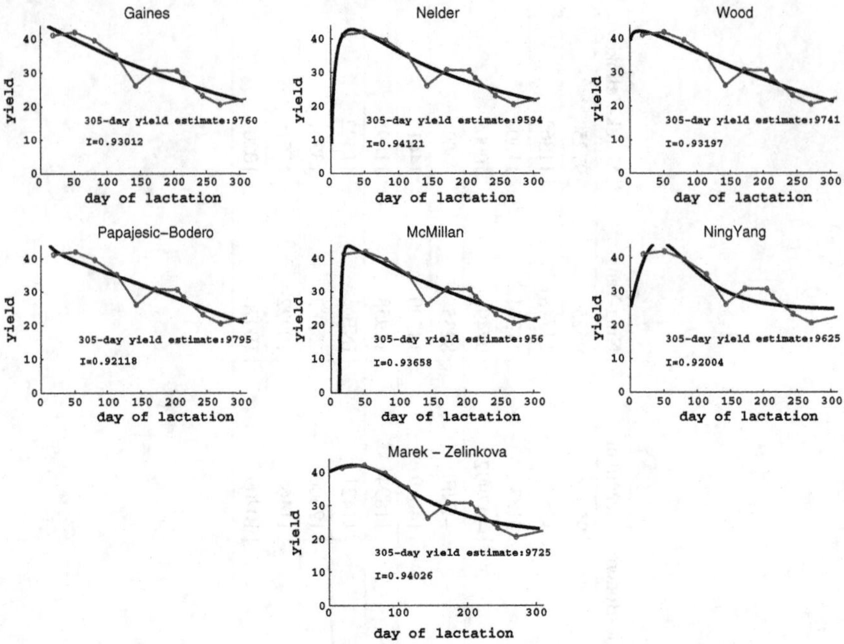

Fig. 1 Cow No 43539, 7*th* lactation

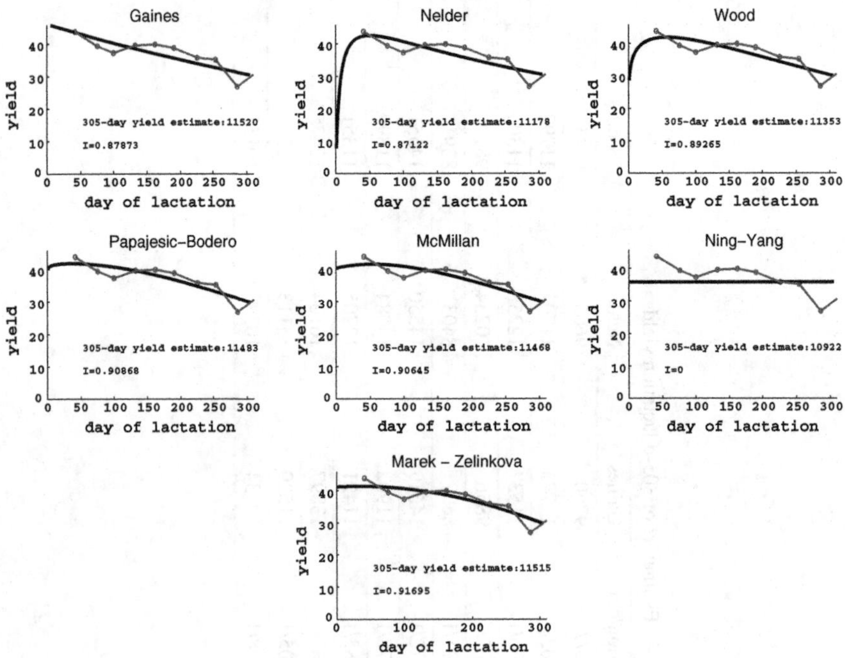

Fig. 2 Cow No 411605, 1*st* lactation

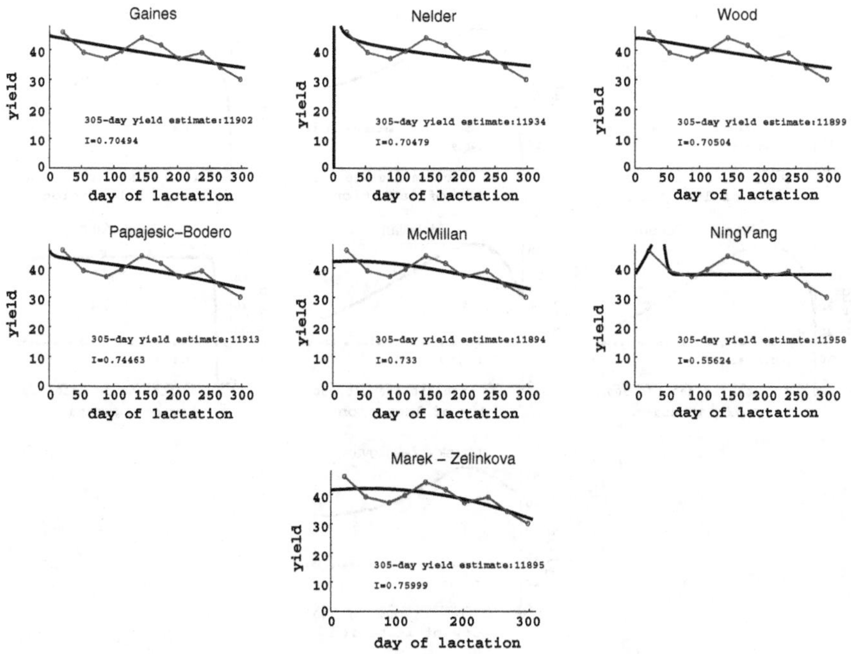

Fig. 3 Cow No 411578, 1*st* lactation

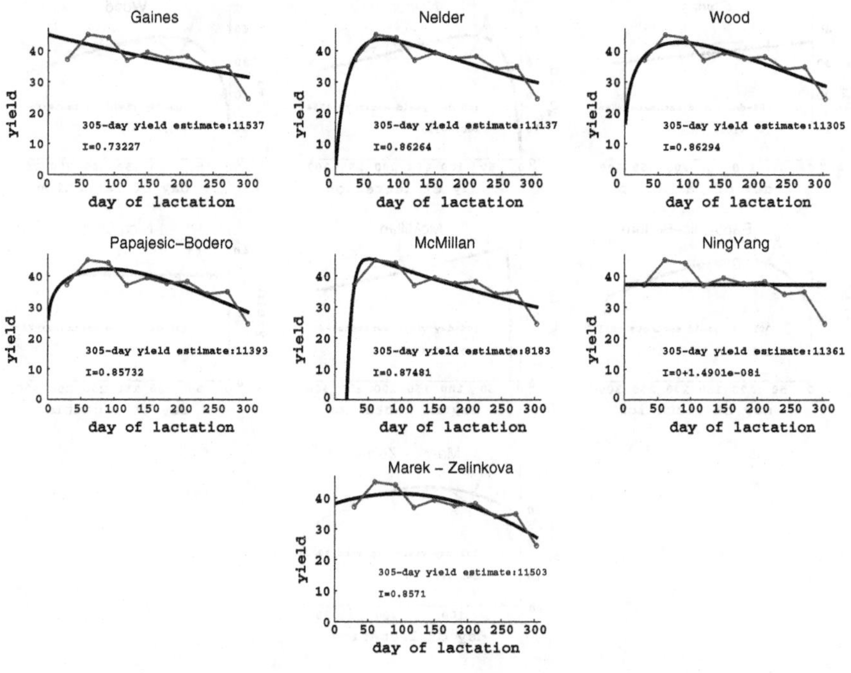

Fig. 4 Cow No 411896, 1*st* lactation

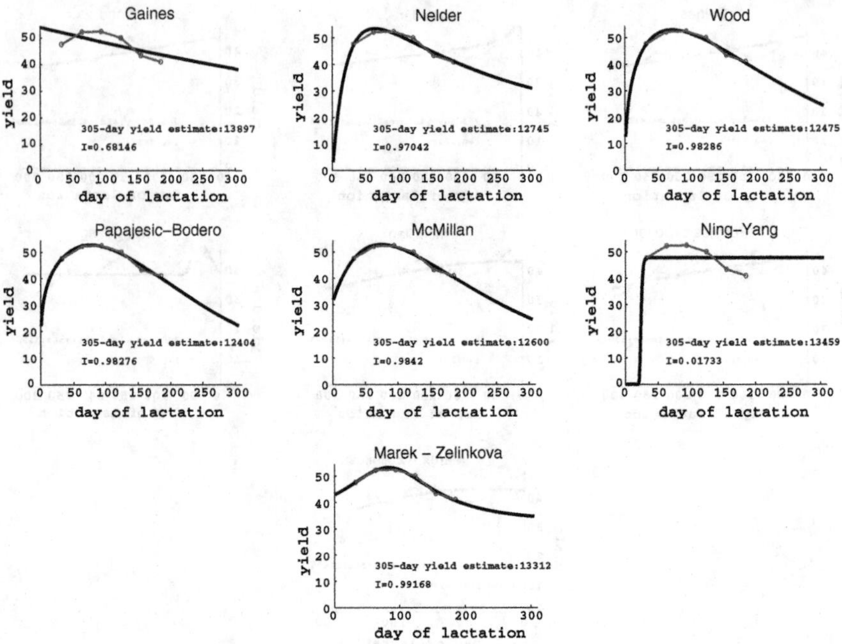

Fig. 5 Cow No 411587, 1*st* lactation

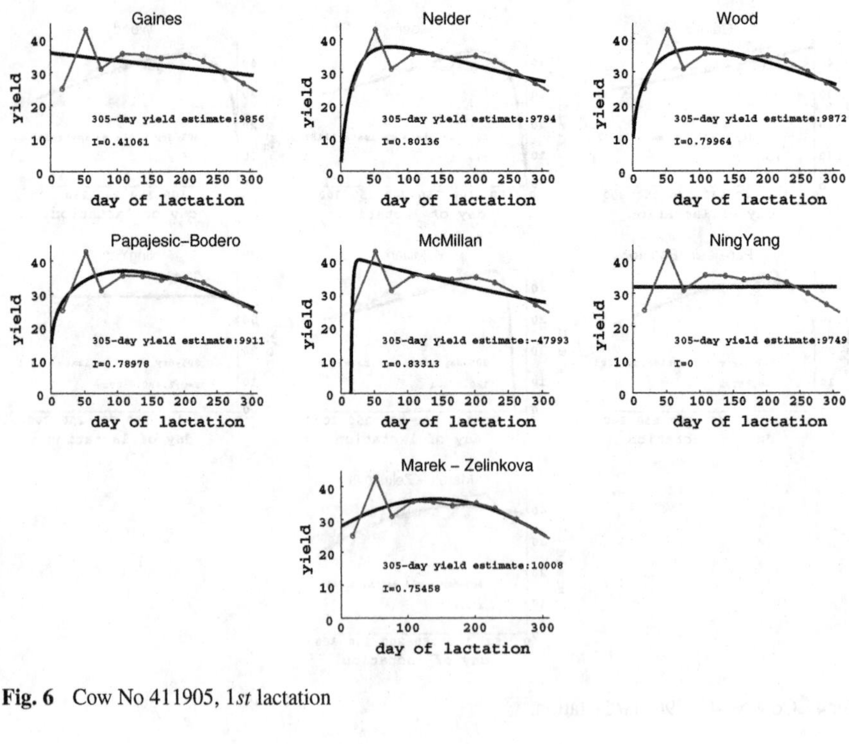

Fig. 6 Cow No 411905, 1*st* lactation

5 Quality of Modelling

Karush in 1963, cf. [3], discussed four interaction options between mathematics and empirical science:

- direct application (using without mathematical innovation, the mathematical model is considered to be an adequate picture of reality),
- invent application (inventive use: derive a new or modification of existing models),
- modelling (in situations when modeling mathematical representation is not known),
- theory construction (the creation of the theory of different degrees of abstractness).

In the third section we presented 7 different nonlinear functions for modeling of day yield over the lactation cycle. We can ask the question of which model is the best. For a choosen cow it can be the second function. But for the same cow on another lactation the fourth function will be better. In the decision process we can compare the values of the determination indexes or residual sum of squares in the proposed models. The results obtained on the data bulk might indicate very good fitting of Nelder, and Wood models. It is a question, whether the curves calculated for individual cows really fitts to biologically expected situation. In our case, we are in a special situation we have seven different models available. All models—hypotheses—will be verified on data sample selected for this purpose.

The role of the test model for all alternatives is not to respond directly to a given problem, it is not in demonstrating the truth, but in calculating probability that is the basis for our decision whether the model will be accepted or rejected. In a strongly negative case, we will reject the model totally, in positive case we will accept the model. But it is important to keep in mind that other data may reject (falsify) our model. This fact makes the truth in the empirical sciences a relative concept. Cf. [14].

Mathematics does not put the truth in the hands of the experimenter. The expert shall decide, whether his hypothesis is acceptable.

The model becomes more likely not only if we verify it inductively on the largest number of cases, but in particular with its incorporation into existing theories, which will arrange their deductive confirmation.

To access the suitability of the proposed models (1–7) we determine the indexes of determination. See Table 5.

Table 5 Indexes of determination for studying functions and data from Tables 1 and 2

Function/ cow no.	Gaines	Nelder	Wood	Papajesic-Bodero	McMillan	Ning-Yang	Marek-Zelinkova
43539/7	0.930	0.941	0.932	0.921	0.937	0.920	0.940
411503/1	0.291	0.693	0.729	0.750	0.737	0	0.761
411896/1	0.732	0.863	0.863	0.857	0.875	0	0.857
411905/1	0.411	0.801	0.800	0.790	0.833	0	0.755
411921/1	0.800	0.800	0.823	0.836	0.817	0.747	0.870
411572/1	0.719	0.714	0.724	0.743	0.719	0.640	0.747
411578/1	0.705	0.705	0.705	0.745	0.733	0.556	0.760
411583/1	0.741	0.728	0.756	0.796	0.741	0.496	0.830
411587/1	0.682	0.970	0.983	0.983	0.984	0.017	0.992
411605/1	0.879	0.871	0.893	0.909	0.907	0	0.917
411898/1	0.861	0.862	0.882	0.894	0.893	0.828	0.896

6 Discussion

We studied different nonlinear functions for 305-day yield approximation. The data included 5854 lactation cycles, 2589, 2080, 1185 cows in the first, second, third and larger lactation, respectively. Estimation of 5854 yields was realized from 76724 day records.

We demonstrated that indexes of determination for cow No 43539 (see Fig. 1) were high in all models. However, in Fig. 1 we can see, that estimated McMillan function takes negative values on interval (0, 15). This fact disqualifies this model, see also Tables 4 and 5. On the left we can see, that Papajesic-Bodero function at the beginning of lactation is a convex. It does not match reality. Always Gaines function is decreasing, but real lactation curve does not peak at the beginning of the cycle. Wood, Nelder and Marek-Zelinkova functions have the highest values of index of determination, see Table 5.

In the next picture—Fig. 2 left we demonstrate, how Ning-Yang's function could inappropriately deal with approximation of lactation curve of cow No. 411605 only by constant function. In Fig. 2 we can see the approximation by Nelder function of lactation of cow No 411578. However, at the beginning of lactation the approximating function greatly exceeds the value of 50 and it gives a greatly deviated estimate of a 305-day yield of 14084 litres, see Tab. 4 and 5. These facts direct to disqualification of both functions.

Analysis of the yields of our dairy cows selected for testing the models and the results of Tables 4 and 5 leads to the decision to apply Wood function or Nelder function for the calculation of the 305-day yield. The alternative could be a function of the Marek-Zelinkova, but it is not generally known.

7 Conclusion

Our study demonstrates that high indexes of determination do not ensure a sufficient fitting of the model from the biological point of view. If we want to use the model for estimation of the situation at any day of lactation it is important – evaluating the individual model quality - to take into account also this aspect. On the basis of our experience, the best results were obtained with Wood, Nelder or Marek-Zelinkova models.

Acknowledgments We thank Zdenek Kovar from Zahori for his support and data preparation.

References

1. Golebiewski M, Brzozowski P, Golebiewski L (2011) Analysis of lactation curves, milk constituents, somatic cell count and urea in milk of cows by the mathematical model of Wood. Acta Vet Brno 73–80
2. Gradiz L, Alvarado L, Kahi AK, Hirooka H (2009) Fit of Wood's function to daily milk records and estimation of environmental and additive and non-additive genetic effects on lactation curve and lactation parameters of crossbred dual purpose cattle. Livestock Sci 124: 321–329
3. Karush W (1963) On the use of mathematics in behavioral reseach. In: Garvin V (ed) Natural language and the computer. McGraw-Hill, New York, pp 67–83
4. Kubacek L, Kubackova L, Volaufova J (1995) Statistical models with linear structures. Publishing House of the Slovak Academy of Sciences, Bratislava, Veda
5. Kubacek L (1995) On a linearization of regression models. Appl Math (1) 40:61–78
6. Leon-Velarde CU, McMillan I, Gentry RD, Wilton JW (1995) Models for estimating typical lactation curves in dairy cattle. J Anim Breed Genet 112(1995):333–340
7. Mellado J, Sepulveda E, Garcia JE, Rodriguez A, De Santiago A, Veliz FG, Mellado M (2014) Milk yield of Holstein cows induced into lactation twice consecutively and lactation curve models fitted to artificial lactations. J Integr Agric 13(6):1349–1354
8. Murphy MD, O'Mahony MJ, Shalloo L, French P, Upton J (2014) Comparison of modeling techniques for milk-production forecasting. J Dairy Sci 97:3352–3363
9. Olori VE, Brotherstone S, Hill WG, McGuirk BJ (1999) Fit of standard models of the lactation curve to weekly records of milk production of cows in a single herd. Livestock Prod Sci 58:55–63
10. Rao CR (1973) Linear statistical inference and its applications, 2nd edn. Wiley, New York
11. Rekik B, Gara AB, Hamouda MB, Hammami H (2003) Fitting lactation curves of dairy cattle in different types of herds in Tunisia. Livestock Prod Sci 83:309–315
12. Seber GAF, Wild CJ (2003) Nonlinear regression. Wiley, New Jersey
13. Silvestre AM, Martins AM, Santos VA, Ginja MM, Colaco JA (2009) Lactation curves for milk, fat and protein in dairy cows: a full approach. Livestock Sci 122:308–313
14. Wimmer G, Altmann G, Hrebicek L, Ondrejovic S, Wimmerova S (2003) Uvod do analyzy textov. Vydavatelstvo Slovenskej akademie vied, Bratislava
15. Wood PDP (1967) Algebraic model of the lactation curve in References cattle. Nature (London) 216:164–165
16. Zavadilova L, Jamrozik J, Schaeffer LR (2005) Genetic parameters for test-day model with random regressions for production traits of Czech Holstein cattle. Czech J Anim Sci 50(4):142–154

17. Zavadilova L, Nemcova E, Pribyl J, Wolf J (2005) Definition of subgroups for fixed regression in the test-day animal model for milk production of Holstein cattle in the Czech Republic. Czech J Anim Sci 50(1):7–13
18. Zelinkova G, Marek J (2010) Dependence of milk yield, fat: protein ratio and somatic cell counts. Proceedings of XI, Middle-European Buiatrics Congress, Brno

Model Predictive Control of Trajectory Tracking of Differentially Steered Mobile Robot

K. Rahul Sharma, Daniel Honc and František Dušek

Abstract The paper is focused on trajectory tracking of differential drive mobile robot. The mathematical model of dynamics and kinematics of the mobile robot is considered, based on first principle approach. The dynamic behaviour of engines and chassis, coupling between engines and wheels and basic geometric dimensions are taken into consideration. Reference tracking of linear and angular velocities are achieved by model predictive control of supply voltage of both the drive motors by considering constraints on controlled variable, manipulated variable as well as state variables. Simulation results are provided to demonstrate the performance of proposed control strategy in MATLAB simulation environment.

Keywords Dynamic system modelling · Advanced control · Mobile robot · MPC

1 Introduction

For complex, constrained, multivariable control problems, Model Predictive Control (MPC) has become a standard control strategy in the process industries [1]. With increase of computational power, the MPC is not only limited to slow dynamics processes, where dynamical optimization is easily possible, but also there are new applications for faster systems. For example, MPC control techniques for trajectory tracking of mobile robots as can be seen in [2–4]. MPC concept is even stronger if future set-points are known (at least on finite future horizon) which is

K. Rahul Sharma (✉) · D. Honc · F. Dušek
Department of Process Control, Faculty of Electrical Engineering and Informatics, University of Pardubice, Pardubice, Czech Republic
e-mail: rahul.sharma@student.upce.cz

D. Honc
e-mail: daniel.honc@upce.cz

F. Dušek
e-mail: frantisek.dusek@upce.cz

A. Abraham et al. (eds.), *Intelligent Data Analysis and Applications*,
Advances in Intelligent Systems and Computing 370,
DOI 10.1007/978-3-319-21206-7_8

usually the case of mobile robots. Review of motion control of Wheeled Mobile Robots (WMRs) using MPC can be found in [5]. Authors classify applications according MPC models, robot kinematic models, and basic motion tasks. Mobile robot trajectory tracking problem with linear and nonlinear state-space MPC is presented in [6]. Experimental overview of WMR is published in [7]—several control strategies for trajectory tracking and posture stabilization in an environment free of obstacles are reviewed and compared.

In contrast to the commonly used WMRs models, we consider dynamics of motors as well, so the controller can respect their state variables constraints. In our previous work [8], we modelled the dynamic behavior of differentially steered robot, where the reference point can be chosen independently and gives us more general formulation. In this paper we propose predictive control of the mobile robot, where the linear and angular velocities are optimally controlled by voltages to the drive motors with constraints on controlled variable, manipulated variable and states (current and wheel speed of the motors).

The remainder of this paper is organized as follows. In Sect. 2, mathematical modelling of dynamics and kinematics of mobile robot is derived. In Sect. 3, the model predictive control method is presented. Simulation results are presented in Sect. 4. Finally, conclusions are drawn in Sect. 5.

2 Mathematical Modelling of Dynamics and Kinematics of Mobile Robot

Mathematical model of the robot consists of three relatively independent parts [8]. Dynamics of DC series motor, chassis dynamics—dependency between translational and rotational velocities of the chassis reference point on moments acting to driving wheels and kinematics—influence of motor speed to translational and rotational velocities.

Dynamics of series DC motor can be expressed in terms balancing of voltages (Kirchhoff's law) and balancing of moments. From Kirchhoff's voltage law, we can derive,

$$Ri_L + R_z(i_L + i_R) + L\frac{di_L}{dt} = u_L U_0 - K\omega_L \tag{1}$$

$$Ri_R + R_z(i_L + i_R) + L\frac{di_R}{dt} = u_R U_0 - K\omega_R \tag{2}$$

where, K is the back EMF constant, ω_P and ω_R are the right and left motor speeds. All the other parameters are shown in Fig. 1.

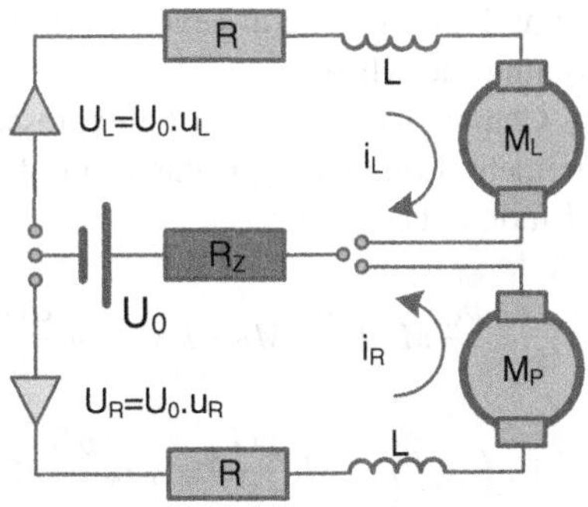

Fig. 1 DC motor wiring

By considering balance of moments we can derive,

$$J\frac{d\omega_L}{dt} + k_r\omega_L + M_L = Ki_L \tag{3}$$

$$J\frac{d\omega_R}{dt} + k_r\omega_R + M_R = Ki_R \tag{4}$$

where, J is moment of inertia of the robot, k_r is coefficient of rotational resistance. M_{L} and M_{R} are load moments on left and right wheel respectively.

Chassis dynamics is defined with vector of linear velocity v_{B} acting on chassis reference point and with rotation of this vector with angular velocity ω_{B} (constant for all chassis points). The chassis reference point B is the point of the intersection of axis joining the wheels and centre of gravity normal projection—see Fig. 2.

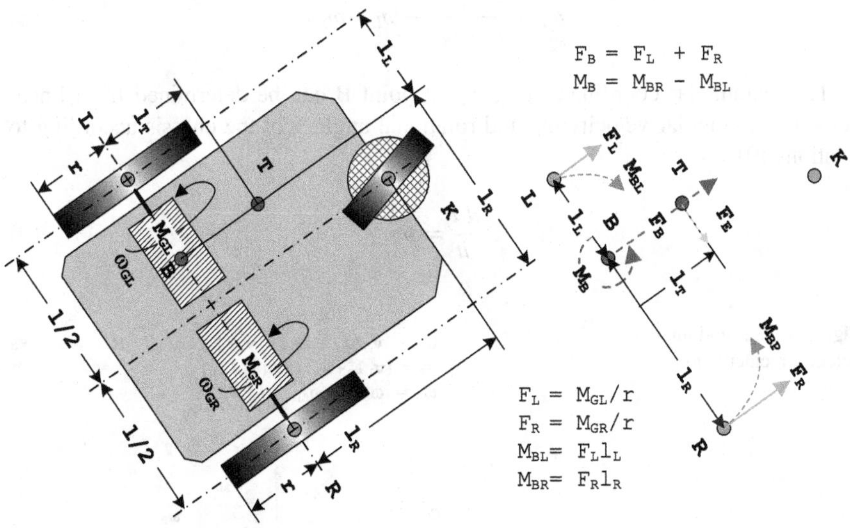

Fig. 2 Chassis scheme and forces

Point T is the general centre of gravity—usually it is placed to centre of the axis joining the wheels.

Chassis dynamics can be expressed by taking balance of force and balance of moment. Equation (5) is the result of applying balance of force and (6) from balance of moments.

$$\frac{p_G}{r}M_L + \frac{p_G}{r}M_P - k_v v_B - m\frac{dv_B}{dt} = 0; M_L + M_P - r_G k_v v_B - r_G m\frac{dv_B}{dt} = 0 \quad (5)$$

$$\begin{aligned} -l_L\frac{pG}{r}M_L + l_R\frac{pG}{r}M_R - k_\omega\omega_B - \left(J_T + ml_T^2\right)\frac{d\omega_B}{dt} &= 0 \\ -l_L M_L + l_R M_R - k_\omega\omega_B - r_G J_B\frac{d\omega_B}{dt} &= 0 \end{aligned} \quad (6)$$

where, p_G is the gear box transmission ratio, k_ω is resistance coefficient against rotational motion. The rest of the parameters are shown in Fig. 2. The parameters r_G and J_B are described as,

$$r_G = \frac{r}{p_G} \qquad J_B = J_T + ml_T^2$$

Kinematics of mobile robot can be derived from theorems of similar triangles, depicted in Fig. 3, we can recalculate the peripheral velocities of the wheels v_L, v_R to the linear velocity v_B and angular velocity ω_B at point B as,

$$v_B = \frac{r_G}{l_L + l_R}(l_R\omega_L + l_L\omega_R) \quad (7)$$

$$\omega_B = \frac{r_G}{l_L + l_R}(-\omega_L + \omega_R) \quad (8)$$

The rectangular coordinates (x_B, y_B) at point B can be determined from linear velocity v_B, angular velocity ω_B and rotational angle α of the chassis according to relations [9].

$$\frac{d\alpha}{dt} = \omega_B \quad (9)$$

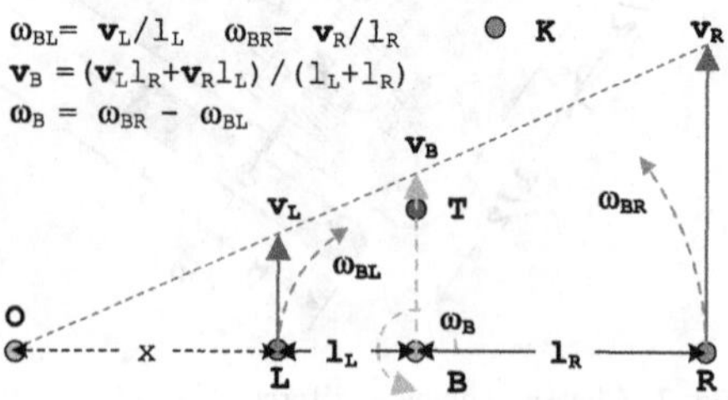

Fig. 3 Linear and angular velocity recalculation

$$\frac{dx_B}{dt} = v_B \cos(\alpha)\,;\ \frac{dy_B}{dt} = v_B \sin(\alpha) \tag{10}$$

These six differential Eqs. (1)–(6), and two algebraic Eqs. (7)–(8) containing eight state variables represent a mathematical description of dynamic behaviour of ideal differentially steered mobile robot with losses linearly dependent on the revolutions or speed. Control signals, u_L and u_R, that control the supply voltages of the motors are input variables and the linear velocity v_B and angular velocity ω_B are output variables. From them, using (9)–(10), we can determine the current coordinates of a point B and the angle of rotation of the chassis.

Calculation of steady-state values for constant engine power voltages are given below. Calculation of steady-state is useful both for the checking of derived equations and for the experimental determination of the values of the unknown parameters. Steady state in matrix representation is,

$$\begin{bmatrix} R+R_z & R_z & K & 0 & 0 & 0 & 0 & 0 \\ R_z & R+R_z & 0 & K & 0 & 0 & 0 & 0 \\ K & 0 & -k_r & 0 & -1 & 0 & 0 & 0 \\ 0 & K & 0 & -k_r & 0 & -1 & 0 & 0 \\ 0 & 0 & 0 & 0 & 1 & 1 & -r_G k_v & 0 \\ 0 & 0 & 0 & 0 & -l_L & l_P & 0 & -k_\omega \\ 0 & 0 & l_P & l_L & 0 & 0 & -\frac{l_P+l_L}{r_G} & 0 \\ 0 & 0 & -1 & 1 & 0 & 0 & 0 & -\frac{l_P+l_L}{r_G} \end{bmatrix} \begin{bmatrix} i_L \\ i_P \\ \omega_L \\ \omega_P \\ M_L \\ M_P \\ v_B \\ \omega_B \end{bmatrix} = \begin{bmatrix} U_L \\ U_P \\ 0 \\ 0 \\ 0 \\ 0 \\ 0 \\ 0 \end{bmatrix} \tag{11}$$

2.1 Computational Form of the Model

Model can be divided into three parts as shown in Fig. 4. Supply voltages, u_L and u_R, are the control inputs. Linear velocity v_B and angular velocity ω_B are output variables from linear part of the model. These variables are the inputs to the consequential non-linear part of the model, whose outputs are the coordinates of

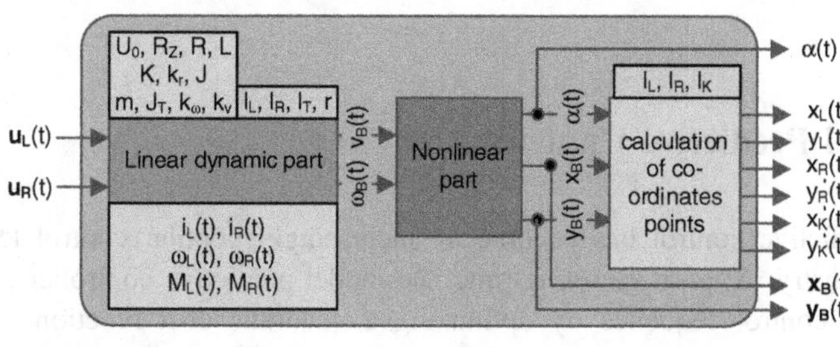

Fig. 4 Model partitioning to linear and nonlinear part

chassis reference point (x_B, y_B) and the rotation angle of the chassis α. The last part is the calculation of coordinates of arbitrary points (co-ordinates of left wheel, right wheel and castor wheel) of the chassis [8].

The Eq. (11) can be reduced to a state space model with four states by introducing the following parameters

$$a_L = k_r + \frac{k_v l_R r_G^2}{l_L + l_R} \quad a_R = k_r + \frac{k_v l_L r_G^2}{l_L + l_R} \quad ; \quad b_L = J + \frac{m l_R r_G^2}{l_L + l_R} \quad b_R = J + \frac{m l_L r_G^2}{l_L + l_R}$$

$$c_L = k_r l_L + \frac{k_\omega r_G^2}{l_L + l_R} \quad c_R = k_r l_R + \frac{k_\omega r_G^2}{l_L + l_R} \quad ; \quad d_L = J l_L + \frac{J_B r_G^2}{l_L + l_R} \quad d_R = J l_R + \frac{J_B r_G^2}{l_L + l_R}$$

The reduced state-space model in matrix form can be written as,

$$\frac{d\mathbf{x}}{dt} = \mathbf{Ax} + \mathbf{Bu} \quad \mathbf{x} = \begin{bmatrix} i_L \\ i_R \\ \omega_L \\ \omega_p \end{bmatrix} \quad \mathbf{u} = \begin{bmatrix} u_L \\ u_R \end{bmatrix} \quad \mathbf{y} = \begin{bmatrix} v_B \\ \omega_B \end{bmatrix} \tag{12}$$

$$\mathbf{y} = \mathbf{Cx}$$

with matrices **A**, **B** and **C** as,

$$A = \begin{bmatrix} -\frac{R+R_z}{L} & -\frac{R_z}{L} & -\frac{K}{L} & 0 \\ -\frac{R_z}{L} & -\frac{R+R_z}{L} & 0 & -\frac{K}{L} \\ \frac{K(d_R + b_R l_L)}{b_L d_R + b_R d_L} & \frac{K(d_R - b_R l_R)}{b_L d_R + b_R d_L} & -\frac{d_R a_L + b_R c_L}{b_L d_R + b_R d_L} & -\frac{d_R a_R - b_R c_R}{b_L d_R + b_R d_L} \\ \frac{K(d_L - b_L l_L)}{b_L d_R + b_R d_L} & \frac{K(d_L + b_L l_R)}{b_L d_R + b_R d_L} & -\frac{d_L a_L - b_L c_L}{b_L d_R + b_R d_L} & -\frac{d_L a_R + b_L c_R}{b_L d_R + b_R d_L} \end{bmatrix}$$

$$\mathbf{B} = \begin{bmatrix} \frac{U_0}{L} & 0 \\ 0 & \frac{U_0}{L} \\ 0 & 0 \\ 0 & 0 \end{bmatrix} \quad \mathbf{C} = \begin{bmatrix} 0 & 0 & \frac{l_R r_G}{l_L + l_R} & \frac{l_L r_G}{l_L + l_R} \\ 0 & 0 & -\frac{r_G}{l_L + l_R} & \frac{r_G}{l_L + l_R} \end{bmatrix} \tag{13}$$

3 Model Predictive Control

Model predictive control has become an increasingly popular control technique used in industry. At each sampling time, the model predictive controller generates an optimal control sequence by optimizing a quadratic cost function. The first control action of this sequence is applied to the system. The optimization problem is solved again at the next sampling time using the updated process measurements and a shifted horizon. The cost function for a Single Input Single Output system is described as by,

$$\mathrm{J}=\sum_{i=1}^{2} r_i \sum_{j=N1}^{N2}\left[\hat{\mathrm{Y}}(k+j)-W(k+j)\right]^2+\sum_{i=1}^{2} q_i \sum_{j=1}^{N3}[\Delta \mathrm{U}(k+j-1)]^2 \tag{14}$$

where $\hat{\mathrm{Y}}(k+j)$ is an optimum j-step ahead prediction of the system outputs, N_1 is the starting point, N_2 is the prediction horizon, N_3 is the control horizon and $W(k+j)$ is future set-points or reference for the controlled variables. The parameters, r_i and q_i are the weighting coefficient for control errors and control increments respectively.

The cost function consists of two parts, mainly costs due to control error during the control error horizon N_2 and costs to penalize the control signal increments N_3. For simplicity in the following text we consider $N_1 = 1$, $N_2 = N_3 = N$.

A general discrete-time state-space model is given as,

$$\begin{aligned}\mathbf{x}(k+1)&=\mathbf{A}\,\mathbf{x}(k)+\mathbf{B}\,\mathrm{U}(k)\\ \mathrm{Y}(k)&=\mathbf{C}\,\mathbf{x}(k)\end{aligned}$$

An incremental state space model [10] can also be used, if the model input is the control increment $\Delta u(k)$ instead of $u(k)$. $\Delta u(k) = u(k) - u(k-1)$

$$\begin{aligned}\underbrace{\begin{bmatrix}\mathbf{x}(k+1)\\ \mathrm{U}(k)\end{bmatrix}}_{\bar{\mathbf{x}}(k+1)}&=\underbrace{\begin{bmatrix}\mathbf{A}&\mathbf{B}\\ \mathbf{0}&\mathbf{I}\end{bmatrix}}_{\mathbf{M}}\underbrace{\begin{bmatrix}\mathbf{x}(k)\\ \mathrm{u}(k-1)\end{bmatrix}}_{\bar{\mathbf{x}}(k)}+\underbrace{\begin{bmatrix}B\\ \mathbf{I}\end{bmatrix}}_{\mathbf{N}}\Delta\mathrm{U}(k)\\ \mathrm{Y}(k)&=\underbrace{\begin{bmatrix}\mathbf{C}&\mathbf{0}\end{bmatrix}}_{Q}\underbrace{\begin{bmatrix}\mathbf{x}(k)\\ \mathrm{U}(k-1)\end{bmatrix}}_{\bar{\mathbf{x}}(k)}\end{aligned} \tag{15}$$

The predicted output representation of state space model in matrix form is,

$$\underbrace{\begin{bmatrix}\hat{\mathrm{Y}}(k+1)\\ \hat{\mathrm{Y}}(k+2)\\ \vdots\\ \hat{\mathrm{Y}}(k+N)\end{bmatrix}}_{\mathbf{Y}}=\underbrace{\begin{bmatrix}\mathbf{QN}&0&\cdots&0\\ \mathbf{QMN}&\mathbf{QN}&\cdots&0\\ \vdots&\vdots&\ddots&\vdots\\ \mathbf{QM}^{N-1}\mathbf{N}&\mathbf{QM}^{N}\mathbf{N}&\cdots&\mathbf{QN}\end{bmatrix}}_{\mathbf{G}}\underbrace{\begin{bmatrix}\Delta U(k)\\ \Delta U(k+1)\\ \vdots\\ \Delta U(k+N-1)\end{bmatrix}}_{\mathbf{U}}+\underbrace{\begin{bmatrix}\mathbf{QM}\\ \mathbf{QM}^2\\ \vdots\\ \mathbf{QM}^N\end{bmatrix}}_{\bar{\mathbf{F}}}\bar{\mathbf{x}}$$

$$\mathbf{Y}=\mathbf{GU}+\bar{\mathbf{F}}_{\mathbf{f}}\bar{\mathbf{x}} \tag{16}$$

3.1 Cost Function

The cost function [11] in (14) can be represented in matrix format as,

$$\mathbf{J} = (\mathbf{Y} - \mathbf{W})^{\mathrm{T}}\mathbf{R}(\mathbf{Y} - \mathbf{W}) + \mathbf{U}^{\mathrm{T}}\mathbf{Q}\mathbf{U} \tag{17}$$

where, **R** and **Q** are diagonal matrices with diagonal elements r_i and q_i respectively and **W** is a column vector of N future set points.

3.2 Constraints

In a long range predictive control, the controller has to anticipate constraint violation and correct control actions in an appropriate way. The output constraints are mainly due to safety reasons and must be controlled in advance, because output variables are affected by process dynamics. Manipulated variables can always be kept in bound, but predictive controller will find optimal solution with respect to constraints. The input, state and output constraints are,

$$\mathrm{U}_{\min} \leq \mathrm{U}(k+i-1) \leq \mathrm{U}_{\max}\ ; \mathrm{X}_{\min} \leq \mathrm{X}(k+i) \leq \mathrm{X}_{\max}$$
$$\mathrm{Y}_{\min} \leq \mathrm{Y}(k+i) \leq \mathrm{Y}_{\max}; i = \{1, N\}$$

The implementation of MPC with constraints involves the minimization of a quadratic cost function subject to linear inequalities: Quadratic Programming (QP) problem. The QP problem can be solved using function *quadprog* in MATLAB. The inequality, equality constraints, lower and upper bounds are given by,

$$\mathbf{AU} \leq \mathbf{b}; \mathbf{A}_{eq}\mathbf{U} = \mathbf{b}_{eq}\ ; \ \mathbf{lb} \leq \mathbf{U} \leq \mathbf{ub}$$

3.3 Predictive Control of Mobile Robot

Considering the robot model (12)–(13), the optimization problem (14) for trajectory tracking is solved by MPC, with a cost function in the form of (17). Model can be divided into three parts as shown in Fig. 5. Supply voltages, u_{L} and u_{R}, are the control inputs. Linear velocity v_{B} and angular velocity ω_{B} are output variables from linear part of the model. These variables are the inputs to the consequential non-linear part of the model, whose outputs are the coordinates of chassis reference point (x_{B}, y_{B}) and the angle of rotation of the chassis α. Constraints are applied to

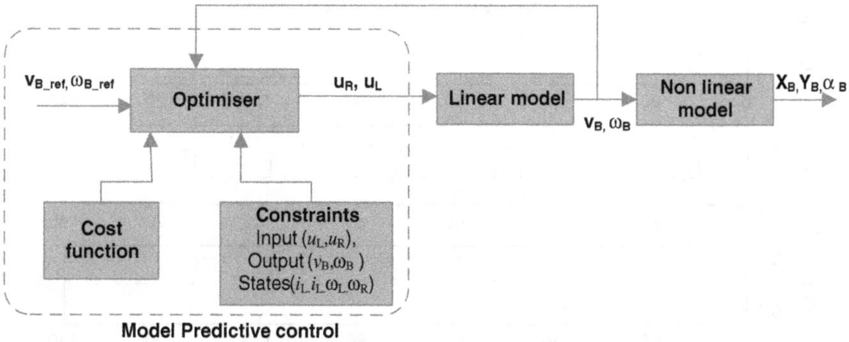

Fig. 5 Predictive control of mobile robot

controlled variable, manipulated variable and state variables (motor current, i_L and i_R, wheel speeds, ω_L and ω_L.). Overall control scheme is shown in Fig. 5.

4 Simulation Results

Chassis parameters and DC motor parameters were chosen as in [8]. These values are chosen so that they roughly correspond to the real physical values of mobile robot.

The mobile robot MPC is simulated in MATLAB simulation environment with a sample time of 0.1 s. With the tuning parameters: $N = 5$, $q_1 = 1$, $q_2 = 10$, $r_1 = 1000$, $r_2 = 100$ and with an initial condition calculated from steady state (11) we have the simulation results shown in Figs. 6 and 7. The matrices **Q** and **R** are diagonal matrices with first N diagonal elements r_1 and q_1 respectively and the consequent N diagonal elements are r_2 and q_2 respectively. Mobile robots initial orientation was considered along y axis.

Constraints where applied to controlled variables (control voltages to right and left wheel), manipulated variables (linear and angular velocities) and state variables (current to the right and left motors and wheel speeds). The constraints of control voltage of the motors where set to [0, 1] since the source voltage is 10 V. The constraints of linear velocity and angular velocity were set to [0, 0.5] m s^{-1} and [−5, 5] rad s^{-1} as these values resembles the physical limits of the robot. The constraints on current and wheel speeds of DC motor were set to [−3, 3] A and [0, 50] rad s^{-1}. The trajectory was chosen in such a way that we can see the response of robot when a sudden change of direction and orientation to the robot and movements of rotation and translation in the same time occurs.

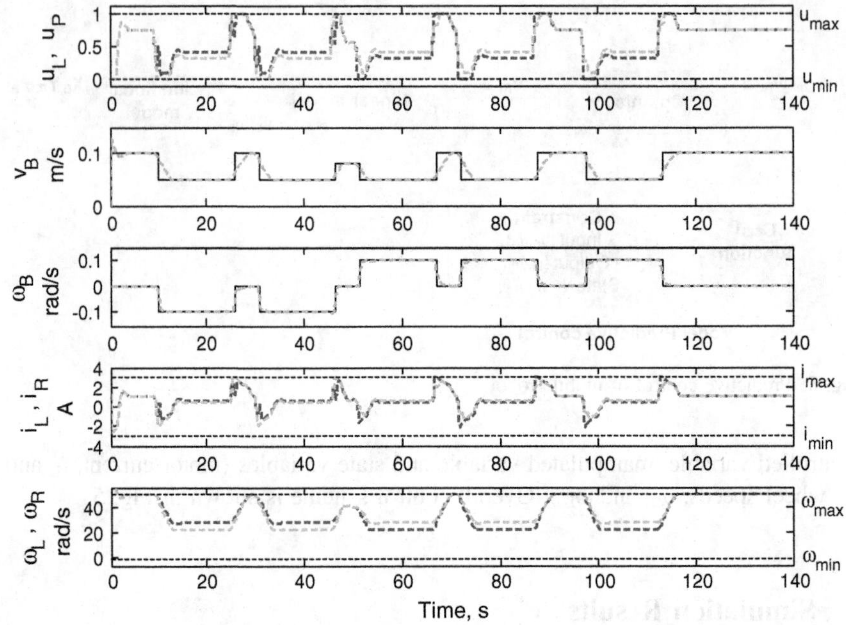

Fig. 6 Control inputs (left and right wheel voltage), responses (linear and angular velocity) and state variables (current and speed of left and right motors). *Red* and *blue dashed lines* indicates right and left wheels respectively. *Solid black* and *dashed green* represents reference and response signals respectively (Colour figure online)

Fig. 7 Trajectory tracking (*Red*—right wheel, *Blue*—left wheel, *Green*—Castor wheel) (Colour figure online)

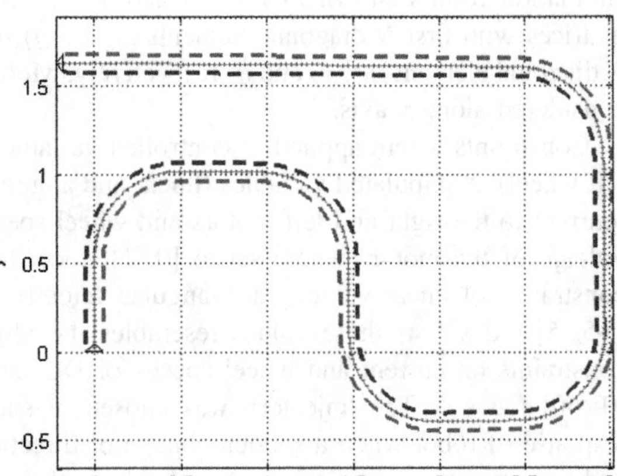

5 Conclusion

In this paper, a model predictive control structure has been proposed for a mobile robot path tracking problem. The use of a simplified model allowed the use of a linear MPC under constraints, which it was developed to control the robot dynamics and kinematics by tracking the reference linear and angular velocities. Simulation results have shown the good performance of the control strategy proposed, besides to be flexible for several applications and to have a vast calibration capacity. For example, the penalization coefficients, prediction and control horizons, and constraints can be adjusted depending on the application and/or type of trajectory.

Acknowledgments This research was supported by project SGSFEI_2015007, New modelling and identification methods for Model Predictive Control at FEI, University of Pardubice. This support is very gratefully acknowledged.

References

1. Camacho EF, Bordons C (2004) Model predictive control, 2nd edn. Springer, London
2. Gu D, Hu H (2006) Receding horizon tracking control of wheeled mobile robots. IEEE Trans Control Syst Technol 14(4):743–749
3. Kuhne F, Lages WF, da Silva Jr JMG (2004) Model predictive control of a mobile robot using linearization. In: Proceedings of mechatronics and robotics
4. Lages WF, Alves JAV (2006) Real-time control of a mobile robot using linearized model predictive control. In: Proceedings of 4th IFAC symposium on mechatronic systems
5. Kanjanawanishkul K (2012) Motion control of a wheeled mobile robot using model predictive control: a survey. KKU Res J 17:811–837
6. Kühne F, Lages WF, da silva JMG (2005) Mobile robot trajectory tracking using model predictive control. In: II IEEE Latin-American robotics symposium
7. De Luca A, Oriolo G, Vendittelli M (2001) Control of wheeled mobile robots: an experimental overview. In: Ramsete. Springer, Berlin, pp 181–226
8. Dušek F, Honc D, Rozsíval P (2011) Mathematical model of differentially steered mobile robot. In: 18th International Conference on Process Control, Tatranská Lomnica, Slovakia
9. Šrejtr J (1954) Technická mechanika II. Kinematika 1. část. SNTL Praha, p 256
10. Honc D, Dušek F (2013) State-Space constrained model predictive control. In: ECMS, pp 441–445
11. Sharma R, Abraham A, Honc D, Dušek F, Pappa N (2015) Predictive control of GUNT RT 010 and RT 050 laboratory systems. In: 20th international conference on process control, Tatranská Lomnica, Slovakia (unpublished)

Generalized Linear Models Applied for Skin Identification in Image Processing

Sebastián Basterrech, Andrea Mesa and Ngoc-Tu Dinh

Abstract We investigate the performance of a hybrid classifier for solving a classic problem in the area of image processing. We analyse the performance of this method for a specific classification task that is detecting skin regions in a picture. Our approach consists in partitioning clustering the input dataset. Then, for each cluster we apply the well-known Generalized Linear Models in order to identify the skin and non-skin points. We evaluate the performance of our approach using several well-known metrics. Besides, we compare the reached performance with the Feed-forward Neural Networks. The reached results prove that the proposed approach is a well-alternative for solving the skin-identification problem.

Keywords Classification · Generalized linear models · Computer vision · Skin detection · Image segmentation

1 Introduction

In computer vision and image processing, the classification of pictorial information is an important problem, which is done using a set of meaningful features in the representation of the image. In the image segmentation problem, one of the most crucial

---

S. Basterrech (✉)
IT4Innovations, National Supercomputing Center, VŠB—Technical University of Ostrava, Ostrava-poruba, Czech Republic
e-mail: Sebastian.Basterrech.Tiscordio@vsb.cz

A. Mesa
Facultad de Ciencias Económicas y de Administración, Universidad de la República, Montevideo, Uruguay
e-mail: andreagmesa@ccee.edu.uy

N-T. Dinh
School of Information and Communication Technology, Hanoi University of Science and Technology, Hanoi, Vietnam
e-mail: dinhngtu@gmail.com

A. Abraham et al. (eds.), *Intelligent Data Analysis and Applications*,
Advances in Intelligent Systems and Computing 370,
DOI 10.1007/978-3-319-21206-7_9

research areas of computer vision, the main task refers of partitioning a digital image into multiple segments. The goal is to separate the image in multiple regions following features that are used by human interpretation of pictorial information. Several works explore the kind of textural features appropriate to classify images [1–3]. Once these features are defined, the blocks of images can be categorized using anyone of a multitude of pattern-recognition methods. The literature about classifier methods is very extensive [4]. For instance, in [5] the authors use a specific type of Neural Network (named Random Neural Network [6]) to classify the pixels into texture classes. This kind of approach is often used for medical imaging, for example in [7] was applied for magnetic resonance imaging and in [8] was applied for ultrasound images. In [9] the authors apply the Bayesian classifiers, Gaussian classifiers and Neural Networks for identifying a skin pixel. In addition, they used different kind of color representations.

In [10] the authors use a large database composed of images from the World Wide Web for generating a classifier. The authors analyse the skin color properties according several color space representations. The skin detection has been also studied in video; for instance in [11] the authors use chrominance, shape, and probability distributions for developing their methods.

The dataset consists of a collection of pixels in the RGB channels with a label that indicates if the pixel represents corresponds to skin or not. We assume that the image has several regions that can be classified into categories. A skin classifier must be able to identify any type of skin (blackish, yellowish, brownish, whitish, etc.) in multiple contexts [9]. Therefore, we are interested in finding a mapping that receives pixels in the RGB space and identifies them as skin points or not. In this article, we are interested in an approach that separates the image in regions and for each partition estimates a classifier. The main goal of this work is analysing the performance of a hybrid method for solving the skin detection problem. The technique combines a linear partition of the RGB space and the *Generalized Linear Model (GLM)* technique using the logit link function. Within the linear methods for classification problems, the logistic regression for the problem of two classes have been widely used being a classic technique that yields very good results despite its simplicity. We use for training the classifiers a popular dataset from the *UCI Repository of Machine Learning* [12].

The article is organised as follows. Section 2 presents an overview of the binary classifiers studied in this paper. This section starts with the specification of the problem. Next, we introduce the Generalized Linear Models and the hybrid classifier. This section ends with the definition of the objective evaluation metrics. Empirical results are presented in Sect. 3, including the dataset description, the model parameter setting and the analysis of the results. Section 4 ends this work providing a conclusion and a discussion about future research directions.

2 Binary Classifier Models

This section is structured as follows. This section begins by defining the problem. We follow by describing the GLM. Next, we introduce our approach that combines a linear partition of the input space and the GLM. After that, the section presents the evaluate metrics used for computing the model performance.

2.1 Problem Description

In this article, the goal is developing a method for identifying skin pixels on color images. For this purpose, we use a dataset of labeled pixels where for each sample there is a label indicating if the pixel is skin or not. The learning dataset was taken from the UCI Machine Learning Repository [12]. Most relevant papers that have already studied this benchmark problem can be seen in [13, 14]. Skin and non-skin dataset is generated using skin textures from face images of many people with diverse age, gender and race. The input features are given by the terna of RGB colors and the output feature is a binary indicating skin or not-skin. The learning models are adjusted using this particular dataset. Formally, the problem is defined as follows. Given a dataset $\mathcal{L} = \{(\mathbf{a}^{(k)}, \mathbf{b}^{(k)}), k = 1, \ldots, K\}$, where $\mathbf{a}^{(k)} \in [0, 255]^3$ represents the input features of a pixel k and $\mathbf{b}^{(k)} \in \{-1, 1\}$ represents if there is skin or not in k, the goal of a learning algorithm consists in inferring a parametric mapping $v(\mathbf{a}, \boldsymbol{\beta}, \mathcal{L})$ in order to predict the values $\mathbf{b}$, such that some distance $d(v(\mathbf{a}^{(k)}, \boldsymbol{\beta}, \mathcal{L}), \mathbf{b}^{(k)})$ is minimised for all $k \in \{1, 2, \ldots, K\}$. The parameters of the mapping are collected in the vector $\boldsymbol{\beta}$. In our specific approach we are interested in a parametric mapping, although exists other approaches which in the mapping is non-parametric. The type of distance $d(\cdot)$ depends of the kind of algorithms used for adjusting $\boldsymbol{\beta}$. There are several kind of distances and algorithms for solving this optimisation problem, most common include: sum of square residuals and cross-entropy [4].

2.2 Generalized Linear Model

The linear models are one of the most applied learning tool during the last 50 years [4]. Let N_a be the dimension of the input space and N_b the dimension of the output space. In our particular problem we have that $N_a = 3$ and $N_b = 1$. A simple linear regression with $N_a + 1$ unknown parameters collected in the vector $\boldsymbol{\beta}$ is defined as follows. Given an input $\mathbf{a}^{(k)}$ a linear classifier has the form [4]

$$v(\mathbf{a}, \boldsymbol{\beta}, \mathcal{L}) = \beta_0 + \beta_1 a_1^{(k)} + \ldots + \beta_{N_a} a_{N_a}^{(k)}. \tag{1}$$

In a classification context when the output is binary, the output response can be coded to −1 and 1 and the fitted values are discretised according to the rule:

$$\nu(\mathbf{a},\boldsymbol{\beta},\mathcal{L}) = \begin{cases} -1, & \nu(\mathbf{a},\boldsymbol{\beta},\mathcal{L}) \leq 0, \\ 1, & \nu(\mathbf{a},\boldsymbol{\beta},\mathcal{L}) > 0. \end{cases}$$

The *Generalized Linear Models (GLMs)* were proposed by Nelder and Wedderburn in 1972 [15, 16]. Let μ be the expected value of the output variable b. A linear model expresses the expected value μ as a linear predictor $\nu(\cdot)$ of the input variable **a**. On the other hand, GLM expresses a relationship between a monotone and differentiable function $g(\cdot)$ of μ and a linear combination of **a**. The model is given by:

$$g(\nu(\mathbf{a},\boldsymbol{\beta},\mathcal{L})) = \beta_0 + \beta_1 a_1^{(k)} + \ldots + \beta_{N_a} a_{N_a}^{(k)}. \tag{2}$$

The function $g(\cdot)$ is called *link function* and depends on the type of problem. In this work, the output variable is binary, therefore we consider the *logit function* $g(p) = \ln\left(\frac{p}{1-p}\right)$, where p is the probability that the response variable takes the value 1 [16].

The parameter estimation is based on both: the Maximum Likelihood Method and in a variation of the Newton-Raphson algorithm called the Fisher Scoring Method [15, 16]. There are several ways to measure the fit of a model, a widely used in GLMs is the deviance of the model [16]. However, in this article is presented the performance using other metrics usually used for binary classification problems.

2.3 A Hybrid Binary Classification Method

The algorithm has two phases. The first step consists in partitioning clustering the input patterns. Partitioning clustering consists in dividing the input dataset into clusters, so for each input pattern is assigned one cluster. The second step consists in estimating a binary predictor for each cluster. The main hypothesis of this method is that the mapping between **a** and **b** can be well-approximated using local information of the neighbours of **a**. The procedure is as follows.

- Partitioning clustering: we create a fix partition of the input space onto clusters: $B_1, B_2, \ldots, B_S$, where each B_j is an hypercube in $[0, 255]^3, j = 1, \ldots, S$. Each input point of the training data can be assigned to one and only one cluster B. Let S be the number of clusters B_j, and R be the edge length of each B_j. The mapping between an input pattern $a_i^{(k)}$ and B_j is given by the module function between $v_i^{(k)}$ and R denoted by $\mod(v_i^{(k)}, R)$.
- Predictor estimation: in each cluster, we study the accuracy when the prediction is given by both the majority vote rule and the GLM with logit link function.

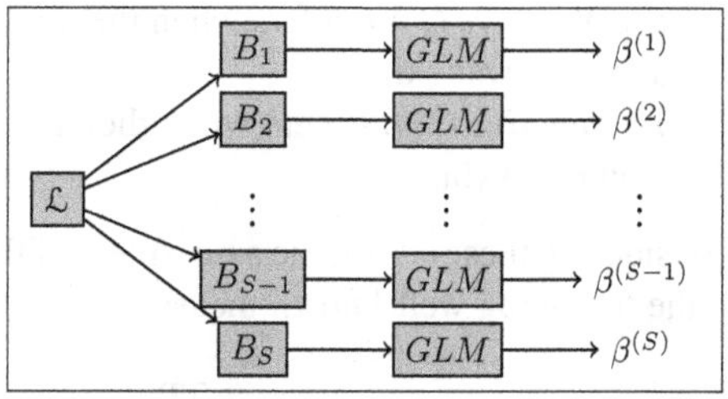

Fig. 1 Schema of the proposed hybrid approach

– The majority vote rule: it assigns a class $C_j \in \{-1, 1\}$ for each cluster B_j according

$$C_j = \text{sign}\left(\sum_{v_i^{(k)} \in B_j} b^{(k)} \right), \tag{3}$$

for each color i. A good property of this rule is that processes and classifies one input pattern in constant complexity time.

– GLM predictor: for each cluster B_j we find the parameters $\boldsymbol{\beta}^{(j)}$ that verifies the expression (2). Due to the fact that the output is binary, it is used the logit function as link function [4]. Figure 1 illustrates the approach.

Note that, the visual perception of an image can change depending of the level of illumination, the background of the image, and even depending of other external factors. Besides, the image appearance depends of several factors that include the technology used for color management. The proposed approach consists of learning a classifier that uses images taken with a specific color management technology. Thus, the learned classifier is dependent of the color gamut and technological devices.

2.4 Performance Evaluation Criteria

Let P and N be the total number of skin and non-skin pixels in the dataset, respectively. In the supervised learning area a common quality measure is the quadratic error function as distance among the predicted values and the target values, defined as

$$MSE = \frac{1}{K} \sum_{k=1}^{K} \left(b^{(k)} - v(\mathbf{a}, \boldsymbol{\beta}, \mathcal{L}) \right)^2. \tag{4}$$

In a binary classification we have four kind of errors that may happen during the estimation. These type of errors are:

- *True Positive (TP)*: a pixel that corresponds to skin was correctly classified.
- *False Positive (FP)*: an incorrect identification, it occurs when the predictor assigns skin when the pixel was non-skin.

- *False Negative (FN)*: refers when the model incorrectly assign non-skin to a pixel that in fact was skin.
- *True Negative (TN)*: refers when the model correctly rejected a pixel, it correctly assign non-skin.

Considering these errors, the assessment of the predictor quality is realised according to the following well-known measures:

$$\text{Sensitivity} = \frac{\text{TP}}{\text{TP} + \text{FN}}, \qquad \text{Specificity} = \frac{\text{TN}}{\text{N}},$$
$$\text{Precision} = \frac{\text{TP}}{\text{TP} + \text{FP}}, \qquad \text{Accuracy} = \frac{\text{TP} + \text{TN}}{\text{P} + \text{N}}.$$

In addition, we use the *Geometric Mean of Accuracy (GMEAN)* over skin and non-skin samples, given by [17]:

$$\text{GMEAN} = \sqrt{\frac{\text{TP}}{\text{TP} + \text{FN}} \frac{\text{TN}}{\text{TN} + \text{FP}}}. \tag{5}$$

3 Empirical Results

This section starts presenting a description of the dataset used for learning the classifiers. Next, the experimental setup for each technique is presented. The section ends with the presentation of the empirical results.

3.1 Dataset Description

The benchmark dataset was taken from [12]. The collection has a dimension of 245057 × 4, where the first three columns are *B*, *G*, *R* values and the fourth column indicates the class label (skin and non-skin). The dataset is imbalanced, there are 50, 859 (20.75 %) observations of skin pixels and 194, 198 (79.25 %) observations of non-skin. It is well-known the difficulties for estimating predictors when the training set is imbalanced. The collection was generated sampling points from face images of various age groups (young, middle, and old), race groups (white, black, and asian), and genders [12]. The methodology describing the dataset collection can be seen in [12].

3.2 Experimental Setup

A random partition for splitting the dataset into two parts is generated: the first one represents the 80 % (196, 045 points) that is used for estimating the parameters (training), the remaining 20 % (49, 012 points) is used for evaluating the estimation of the training phase (testing). We compare the reached performance by the Hybrid Classifier with the reached one when is used a Feedforward Neural Network (FFNN). For setting the weights of the network we use a back propagation algorithm with learning rate equal to 0.001. The neurons has associated a sigmoid function as activation function. The performance is evaluated on several topologies, we experiment with 3, 6, 10, 20 and 30 hidden neurons. The Hybrid Classifier has a main parameter that is the length range, we evaluate the model for the range values: $R = 4$, 8, 16 and 32. A pixel has G, B, R values in [0, 255], then the number of segments for these ranges were: $S = 32$, 16, 8 and 4.

3.3 Results

Table 1 shows the values of the confusion matrix, this is the TP, FP, TN and FN values. In addition the last columns present the MSE for the training and test dataset and the GMEAN. These metrics are shown for the hybrid classifier in both cases, when the GLM and the majority vote rule (MVR) are used. For each case, it is presented the performance of the model for different number of clusters. The results show that the hybrid classifier reached the best performance when the number of clusters is 32. Although, in all cases the method performs better when is employed the GLM

Table 1 Values of the confusion matrix obtained with the hybrid classifier method when the majority vote rule (MVR) and the GLM are used

Method	Parameter (S)	TP	FP	TN	FN	Train MSE	Test MSE	GMEAN
GLM	4	10141	93	38762	15	0.001438	**0.205899**	0,998065
	8	**10151**	33	38822	**5**	**0.000332**	0.207273	0,999329
	16	10144	28	38827	12	0.000598	0.208189	**0,999049**
	32	10098	**11**	**38844**	58	0.002541	0.210936	0.996999
MVR	4	10051	754	38101	105	0.017521	0.220461	0.985117
	8	10117	329	38526	39	0.007340	0.213136	0.993844
	16	10121	81	38774	35	0.002693	0.208157	0.997234
	32	**10126**	**43**	**38812**	**30**	**0.000821**	**0.207484**	**0.997969**

The second column presents the number of clusters (S). Columns 3, 4, and 5 present the True Positive (TP), False Positive(FP), True Negative (TN) and False Negative (FN) values. The last three columns present the MSE computed with the training dataset and the MSE computed with the testing dataset. Last column presents the GMEAN

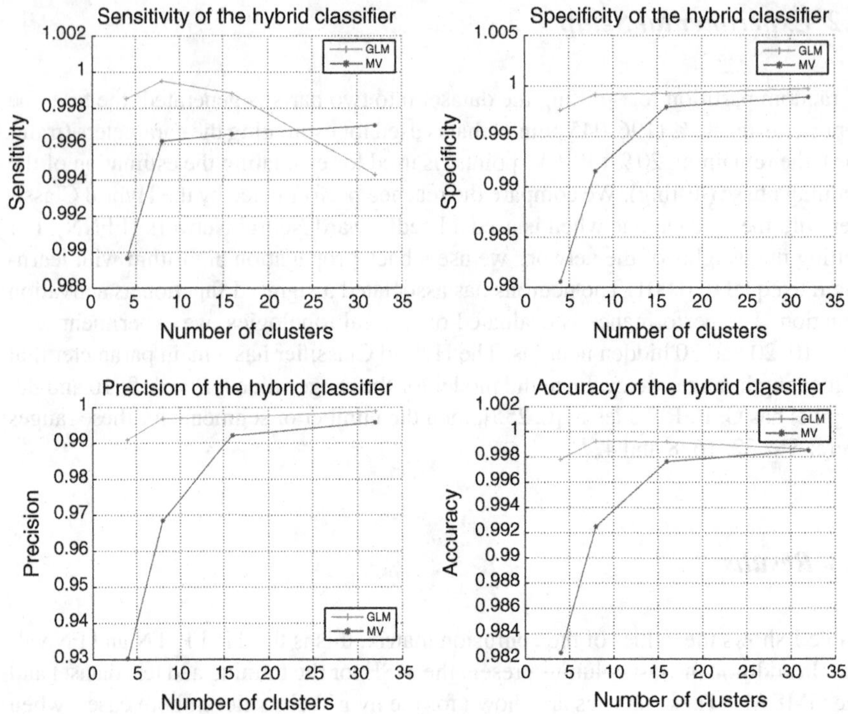

Fig. 2 Impact of the number of clusters in the hybrid classifier on the performance of the algorithms. *Horizontal axe* is the number of clusters 4, 8, 16 and 32, the *vertical axes* are sensitivity, specificity, precision and accuracy metrics. Each figure has two *curves*, the *red* one with (+) *dots* presents the performance of the classifier when the GLM is applied, and the *blue curve* with (*) *dots* presents the performance when the majority vote rule (MVR) is applied

classifier. According with the GMEAN metric, it is enough to consider 4 clusters using GLM for reached a better GMEAN than the method with the majority vote rule and 32 clusters.

Figure 2 presents the evolution of the sensitivity, specificity, precision and the accuracy for the hybrid method when is used GLM and the majority vote rule. The evolution is presented in respect of the number of clusters. There are 4 subfigures, each one corresponds to different metric. In all cases, the performance of GLM outperforms the majority vote rule. There is only one exception that is in the sensitivity case when the number of clusters is 34 for both predictors (MVR and GLM) clusters.

Table 2 shows the sensitivity, specificity, precision and the accuracy for the hybrid method and for the FFNN. The first 5 rows present the results for the neural network with different number of hidden units. The FFNN reaches a higher accuracy when the hidden layer is large. The highest values of sensitivity and accuracy are reached when the network has 30 hidden units. On the other hand, the best specificity and precision are reached when the network has few hidden neurons. A second block of

Table 2 Performance metrics for the classifiers

Method	Parameter	Sensitivity	Specificity	Precision	Accuracy
FFNN	$h = 3$	0.842178	**0.939313**	0.783908	0.919183
	$h = 6$	0.913656	0.937846	**0.793501**	0.932833
	$h = 10$	0.945456	0.931592	0.783215	0.934465
	$h = 20$	0.950773	0.930279	0.780932	0.934526
	$h = 30$	**0.951659**	0.930305	0.781154	**0.934730**
GLM	$S = 4$	0.998523	0.997606	0.990913	0.997796
	$S = 8$	**0.999508**	0.999151	0.996760	**0.999224**
	$S = 16$	0.998818	0.999279	0.997247	0.999183
	$S = 32$	0.994289	**0.999717**	**0.998912**	0.998592
MVR	$S = 4$	0.989661	0.980595	0.930217	0.982473
	$S = 8$	0.996160	0.991533	0.968505	0.992491
	$S = 16$	0.996554	0.997915	0.992060	0.997633
	$S = 32$	**0.997046**	**0.998893**	**0.995771**	**0.998511**

The first block of rows corresponds to the FFNN with different number of hidden neurons (h). Second blocks corresponds to the hybrid classifier with GLM predictor, and the last block of rows corresponds to the hybrid classifier with the majority vote rule

rows presents the performance of the hybrid model when the GLM is used as binary classifier. The best sensitivity and accuracy are reached when is used only 8 clusters. To obtain the best specificity and precision is necessary a larger number of clusters, they are obtained when the partition generates 32 clusters. The last block of rows shows the performance when the hybrid classifier is applied with the majority vote rule. In all cases, the best situation is produced when the method has 32 clusters.

Even though the simple majority vote is applied, the hybrid classifier has better performance than the arched by the neural network. Although, the best case is presented when a GLM is applied for each cluster.

A remark, in the case that a partition has not input training patterns (the potential cluster is empty), therefore we assign a value according to the majority vote rule over the whole training set. The same criteria is used for estimating the parameters of the GLM, in the case that a partition region has not got training patterns, the GLM is replaced by the majority vote rule using all the training samples.

4 Conclusions and Future Work

In this article, we study the performance of binary classifiers for detecting skin pixels in images. In particular, we analyse the performance of a method based in the locality of the inputs points. We introduce a hybrid method that consists of two phases. In a first step the method partitions the input dataset onto several clusters. Next, for each cluster is estimated a binary classifier. The hypothesis of this approach is that any input pattern can be approximated using local information. The proposed method use a popular dataset from the *UCI Repository of Machine Learning* [12] for training the classifiers.

We analyse this approach defining two type of classifiers. In one case, it is used the majority vote rule among the points inside each cluster. On the another case, it is estimated a Generalized Linear Model with logit link function using the training points of each cluster. The proposed model is evaluated using several types of well-known metrics. In addition, the model performance is compared with the performance reached by a Feedforward Neural Networks. In spite of the simplicity, clustering the dataset and applying the majority vote rule performs better than a feedforward neural network. On the other hand, the best results are reached when a GLM with logit link function in each cluster is used. In the near future, another kind of color representation can be used, and the color management technology can be inserted as input feature of the learning predictor. In addition, we are interested in applying our learning predictor on real images, as well as to apply the algorithm for skin detection for frames of video.

Acknowledgments This article has been elaborated in the framework of the project *New creative teams in priorities of scientific research*, reg. no. CZ.1.07/2.3.00/30.0055, supported by Operational Programme Education for Competitiveness and co-financed by the European Social Fund and the state budget of the Czech Republic and supported by the IT4Innovations Centre of Excellence project (CZ.1.05/1.1.00/02.0070), funded by the European Regional Development Fund and the national budget of the Czech Republic via the Research and Development for Innovations Operational Programme, and by the Project SP2015/105 *DPDM—Database of Performance and Dependability Models* of the Student Grand System, VŠB—Technical University of Ostrava.

References

1. Haralick RM, Shanmugam K, Dinstein I (1973) Textural features for image classification. IEEE Trans Syst Man Cybern 3(6):610–621
2. Rosenfeld A, Troy EB (1970) Visual texture analysis. Technical report TR-70-116, Maryland University, College Park, USA. Computer Science Center, USA, June 1970. OSTI Identifier: OSTI ID: 6918260. Legacy ID: DE87008108. ORO-3662-10
3. Russ JC (2011) The image processing handbook 6th edn. CRC Press, Taylor and Francis Group, USA
4. Hastie T, Tibshirani R, Friedman J (2001) The elements of statistical learning. Spring Series in Statistics. Springer, New York
5. Teke A, Atalay V (2006) Texture classification and retrieval using the random neural network model. Comput Manage Sci 3(3):193–205
6. Gelenbe E (1989) Random neural networks with negative and positive signals and product form solution. Neural Comput 1(4):502–510
7. Gelenbe E, Feng Y, Ranga K, Krishnan R (1996) Neural networks for volumetric magnetic resonance imaging of the brain. In: International workshop on neural networks for identification, control, robotics, and signal/image processing (NICROSP'96), IEEE computer society, vol 0, Venice, Italy, pp 194–202, Aug 1996
8. Lu R, Shen Y (2005) Image segmentation based on random neural network model and gabor filters. In: 27th annual international conference of the engineering in medicine and biology society, 2005. IEEE-EMBS 2005, pp 6464–6467, Jan 2005

9. Phung SL, Bouzerdoum A, Chai S, Skin D (2005) Segmentation using color pixel classification: analysis and comparison. IEEE Trans Pattern Anal Mach Intell 27(1):148–154
10. Jones MJ, Rehg JM (1999) Statistical color models with application to skin detection, In: Proc. IEEE Conf. Comput. Vis. Pattern Recognit., Vol. 1, pp. 274–280
11. Wang H, Chang S-F (1997) A highly efficient system for automatic face region detection in MPEG video. IEEE Trans Circ Syst Video Technol 7(4):615–628
12. Bhatt R, Dhall A (2015) Skin segmentation dataset. UCI machine learning repository. http://archive.ics.uci.edu/ml/datasets/Skin+Segmentation. Accessed 20 April 2015
13. Bhatt RB, Sharma G, Dhall A, Chaudhury S (2009) Efficient skin region segmentation using low complexity fuzzy decision tree model. In: Proceedings of the IEEE-INDICON, Ahmedabad, India, pp 1–4, Dec 2009
14. Dhall A, Sharma G, Bhatt R, Khan GM (2009) Adaptive digital makeup. In: Bebis G, Boyle R, Parvin B, Koracin D, Kuno Y, Wang J, Pajarola R, Lindstrom P, Hinkenjann A, Encarnação ML, Silva CT, Coming D (eds) ISVC 2009, part II, vol 5876. LNCS, Springer, Heidelberg, pp 728–736
15. Nelder JA, Wedderburn RWM (1972) Generalized linear models. J Roy Stat Soc Ser A Gen 135:370–384
16. McCullagh P, Nelder JA (1989) Generalized linear models, 2nd edn. Chapman and Hall, New York
17. Zong W, Huang G-B, Chen Y (2013) Weighted extreme learning machine for imbalance learning. Neurocomputing 101:229–242

A Survey on Big Data, Mining: (Tools, Techniques, Applications and Notable Uses)

Nour E. Oweis, Suhail S. Owais, Waseem George, Mona G. Suliman and Václav Snášel

Abstract Big Data is a massive set of data that is so complex to be managed by traditional applications. Nowadays, it includes huge, complex, and abundant structured, semi-structure, and unstructured data as well as hidden data that are generated and gathered from several fields and resources. The challenges for managing Big Data include extracting, analyzing, visualizing, sharing, storage, transferring and searching such data. Currently, the traditional data processing tools and its applications are not capable of managing such revolutionized data. Therefore, there is a critical need to develop effective and efficient Big Data Mining techniques. This, in turn, has opened opportunities for research frontiers by using the exploiting artificial intelligence techniques for Big Data management. This study investigates the most effective Big Data Mining techniques and their rationale applications in various social, medical and scientific fields.

Keywords Big data · Data mining · Smart devices

N.E. Oweis (✉) · W. George · V. Snášel
Department of Computer Science, VŠB-Technical University of Ostrava, Ostrava, Czech Republic
e-mail: nour.easa.oweis.st@vsb.cz

S.S. Owais
Deptartment of Computer Science, FIT, Applied Science University, Amman, Jordan

M.G. Suliman
Faculty of Information Technology, MEU-Middle East University, Amman, Jordan

N.E. Oweis · V. Snášel
IT4Innovations, VŠB-Technical University of Ostrava, Ostrava, Czech Republic

A. Abraham et al. (eds.), *Intelligent Data Analysis and Applications*,
Advances in Intelligent Systems and Computing 370,
DOI 10.1007/978-3-319-21206-7_10

1 Introduction

From the ancient graphics symbols, proto-writing age to our new and modern digital data age, the data generation has not been stopped and the amount of digital data has been exploding to an unlimited rate.

Small data has a limited store schema, well structured, and relational database. The traditional techniques for analysis small data is not so complex and built on the relational database model between subjects [1].

Generally, the small data which are stored in the data warehouse is mostly understood. On the other hand after data prevalence our digital world with its large, complex, non-relational and unstructured amount of data that comes from several fields and resources starting from social network, medical science, commercial, industrial, scientific until many more, lead to realize the term of Big Data, were the main purposes of the big data is to generate small data to be understood.

Big data gives a lot of opportunities to make great development in many fields. Generally, data and specially big data techniques is the complement of the traditional data tools were the big data is still the main and one of the newest term of contemporary debate in the whole of data word [2].

In our fields as a scientists researchers, the massive amount of data has altered the way that we assumes during the past research implementation, analysis to control huge complicated, hidden, and some time unavailable data. This data-intensive has been improved to serve us as a new scientific discoveries called big data mining.

The rest of the paper is organized as follows. In Sect. 2, we review the main big data definitions and characteristics including the 5 V's. In Sect. 3, we survey the big data sources including the smart hardware devices and system software resources. In Sect. 4, we present the main big data tools. In Sect. 5, data mining concept and types are discuss comprehensively. In Sect. 6, data mining techniques are briefly described. In Sect. 7, the big data application and its notable uses in company, healthcare, financial, telecommunication, marketing, and industrial. In Sect. 8 we give the conclusion and future work.

2 Big Data Definitions and Characteristics

After we inclined to accept the huge data alteration; Big Data appear to be the biggest issue and the next research innovations [3].

One of the most popular definition of the big data were defined by Gartner is, "Big data is high-volume, high-velocity and high-variety information assets that demand cost-effective, innovative forms of information processing for enhanced insight and decision making" [4].

After the 3 V's of Gartner (Volume, Velocity, Variety); IBM data scientists introduced the 4th V of big data so-called "Veracity" [5] for ambiguity and incompleteness data which leads to another challenge for keeping big data organized [6].

To discover and analyze the hidden valuable data from these 4 V's (Volume, Velocity, Variety and Veracity) of big data leads to the fifth V so-called "Value" which is the main opportunities that most of organizations are looking for.

The other definition of the big data is Matt Aslett definition, he define the term big data as: "Big data is now almost universally understood to refer to the realization of greater business intelligence by storing, processing, and analyzing data that was previously ignored due the limitation of traditional data management technologies".

Each of the 5 V's (Volume, Velocity, Variety, Veracity, and Value) figures out as part of the big data scope. These 5 V's are now the main characteristics terms of the big data [1, 7] as listed below:

1. Data Volume is the size of the dataset, by Terabytes, Exabyte, and Zettabyt age.
2. Data Velocity indicates the speed of data for in and out process in a real time. Data is begin generated fast and need to be processed fast, this means the data velocity measure how often the data is generated in time.
3. Data Veracity describes the ambiguity and Incompleteness data.
4. Data Variety describes the scope of several data types that are comes from deferent fields and resources with structured, semi-structure, and also unstructured data.
5. Data Value to explore, discover and analyze the most valuable information from the dataset.

3 Big Data Sources

Nowadays among more than two billion users are using the newest smart technology that are daily generating data, starting from smart home, smart city, smart business, smart posters, up-to-the-minute entertainments applications, and even the great new facilities that allow machine to machine communication without user in between, which is known as the Internet Of Things (IoT) [8]. This smart devices are widely used in our world with huge amount of data daily generated, this lead to increase to data capacity and variety. We classify these data sources into two main parts: the first is the hardware, and the second, is the system software.

1. **System Hardware**
   The system hardware sources including the smart devices that generating and collecting daily data, such as Data Center, Wireless Sensors Network (WSN), Radio Frequency Identification (RFID), Near Field Communication (NFC), Machine Log Data, Cloud Computing, Smart Phone, Smart Poster and many more [8–10].

2. **System Software**
   The system software including, many software and platforms developed specially for the big data, like big data platforms (IBM, SAP, SAS Microsoft, Intel, Infobright, Hortonwork, Kognitio, ORACLE, and Amazon); Internet of Things such as (Internet Protocol version 6 (IPv6), Machine to Machine (M2M) protocol; Service Oriented Architecture (SOA), and Representational State Transfer (REST); Web of Things; Cloud platform such as(Google cloud, and Dropbox); Social Network such as (Facebook, LinkedIn, and Twitter); Search Engine such as (Google BigQuery) [8–10].
   All of these data sources and technologies are playing an important and critical impact in many in several fields like medical science, social network, commercial, industrial, scientific, and many more.

4 Big Data Tools

Storing huge amount of data is not a big challenge, but one of the most challenge of the big data deal with how to figuring out this huge amount of data to make it valuable and more senses. Big data deals with multi data types, such as structure, semi-structure, and even unstructured and untraditional databases. These types of data can reach a high capacity of storage media, scaled today by petabytes, Exabyte or zettabytes that require a specific tools to handle this huge capacity.

Nowadays, many big company like Google are dealing with big data doesn't use the traditional techniques to process their data, its use a special big data tools to manipulate, store, and analyze their data stream. Also there are several big data tools used to extract, analyze, and visualize the complex and different data types. In this section we will introduce briefly the four main open source big data tools, such as; NoSQL; MapReduce; Hadoop, and R Language [11].

1. **NoSQL**
   NoSQL refers to "Not Only SQL", that means, this tool have combine two parts, the traditional SQL techniques with additional other new and alternative techniques used for querying and access the large, complex, unstructured, and non-relational dataset [12] that can be stored remotely on multiple virtual services in cloud dataset.
   NoSQL is an open source database software that is useful for big data management. NoSQL is combine with other tools like massive parallel processing, columnar-based databases and database-as-a-service (Dass), and most of the recent social network such Facebook, LinkedIn and Twitter are using Apache Casandra NoSQL database tool [13].

2. **MapReduce**
   MapReduce is one of the most open source data mining techniques model that allows programmers to implement, processing, and develop large dataset with

parallel and distributed algorithm on a cluster by using several programming languages like C, C++, Java, Perl and more by using several MapReduce libraries [13, 14].
MapReduce is inspired by Google Company, consisting of two parts, Map and Reduce. The Map is a procedures that divided, filter, and sort in the distributed cluster, while the Reduce is another procedures that summarize the results into a single mode at a time [11].
MapReduce can be applied to large volumes of data that can be processed by a large number of servers. MapReduce can be used to sort a petabyte of data, with only a few hours. Parallelism also gives some possibilities partial recovery server failures: if the operating portion [12], which produces a preprocessing operation or convolution fails, its operation may be transferred to another working unit (assuming that the input data for the ongoing operation are available) [15]. The most popular open source implementation of the MapReduce is the Apache Hadoop software.

3. **Hadoop**
Hadoop is a software with an open source, freely available set of tools and libraries, based on Java software framework for processing, development, and execution the large volume of distributed datasets and application [12]. Hadoop can execute thousands terabytes of large, complex and non-relational dataset under several operating system like Windows, Linux, BSD (UNIX), and OS X for Apple Macintosh [11]. Hadoop framework is used by several online search engines like Google and Yahoo [13, 15].

4. **R-Language**
R is a programming language for statistical data processing and graphics, free software environment computing, and open source project GNU developed at Bell Labs. R-Language is an implementation of S programming language that is used for processing large amount of data.
Big Data has altered the way that we espouse in doing scientific research trends. Several analysis techniques for extract and visualize this complex amount of big data and being used for optimization processes, decisions, design and implementation by using data mining and machine learning [16].

5 Data Mining Concept and Types

The idea of parallel data mining has been emerged with big data to improve the usage of huge, complex amount of dataset by using Artificial Intelligence (AI) system that makes the computer thinks and operates like a human being [17].

Data mining refers to the steps of searching, analyzing and extracting the valuable needed data from a data warehouses to exploit problem-solving and decision-making. This is known as Knowledge Data Discovery (KDD) [18].

Data mining technique contains a variety of applications and notable uses which are designed to work skillfully with a huge amount of datasets. These applications and their notable uses cover a colossal domain of our life, and most of them will be presented briefly in this study.

There are two main types of data mining models: The first type is descriptive data mining, which used for summarization analysis tools for re-organizing and extracting the basic structure and interconnections between data, this model is commonly used in marketing advertising such as: summarization, clustering, and interconnection data mining and the second type is the predictive data mining, which used for development model based on existing data that carries out the analysis and extraction in more specifications and classifications. This model is commonly used in marketing predictions, like which new products may be popular in the future such as: classification, specification, and prediction [19]. This model is commonly used by big data for awfully valuable techniques to produce productivity in deferent fields.

6 Data Mining Techniques

To optimize the available and suitable data needed from a datasets, parallel data mining tools have been developed to use for solving a lot of problems using several techniques like, artificial neural network, decision tree, rule induction, genetic algorithms, nearest neighbor and many more. Some of these tools are descript briefly in this section.

Artificial Neural Network

Artificial Neural Networks are relatively new technology in computing, which is inspired by the workings of the human brain. Neural networks consist of a large number of processing unites and neurological that are interlocking and interconnected with each other to be able to treat certain types of problems [20].

Neural networks are characterized by its ability for extracting and predicting data from complex input or inaccurate. It can also extract patterns and detect trends of complexity that cannot be observed by a human or by other computer technology [21].

Decision Trees

There are tree-shaped structures of decision tree that perform the sets of decisions commonly used in operation research, specifically in decision analysis. Decision tree consists of three main types, the first, is the decision nodes represented as square, the second, is the chance nodes represented as circle, and the third is the end nodes represented as a triangles.

Genetic Algorithms (GA)

Genetic Algorithms represented one of the artificial intelligence algorithms that are attractive paradigm to improve performance in information retrieval system that are natural evolution techniques for optimizing and searching problems.

There are several applications used genetic algorithms such as bioinformatics, airlines revenue management, artificial creativity, clustering, biology and chemistry, electrical circuit design, financial mathematics, software engineering and many more.

Also there are several techniques for data mining such as, Nearest Neighbor and Rule Induction. The Nearest Neighbor, sometimes called the k-Nearest Neighbor (KNN) used a classification technique that classifies each record depend on the records most related to it in a historical database and the Rule Induction for extraction of useful if-then rules from dataset based on statistical impact [22].

7 Big Data Applications and Notable Uses

There are several varieties of big data notable uses applications that exist in current information age, these applications are available to cover our scope for extracting and analyzing data from several data resources. This section is briefly introduce most of each applications and its notable uses.

7.1 Companies

Most of the big companies have been using big data applications. Big Data mining is a powerful new technology, with great potential to help companies focus on the most important information in the data they have collected, behavior and potential customers. Big data mining techniques can help companies with more vigorously promote to their marketing programs and pricing to retain existing customers and attract new ones.

An example of a company that already deals with big data and the new techniques like IoT is the Infobright Company [23], they provide a high quality, complex queries resulting in faster rapid analysis techniques software and services for machine generated data.

One of the most attractive software that built by Infobright company with (SaaS) platform is called the ERZ-1 software. ERZ-1 provides the optimal solutions that allow clients to visualize distribute of their equipment, connect to the Intermodal marketplace, execute efficient, and accurate financial services, this help them to get the most out of their assets. REZ-1 provide tailored solutions that utilize most of product areas enhanced to use the website such as tracking that provides real time visibility and location for drive movement, inventory management that supplies of assets complex bossiness at any location, and financial management.

There are several services and applications for the company and its notable uses area, some of them are listed below:

7.1.1 Credit Company

Credit Company detects of information fraud that typical behavior credit card holder, reliability analysis of customer accounts, and cross-selling program. An example of financial management company is the Infobright ERZ-1 software that supports billing rules, invoices, dispute and other business services.

7.1.2 Heath Care and Medicine Company

Big Data created a link between life sciences and the medical term to predict, expedite more interactions between patients, doctors, diagnosis, choice of treatment modalities, and predicting the outcome of surgery, and pharmaceutical companies [15, 24].

In addition there are many other company types are also dealing with big data techniques, such as, airline, and insurance company.

7.2 Financial Banking

Big data techniques help the financial banks to build many analytical techniques, especially in predictive area to their customer's behavior, and appropriately serve each category.

There are several big data tools that are available in the business application especially for financial banking fields [25], most of them are based on the classification and prediction techniques to support an intelligent system which is known as Business Intelligence (BI) system [26]. The banking trade application use, such as [15, 27], they use the big data mining technology to perform several common tasks like identification of credit card fraud by analyzing the historical transactions, marketing policy, prediction of customer changes, credit risk analysis, and customer acquisition.

7.3 Telecommunications

Nowadays; Telecommunication companies are dealing with big data mining techniques, they are involves in many records and calls analysis, pricing, failure analysis, attracting customers, prediction of funds, and Identification of customer loyalty.

One of the most modern tools in the telecommunication company is the Mobile user data mining which is fast becoming the most important communication way in our work and life. Mobile user data mining is aims to analyze and predict behavior of mobile users from the data collected, and one of the advantages of possessing mobile data based on real user behavior [28].

7.4 Marketing and Retail Big Data

Marketplace is one of the most business big data applications trend. Today retailers gather detailed information on each individual purchase using credit cards and store-brand computerized control system. Here are the typical problems that can be solved with the data mining algorithms in the retail sector such as [29], the analysis of the shopping cart (similarity analysis) is designed to identify products that customers tend to buy together. Knowledge of the shopping cart need to improve advertising, develop a strategy of stockpiling goods, and ways of their layout in the sales places.

7.5 Industrial

Data mining tools include several applications to automatically control the industrial production management such as, quality control, logistics, and process optimization [30].

After we mentioned all of the above applications, big data is also impact many other areas like science and engineering, text and web data mining, and many more.

8 Conclusion and Future Work

In this survey reported on the current state of Big Data and Data mining, we introduce a short history of the big data, definitions, characteristics and tools. We mentioned the traditional small data and the demand necessity for the big data value in several scientific researches. We also propose briefly the data mining types, techniques that support searching, extracting, and analyzing different data types.

Finally we introduce the most big data opportunities that serve a large number of notable uses areas like (Financial, Telecommunication, Company, Science and Engineering and Industrial). So finally we can conclude that, big data which are flying over the world is the master key for any new digital decorations, and parallel algorithm is the suitable solution for the big data mining techniques.

In the soon future, the next generation of big data is the; Data Fusion and; Data Binding to integrate both services and data from multiple resources with remotely access process and execute, both for providing a better analysis and decision make.

Acknowledgments This work was supported by the IT4Innovations Centre of Excellence project (CZ.1.05/1.1.00/02.0070), funded by the European Regional Development Fund and the national budget of the Czech Republic via the Research and Development for Innovations Operational Programme and by the SGS in VSB—Technical University of Ostrava, Czech Republic, under the grant No. SP2015/146.

References

1. Kudyba S (2014) Big data, mining, and analytics: components of strategic decision making. Boca Raton, CRC Press
2. Schroeder R, Cowls J (2014) Big data, ethics, and the social implications of knowledge production
3. Manyika J, Chui M, Brown B, Bughin J, Dobbs R, Roxburgh C, Byers A (2011) Big data: the next frontier for innovation, competition, and productivity [Kindle edition]. McKinsey Global Institute. Accessed 11 June 2012
4. Gartner IT Glossary. http://www.gartner.com/it-glossary/big-data/. Accessed 05 April 2015
5. The Four V's of Big Data—IBM. http://www.ibmbigdatahub.com/infographic/four-vs-big-data. Accessed 05 April 2015
6. Elorie K (2015) The 5 V's of big data. Avnet advantage: the blog, solution-focused insight for growth-minded VARs. http://blogging.avnet.com/ts/advantage/2014/07/the-5-vs-of-big-data/#comment-474. Accessed 05 April 2015
7. Gupta R (2014) Journey from data mining to web mining to big data. arXiv preprint arXiv:1404.4140
8. Gubbi J, Buyya R, Marusic S, Palaniswami M (2013) Internet of things (IoT): a vision, architectural elements, and future directions. Future Gener Comput Syst 29(7):1645–1660
9. Domingo MC (2012) An overview of the internet of things for people with disabilities. J Netw Comput Appl 35(2):584–596
10. Whitmore A, Agarwal A, Da Xu L (2014) The internet of things—a survey of topics and trends. Inf Syst Front 1–14
11. Wu X, Zhu X, Wu GQ, Ding W (2014) Data mining with big data. IEEE Trans Knowl Data Eng 26(1):97–107
12. Barbierato E, Gribaudo M, Iacono M (2014) Performance evaluation of NoSQL big-data applications using multi-formalism models. Future Gener Comput Syst 37:345–353
13. Lee KM, Park SJ, Lee JH (2014) Soft computing in big data processing
14. Koch C (2013) Compilation and synthesis in big data analytics. In: Big data. Springer, Berlin, pp 6–6
15. Srinivasa S, Bhatnagar V (eds) (2012) Big data analytics: first international conference, BDA 2012, New Delhi, India, 24–26 December 2012: Proceedings (vol 7678). Springer
16. Verzani J (2014) Using R for introductory statistics. CRC Press
17. Jain N, Srivastava V (2013) Data mining techniques: a survey paper. IJRET: Int J Res Eng Technol
18. Sayad S (2012) Data mining map, an introduction to data mining. http://www.saedsayad.com/. Accessed 05 April 2015
19. Zaki MJ, Meira Jr W (2014) Data mining and analysis: fundamental concepts and algorithms. Cambridge University Press
20. Ghore S (2014) Data mining used of neural networks approach, Department, CSE, Govt. Engg. College Bilaspur, Chhattisgarh, India. ISSN: 2348 – 7968
21. Singh Y, Chauhan AS (2009) Neural networks in data mining. J Theor Appl Inf Technol 5 (6):36–42
22. Lahoti AA, Ramteke PL (2014) Data mining technique its needs and using applications. IJCSMC 3(4):572–579

23. Infobright, Data analysis institute. https://www.infobright.com/index.php/case-study/rez-1-ad-hoc-reporting-reduced/#.VE5MOiLF98F. Accessed 05 April 2015
24. Wang Y, Kung L, Ting C, Byrd TA (2015) Beyond a technical perspective: understanding big data capabilities in health care. In: Proceedings of 48th annual Hawaii international conference on system sciences (HICSS), Kauai, Hawaii
25. Akerkar R (2014) Big data computing, Chapman & Hall Book, CRC Press Western Norway Research Institute Sogndal
26. Kambatla K, Kollias G, Kumar V, Grama A (2014) Trends in big data analytics. J Parallel Distrib Comput 74(7):2561–2573
27. Saraswathi K, Ganesh Babu V (2015) A survey on data mining trends, applications and techniques. History 30(135):383–389
28. Du R, Huang J, Huang Z, Wang H, Zhong N (2014) A system to generate mobile data based on real user behavior. In: Web information systems engineering–WISE 2013 workshops. Springer, Berlin, pp 48–61
29. Feinleib D (2014) Doing a big data project. In: Big data Bootcamp. Apress, New York, pp 103–123
30. Johnston WJ (2014) The future of business and industrial marketing and needed research. J Bus Mark Manag 7(1):296–300

Modeling of Acoustic Short Circuits

Stanislav Misak, Viktor Pokorny and Petr Kacor

Abstract The team of authors of this article have been involved in the issue of eliminating noise emissions by electronic equipment from the year 2009. This theme is extremely relevant at present since the health of the human population is a definite priority in human efforts. The first functional prototype for an Antiphase Eliminator, which was installed on a power transformer, was successfully constructed in 2012 on the basis of research. The team develops anti phase eliminator for over 6 years. Now after completion of analog prototype began with digitization and thus building another prototype. This new prototype will processes the signals into digital form. Because the efficiency of the analog prototype is relatively low, the team began together with the digitization also work on computer models that would reveal possible causes of low efficiency. These models already subtracted signals only, as it was on the start of research. These models simulate real acoustic propagation in space, his reflections and the resulting impact of the action of antiphase eliminator.

Keywords Anti-phase · Noise · Acoustic signal · Modeling · ANSYS · Boundary conditions

S. Misak
Faculty of Electrical Engineering and Computer Science, Department of Electrical Power Engineering and IT4Innovations, VŠB-Technical University of Ostrava, Czech, Republic
e-mail: stanislav.misak@vsb.cz

V. Pokorny (✉)
IT4Innovations, VŠB-Technical University of Ostrava, Ostrava, Czech Republic
e-mail: viktor.pokorny@vsb.cz

P. Kacor
Faculty of Electrical Engineering and Computer Science, Department of Electrical Power Engineering, VŠB-Technical University of Ostrava, Ostrava, Czech Republic
e-mail: petr.kacor@vsb.cz

A. Abraham et al. (eds.), *Intelligent Data Analysis and Applications*,
Advances in Intelligent Systems and Computing 370,
DOI 10.1007/978-3-319-21206-7_11

1 Introduction

Currently, the health of human population in any activity is a priority of human pursuit. Indivisible factor having an adverse impact on the human organism is a parasitic noise around us. Human physiology is not adapted to deal with noise emissions, which currently man produces, therefore are issued orders, regulations and directives governing the noise level to human in various activities in various areas. The aim of designers is the parasitic noise as possible to eliminate [1].

2 Antiphase

To understand the system is important to know basic, namely waves alone. Phase of the wave is a dimensionless quantity that determines the relation variable waves, (e.g. displacement noise) at that place and time and to the state variables characteristic waves in temporal and spatial origin. Dependence characteristic of variables determines the shape of "waves" regardless of its dissemination. Phase is a parameter, which depends on the timing characteristic values in a fixed location, which the wave passes, respectively spatial field characteristic values for a fixed moment in time. Noise so we can capture by the curve demonstrate displacement in time. Anti-phase is turning the current signal o 180°.

The result of the exact anti-phase is an absolute deduction of both signals and therefore their complete elimination [1, 2] (Fig. 1).

3 The Current State of Research

The first experiments with an acoustic antiphase wave began to be carried out at VŠB-Technical University of Ostrava in 2008. This impulse arose after measuring the noise of wind power plants as a result of people in their surroundings complaining about the noise. The commissioner of the measuring required values after the analysis as well as a plan for improving the current problematic state of affairs. The idea of making use of an antiphase wave consequently came about as a way of

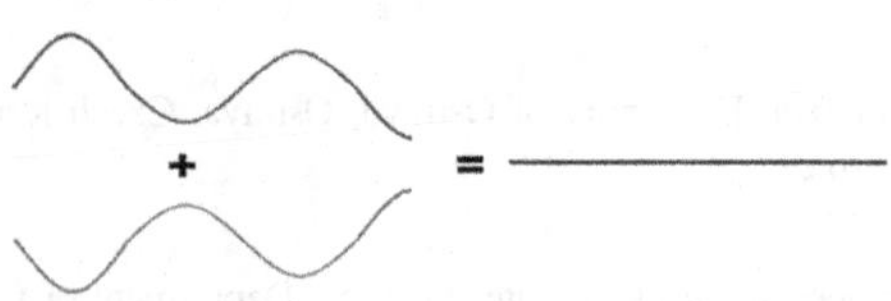

Fig. 1 Generating of antiphase wave

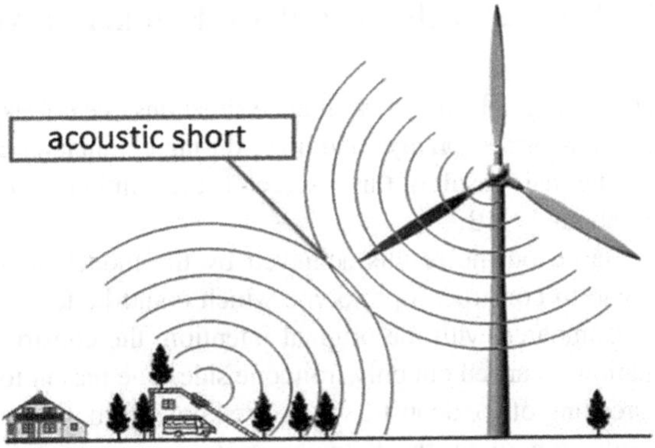

Fig. 2 Application of the Eliminator to buildings, protective walls and noise barriers

actively counteracting this parasitic noise. The following figures demonstrate two possible solutions [1, 3].

The following pictures indicate how the antiphase elimination can be used for elimination of the noise of wind power plants.

The first possibility involves speakers which will serve as an acoustic barrier as in Fig. 2. This consists of the possibility of application of the Eliminator to buildings, protective walls and noise barriers.

The second possibility involves the application of speakers, with the Antiphase Eliminator directly on the walls of buildings (Fig. 3).

The first simulation in a computer environment was carried out on the basis of these considerations. The influences of the antiphase waves in relation to parasitic sounds were tested. The measured signals were compared and the required exactness for the successful elimination was tested.

A decision was made to create a model for the equipment, which will be involved in the elimination of the sound of less complicated equipment in terms of noise, on the basis of these computer simulations. This was done because, for example, a wind power station often changes the spectrum and intensity of its emitted sound. This spectrum is measured on the basis of the intensity of the wind and the overall wind conditions, these being pressure, temperature and air humidity. All of these parameters have an influence on both the parasitic, as well as the potential antiphase (anti-noise) acoustic waves [5].

Fig. 3 Application of the Eliminator to buildings, protective walls and noise barriers

4 Results of the Antiphase Eliminator Model

The polar graph compares all three situations, the noise of the room (blue), the noise of the transformer (orange) and noise during elimination by antiphase (violet) (Fig. 4).

The main result of this research is the elimination of the noise of the transformer by about 17 dB.

Based on the results achieved by the model of the Antiphase Eliminator, we began to construct a prototype which could be tested on an actual transformer.

Compared with the original intention, the construction is different as the elimination is carried out only from one side. The reason for this is the complexity of the spreading of the acoustic wave. We know from the measuring on the model that the production of feedback in the system is a serious problem. The effect of two speakers against one other would cause serious problems with this phenomenon. The entire system will be one-sided after the function verification (Fig. 5).

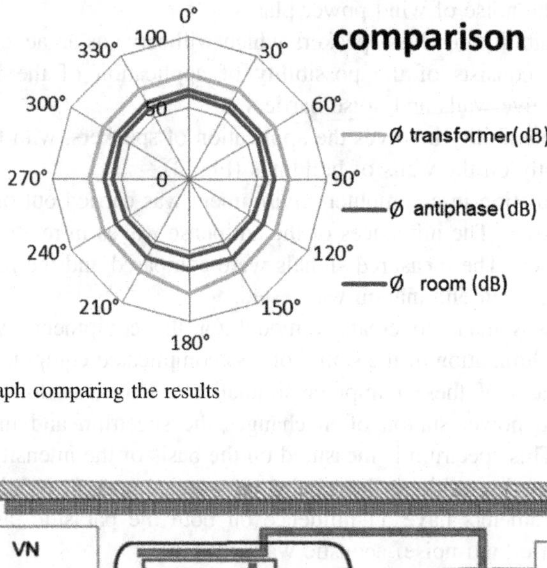

Fig. 4 Polar graph comparing the results

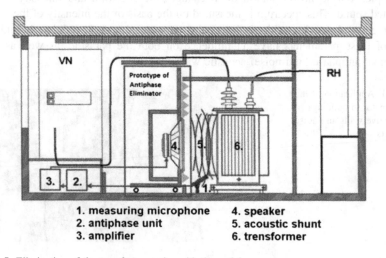

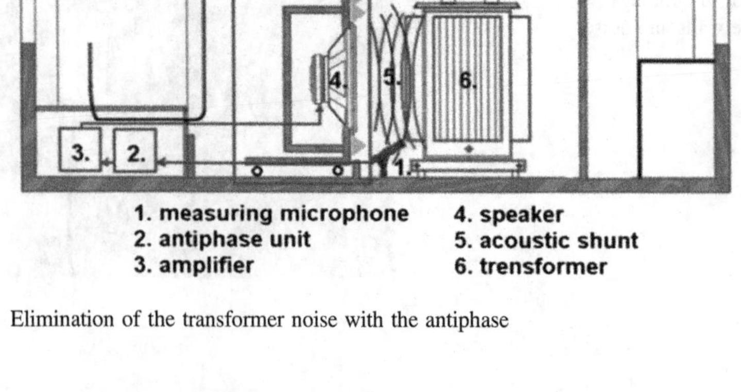

Fig. 5 Elimination of the transformer noise with the antiphase

5 Results of the Antiphase Eliminator Model

During all connections and measurements, a spectral analysis of transformer noise was continuously performed both in the regular operation and in elimination.

From the graphs of spectral analysis of transformer noise both during elimination and without the elimination, it is clear that the elimination was also successful in a real transformer. Frequency 101 Hz was suppressed by 131 dB and also the remaining frequencies ranged below-60 dB.

The graph on the Fig. 6 presents the spectral analysis of noise transformer T616; it is clearly evident that the carrier frequency 101 Hz stands above the rest of the spectrum. In addition, the multiples of this frequency of 200, 300 and 500 Hz are evident. All these frequencies exceed level −50 dB (Fig. 7).

All of the achieved results indicate the success of this method in eliminating noise. The actual core of the prototype is currently established as only analogical equipment which is additionally made up of elements which are far from sufficient for the eventual requirements. These elements are unnecessarily uneconomical, both in terms of their purchase price and in terms of their operations.

The prototype supplied with the protection of industrial ownership, at this point an Applied Model, was created on the basis of these successful tests. This Applied Model was registered with the Office of Industrial Property under the registration number UPV no.25699 and further as the functional sample no.:039/22-05-2012_F. There will be efforts carried out in order to obtain patent protection.

A consequent step in this research will be the digitalization and specification of the entire Antiphase Eliminator [1, 4, 6].

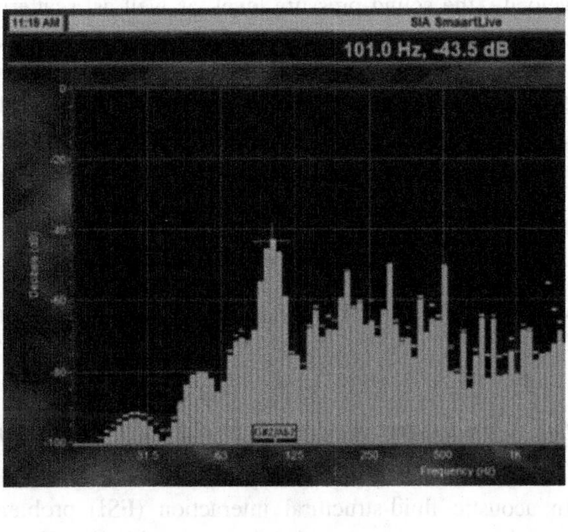

Fig. 6 Spectral analysis of the noise of transformer T616

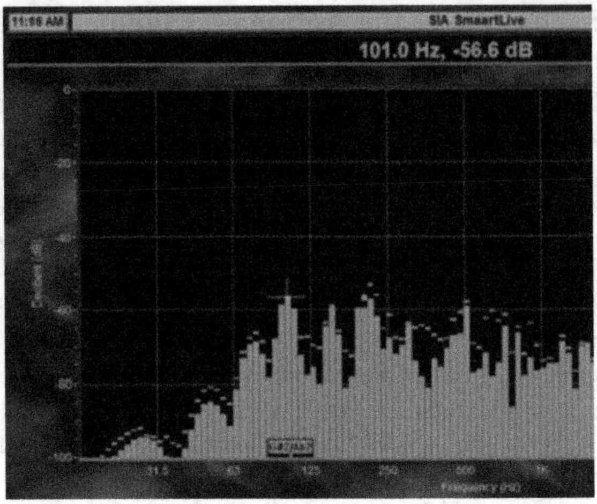

Fig. 7 Spectral analysis of the noise of transformer T616 during elimination

6 Modeling of Acoustic Short

With the idea to make the whole system modeling came the team together with the idea to continue the development of a prototype and make digitization of the prototype. If the prototype is to be more effective, they must find out exactly how acoustic waves propagate and what occurs when exposed to parasitic antiphase wave acoustic signals from the madding sources [7].

Acoustics is the study of the generation, propagation, absorption, and reflection of sound pressure waves in a fluid medium. Typical quantities of interest are the pressure distribution in the fluid at different frequencies, pressure gradient, particle velocity, the sound pressure level, as well as, scattering, diffraction, transmission, radiation, attenuation, and dispersion of acoustic waves.

In normal gases, at audible frequencies, the pressure fluctuations occur under essentially adiabatic conditions (no heat is transferred between adjacent gas particles). Speed of sound then becomes

$$c = \sqrt{\frac{\gamma P}{\rho}} \tag{1}$$

where: $\Upsilon = Cp/Cv = 1.4$ for air and $P = \rho RT$ (Ideal Gas Law)

6.1 Conditions Acoustic Field and Boundary Conditions

In acoustic fluid-structural interaction (FSI) problems, the structural dynamics equation must be considered along with the Navier-Stokes equations of fluid momentum and the flow continuity equation. From the law of conservation of mass law comes the continuity equation:

$$\frac{\partial \rho}{\partial \rho} + \frac{\partial(\rho Vx)}{\partial x} + \frac{\partial(\rho Vy)}{\partial y} + \frac{\partial(\rho Vz)}{\partial z} = 0 \tag{2}$$

where: vx, vy and vz = components of the velocity vector in the x, y and z directions, respectively, ρ = density, x, y, z = global Cartesian coordinates, t = time [7].

The Navier-Stokes equations are as follows:

$$\begin{aligned}&\frac{\partial \rho Vx}{\partial t} + \frac{\partial(\rho VxVx)}{\partial x} + \frac{\partial(\rho VyVx)}{\partial y} + \frac{\partial(\rho VzVx)}{\partial z} = \rho gx - \frac{\partial P}{\partial x}\\ &+ Rx + \frac{\partial}{\partial x} + \left(\mu e\frac{\partial Vx}{\partial x}\right) + \frac{\partial}{\partial y}\left(\mu e\frac{\partial Vx}{\partial y}\right) + \frac{\partial}{\partial z}\left(\mu e\frac{\partial Vx}{\partial z}\right) + Tx\end{aligned} \tag{3}$$

$$\begin{aligned}&\frac{\partial \rho Vy}{\partial t} + \frac{\partial(\rho VxVy)}{\partial x} + \frac{\partial(\rho VxVy)}{\partial y} + \frac{\partial(\rho VxVy)}{\partial z} = \rho gy - \frac{\partial P}{\partial y}\\ &+ Ry + \frac{\partial}{\partial x} + \left(\mu e\frac{\partial Vy}{\partial x}\right) + \frac{\partial}{\partial y} + \left(\mu e\frac{\partial Vy}{\partial y}\right) + \frac{\partial}{\partial z} + \left(\mu e\frac{\partial Vy}{\partial z}\right) + Ty\end{aligned} \tag{4}$$

$$\begin{aligned}&\frac{\partial \rho Vz}{\partial t} + \frac{\partial(\rho VxVz)}{\partial x} + \frac{\partial(\rho VyVz)}{\partial y} + \frac{\partial(\rho VzVz)}{\partial z} = \rho gz - \frac{\partial P}{\partial z}\\ &+ Rz + \frac{\partial}{\partial x} + \left(\mu e\frac{\partial Vz}{\partial x}\right) + \frac{\partial}{\partial y} + \left(\mu e\frac{\partial Vz}{\partial y}\right) + \frac{\partial}{\partial z} + \left(\mu e\frac{\partial Vz}{\partial z}\right) + Tz\end{aligned} \tag{5}$$

where:

gx, gy, gz = components of acceleration due to gravity, ρ = density, μe = effective viscosity, Rx, Ry, Rz = distributed resistances, Tx, Ty, Tz = viscous loss terms The fluid momentum (Navier-Stokes) equations and continuity equations are simplified to get the acoustic wave equation using the following assumptions:

1. The fluid is compressible (density changes due to pressure variations).
2. There is no mean flow of the fluid [7].

The acoustic wave equation is given by:

$$\begin{aligned}&\nabla\left(\frac{1}{\rho 0}\nabla \rho\right) - \frac{1}{\rho 0 c^2}\frac{\partial^2 \rho}{\partial t^2} + \nabla\left[\frac{4\mu}{3\rho 0}\nabla\left(\frac{1}{\rho 0 c^2}\frac{\partial \rho}{\partial t}\right)\right]\\ &= -\frac{\partial}{\partial t}\left(\frac{Q}{\rho 0}\right) + \nabla\left[\frac{4\mu}{3\rho 0}\nabla\left(\frac{Q}{\rho 0}\right)\right]\end{aligned} \tag{6}$$

where:

c = speed of sound ($\sqrt{K/\rho_0}$) in fluid medium, ρo = mean fluid density, K = bulk modulus of fluid, μ = dynamic viscosity, p = acoustic pressure (=p(x, y, z, t)), Q = mass source in the continuity equation, t = time

Harmonically varying pressure is given by:

$$p\left(\vec{r},t\right)=\mathrm{Re}\left[p\left(\vec{r}\right)e^{j\omega t}\right] \tag{7}$$

6.2 Harmonic Response Analyses

The objective of harmonic analyses is to calculate response of system as a function of frequency based on volumetric flow rate or pressure excitation. Harmonic analysis is a technique used to determine the response of a linear structure to loads that vary sinusoidally (harmonically) with time. The wave equation resolved in acoustic simulation requires mass density ρ and sound velocity c of the fluid media [7].

In harmonic response analyses the following equation is resolved for pure acoustic problems:

$$\left(-\omega^2[M_a]+J\omega[C_a]+[K_a]\right)\{p\}=\{f_F\} \tag{8}$$

[Ma]—acoustic fluid mass matrix, [Ca]—acoustic fluid damping matrix, [Ka]—acoustic fluid stiffness matrix, {p}—acoustic pressure, {ff}—acoustic fluid load vector.

6.3 Wave Absorption Conditions (Surface of Model)

The exterior structural acoustics problem typically involves a structure submerged in an infinite, homogeneous, inviscid fluid. In FEA we need to truncate the domain. Wave absorption conditions allow us to model a smaller portion of the domain and assume that outgoing waves keep propagation outwards and do not reflect back. For that we use Perfectly Matched Layers (PMLs). PMLs are artificial anisotropic materials that absorb all incoming waves without any reflections, except for the gazing wave that parallels the PML interface in the propagation direction. PMLs are currently constructed by the propagation of acoustic waves in the media for harmonic response analysis [7].

In harmonic analysis the governing lossless and source-free momentum and conservation are given by:

$$\nabla p=-j\omega p_0\bar{v} \qquad \nabla p=-\frac{j\omega p_0}{p_0c^2} \tag{9}$$

All the above formulas were used for subsequent models as boundary conditions or formulas for conduct acoustic variables in the model.

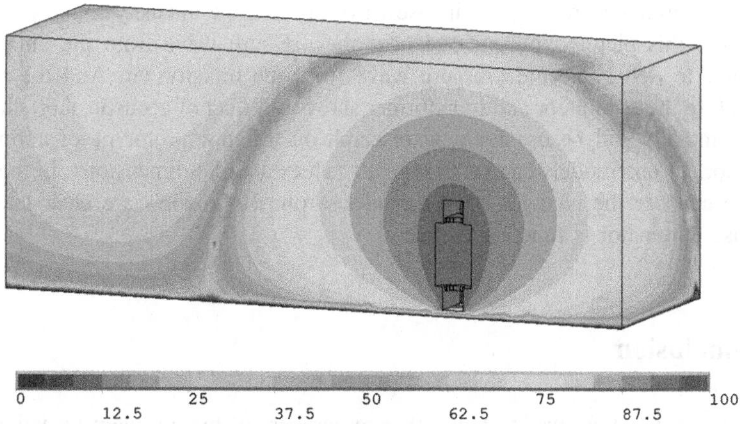

Fig. 8 Model of the spread of parasitic noise of the transformers

6.4 Models of Acoustic Antiphase

Modeling was performed using ANSYS software. On the Fig. 8 you see the model transformer, which makes a noise of 1 Pa = 90 db. On the model we can observe how the acoustic quantities kept in the air flow.

On the surface of the transformer has commissioned a pressure boundary condition (overloaded).

On the outside of surface was applicated Perfectly Matched Layers this enabled modeling only actual element of model (transformer, anti-phase eliminator) and do not infinitely large area. All the models are listed in the SPL = Sound Pressure Level.

The scales on both these models are in dB.

On the Fig. 9 you can see effects of prototype of antiphase Eliminator on the transformer. Both elements against radiate the pressure wave (acoustic signal). The

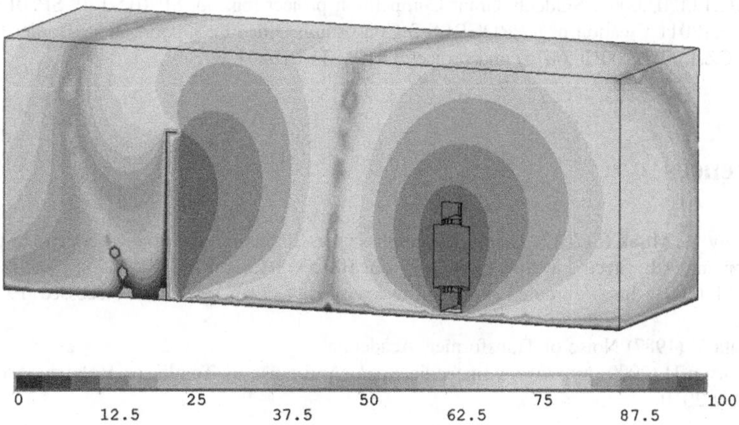

Fig. 9 Model of antiphase wave action

antiphase eliminator produces antiphase wave of the same intensity sound pressure 90 db. On this picture you can see how the pressure wave from the antiphase eliminator to deformate the pressure wave from the transformer. And the green color before the eliminator and transformer show the effect of acoustic short circuit.

Both models will be use for next research on the new prototype of antiphase eliminator. In this models can be change all values and the dimensions. In the real time we can see the changes on the model. From this models are clear that the antiphase eliminator is function device.

7 Conclusion

The aim of this article was to help other researchers of this problem to define the conditions for modeling antiphase acoustic short. A decision was made to continue with the research on the basis of the extremely successful completion of the first phase of the research, wherein the functional prototype of the Antiphase Eliminator was created. The next step will involve the actual digitalization of the instrument and the overall optimization of the method. The optimization will consist of a simplified, for the current time, analogical chain providing the process of treating the actual parasitic sound signal. An algorithm has been created from the acoustic equations which will make up the mathematical basis for adding up the two acoustic signals. A completely new prototype created on the basis of this process will be more economical, more effective and which will have a wider field of application.

Acknowledgments This paper has been elaborated in the framework of the project New creative teams in priorities of scientific research, reg. no. CZ.1.07/2.3.00/30.0055, supported by Operational Programme Education for Competitiveness and co-financed by the European Social Fund and the state budget of the Czech Republic and supported by the IT4Innovations Centre of Excellence project CZ.1.05/1.1.00/02.0070, funded by the European Regional Development Fund and the national budget of the Czech Republic via the Research and Development for Innovations Operational Programme. This paper was conducted within the project ENET—CZ.1.05/2.1.00/03.0069, Students Grant Competition project reg. no. SP2015/170, SP2015/178, project LE13011 Creation of a PROGRES 3 Consortium Office to Support Cross-Border Cooperation (CZ.1.07/2.3.00/20.0075) and project TACR: TH01020426.

References

1. Pokorny V, Misak S (2013) Using the antiphase for elimination the noise of electrical power equipment. Adv Electr Electron Eng 11(3). doi:10.15598/aeee.v11i3.794
2. Chu M (2001) http://gilmore2.chem.northwestern.edu/tech/anr_tech.htm. Accessed 19 April 2010
3. Hamata V (1987) Noise of Transformer, Academia
4. Tomlinson H (2009) http://www.mefeedia.com/– Anti-noise by Tomlinson Holman. Accessed 15 May 2010

5. Qiu X, Hansen CH (2001) An algorithm for active control of transformer noise with on-line cancellation path modelling based on the perturbation method. J Sound Vib 240:647–665. http://dx.doi.org/10.1006/jsvi.2000.3256
6. Hinamoto Y, Sakai H (2007) A filtered-xLMS algorithm for sinusoidal reference signals—effects of frequency mismatch. IEEE Signal Process Lett 14:259–262. doi:10.1109/LSP.2006.884901
7. ANSYS Theory Manual (2015)

A Novel Method for Detection of Covered Conductor Faults by PD-Pattern Evaluation

Tomas Vantuch, Tomas Burianek and Stanislav Misak

Abstract The partial discharge activity as a side effect of the current conductor's disorder was taken into analysis to develop the correct classification of its behavior. This derived knowledge can decrease the risk of the possible damage on the environment caused by uncontrolled conductor's failure. The preprocessing part of the experiment was the synthesis of non-linear features by genetic programming (GP). The inputs for GP were obtained by discrete wavelet transformation (DWT) of the signal data. This preprocessing phase was aimed to create the input values for the classification algorithm which was based on the artificial neural network (ANN).

Keywords Signal classification · Discrete wavelet transformation · Genetic programming · Artificial neural network

1 Introduction

The medium voltage (hereinafter MV) overhead lines with covered conductors (hereinafter CC) are special conductors with additional insulation system. These CC are specified by higher operational reliability and safety in comparison to the ACSR conductors (Aluminium-Conductor Steel-Reinforced) which is given by additional insulation system (SAX-W CC) [26] and by additional insulation system combined with semi-conducting layer (BLT CC) [26]. Therefore CC can be placed on a post

T. Vantuch (✉) · T. Burianek
Department of Computer Science, VŠB-Technical University of Ostrava, 17.
Listopadu 15, 708 33 Ostrava-Poruba, Czech Republic
e-mail: tomas.vantuch@vsb.cz

T. Burianek
e-mail: tomas.burianek.st1@vsb.cz

S. Misak
Faculty of Electrical Engineering and Computer Science, Department of Electrical
Power Engineering and IT4Innovations, VŠB-Technical University of Ostrava, 17.
Listopadu 15 , 708 33 Ostrava-Poruba, Czech RepublicS. Misak
e-mail: stanislav.misak@vsb.cz

A. Abraham et al. (eds.), *Intelligent Data Analysis and Applications*,
Advances in Intelligent Systems and Computing 370,
DOI 10.1007/978-3-319-21206-7_12

in smaller interphase distance and built up area is smaller too. In case of interphases touch of individual conductors of CC or for CC contact with branches of tree does not arise interphase short-circuit, this is a main advantage of MV overhead lines with CC. For this advantage are CC mostly installed in the forested and broken terrain.

However, in case of CC rupture with subsequent CC downfall, this CC fault is not possible to detect by standard digital relays because the standard earth fault does not arise [7]. The low-energy current passes through the fault point and therefore standard digital relays working on current principle cannot detect this fault. Nevertheless, in fault point there is possible to detect partial discharges (hereinafter PD) activity which is generating inhomogeneous electric field in round of degradation of insulation system by fault. The evaluation of PD activity is the basic principle of some methods for CC fault detection. These methods are possible to divide into two categories: (i) methods evaluating PD activity as a low-energy current signal [6, 8, 28]; (ii) methods evaluating PD activity as a voltage signal measured in CC vicinity [13].

In the case of the second approach there is the electric field evaluated in vicinity of CC measured as the voltage signal. The high-frequency component (hereinafter impulse component) within the limits (1–10) MHz of voltage signal is obtained from this voltage signal as a so called PD-Pattern. For no-fault state of CC, the impulse component approximates to zero value and PD-Pattern has created by high-frequency disturbances of external radio transmitters. For fault state, impulse component is generated by PD activity and shape of PD-pattern corresponds to this impulse component change. However, except of impulse component generated by PD activity is PD-Pattern distorted above mentioned disturbances.

The main goal of this paper was detection of CC fault by evaluation of PD-Pattern of voltage signal with using artificial intelligence methods.

2 Experiment Design

The entire work-from of the experiment contained more stages covering analysis, preprocessing and the classification itself. At the first stage of the process, the signal was split into seven parts which were overlapped by 50 % to avoid the potential lost of some hidden patterns. This sub-signals were then transformed by DWT.

2.1 Discrete Wavelet Transform

Wavelet transform (WT) [3, 18] has been widely applied in many engineering fields for solving various real-life problems, not only monitoring of power quality disturbance [4, 11] but in EEG signal classification [5] or image processing too [27].

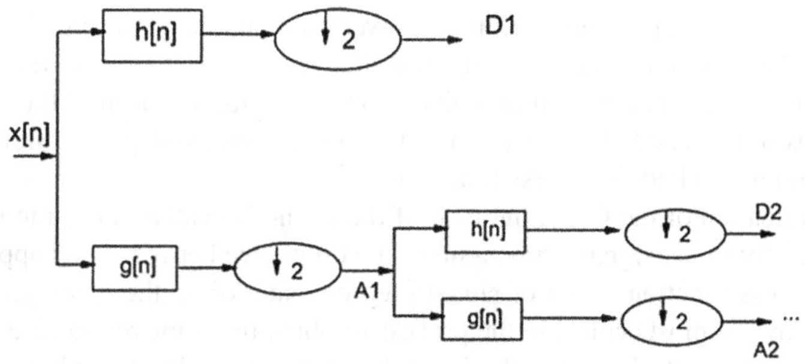

Fig. 1 Block diagram of filter analysis

In this area of study, the wavelet transform comes with better results compared to Fourier transform [22, 25] due to the more flexible way of time frequency representation of a signal by allowing the use of variable sized analysis windows.

The wavelet transform is the projection of a discrete signal into two spaces: the set of approximation coefficients and the set of detail coefficients. There is an effective implementation for obtaining these coefficients, developed by Mallat [12] and it is working by passing the signal through the set of low-pass and high-pass filters, as it is show in Fig. 1.

$$y_{low}[n] = \sum_{k=-\infty}^{\infty} x[k]h[2n-k] \quad (1)$$

$$y_{high}[n] = \sum_{k=-\infty}^{\infty} x[k]g[2n-k] \quad (2)$$

Various high quality software packages are available for wavelet analysis and by this experiment the MATLAB wavelet toolbox was used.

In case of this experiment the previously mentioned parts of the signal went trough the low and high pass filters to obtain a first level of their coefficients (A_1, D_1) and then by the repeating of this procedure on A_1 coefficients, there was obtained the second level of coefficients (A_2, D_2). Each of the series of the coefficients was used as the inputs for the next phase of the experiment, the GP's driven synthesis of the nonlinear features.

2.2 Genetic Programming

The Genetic programming is an evolutionary based algorithm used to create approximation to the ideal solution for a given problem [10, 23]. This approach is inspired by the Evolution Theory. The main part of the GP is a population of individuals based on given rules. The individual is mostly represented by vector or tree structure, which brings the possibility of its crossover and mutation in the breeding phase. The set of the best individuals from the population is chosen to create next generation and this process is repeated in a given count of iterations or until the stopping condition is not reached. The selection of the best individual depends on the given problem and it is called the fitness function.

The application of the GP in the field of the feature's synthesis is quite wide [5, 15, 24], mostly in the signal's processing. The difference between used approaches is in the fitness functions, data or signal preprocessing or on the given grammars. This parts are the most critical in the process of obtaining some reasonable results.

In this experiment, there was the individual represented by the polynomial formula, that was able to compute single value for a given part of the signal. This value represented in this experiment some non-linear feature evaluated by the Information Gain [14]. The IG criteria is used in creating C4.5 Decision tree [17] and the attribute with the highest IG is taken as splitting attribute. In this experiment the IG was used as a fitness function to synthesize the set of non-linear attributes with best splitting ability.

The information gain is calculated as a difference of the entropy and average entropy (information), where entropy of the attribute is simple a measure of disorder of data.

$$Entropy(S) = -\sum_{j=1}^{n} P(value_j)log(P(value_j)) \tag{3}$$

$$Information(S|A) = \sum_{i} \frac{|S_i|}{|S|} \times E(S_i) \tag{4}$$

$$Gain(S|A) = Entropy(S) - Information(S|A) \tag{5}$$

The individual's construction was driven by the created grammar (Backus-Naur form) which supported some useful mathematical operation to chose (see in Table 1).

As an input to the synthesized polynomial there were given the A_1, A_2, D_1, D_2 coefficients and the unprocessed sub-signal's data. The polynomial could be made up on the only one of the vectors and only the polynomials with highest fitness values were kept. As an output from the synthesized polynomials, there was only vectors of the double values of the length of 35 (7 subsignals $\times$ 5 chosen polynomial features) for each of the signal and this comes into next phase as the input for the artificial neural network.

2.3 Neural Network

The artificial neural network (ANN) is a computational model structurally and functionally inspired by the human brain [19]. The most common form of the ANN is a feed-forward multilayer neural network, which is often employed in the tasks of the

Table 1 Operations supported by the adjusted grammar for the GP

Operation	Symbol	Protection
Addition	+	N/A
Subtraction	−	N/A
Multiplication	×	N/A
Division	÷	Output 0 when denominator input is zero
Square root	$\sqrt{}$	Apply an absolute value operator before radical
Natural logarithm	log	Output zero for an argument of zero; and apply an absolute value operator to negative arguments
Sine	sin	N/A
Cosine	cos	N/A
Natural exponential function	e^x	N/A
Maximum	max	N/A
Minimum	min	N/A
Summation	Σ	N/A
Product	Π	N/A

approximation of non-linear functions [9], pattern recognition [1] or signal classification.

Optimization of this kind of the network is provided by a supervised learning with a training set of predetermined knowledge.

Structure of the ANN is performed by a directed graph of the neurons organized into the recognizable layers. The directed graph of a network contains at least recognizable 3 layers: one input layer, at least one hidden layer and one output layer. The number of the hidden layers is optional and depends on designed problem. Neurons between layers are fully connected by weighted connections in a way that each neuron in the previous layer has a connection with each neuron in the next layer. The feed-forward propagation of the input signal starts on the input layer, where the signal is presented as an input and brought by synapsis (connections) to the next layers until the output layer is not reached. Each neuron of the hidden and output layer proceeds the summation of its all input signals x_i multiplied by the particular connection's weight coefficient w_i as it is show in the equation:

$$z = \sum_{i}^{n} w_i x_i. \tag{6}$$

The excitation value of the neuron is computed by its sigmoid activation function presented in equation:

$$y = \frac{1}{1 + e^{-\lambda z}}, \quad (7)$$

where λ is the slope of the sigmoid function. Final output values of the neural network are the excitations of neurons of the output layer.

Due to supervised learning mechanism, there is an comparison of gained and desired outputs by the error function, which leads to a kind of optimization (ANN's learning). The learning of the feed-forward neural network is processed by updating of the connection weights between neurons based on the amount of the error between desired and computed errors. The most popular method for neural network adaptation is a backpropagation.

2.4 Backpropagation Learning Algorithm

The backpropagation (BP) algorithm is based on a gradient descent method. Adaptation of weights is processed by propagation of output errors back through the network from the upper layers to the lower layers. The goal is to minimize the error function:

$$E = \frac{1}{2} \sum_{p=1}^{P} \sum_{j=1}^{M} (y_j - d_j)_p^2, \quad (8)$$

where error between real output from network y_j and desired output d_j is summed for P patterns of training set and for all m output neurons in the output layer. Error minimization is done by adaption of the weights between neurons. Change of weight is done by equation:

$$\Delta w_{ij} = -\eta \frac{\partial E}{\partial w_{ij}}, \quad (9)$$

where η is a learning coefficient. Then partial derivation of error E based on connection weight w_{ij} is obtained:

$$\frac{\partial E}{\partial w_{ij}} = \frac{\partial E}{\partial y} . \frac{dy}{dz} . \frac{\partial z}{\partial w_{ij}}. \quad (10)$$

Based on Eqs. 6 and 7 previous equation is simplified by:

$$\frac{\partial z}{\partial w_{ij}} = x_i \quad \text{and} \quad \frac{dy}{dz} = y(1 - y)\lambda. \quad (11)$$

More details about solution of partial derivations are described in [20, 21].

3 Results

In this section, there is a review of the outputs from all the steps of the previously mentioned work-flow. There was created seven sub-signals from each of the signal by simple splitting and 50 % overlap. From every sub-signal there were obtained A_1, A_2, D_1, D_2 coefficient vectors by the DWT first and second level of filtering. Parts of the signals and their coefficient vectors served as the inputs for the GP and because of the variability of inputs, more than ten non-lineary synthesized features was created with non-zero IG. The top five indicators with highest IG were chosen to build the input data set for the ANN. The example of the indicators are shown below as polynomial 12 and polynomial 13.

$$i_1 = max(\sqrt{\sum_{n=0}^{4} sin(sin(x_n))}, \sqrt{\sum_{n=1}^{5} sin(sin(x_n))}, \ldots, \sqrt{\sum_{n=N-4}^{N} sin(sin(x_n))}) \quad (12)$$

$$i_3 = \frac{\sqrt{min(x_0, x_1, x_2, \ldots x_N)}}{log(\sum_{n=0}^{N} sin(x_n))} \quad (13)$$

In the Fig. 2, there is a scatter plot of signal's indicator's values. This two indicators are transforming the searched space of samples into the recognizable zones due to their significant IG value (IG(i_1) = 0.757131, IG(i_3) = 0.607786).

In the phase of the creation of the final classification algorithm, there was obtained the neural network with one hidden layer. The number of neurons of the hidden layer was adjusted to an intrinsic dimension of the training input data set. This intrinsic

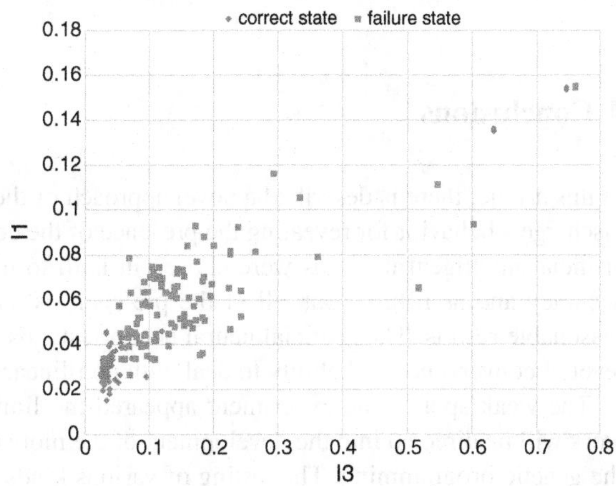

Fig. 2 The *scatter plot* of the i_1 and i_3 indicator's values

Table 2 Table of performance

	Class 0	Class 1	Total performance
Training	46 (44 | 2) 95.7 %	59 (51 | 8) 86.4 %	105 (95 | 10) 90.5 %
Validation	15 (15 | 0) 100 %	9 (8 | 1) 88.9 %	24 (24 | 1) 95.8 %
Testing	16 (14 | 2) 87.5 %	16 (15 | 1) 93.8 %	22 (19 | 3) 90.6 %
Total	77 (73 | 4) 94.8 %	84 (74 | 10) 88.1 %	161 (147 | 14) 91.3 %

dimension is a number of principal components needed to capture 80 % of the variance in the input data set [2].

The data set was then divided into three parts. The first part consisted of 70 % of the data set for training phase, the second part of 15 % for validation phase and third part of 15 % for testing phase. Validation set was used during training phase to avoid over-fitting by early stopping of training [16]. The learning coefficient η and the slope of the sigmoid function λ were experimentally set to 0.4 and 0.6 respectively.

The success rate of the classifications in all the ANN's phases is shown in Table 2 below.

Signals labeled as a "class 1" are the signals containing the PD's character indicating the CC's failure and the signals labeled as a "class 0" mean that measured behavior of the CC reveals the normal state of the system. The bottom line of the Table 2 stands for the success rates of the classification of the both classes and the total performance. The performance of the classification of the "class 0" is reported by values 77 (73 | 4) 94.8 % which means that there were 77 signals indicating the correctness of the system and 73 of these signals were correctly identified as the "class 0" and 4 of them incorrectly as the "class 1" with the success rate close to 94.8 %. There was achieved the similar precision in case of classification of the "class 1" where the performance went to 88.1 %. Total performance of all phases in the proposed model was 91.3 %

4 Conclusions

In this article, there is described a novel approach of the classification of the partial discharge's behavior for revealing the presence of the covered conductors fault. The artificial intelligent methods were chosen in faith to maximize the classification's accuracy and as it turns out, all of the pre-analysis and pre-processing led to the reasonable results. The artificial neural network stands its place too, due to its high level of convergence and ability to deal with non-linear problems.

The weak spot of the experiment appeared the limited data set and our future work will be directed into the development of the more complex fitness function for the genetic programming. The testing of various kinds of signal transformations is planed for improve pre-analysis as well.

Acknowledgments This paper was conducted within the framework of the IT4Innovations Centre of Excellence project, reg. no. CZ.1.05/1.1.00/02.0070, project ENET CZ.1.05/2.1.00/03.0069, Students Grant Competition project reg. no. SP2015/142, SP2015/146, SP2015/170, SP2015/178, project LE13011 Creation of a PROGRES 3 Consortium Office to Support Cross-Border Cooperation (CZ.1.07/2.3.00/20.0075) and project TACR: TH01020426.

References

1. Bishop CM (1995) Neural networks for pattern recognition. Oxford University Press Inc, New York
2. Boger Z, Guterman H (1997) Knowledge extraction from artificial neural network models. In: IEEE International Conference on Systems, Man, and Cybernetics. Computational Cybernetics and Simulation, vol 4, pp 3030–3035
3. Burrus CS, Gopinath RA, Guo H (1997) Introduction to wavelets and wavelet transforms 1st edn. Prentice Hall, Englewood Cliffs
4. Gawre K, Patidar NP, Nema RK (2012) Article: application of wavelet transform in power quality: a review. Int J Comput Appl 39(18):30–36
5. Guo L, Rivero D, Dorado J, Munteanu CR, Pazos A (2011) Automatic feature extraction using genetic programming: an application to epileptic EEG classification. Expert Syst Appl 38(8):10425–10436. http://www.sciencedirect.com/science/article/pii/S0957417411003253
6. Hashmi GM, Lehtonen M (2009) Effects of Rogowski Coil and covered-conductor parameters on the performance of pd measurements in overhead distribution networks. Int J Innovations Energy Syst Power 4(2):14–21
7. Hashmi G, Lehtonen M, Nordman M (2010) Modeling and experimental verification of on-line pd detection in mv covered-conductor overhead networks. IEEE Trans Dielectr Electr Insul 17(1):167–180
8. Hashmi G, Lehtonen M, Nordman M (2011) Calibration of on-line partial discharge measuring system using Rogowski Coil in covered-conductor overhead distribution networks. Sci Meas Technol IET 5(1):5–13
9. Iannella N, Back AD (2001) A spiking neural network architecture for nonlinear function approximation. Neural Netw 14(67):933–939
10. Koza JR (1992) Genetic programming: on the programming of computers by means of natural selection. MIT Press, Cambridge
11. Littler T, Morrow D (1999) Wavelets for the analysis and compression of power system disturbances. IEEE Trans Power Delivery 14(2):358–364
12. Mallat S (1989) A theory for multiresolution signal decomposition: the wavelet representation. IEEE Trans Pattern Anal Mach Intell 11(7):674–693
13. Misak S, Pokorny V (2014) Testing of a covered conductor's fault detectors. IEEE Trans Power Delivery PP(99):1–1
14. Mitchell TM (1997) Machine learning, 1st edn. McGraw-Hill Inc, New York
15. Muharram MA, Smith GD (2004) Evolutionary feature construction using information gain and Gini index. In: Keijzer M, O'Reilly U-M, Lucas S, Costa E, Soule T (eds) EuroGP 2004, vol 3003. LNCS, Springer, Heidelberg, pp 379–388
16. Prechelt L (1998) Automatic early stopping using cross validation: quantifying the criteria. Neural Netw 11(4):761–767. http://www.sciencedirect.com/science/article/pii/S0893608098000100
17. Quinlan JR (1993) C4.5: programs for machine learning. Morgan Kaufmann Publishers Inc., San Francisco
18. Ribeiro PF (1994) Wavelet transform: an advanced tool for analyzing non-stationary harmonic distortions in power systems. In: Proceedings IEEE ICHPS VI, pp 365–369

19. Rosenblatt F (1962) Principles of neurodynamics: perceptrons and the theory of brain mechanisms. Spartan Books, Washington
20. Rumelhart DE, Hinton GE, Williams RJ (1986) Learning internal representations by error propagation. In: Rumelhart DE, McClelland JL, PDP Research Group (eds) Parallel distributed processing: explorations in the microstructure of cognition, vol 1. MIT Press, Cambridge, MA, USA, pp 318–362
21. Rumelhart DE, Hinton GE, Williams RJ (1988) Learning representations by back-propagating errors. In: Anderson JA, Rosenfeld E (eds) Neurocomputing: foundations of research. MIT Press, Cambridge, pp 696–699
22. Santoso S, Grady W, Powers E, Lamoree J, Bhatt S (2000) Characterization of distribution power quality events with fourier and wavelet transforms. IEEE Trans Power Delivery 15(1):247–254
23. Sette S, Boullart L (2001) Genetic programming: principles and applications. Eng Appl Artif Intell 14(6):727–736. http://www.sciencedirect.com/science/article/pii/S0952197602000131
24. Smart O, Firpi H, Vachtsevanos G (2007) Genetic programming of conventional features to detect seizure precursors. Eng Appl Artif Intell 20(8):1070–1085. http://www.sciencedirect.com/science/article/pii/S0952197607000127
25. Vega V, Kagan N, Ordonez G, Duarte C (2009) Automatic power quality disturbance classification using wavelet, support vector machine and artificial neural network. In: 20th international conference and exhibition on electricity distribution—part 1, CIRED 2009, pp 1–4 (June 2009)
26. Wareing JB (2005) Covered conductor systems for distribution. Technical report 70580, EA Technology Ltd, Capenhurst Technology Park, Capenhurst, Chester, CH1 6ES (December 2005)
27. Xizhi Z (2008) The application of wavelet transform in digital image processing. In: International conference on multimedia and information technology. MMIT '08, pp 326–329 (Dec 2008)
28. Zhang W, Hou Z, Li HJ, Liu C, Ma N (2014) An improved technique for online pd detection on covered conductor lines. IEEE Trans Power Delivery 29(2):972–973

Application Tool for Prediction and Implementation of QoS in IP Based Network

Jaroslav Frnda, Miroslav Voznak, Martin Hlozak, Jiri Slachta and Jerry Chun-Wei Lin

Abstract This paper deals with the application able to predict the quality of Triple play services in the network. Our previous results led to the creation of mathematical models for each service type (voice, video and data), and the application offered fast calculation of reached quality according to the objective evaluation methods. Because our application is still being developed and tested, the version described here offers a new features and measurement parameters. The verification of results was performed on two the most common network infrastructure vendors (Cisco and Huawei) and new codecs were included too. The application can not only predict QoS parameters, but also generate the source code of particular QoS policy setting according to the user interaction and apply the selected policy to the routers in the network. The contribution of this paper lies in designing an extended version of SW tool capable of predicting the quality of Triple-play services in networks and configuring QoS policies on routers.

Keywords Application · Delay · E-Model · Network performance monitoring · Packet loss · Qos · Triple play · Voice service

J. Frnda · M. Voznak (✉) · M. Hlozak · J. Slachta
Department of Telecommunications, VSB—Technical University of Ostrava, 17. Listopadu 15, 70833 Ostrava, Czech Republic
e-mail: miroslav.voznak@vsb.cz

J. Frnda
e-mail: jaroslav.frnda@vsb.cz

M. Hlozak
e-mail: martin.hlozak@vsb.cz

J. Slachta
e-mail: jiri.slachta@vsb.cz

J.C.-W. Lin
School of Computer Science and Technology, Innovative Information Industry Research Center, Harbin Institute of Technology Shenzhen Graduate School, Shenzhen 518055, China
e-mail: jerrylin@ieee.org

A. Abraham et al. (eds.), *Intelligent Data Analysis and Applications*, Advances in Intelligent Systems and Computing 370,
DOI 10.1007/978-3-319-21206-7_13

1 Introduction

Network convergence that took place during the early 90s of the 20th century brought the new network concept based on IP called NGN (Next Generation Network). This concept allowed the transfer of formerly separate services (voice, video and data) by one common network infrastructure. However, this transition had to deal with some difficulties because packet networks based on IP protocol had not been designed to transfer delay-sensitive traffic. The difficulties appeared especially at the transfer of voice. Packet network has to use supplementary mechanisms securing the quality of service during the transmission over the network, able to provide a high-quality interactive communication similar to standard fixed lines (PSTN).

QoS policies used for prioritization of time-sensitive data streams on routers seems to be a one of the techniques for securing at least minimal QoS level in a packet network. The second important factor is network monitoring. Constant network performance monitoring and intervening as needed, keep the impact of network behavior on the acceptable level. Therefore, the purpose of the application described in this paper is to provide an effective monitoring tool capable of predicting qualitative QoS parameters according to network status. The application aims to be an alternative to expensive monitoring tools, as well as a helpful tool for designing the network infrastructure and implementation of QoS policies on selected routers.

2 State of the Art

Last decade brought growing interest in multimedia services transfer through packet networks based on IP protocol. Nowadays, voice and video service as a component of Triple play make a significant part of total data sended via the network. Voice as the most sensitive service to an overall network status, depends on many QoS parameters. Particularly, jitter and packet loss have an important impact on voice quality [1, 2]. To achieve the satisfactory voice quality, the network architecture must be designed by using representative congestion management techniques. Congestion management features allow to control congestion by determining the order in which packets are sent out an interface based on priorities assigned to those packets. The main contribution of these works [1–4] is to analyse of the impact of impairments mentioned above and effectiveness of particular QoS policy implementation. In order to implement methods for packet classification and prioritization, it is necessary to monitor the QoS parameters, by means network measurements. Evaluation of voice service according to the network behavior is based on the simplified version of calculation model based on recommendation ITU-T G.107 (also known as E-model) [5]. This model allows to evaluate the quality of speech, adjusting the model to be suitable especially for packet networks.

The reason why network monitoring plays an important role in securing the QoS is depends both on a delay and jitter factor on network utilization. Works [6, 7] analyse in detail the impact of network utilization on the variable component of total delay and packet loss.

This paper follows directly our previous experiments studying the impact of full network utilization and performance of data prioritization on the final quality of service [3]. For further upgrade of the prediction model, it was essential to focus on several new factors. The extended version of our application offers QoS prediction for additional voice codecs and the possibility of QoS policy implementation. We used our measurements results of two most commonly used network vendors too. According to the reliable information on Triple play quality in network calculated by our tool, the user can select and generate the implementation code for particular QoS policy for both vendors and send this configuration on the router. Because the inclination to NGN concept is huge, that is why monitoring of impacts on QoS efficiency and evaluation is highly important and necessary in order to secure the competitiveness of any multimedia services provider.

3 Methodology

Nowadays, practically, computer networks are not built only on a homogeneous infrastructure, but they use heterogeneous devices. The first version of our mathematical prediction model was based on Cisco devices, and we wanted to know if it is possible to use this model for the network devices of Huawei company. Our testing scenarios brought the verification of mathematical model used in the application by performing measurements in a topology with Huawei network items. The next part of measurements was dedicated to analyse packet loss tolerance on additional voice codecs. Implementation of the computational model was carried out in programming language C#.

3.1 Measured Parameters

E-model, defined by Recommendation G.107 by ITU-T [5], was used as an objective evaluation method for voice service. This recommendation also includes a set of recommended values which enable to simplify the calculation so that it corresponds with packet networks as follows:

$$R = 93.35 - I_D - I_{E-EFF} \tag{1}$$

Regarding of QoS in IP networks, transmission bandwidth, packet loss, network delay and variation in latency (jitter) are key parameters [2]. The goal of the QoS is to improve values of these transmission parameters. These parameters can be express in two factors used for R-factor calculation, namely in I_D (2) and I_{E-EFF}.

$$I_D = I_{DTE} + I_{DD} \tag{2}$$

The parameter I_{DTE} represents the factor of impairment caused by echo (Echo-cancellation has been solved in ITU-T G.168 recommendation), and I_{DD} represents the factor of impairment caused by the too long transfer delay. In generally, the factor I_{e-eff} represents impairment caused by low bit-rate codecs and actual packet loss ratio of the network. Typical score range of R-factor is from 50 (bad) to 100 (excellent). By keeping all the default values during the calculation, the R-factor reaches a final value of 93.35 [3]. In order to meet user's expectations, the value of 70 or more is needed [5]. Except from the R-factor values, a rating scheme of 1–5 (5 is the best), called MOS (Mean Opinion Score), can be used as an evaluation scale. Conversion of the R-factor values to MOS scale values is also described in G.107 recommendation.

In our case, network delay is an important component of end to end delay in network. It represents queuing (processing) delay on routers and it is directly influenced by using QoS policies. Delay variation (Jitter) describes variability in the packet delivery time to the target node causes incorrect order. Packet loss occurs when one or more packets of data traveling across a computer network fail to reach their destination. It is most often expressed as a percentage. As it can be seen in Fig. 1, we created network topology, which can be used to obtain the appropriate values of important parameters.

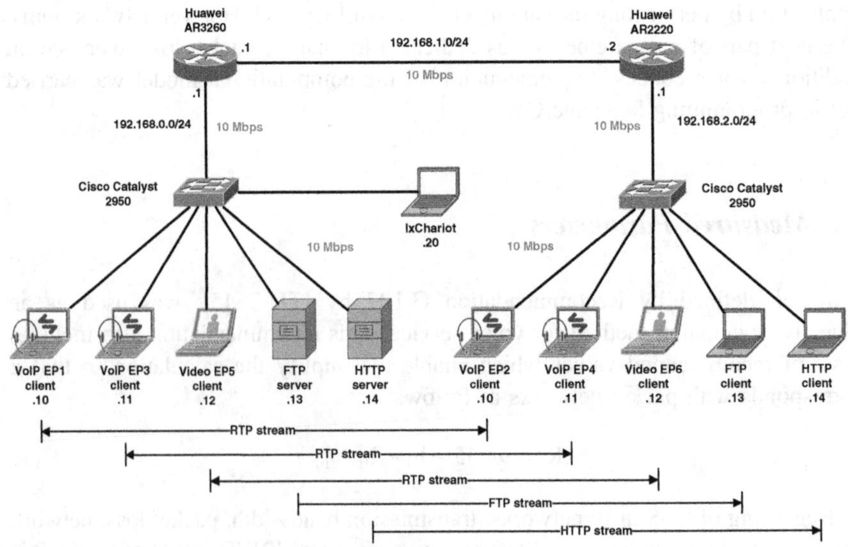

Fig. 1 Network topology for QoS testing

Our network topology consists of two Huawei routers (AR3260 and AR2220), two Cisco switches (Catalyst 2950), four VoIP clients, two Video clients, FTP client and server. During testing, all interfaces used only Ethernet-based protocol and speed of 10 Mbps for better full network utilization.

In order to simulate voice and video traffic, IxChariot tool from Ixia company was used. Simultaneously, the first VoIP traffic stream was transmitted between VoIP endpoint EP1 and VoIP endpoint EP2, the second between VoIP EP3 and VoIP EP4. Video stream was transmitted between the EP5 and EP6 and the data service was generated by using FTP service. QoS policies were implemented on both routers, and this topology served us for verification of our prediction model for Huawei devices. If we wanted to add new voice codecs into the application, we had to make tests of packet loss robustness. For emulation of packet loss in the computer network, we used Linux tool, called Network Emulator (NetEm). Due to this Network Emulator, we could change values of packet loss using individual stepwise test. For every scenario, we performed ten individual tests of packet loss from 0 to 10 % while each test was performed five times. For audio traffic, we chose the most used codecs G.711, G.711 with PLC, G.729, and AMR 12.2 kbps. G.711 is a primary codec used in telephony services using 64 kbps of network bandwidth. Packet Loss Concealment (PLC) algorithm is an appendix to G.711, and it can help hide transmission losses in a packetized network based on CELP. G.729 is low bandwidth codec using 8 kbps of bandwidth and mostly used as an alternative to G.711 codec for VoIP communication. AMR codec is a hybrid speech codec using LPC and ACELP and nowadays is widely used in UMTS and LTE networks (Table 1).

At first, we tested performance of QoS policies settings and compared to the results obtained if no policy was set. For voice flow, we chose Low-latency queuing (LLQ), which is useful for delay-sensitive and gave absolute preference over the other traffic. Video and FTP data flow used Class-Based Weighted Fair Queuing (CBWFQ), which can guarantee percentage transmission rate of the interface for every CBWFQ queue. The rest of the data flows (with no classification) were inserted into the special queue, called Weighted Fair Queuing (WFQ). For every unclassified data flows were automatically created an individual queue. For CBWFQ with priority queue LLQ were used two scenarios. Voice service got 7.5 and 10 % of total transmission speed video 40 and 30 % respectively. The process of classifying packet and the subsequent use appropriate queue using Huawei devices is very similar to classifying packet of Cisco router. The input interface of this router marks every incoming packet. Based on IP address, the IP packets get the appropriate value of DSCP (Differentiated services code point). According to this classification in IP header were the relevant IP packets queued and sent to the output interface.

We found the different behavior and packet processing compared to Cisco. Huawei routers had smaller buffers for processing of IP packets. During the buffer overflow, packets were dropped. The disadvantage of this behavior was a high packet loss. Cisco routers have a much bigger buffer for processing of packets. Therefore, more packets are able to queued much longer. This behavior caused very low or no

Table 1 Different QoS configuration commands with descriptions

<table>
<tr><th>Description</th><th>Huawei configuration commands</th><th>Cisco configuration commands</th></tr>
<tr><td>Creating classes for defining appropriate traffic</td><td>[Huawei]**traffic classifier** <name of class></td><td>Cisco(config) #**class-map** <name of class></td></tr>
<tr><td>The definition of the type of traffic</td><td>[Huawei-classifier-VOICEin] **if-match** {acl <ACL number> | **dscp** <value> }</td><td>Cisco(config-cmap)#**match** {**access-group** <ACL number> | **ip dscp** <value> }</td></tr>
<tr><td>Creating appropriate traffic behavior</td><td>[Huawei]**traffic behavior** <name of traffic behavior></td><td></td></tr>
<tr><td>Creating of traffic policy</td><td>[Huawei]**traffic policy** <name of policy traffic></td><td>Cisco(config) **#policy-map** <name of policy traffic></td></tr>
<tr><td>Assigning class to policy-map</td><td></td><td>Cisco(config-pmap) **#class** <name of created class></td></tr>
<tr><td>Setting appropriate policy</td><td>[Huawei-behavior-VOICEin] **remark dscp** | **queue llq bandwidth** {[**pct** <value>] | [**transmission rate** <value>] } | **bandwidth** {[**percent** <value>] | [**transmission rate** <value>] }</td><td>Cisco(config-pmap-c)#**set dscp** | **priority** {[**percent** <value>] | [**transmission rate** <value>] } | **bandwidth** {[**percent** <value>] | [**transmission rate** <value>]}</td></tr>
<tr><td>Assigning class and behavior of traffic under a common policy</td><td>[Huawei-trafficpolicy-QueueOut] **classifier** <name of class> **behavior** <name of traffic behavior></td><td></td></tr>
<tr><td>Assigning a common policy to the interface</td><td>[Huawei-FastEthernet0/0/0] **traffic-policy** <name of policy traffic> {**inbound** | **outbound**}</td><td>Cisco(config-if)#**service-policy** {**inbound** | **outbound**} <name of policy traffic></td></tr>
</table>

packets loss, but processing delay on the router was higher. QoS configuration commands are very similar to Cisco devices. Following table show up differences between this two big vendors. During the measurements, we mainly focused on critical parameters called Packet loss, Network Delay and Jitter. These calculated values were obtained from IxChariot tool and described below. As a default QoS policy on Ethernet interface, Cisco routers use Best effort policy based on FIFO (First In First Out) [8]. It is a connectionless model of delivery that provides no guarantees for reliability, delay or other performance characteristics. Routers from Huawei company use Custom Queuing (CQ) as a default QoS policy on Ethernet interface [1]. This different attitude causes that we must implement the configuration of CQ for Cisco, but for Huawei it is sufficient to create a packet classification by using DSCP and this policy will be set automatically. With CQ, a network administrator can control the available bandwidth on an interface when it is unable to accommodate the aggregate traffic enqueued. Associated with each output queue is a configurable byte count, which specifies how many bytes of data should be delivered from the current queue by the system before the system moves on to the next queue (maximum 16 queues). We used for each service type separate queue.

4 Results

As it was mentioned above, simplified E-model consist of two factors influence the final score of R-factor. Parameter I_D can be expressed like function between the absolute delay and obtained R-factor value [9]. When the value of absolute delay is higher than the recommended boundary 150 ms, voice quality goes rapidly down [9]. This relation is the same for all voice codes. For adding new voice codecs, we need to focus on the packet loss impact on voice codec. Figure 2 shows the packet loss impairment factor I_{E-EFF} on voice codecs G.711 and G.729 achieved from previous work [3], and newly added codecs G.711 with PLC and AMR codec. The results of measurements showed following regressive equations:

For codec G.711 PLC

$$Y = a + (b^*X), \quad R^2 = 99.13\,\% \tag{3}$$

$Y = I_{E-EFF}$ $\quad X = \text{packet loss}(\%)$
$a = 0.0789377$ $\quad b = 2.38358$

For codec AMR 12.2 kbps

$$Y = \sqrt{(a + (b^*X))}, \quad R^2 = 99.51\% \tag{4}$$

$Y = I_{E-EFF}$ $\quad X = \text{packet loss}\ (\%)$
$a = 35.3068$ $\quad b = 225.352$

The measurements of QoS Policy effectiveness on Huawei routers show, that automatically set CQ policy does not achieve the same level of performance than on Cisco routers. Especially packet loss rate is too high for securing a good voice quality. On the other side, different Huawei behavior described in Sect. 3.1 reaches

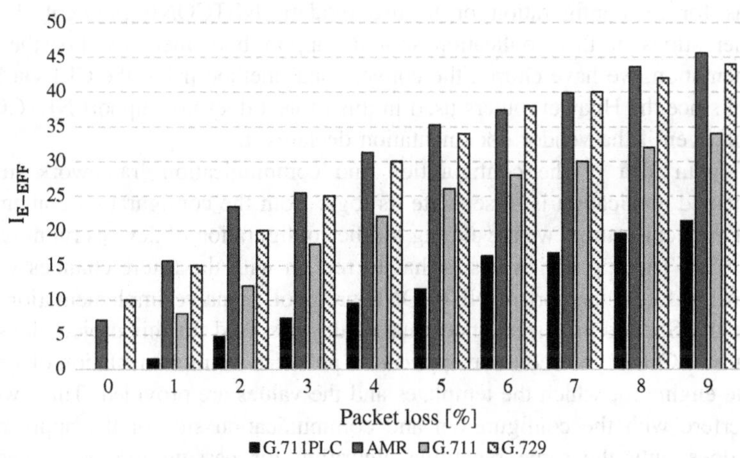

Fig. 2 I_{E-EFF} factor for voice codecs

Table 2 Comparison of results between Cisco, Huawei and prediction model

Name of Policy	Huawei			Cisco			Application	
	Network delay	Jitter	Packet Loss [%]	Network delay	Jitter	Packet Loss [%]	Network delay	Jitter buffer
CQ	14.8	2.9	6.1	9.3	5.5	0.6	10	10
CBWFQ	11.9	3.6	0.04	17.2	18	0.02	15	30
CBWFQ1	12	5	0.07	14.39	23.5	0.01	15	35

Note Delay and jitter values are expressed in milliseconds

better results for both CBWFQ policy scenarios. Actually, CBWFQ is very preferred QoS policy and our results confirmed that Huawei is ready to be a full-fledged competitor for the Cisco and it can be a part of heterogeneous network cooperating with Cisco routers.

Table 2 also shows predicted values from our application and as it can be seen, prediction of network delay is similar with real values. Jitter buffer represents recommended size of De-jitter buffer on receiving side, that should be at least 1.5 bigger than average measured jitter value [10]. Because Huawei routers need less buffer size than routers from Cisco, we must edit and calculate the different size for Cisco router and Huawei. These results will help us to prepare better model will correspond with real behavior in the network.

4.1 Configuration of Network Components from Implemented SW

For the purpose of the configuration of network devices from developed application, it was necessary to decide whether the application should use conventional methods for its configuration or to use modern NETCONF protocol. Future implementations of this application should support both methods. For the first implementation, we have chosen the conventional method using the CLI via SSH protocol since the Huawei routers used in this project does not support NETCONF protocol, even if the vendor documentation declares it.

The main idea of the configuration and communication framework in the implemented application is to separate its logic from the configuration content. In the first project iteration, we are aiming for the configuration of devices running any remote CLI. The application design has to reckon with the future changes which assumes the future support of NETCONF protocol. A naive implementation that uses Secure Shell is preferred since Huawei devices used for this project does not support NETCONF protocol. The application provides an implementation of simple template engine for which the templates and the values are provided. Thus, we do not interfere with the configuration and communication side of the application, which does only the same repetitive algorithm for passing through generated configuration to remote CLI.

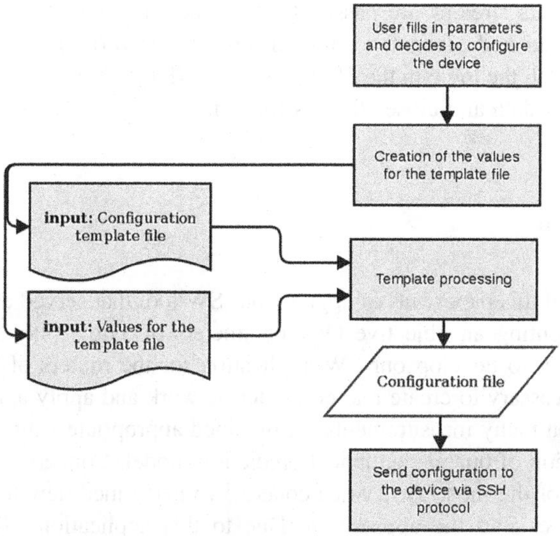

Fig. 3 Diagram of communication between the application and router

Currently, besides the mentioned implementation, the application is prepared to support editing configuration on remote devices. The basic approach is depicted in Fig. 3. Architecture is divided into two separate parts–template and communication part. The template part accepts data streams in a form of template file and values

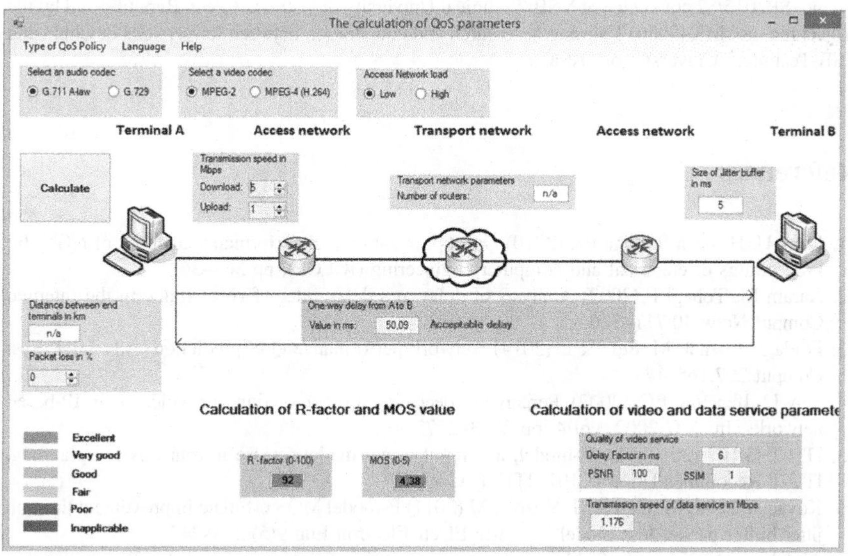

Fig. 4 Graphical design of the application

file. When the data streams are passed through the template engine, the configuration is generated and given to communication framework. The communication framework does all the logic in the SSH session or NETCONF session, transmits all the configuration data and closes the session (Fig. 4).

5 Conclusion

The main goal of this paper was to upgrade our SW tool that served as a simulation model for computing an objective QoS parameters of Triple play services in IP network. In order to develop our SW application for the routers of Huawei company, it was necessary to create real computer network and apply appropriate QoS policy. Based on many measurements we obtained appropriate values, which were used for extension of our mathematical prediction model. Concerning to cover the wide spectrum of the most used voice codecs, we performed new tests and added new voice codecs into the application. Due to this application ISP can predict objective QoS parameters of triple play services and help to him in network design. The application also allows to generate the configuration of selected QoS policy and saves time during the implementation of QoS techniques on routers in the network.

The next step should be a performance analysis of the new video codec H.265 (HEVC) that is the successor of today used codecs MPEG-2 and MEPG-4 (H.264). Our goal is to predict QoS not only for the most used voice codecs but also for all video codes typically used for video streaming.

Acknowledgments The research leading to these results has received funding from the grant SGS reg. no. SP2015/82 conducted at VSB-Technical University of Ostrava, Czech Republic and by the project reg. no. MK4005611 aiming at research and cooperation between universities in China and VSB-Technical University of Ostrava.

References

1. Hany U, Hossain S, Saha PK (2010) QoS optimization and performance analysis of NGN. In: Proceedings of electrical and computer engineering (ICECE), pp 364–367
2. Karam M, Tobagi F (2002) Analysis of delay and delay jitter of voice traffic in the Internet. Comput Netw 40:711–726
3. Frnda J, Voznak M, Sevcik L (2014) Network performance QoS prediction. Adv Intell Syst Comput 297:165–174
4. Sun L, Ifeachor EC (2002) Perceived speech quality prediction for voice over IP-based networks. In: ICC 2002, vol 4, pp 2573–2577
5. ITU-T G.107 (2010) The E-model, a computational model for use in transmission planning, ITU-T Recommendation G.107, ITU-T, Geneva
6. Kovac A, Halas M, Orgon M, Voznak M (2011) E-model MOS estimate improvement through jitter buffer packet loss modelling. Adv Electr Electron Eng 9(5):233–242
7. Abdelkefi A, Jiang Y (2011) A structural analysis of network delay. In: Proceedings of communication networks and services research conference (CNSR), pp 41–48

8. Mancas C, Mocanu M. Mancas D (2013) Congestion avoidance in multimedia networks. In: Proceedings of 11th Roedunet international conference, pp 1–5
9. Voznak M (2011) E-model modification for case of cascade codecs arrangement. Int J Math Models Methods Appl Sci 5(8):1439–1447
10. Kyrbashov B, Baronak I, Kovacik M, Janata V (2011) Evaluation and investigation of the delay in voip networks. Radioengineering 20(2):540–547

8. Manas C, Moraru V, Mancini D (2012) Congestion avoidance in multimedia networks. In: Proceedings of 11th Roedunet international conference, pp 1–3
9. Voznak M (2011) E-model modification for case of cascade codecs arrangement. Int J Math Models Methods Appl Sci 5(8):1439–1447
10. Kyrbashov B, Baronak I, Kovacik M, Janata V (2011) Evaluation and investigation of the delay in voip networks. Radioengineering 20(2):540–547

Procedure for Mapping Objective Video Quality Metrics to the Subjective MOS Scale for Full HD Resolution of H.265 Compression Standard

Miroslav Uhrina, Miroslav Voznak, Martin Vaculik and Michal Malicek

Abstract The article deals with the correlation between objective and subjective methods and the subsequent mapping of the selected objective metric to MOS scale. The theoretical part of the paper focuses on objective and subjective metrics applied for testing. The practical part describes the measurements and the experimental results. The results are implemented into the new model. The developed model will facilitate predicting the quality in networks based on IP.

Keywords ACR · DSIS · DSCQS · MOS · Qoe · PSNR · SSIM · VQM

1 Introduction

Recently, the demand for the multimedia services, i.e. broadcasting, transmission and receiving the video, audio and other data in a single stream, referred to as the multimedia stream, has increased. As a result, the assessment of the quality of the video as a part of the multimedia technology has gained importance. The quality of the video is affected by many factors, especially by compression technology and transmission link imperfections. Nowadays many new compression standards are being developed, e.g. H.265/HEVC, VP9 or DAALA. Accordingly, the assessment of the video quality maintains its importance in the field of research.

M. Uhrina · M. Vaculik · M. Malicek
Department of Telecommunications and Multimedia, University of Zilina,
Univerzitna 1, 01026 Zilina, Slovak Republic
e-mail: miroslav.uhrina@fel.uniza.sk

M. Vaculik
e-mail: martin.vaculik@fel.uniza.sk

M. Voznak (✉)
Department of Telecommunications, VSB—Technical University of Ostrava,
17. Listopadu 15, 70833 Ostrava, Czech Republic
e-mail: miroslav.voznak@vsb.cz

A. Abraham et al. (eds.), *Intelligent Data Analysis and Applications*,
Advances in Intelligent Systems and Computing 370,
DOI 10.1007/978-3-319-21206-7_14

2 State of the Art

The growing interest in multimedia services poses new requirements on the assessment of video quality. There are many factors that influence the quality of the video. The most important ones include the compression technology and transmission link imperfections. Consequently, the need to create a new model that will predict the quality has arisen. Papers [1–3] discuss the impact of bitrate on the video quality using the objective metrics; and paper [4] discusses the impact of bitrate on the video quality using the subjective metrics. The measurements in these papers show the relation between the resolution, bitrate and the type of content and the quality of the video. References [5–7] focus on the degradation of quality due to delays and packet losses. Paper [8] analyses in detail the impact of network utilization and set policies on the variable component of the total delay. Since the final delay and packet loss depend on the full network utilization and QoS policy applied to prioritized data flow processing by the routers, it is also necessary to consider this relationship. What is yet missing is the model for predicting the video quality that is influenced by the compression and the network impact. It is necessary to mention that such a model should correlate well with subjective feelings. Accordingly, it is necessary to examine the correlation between the objective and subjective metrics used to assess the quality of the video and the subsequent mapping methodology of the selected objective metric to the MOS scale. Some papers [9–11] focused on such issues. However, no study has yet been undertaken in the area of Full HD resolution. Therefore this paper addresses the impact of bitrate on the video quality for Full HD resolution using both objective and subjective methods, the correlation between various metrics with the subsequent mapping of the selected objective metric to the MOS scale.

3 Video Quality Assessment

The quality of a video can be measured directly in the studio chain or after it has been compressed and/or transmitted via the network. In the former case, the vectorscope and waveform monitor is applied, the evaluation methods are used for the latter. These methods are divided into the objective and subjective assessment.

3.1 Objective Assessment

The objective video quality assessment applies the computational methods referred to as the "metrics". These return values that enable classifying the video quality. They measure the physical characteristics of a video signal such as the signal amplitude, timing and signal-to-noise ratio. Their biggest advantage is their repeatability. The best-known and most frequently used objective metrics are Peak Signal-to-Noise Ratio (PSNR), Video Quality Metric (VQM) and Structural Similarity Index (SSIM).

PSNR (Peak Signal-to-Noise Ratio)
The PSNR in decibels is defined as:

$$\mathrm{PSNR} = 10\log_{10}\left(\frac{m^2}{\mathrm{MSE}}\right)\ [\mathrm{dB}] \tag{1}$$

where *m* is the maximum value that pixel can take (e.g. 255 for 8-bit image) and *MSE* (Mean Squared Error) is the mean of the squared differences between the gray-level values of pixels in two pictures or sequences I and $\tilde{I}$:

$$\mathrm{MSE} = \frac{1}{\mathrm{TXY}} \sum_t \sum_x \sum_y \left[\mathrm{I}(t,x,y) - \tilde{\mathrm{I}}(t,x,y)\right]^2 \tag{2}$$

for pictures of size $X \times Y$ and T frames. Technically, MSE measures the image difference, whereas PSNR measures image fidelity. The biggest advantage of the PSNR metric is that it can be computed easily and fast [12].

SSIM (Structural Similarity Index)
The SSIM metric measures three components—the luminance similarity, the contrast similarity and the structural similarity and combines them into one final value, which determines the quality of the tested sequence. This method differs from the methods described above (all of which are error-based) since it uses the structural distortion measurement instead of the error one. This is due to the human vision system that is highly specialized in extracting structural information from the viewing field and it is not specialized in extracting errors. Owing to this, the SSIM metric achieves a good correlation with the subjective impression [13]. The results are in interval [0,1], where 0 is the worst and 1 the best quality.

VQM (Video Quality Metric)
The VQM metric computes the visibility of artefacts expressed in the DCT domain. The input of the metric is a pair of colour image sequences—the reference one and the test one. Both sequences are cropped, then converted from the input colour space to the YOZ colour space, then transformed to blocked DCT and afterwards converted to units of local contrast. Next, the input sequences are subjected to temporal filtering, which implements the temporal part of the contrast sensitivity function. The DCT coefficients, expressed in a local contrast form, are then converted to just-noticeable-differences (jnds) by dividing their respective spatial thresholds. This implements the spatial part of the contrast sensitivity function. Subsequently, after the conversion to jnds, the two sequences are subtracted to produce a difference sequence. Afterwards, the contrast masking operation to the difference sequence is performed. Finally, the masked differences are

weighted and pooled over all dimensions to yield summary measures of the visual error [14]. The output value of the VQM metric indicates the level of sequence distortion—for no impairment the value equals zero and with the rising level of impairment the output value also rises.

3.2 Subjective Assessment

The subjective assessment involves human observers (people) who score the quality of the video. It is the most reliable way to determine video quality. The disadvantage of this method is that it is time consuming and that human resources are needed.

The well-known subjective methods are:

- Double Stimulus Impairment Scale (DSIS) also known as Degradation Category Rating (DCR),
- Double Stimulus Continuous Quality Scale (DSCQS),
- Single Stimulus Continuous Quality Evaluation (SSCQE),
- Absolute Category Rating (ACR) also known as Single Stimulus (SS),
- Simultaneous Double Stimulus for Continuous Evaluation (SDSCE).

To achieve reliable results, a minimum of 15 observers is required. They should not be experts, in the sense that they are not directly concerned with television picture quality as a part of their daily routine and they are not experienced assessors. The number of assessors needed for the tests depends upon the sensitivity and reliability of the test procedure adopted and upon the anticipated size of the effect sought. Before the testing session, the assessors should be introduced to: the assessment method, the types of impairments, the grading scale, the sequence, the timing (the reference and the test sequence time duration, the time duration for voting). The source signal provides the reference sequence directly and the input for the tested system. It should be of optimum quality (faultless). To obtain stable results, the absence of defects in the reference sequence is crucial. The training sequences demonstrate the range and the type of the impairments that are assessed. They should be used with illustrating pictures other than those used in the test but of comparable sensitivity. The whole session should not exceed 30 min. At the beginning of the first session, some sequences (from three to five) should be shown to consolidate observers' opinions. The data obtained from these presentations should not be taken into account in the results of the test. A random order should be used for the presentations but the test condition order should be arranged so that any effects on the grading of tiredness or adaptation are balanced out from session to session. Some of the presentations can be repeated from session to session to check coherence.

Finally after the test session the mean score (MOS) is determined as follows:

$$\overline{\mathrm{u_{jkr}}} = \frac{1}{\mathrm{N}} \sum\nolimits_{\mathrm{i}=1}^{\mathrm{N}} \mathrm{u_{ijkr}} \tag{3}$$

where: u_{ijkrs}: score of observer i for test condition j, sequence k, repetition r,
N: number of observers

and also the confidence interval which is derived from the standard deviation and size of each sample. It is proposed to use the 95 % confidence interval which is given by:

$$\left[\overline{\mathrm{u}}_{\mathrm{jkr}} - \delta_{\mathrm{jkr}}; \overline{\mathrm{u}}_{\mathrm{jkr}} + \delta_{\mathrm{jkr}}\right] \tag{4}$$

$$\delta_{jkr} = 1.96 \frac{S_{jkr}}{\sqrt{N}} \tag{5}$$

$$S_{jkr} = \sqrt{\sum\nolimits_{i=1}^{N} \frac{\left(u_{jkr} - u_{ijkr}\right)^2}{(N-1)}} \tag{6}$$

DSIS (The Double-Stimulus Impairment Scale Method)

In this method two sequences are shown to the assessor—the unimpaired (the reference) sequence and the same sequence impaired (the tested one). The reference sequence is shown before the tested one and the viewer knows which one is the reference and which one the tested. After watching both sequences the assessor is asked to rate the second, keeping in the mind the first. The five-grade impairment scale is used: 5 imperceptible, 4 perceptible, but not annoying, 3 slightly annoying, 2 annoying, 1 very annoying.

DSCQS (The Double-Stimulus Continuous quality-Scale Method)

In this method, the assessors are also shown two sequences—the unimpaired (the reference) sequence and the same sequence impaired (the test one) but the viewer is not informed which one is the reference and which one is the tested sequence. The position of the reference sequence is changed in pseudo-random fashion. The participant sees sequence A then sequence B twice and then is asked to rate both sequences. The grading scale is divided into five equal intervals: excellent (80–100) = 1, good (60–79) = 2, fair (40–59) = 3, poor (20–39) = 4, bad (10–19) = 5.

ACR (The Absolute Category Rating Method)

This method is also called the single stimulus method (SS). This time, the assessors are only shown the impaired (the tested) sequences and the viewers do not know the quality of the reference sequence. After watching the sequence, the assessor is asked to rate the quality of the test sequence based on the level of the quality. The five-level grading scale is used: 5 excellent, 4 good, 3 fair, 2 poor, 1 bad [13, 15–18].

4 Measurements

Two types of tested sequences depending on content were used in these experiments—one with a dynamic scene—called "Basketball" and one in a slow motion—called "Train". Both sequences were in the FullHD resolution, i.e. 1920 × 1080 pixels and 16:9 aspect ratio with 50fps (frames per second). The length of these sequences was 500 frames, i.e. 10 s. The measurement procedure consists of four steps:

- First, both sequences were downloaded from [19] in the uncompressed format (*.yuv) and used as the reference sequences.
- Afterwards, they were encoded to both H.265/HEVC compression standard using ×265 tool [20]. The target bitrates were ranged from 1 Mbps to 10 Mbps and changed in 1 Mbps steps. For the subjective assessment, only sequences coded 1, 3, 5, 7, 9 Mbps were used. The parameters of the encoded sequences were set to the Main Profile, Level 4. The GOP parameter was set to N = 12 and M = 3 which means the GOP length was 12 and two B frames between two successive P frames were stored.
- Then, the sequences using the same ×265 tool were decoded back to the format *.yuv.
- Finally, the video quality was evaluated. It was done using the MSU Measuring Tool Pro version 3.0 [21]. For the objective assessment PSNR, SSIM and VQM metrics were used, for the subjective DSIS, DSCQS and ACR metrics were used.

5 Experimental Results

5.1 Objective Assessment

Figure 1 shows the measurements' results which were done using the objective metrics. The results show the impact of restriction by bitrate on the video quality for both test sequences coded into H.265/HEVC compression standard.

According to the graphs—the video quality rises exponentially with the increasing bitrate. The plots also support this– the quality depends on the sequence content. Bigger difference in quality between these two tested sequences occurs in lower bitrates—with increasing bitrate the sequences tend to correlate.

5.2 Subjective Assessment

The subjective assessment involved 30 assessors (24 men and 6 women). Their age ranged from 20 to 28 years, the average age was 22 years. Figure 2 shows the

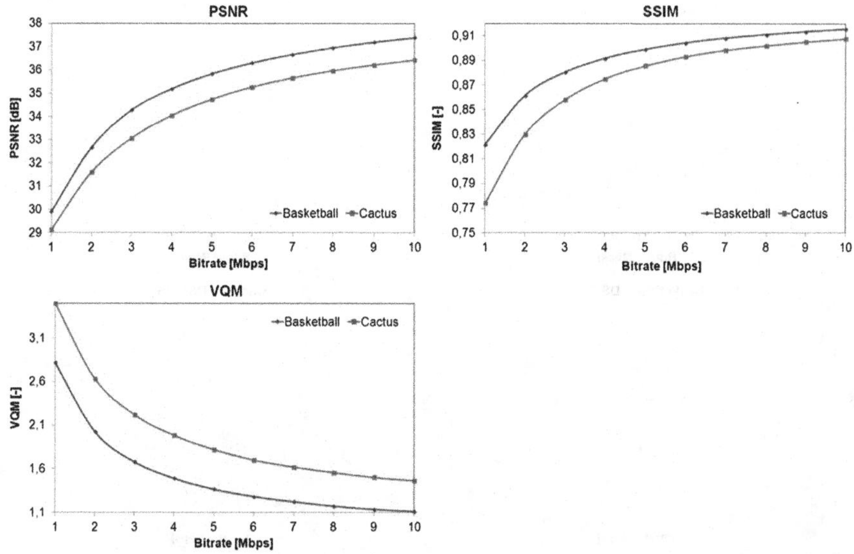

Fig. 1 The relationship between video quality measured by objective metrics and bitrate of both tested sequences

results of the measurements performed using the subjective metrics. The results show the impact of a restriction by bitrate on quality of experience for both tested sequences coded into H.265/HEVC compression standard. The dotted lines show the confidence interval.

As shown on the graphs—the observers recognized the differences in the quality of both tested sequences, especially in lower bitrates. The plots also show that the assessors did not rate the video quality with extremities which confirm the fact that people in general do not like to give the extreme values.

The results from both types of measurements, i.e. objective and subjective ones, were used to compute the correlation coefficients and subsequently for the mapping of the selected objective metric to the MOS scale. For both tested sequences, the Pearson correlation coefficients between particular objective and subjective methods were calculated. It was done for tested sequences coded to 3, 5, 7, 9, 11 Mbps using the formula:

$$r_{xy} = k_{xy}/(d_x d_y) \qquad (7)$$

where k_{xy} is the covariance and d_x and d_y are the standard deviations of the two variables. The correlation coefficients for both tested sequences are shown below in Table 1.

The results show that the differences between the correlation of the applied objective and subjective metrics are not big. The SSIM objective metric was applied

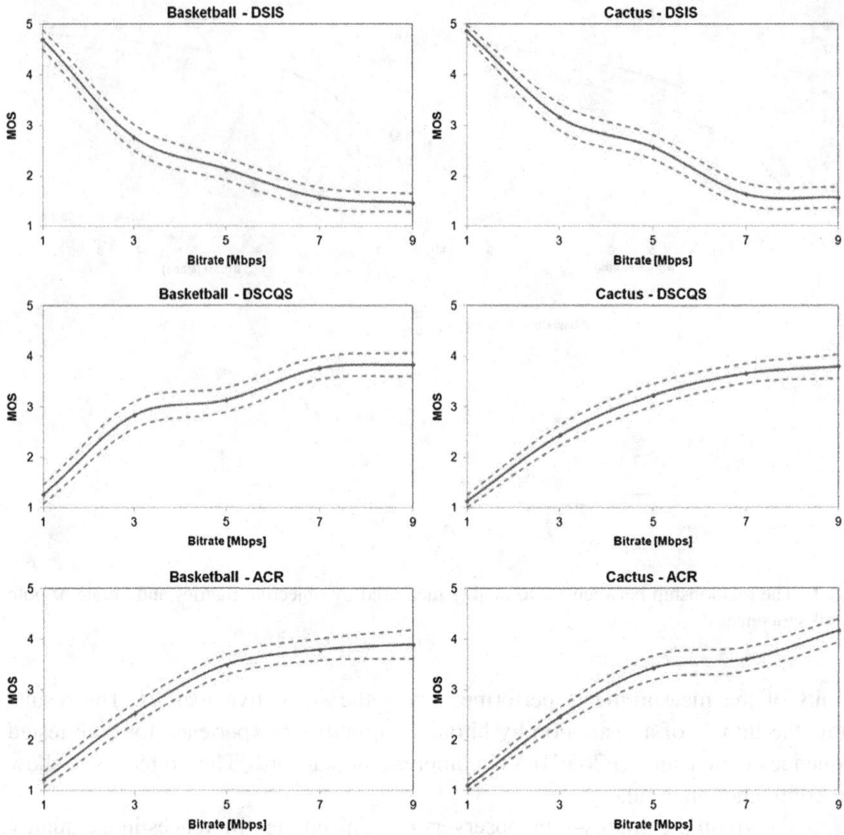

Fig. 2 The relationship between video quality measured by subjective metrics and bitrate of both tested sequences

Table 1 Correlation coefficients between various objective and subjective metrics for "Basketball" (*left*) and "Cactus" (*right*) test sequence

Basketball				Cactus			
	PSNR	SSIM	VQM		PSNR	SSIM	VQM
DSIS	−0.99855	−0.996967	0.993836	DSIS	−0.99053	−0.97875	0.979606
DSCQS	0.992754	0.990373	−0.98701	DSCQS	0.996596	0.986326	−0.986178
ACR	0.989899	0.985707	−0.97913	ACR	0.99089	0.97815	−0.978439

for the subsequent mapping. It was done because the results performed only by this metric are in an interval and thus they can be weighted more easily.

According to the graphs—where the observers considered both sequences by the subjective methods as good, i.e. the score given was at least three, the video quality should be evaluated 0.90 by the objective metric SSIM (Fig. 3).

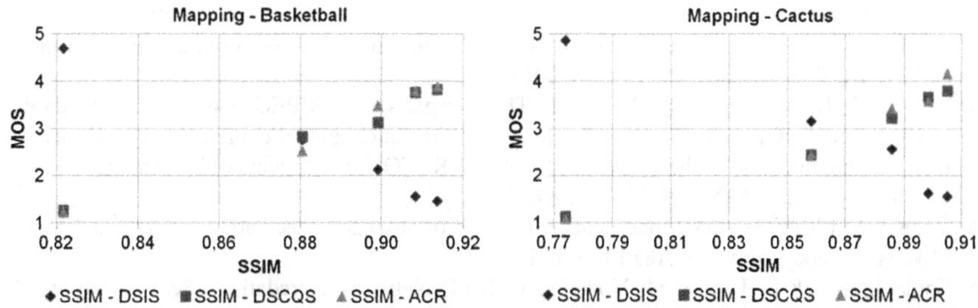

Fig. 3 Mapping of the objective SSIM method to the subjective MOS scale for the "Basketball" (*left*) and "Cactus" (*right*) sequence

6 Conclusion

This paper dealt with the correlation between the objective and the subjective methods with the subsequent mapping of the selected objective metric to the MOS scale. The aim of this paper was to look into the impact of bitrate on the video quality for Full HD resolution using both the objective and subjective methods and afterwards calculating Pearson correlation coefficients between all metrics. The main goal and added value of this paper was the mapping of the selected objective metric SSIM to the MOS scale which—all subjective metrics. The results showed us that the video quality measured by SSIM objective metric should reach at least 0.90 when the observers will consider the video quality using the subjective methods as a good one.

The next step should be to analyse other compression standards with Full HD or 4 K resolutions using both objective and subjective assessment with the subsequent mapping of the obtained results.

Acknowledgments The research leading to these results has received funding from the grant SGS reg. no. SP2015/82 conducted at VSB-Technical University of Ostrava, Czech Republic and by the project reg. no. ITMS 22420320001 within crossborder cooperation between the Czech and Slovak Republic.

References

1. Uhrina M, Hlubik J, Vaculik M (2012) Impact of compression on the video quality. In: 12th international conference on knowledge in telecommunication technologies and optics—KTTO 2012, Malenovice, Czech Republic, 14–16 Nov 2012. Advances in electrical and electronic engineering, vol 10, issue 4, pp 251–258. ISSN 1336-1376
2. Uhrina M, Frnda J, Sevcik L, Vaculik M (2014) Impact of H.264/AVC and H.265/HEVC compression standards on the video quality for 4 K resolution. In: 14th international conference on knowledge in telecommunication technologies and optics—KTTO 2014, Malenovice, Czech Republic, 3–5 Sept 2014. Advances in electrical and electronic engineering, vol 12, issue 4, pp 368–376. ISSN 1336-1376

3. Uhrina M, Sevcik L, Frnda J, Vaculik M (2014) Impact of H.265 and VP9 compression standards on the video quality for 4 K resolution. In: 22nd telecommunications forum TELFOR 2014, Belgrade, 25–27 Nov 2014
4. Uhrina M, Hlubik J, Vaculik M (2011) The comparison of MPEG compression standards using different subjective video quality methods. 11th international conference on knowledge in telecommunication technologies and optics—KTTO 2011, Szczyrk, Poland, pp 56–59, 22–24 June 2011. ISBN 978-80-248-2399-7
5. Voznak M (2011) E-model modification for case of cascade codecs arrangement. Int J Math Models Methods Appl Sci 5(8):1439–1447
6. Frnda J, Sevcik L, Uhrina M, Voznak M (2014) Network degradation effects on different codec types and characteristics of video streaming. J Adv Electr Electron Eng 12(4):377–383. ISSN 1336-1376
7. Frnda J, Voznak M, Sevcik L (2014) Network performance QoS prediction. Adv Intell Syst Comput 297:165–174. doi:10.1007/978-3-319-07776-5_18 (expecting WoS, SCOPUS, SJR = 0.135)
8. Frnda J, Voznak M, Rozhon J, Mehic M (2013) Prediction model of QoS for triple play services. In: 21st telecommunications forum TELFOR
9. Uhrina M, Hlubik J, Vaculik M (2012) Correlation between objective and subjective methods used for video quality evaluation. In: 9th international conference—Elektro 2012, Rajecké Teplice, Slovak Republic, 21–22 May 2012. ISBN 978-1-4673-1180-9
10. Moldvan A-N, Ghergulescu I, Muntean CH (2014) Novel methodology for mapping objective video quality metrics to the subjective MOS scale. In: 2014 IEEE international symposium on broadband multimedia systems and broadcasting (BMSB), Beijing, 25–27 June 2014. INSPEC Accession number: 14515368
11. Wang W (2012) A video quality assessment method using subjective and objective mapping strategy. In: 2012 IEEE cloud computing and intelligent systems (CCIS), pp 514–518, 30 Oct–1 Jan 2012. ISBN: 978-1-4673-1855-6
12. Wooton C (2005) A practical guide to video and audio compression. Elsevier Inc., Oxford, p 787. ISBN 0-240-80630-1
13. Wu HR, Rao KR (2006) Digital video image quality and perceptual coding. Taylor and Francis Group LLC, Boca Raton, p 594. ISBN 0- 8247-2777-0
14. Loke HM, Ong PE, Lin W, Lu Z, Yao S (2006) Comparison of video quality metrics on multimedia videos. In: Image processing IEEE 2006, pp 457–460. ISSN 1522-4880
15. Winkler S (2005) Digital video quality: vision models and metrics. Wiley, Chichester, p 175, ISBN 0-470-02404-6
16. Recommendation ITU-R BT.500-13 (2013) Methodology for the subjective assessment of the quality of television pictures
17. http://www.irisa.fr/armor/lesmembres/Mohamed/Thesis/node2.html (2010)
18. Recommendation ITU-T P.910 (2008) Subjective video quality assessment methods for multimedia applications, April 2008
19. Test Sequence. http://ultravideo.cs.tut.fi/#testsequences
20. H.265/HEVC Encoder, ×265 tool (2013). http://www.videolan.org/developers/x265.html
21. MSU Measurement Tool Pro version (2012). http://compression.ru/video/quality_measure/vqmt_pro_en.html#start

SOM on Interval Variables for Mobile Emergency Call Positioning

Petr Klement and Václav Snášel

Abstract Emergency call localization is an important step in the emergency call taking process. As more than 80 % of emergency calls are dialled from mobile phones, the localization of calls in mobile networks deserves highest attention. In this paper we describe a model-based localization method which do not impose any special investments or modifications on the mobile network infrastructure. We create a model of the mobile network environment using the geographically referenced Self Organizing Map with interval values, trained on common measurements regularly performed by mobile network operator. Localization is then performed by finding the node of the model best matching to the characteristic vector of the call.

Keywords Emergency call · Localization · Self-organizing map · Interval variables · Knowledge discovery in databases · Data clustering

1 Introduction

Emergency call localization is an important step in the emergency call taking process. It can speed up the emergency response significantly, helping to decide which resources are to provide assistance with respect to their position and route of access to the location of the emergency. Exact call localization helps avoiding duplicities in responding the same event. It is also useful in identification of false

P. Klement
Pit Software s.r.o, Nadrazni 140, 702 00 Ostrava, Czech Republic
e-mail: petr.klement@pitsoftware.cz

V. Snášel (✉)
Faculty of Electrical Engineering and Computer Science, Department of Computer Science and IT4Innovations, VŠB-Technical University of Ostrava, Ostrava, Czech Republic
e-mail: vaclav.snasel@vsb.cz

A. Abraham et al. (eds.), *Intelligent Data Analysis and Applications*,
Advances in Intelligent Systems and Computing 370,
DOI 10.1007/978-3-319-21206-7_15

calls as malicious callers usually report places of emergency far different from their real position.

The localization of calls in mobile networks deserves highest attention as more than 80 % of emergency calls are dialled from mobile phones.

Various methods are available to determine a mobile phone location [10, 17]. Widespread usage of smart mobile phones with integrated GNSS (Global Navigation Satellite System, for instance GPS) receiver could suggest that the problem of mobile call localization is solved. Yet the opposite is the truth. Satellite navigation does not work in indoor areas, tunnels, subways etc. and even in the outdoor places as urban canyons—streets where high buildings surrounding the mobile caller hinder receiving signal from satellites to determine the position. Taking into account that the GNSS receiver module is usually switched off in smart phones, relatively long interval in order of minutes of initialization of the receiver creates another serious problem for fast localization of emergency call. Although the latter issue is partially improved by Assisted GPS (AGPS) service which allows the mobile terminal to receive information on satellites available in place from the mobile network, AGPS service availability depends on the mobile network and mobile terminal technology. From reasons like these positioning methods based on physical characteristics and configuration of the mobile network play irreplaceable role in localization of emergency calls.

Localization methods, according to the traditional scheme, fall in three categories of network-based, handset-based and combined methods [16]. From another point of view, these methods can be seen as computational methods and model-based ones.

Computational methods work on physical parameters measured in the course of the call. These are for example signal strength, time of signal arrival, angle of the received signal with respect to the receiving base station etc. From these values the position is computed using triangulation or trilateration formulas. The point here is the need of additional measuring equipment (Location Measurement Unit, LMU) in the network infrastructure, which increases capital expenses of the network significantly.

Model-based methods determine calling party position by comparing information included in the standard call protocol and thus available in various sorts of mobile networks with the model of the environment a priori created from previous measurements and computations. Model-based methods do not impose additional investments in the mobile network infrastructure. On the other hand the quality of results depends on the quality and adequacy of the model preparedness and on regular upgrading of the model to reflect changes in infrastructure and configuration of mobile networks.

Model-based localization can be transformed into the task of finding the best position of the vector of information describing the call in the vector space of the model. Dimensions of vectors of the model bear identical type of information as the descriptive vector of the call. Above this, the model is geographically referenced.

In [14] two possible approaches to this task were presented. In the first one, the model was created from vectors of georeferenced measurements of physical and configuration parameters of the mobile network indexed by suffix tree data structures, where the localization was performed by finding nearest neighbour of the characteristic vector of the call in the suffix tree index. The second approach used the classical SOM trained on measurement vectors to find clusters of measured samples similar in physical and configuration parameters of the mobile network. Members of resulted clusters showed good results in geographic proximity, thus the localization would be performed by finding the nearest cluster or the best-matching node of the SOM to the characteristic vector of the call.

In this paper we propose an extended approach to the model building using the geographically referenced Self Organizing Map with interval values. We collect georeferenced measurements into clusters represented by nodes of the SOM trained on the measurements. Components of weight vectors of the SOM nodes are intervals of values rather than single values used in traditional SOM applications. It is shown that the data preparation as well as competition and adaptation phases of the SOM algorithm using interval values can be successfully tailored to the specifics of the task of localization. Moreover this approach is suitable for extending the model to the areas where it could not be otherwise satisfactorily created because of sparse or lack of measurements.

In Sect. 2 we briefly describe principles of the Kohonen SOM algorithm adopted to interval variables; in Sect. 3 we define attributes we used to create the localization model and characteristic vector of the mobile call to make queries towards the model and describe transformations of these attributes to prepare training and testing sets of data suitable to the task. In Sect. 4 we describe the implementation of the SOM algorithm on interval variables. In Sect. 5 we present experiments showing results of the SOM algorithm on interval variables. In Sect. 6 we discuss the results and in Sect. 7 we draw conclusions.

2 Self-organizing Maps on Interval Variables

The localization model presented in this paper is built on the principle of clustering, or grouping objects (data records) into classes (clusters) in such a way that objects in the same cluster are very similar, while objects in different classes are quite distinct.

One of the possible clustering methods is competitive learning [4, 18]. Given the training set of objects, competitive learning finds an artificial object (representative) most similar to the objects of a certain cluster.

A commonly used application of competitive learning is the Kohonen Self Organizing Map [12], or SOM, described by Teuvo Kohonen in 1982.

SOM is a type of artificial neural network with two fully interconnected layers of neurons, the input layer and the output or Kohonen layer. State of each neuron of the Kohonen layer is described by its weight vector with components of the real numbers domain: $\boldsymbol{w}_j = (w_{j1}, w_{j2}, \ldots, w_{jm}), \quad w_{jk} \in \boldsymbol{R}$.

In the **competition** phase of Kohonen learning every neuron N_j $j = 1, 2, \ldots, n$ of the output layer is excited by the dot product of the input vector $\boldsymbol{x} = (x_1, x_2, \ldots, x_m)$ and the weight vector $\boldsymbol{w}_j$. The largest value of the product is equivalent to the minimum of Euclidean distances $\|x - w_j\| \forall j = 1, 2, \ldots, n$ [8]. The smallest distance of the input vector from the weight vector of a neuron denotes the respective neuron as the winner of the competition.

For interval variables the SOM competition phase must be modified in terms of the distance calculation. Instead of a single value, the k-th component of the weight and training vectors is a real interval described by lower and upper bounds $l_k, u_k; l_k \leq u_k$, midpoint $\mu_k = \frac{u_k + l_k}{2}$ and sometimes by the radius $r_k = \frac{u_k - l_k}{2}$. Various distance measures between intervals are presented in the literature. The Euclidean distance between midpoints [1] is simple, yet the weak one as it completely neglects the length of intervals. Sum of distances of corresponding vertices between m-dimensional hyperrectangles formed by the component-wise intervals, also mentioned in [1] is proved to be equivalent to the Euclidean distance between midpoints plus Euclidean distance between radiuses in [2]. Of the same meaning is the L2 distance measure proposed in [6] or the city block distance applied in [3]. In [13] the sum of Euclidean distances between midpoints and radiuses is supplemented with the dot product of vectors of differences in midpoints and radiuses. The Hausdorff distance applied on intervals [5, 1] is computationally efficient as summing the greater of differences between upper and lower bounds does not require multiplications.

Taking into account specifics of the application domain of this work we proposed algorithmic determination of the distance between two interval-valued vectors. To find the best match for the mobile equipment receiving radio signals from various base stations at the time, the number of somehow corresponding attributes is more significant than the accuracy of the correspondence. Principles of the proposed BHN (Best Hit Node) algorithm are described as follows, given that $(l_k(\boldsymbol{x}), u_k(\boldsymbol{x}))$ and $(l_k(\boldsymbol{w}), u_k(\boldsymbol{w}))$ denote the intervals of k-th components of training and weight vectors and $d(x_k, w_k)$ denotes distance contribution on k-th component:

(i) If there is a non-empty intersection between $(l_k(\boldsymbol{x}), u_k(\boldsymbol{x}))$ and $(l_k(\boldsymbol{w}), u_k(\boldsymbol{w}))$, then $d(x_k, w_k) = 0$ and the k-th component of the weight vector was hit by corresponding component of the training vector

(ii) If for two nodes N_i, N_j holds

$$\sum_{k=1}^{m} d(x_k, w_{ki}) = \sum_{k=1}^{m} d(x_k, w_{kj})$$

then N_i wins if and only if the number of hits on its components is greater than that one of N_j

$$\sum_{\substack{k=1 \\ d(x_k, w_{ki})=0}}^{m} (1) > \sum_{\substack{k=1 \\ d(x_k, w_{kj})=0}}^{m} (1) \tag{2}$$

(iii) If for two nodes N_i, N_j holds (1) and both sides of (2) are equal, then N_i wins if and only if

$$\sum_{\substack{k=1 \\ d(x_k, w_{ki})=0}}^{m} (u_{ki}(w) - l_{ki}(w)) \leq \sum_{\substack{k=1 \\ d(x_k, w_{kj})=0}}^{m} \left(u_{kj}(w) - l_{kj}(w)\right) \tag{3}$$

Or the sum of lengths of matching intervals of N_i is lower than that one of N_j.

The second phase of the SOM learning algorithm is **adaptation**. The neurons of the Kohonen layer are organised mostly in a two-dimensional lattice. A topological neighbour-affecting function is defined on the Kohonen layer, assigning a degree of participation in the learning process to the neurons neighbouring the neuron which won the competition phase. In every learning step the weight vectors of the winning neuron and its neighbours are adjusted to move closer to the input training vector.

The trained network finally sets its weights in such a way that the topologically near neurons represent similar training cases, while distant ones reflect different cases. The topology of a trained SOM forms an inherently useful base for clustering and in case that the nodes of the SOM are bound to geographic coordinates, it suits very well to the task of creating localization models.

Although in simplified cases, as mentioned in [1], adapting single values of midpoints of intervals is possible, the adaptation of interval weight vectors comprises of adaptation of the lower and upper bounds of the intervals. Gaussian neighbourhood function $\theta(c,i,t)$ is used to affect neurons N_i in the neighbourhood $O(N_c)$ of the winning neuron N_c in a learning cycle t

$$\theta(c,i,t) = \exp\left(-\frac{d(N_c,N_i)^2}{h(t)\sigma^2}\right) \qquad i \in \{\arg N_j, N_j \in O(N_c)\} \tag{4}$$

where $d(N_c, N_i)$ means the topological distance between neurons in the SOM, $h(t)$ stands for the width of the Gaussian curve decreasing in time and σ^2 is the variance of $d(N_c, N_i)$ on $O(N_c)$.

Lower and upper bounds of weight vector components of neuron N_i in a learning cycle t are adapted by the learning vector $\boldsymbol{x}$ respective to their values in the preceding cycle $t-1$ according to the formulas

$$l_k(\boldsymbol{w}_i)^{(t)} = l_k(\boldsymbol{w}_i)^{(t-1)} + \alpha(t)\theta(c,i,t)\left(l_k(\boldsymbol{x}) - l_k(\boldsymbol{w}_i)^{(t-1)}\right) \qquad \forall N_i \in O(N_c) \tag{5}$$

$$u_k(\boldsymbol{w}_i)^{(t)} = u_k(\boldsymbol{w}_i)^{(t-1)} + \alpha(t)\theta(c,i,t)\left(u_k(\boldsymbol{x}) - u_k(\boldsymbol{w}_i)^{(t-1)}\right) \qquad \forall N_i \in O(N_c) \tag{6}$$

where $\alpha(t)$ is a coefficient of the learning speed, decreasing in time.

Thanks to the monotonically decreasing functions $\alpha(t)$ and $\theta(c,i,t)$ in Eqs. (5) and (6), the learning is faster and rather rough in first iterations, affecting wider areas of the SOM, while the neighborhood as well as the learning increment are small during final iterations, allowing fine-tuning of weight vectors of winning nodes. Finding the best parameters for the adaptation phase to cope with the speed and accuracy of learning is either a matter of experiment, or a product of superior optimizing algorithm, the latter staying beyond the scope of this paper.

3 Data Description and Preprocessing

Data used in building the mobile network environment model were obtained from routine measurement of the mobile network operator. The measurement is performed regularly in the course of the network optimization task. A car with the mobile measuring equipment cruises through the monitored area and records timestamp, geographic location and various parameters characterising the status of the network in respective time and position. From these Table 1 shows parameters used for creation of the model.

To share transmitting channel between multiple users, GSM mobile networks use Time Division Multiple Access (TDMA) communication scheme. Timing Advance (TA) parameter is used to determine a time slot in TDMA scheme. In this time slot the radio signal covers the distance of roughly 550 m. According to the GSM definition, the TA can reach values between 0 and 63. So for the purpose of our work the TA can be passed as a discrete distance measure with step of 550 m.

Each mobile terminal knows the identification of the serving cell, its TA, broadcasting channel and the signal strength of the broadcasting channel of the serving cell. The terminal also maintains a list of maximum 6 base stations with the strongest signal in its vicinity to be able to switch the serving cell in case of moving to the area covered by another cell.

Table 1 Description of attributes used in building the mobile network environment model

Name	Type	Description
Latitude	Numeric	Geographical latitude of measurement
Longitude	Numeric	Geographical longitude of measurement
CellID	Categorical	Serving cell identity code
TA	Numeric	Serving cell timing advance
BCCH_0	Categorical	Serving cell broadcast control channel
BSIC_0	Categorical	Serving cell base station identity code
RxLev_0	Numeric	Serving cell received signal level
BCCH_1	Categorical	1st neighbouring cell broadcast control channels
BSIC_1	Categorical	1st neighbouring cell base station identity code
RxLev_1	Numeric	1st neighbouring cell received signal level
…	…	BCCH, BSIC and RxLev of the second till the fifth neighbouring Cell
BCCH_6	Categorical	6th neighbouring cell broadcast control channel
BSIC_6	Categorical	6th neighbouring cell base station identity code
RxLev_6	Numeric	6th neighbouring cell received signal level

As the CellID, BSIC and BCCH are categorical attributes, transformation described in [11] was applied to obtain numerical values suitable for the SOM training. Here we used the fact, that each of the categorical attributes has a corresponding numerical value. To the CellID there exists corresponding TA-value, to each of the [BCCH, BSIC] pairs there is corresponding signal strength (RxLev). In the set of measurements there was 64 distinct CellID values and 260 [BCCH, BSIC] pairs. Input vectors of the training set have 324 dimensions plus 2 dimensions storing geographic coordinates of the measurement. These vectors are sparse, most of their values are undefined or empty. Empty TA values were set to 128, as the real TA do not exceed 63. The signal strength is measured in units of decibel (dB) which are represented by negative numbers. Minimal signal strength (maximal attenuation) when the mobile terminal switches to another cell is around −108 dB. Undefined RxLev values were therefore replaced by the negative value of −256 which did not exist in the set of measurements.

4 Implementation of the SOM and Localization Algorithm

We implemented the interval-valued version of the SOM learning algorithm on principles described above. Before the learning itself, initialization of the weight vectors was performed. The SOM was spread over the geographic area in a form of regular grid. Thus every node got its steady geographic position. Measurements (training vectors) were distributed between the nodes of the SOM with respect to the shortest geographic distance of the point of measurement and the geographic position of the node. Then upper and lower bounds of intervals of components of the weight vectors were set to the maximum and minimum of values of the respective components of training vectors assigned to the node. Besides this managed initialization we used also random initialization of weight vectors the SOM nodes. In this case the upper and lower bounds l_k, u_k were set to random values between 0 and 1 on condition that $l_k \leq u_k$.

The geographic latitude and longitude became part of weight vectors of the SOM nodes and were modified by the learning process in some experiments, while in the others they served as anchor points binding SOM nodes to fixed geographic positions not affected by the learning.

We experimented with various interval distance measures mentioned in the paragraph 2. We also experimented with the SOM learning parameters, diameter of the neighbourhood, steepness of the Gaussian curve, affecting intensity of changes of neighbouring nodes and with the learning speed factor. These parameters were decreasing in time so that in the initial learning iterations large areas in terms of both geographic meaning as well as of the number of SOM nodes were affected and in the final iterations only the winning nodes were fine-tuned.

The trained SOM with georeferenced nodes creates the desired model of the mobile network environment which can be used for localization of mobile calls. The localization function is based on the SOM theory again, being nothing else than the competition step of the SOM algorithm where only the non-geographic attributes take part. The geographic position of the testing case is approximated by geographic coordinates of the winning node of the competition. Correctness of the model was tested on a subset of non-geographic components of training vectors which did not take part in forming the model. For visualization of tests we created a light GIS application which interpreted the trained SOM characteristics and the localization function. Various distance measures, identical to those used in the SOM training, were implemented in the localization function.

5 Experiments

We present a brief description of our experiments in four pictures. Figure 1 shows the initialization phase of the model. The SOM nodes are set on fixed geographic positions to cover the area of interest evenly. Intervals of components of weight vectors are initialized to the maximum and minimum of values of the respective

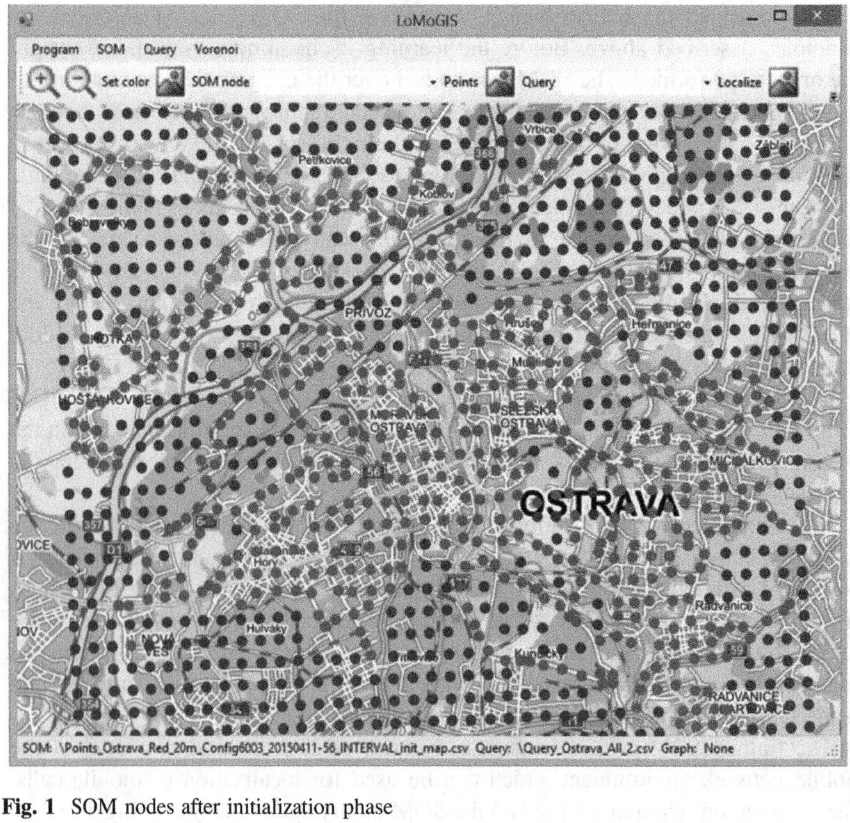

Fig. 1 SOM nodes after initialization phase

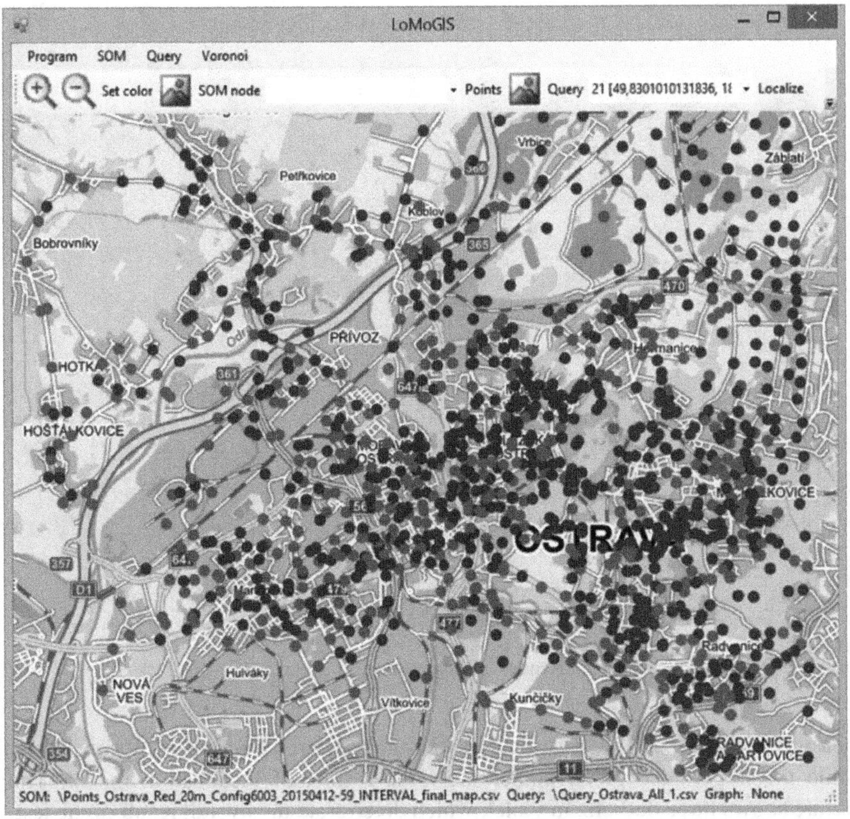

Fig. 2 Trained SOM

components of training vectors geographically closest to the node. Red points show nodes with measurements assigned, blue points are empty nodes, initialized by marginal values.

Trained localization model is shown in Fig. 2. Nodes of the SOM moved from initial positions towards places of higher density of training vectors reflecting the geographic distribution of measurements. Non-geographic components of weight vectors of nodes of the SOM also moved in the n-dimensional space of physical characteristics of the mobile network. Thus the SOM nodes were transformed from centroids of clusters of measurements into codebook vectors of the network-specific subspace of the localization model.

The task of localization is to find the codebook vector closest to the characteristic vector of the mobile call in the network-specific subspace. Geographic position of the codebook vector then approximates position of the mobile call.

The distance in network-specific subspace can be determined in various ways with different results. Figure 3 displays set of 22 test cases presented to the trained model. Enhanced vertex distance described in Sect. 2 was used as the proximity

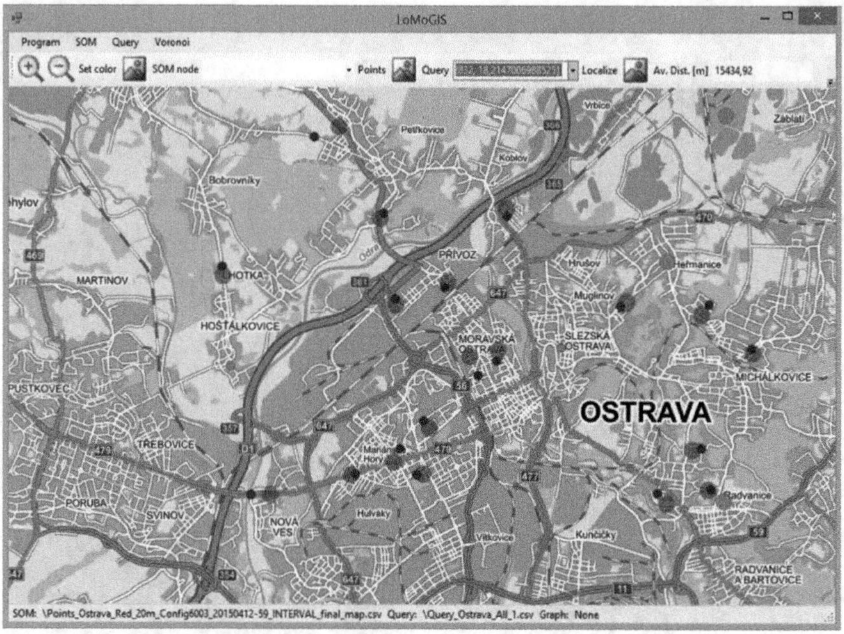

Fig. 3 Localization using interval vertex distance

measure in this case. True positions of testing cases are represented by small dots, while the larger pale green points show the results of localization model. The picture shows three outliers, in violet, which deteriorate the overall result of the test. The mean error, quantified as the mean of differences in geographic distances of the testing cases and their localization points, was 1 115 m.

Results of localization of the same test set using the BHN distance measure as described in Sect. 2 is shown in Fig. 4. The mean error now was 248 m, fairly better than in the previous case.

6 Discussion

The work presented in this paper was done having in mind the concrete application, which was localization of mobile calls. Creating a geographically referenced model of behaviour of the mobile network using Self Organizing Map was driven by two inherent characteristics of the SOM paradigm. The first is the ability of mapping high dimensional input space with nonlinear dependencies into a low dimensional discrete output space. In our application domain the input space consisting of hundreds of attributes varying according to the physics of radio signal transmission

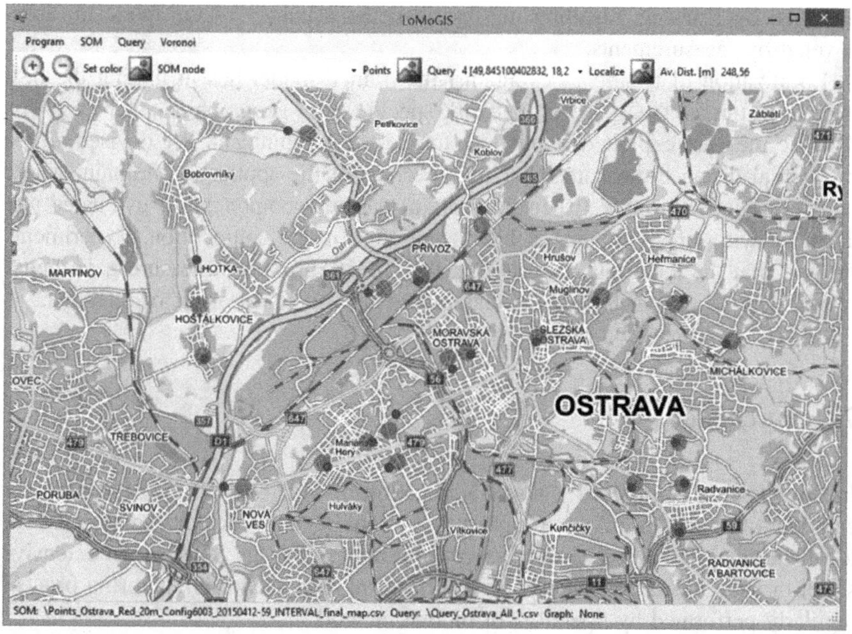

Fig. 4 Localization using best hit node distance measure

and rules of configuration and communication in mobile networks present a stochastic environment for which the discussed property of SOM suits very well.

The second positive feature of using SOM in this application is its ability of approximation. Through the influence on the nodes in a neighbourhood of the best matching node, the resulting model can diffuse itself, covering areas with sparse or none measurements otherwise necessary to create the model.

In components of weight vectors of the SOM nodes we used interval values instead of the traditional SOM with single-valued weights. This again comes out of the application domain, as the network characteristics in a given place vary in time, depending on weather conditions, load of the serving cell and even on temporary obstacles in radio signal transmission.

Nodes of the SOM in our application were geographically referenced. After the SOM was trained, its nodes became codebook vectors for the localization task and their geo-coordinates determined the location of the testing call. In [9] geographic coordinates were used to find the geographically best matching node and then to perform competition on the non-geographic attributes in the restricted neighborhood of this geo-best node. We didn't use geo coordinates in the competition phase, instead we used them in the initialization of the weight vectors and in the learning phase of the SOM algorithm. In some experiments the initial regular geographical grid of nodes stayed preserved in the course of training, which means that the geo coordinates were not involved in learning, while in other experiments they were modified by learning, resulting in slightly better accuracy of the model in areas covered by measurements.

Localization of testing cases against the trained model lies in finding the best matching unit in the space of network-specific (non-geographic) attributes. The distance measure exploited in this task turned to be significant. We devised a type of interval distance measure reflecting specifics of the application domain, where number of interactions between intervals of various components is evaluated and the node with larger number of hits is preferred in the competition. Experiments showed that the localization exploiting this "Best Hit Node" measure gave better results than those which used the classical vertex—type distances of intervals.

7 Conclusion

In this paper we proposed a novel application of Self Organizing Map used for creation of a model of environment of a mobile network for the purpose of localizing mobile calls. The model was built by training the SOM on an extensive set of measurements collected in the course of mobile network monitoring process regularly performed by operators of mobile networks.

The traditional SOM approaches were combined and tailored to the specifics of the application. Namely we used geo-referencing of SOM nodes and interval variables as the components of weight vectors of neurons. We also proposed an interval distance measure suitable to the positioning of the characteristic vector of the mobile call in the multidimensional subspace of physical parameters of the network.

Geo-referenced measurements were grouped into clusters represented by nodes of the SOM in the learning process and the weight vectors of the trained SOM nodes were transformed from centroids of clusters of measurements into codebook vectors of the network-specific subspace of the localization model. Geographic position of the codebook vector closest to the characteristic vector of the mobile call in the network-specific subspace then approximated position of the mobile call.

As the accuracy of the localization model depends on the quality and frequency of measurements of network parameters, the ability of SOM to approximate white areas in the input space is deemed one of the most important feature. It turned out that the competitive learning on interval variables was a suitable approach with respect to the specifics of the task of localization in mobile networks.

Acknowledgments This work was supported by the IT4Innovations Centre of Excellence project (CZ.1.05/1.1.00/02.0070), funded by the European Regional Development Fund and the national budget of the Czech Republic via the Research and Development for Innovations Operational Programme and by Project SP2015/146 "Parallel processing of Big data 2" of the Student Grand System, VŠB—Technical University of Ostrava.

References

1. Bock HH (2003) Clustering algorithms and Kohonen maps for symbolic data. New trends in computational statistics with biomedical applications. J Jpn Soc Comp Statist 15.2:217–229
2. D'Urso P, DeGiovanni L (2011) Midpoint radius self-organizing maps for interval-valued data with telecommunications application. Appl Soft Comput 11:3877–3886
3. De Carvalho F, Bertrand P, De Melo F (2012) Batch self-organizing maps based on city-block distances for interval variables, p 15 <hal-00706519>
4. Gan G, Ma C, Wu J (2007) Data clustering: theory, algorithms and applications. SIAM, Philadelphia
5. Hajjar C, Hamdan H (2011) Self-organizing map based on Hausdorff distance for interval-valued data. In: IEEE international conference on systems, man, and cybernetics (SMC), Anchorage, pp 1747–1752
6. Hajjar C, Hamdan H (2011) Self-organizing map based on L2 distance for interval-valued data. In: 6th IEEE international symposium on applied computational intelligence and informatics, Timişoara, 19–21 May 2011
7. Hajjar C, Hamdan H (2013) Interval data clustering using self-organizing maps based on adaptive Mahalanobis distances. Neural Netw 46:124–132
8. Haykin S (1999) Neural networks: a comprehensive foundation, 2nd edn. Prentice-Hall, Upper Saddle River
9. Henriques R, Bacao F, Lobo V (2012) Exploratory geospatial data analysis using the GeoSOM suite. Comput Environ Urban Syst 36:218–232
10. Kapicak L, Sebesta R, Michalek L, Dvorsky M (2011) Comparison method for mobile location techniques. In: 13th international conference on research in telecommunication technologies 2011, Techov, Czech Republic
11. Klement P, Snášel V (2009) Anomaly detection in emergency call data—the first step to intelligent emergency call system management. In: Proceedings of the 1st international conference on intelligent networking and collaborative systems (INCoS 2009), Barcelona, Spain, Nov 2009
12. Kohonen T (1995) Self-organizing maps. Springer, Berlin
13. Liu L, Xiao J, Yu L (2008) Interval self-organizing map for nonlinear system identification and control. Advances in neural networks—ISNN 2008. In: 5th international symposium on neural networks, ISNN 2008, Beijing, China, 24–28 September 2008, Proceedings, Part I, pp 78-86
14. Martinovič J, Novosád T, Snášel V, Scherer P, Klement P, Šebesta R (2012) Clustering the mobile phone positions based on suffix tree and self-organizing maps. Neural Netw World 22:371–386
15. Moore RE, Kearfott RB, Cloud MJ (2009) Introduction to interval analysis Society for industrial and applied mathematics, Philadelphia
16. Novosád T, Martinovič J, Scherer P, Snášel V, Klement P, Šebesta R (2011) Mobile phone positioning in GSM networks based on information retrieval methods and data structures. in: Proceedings of the international conference on digital information processing and communications (ICDIPC), published in Communications in computer and information science series of springer LNCS
17. Sebesta R, Dvorsky M, Kapicak L, Michalek L, Martinovič J, Scherer P (2011) Visualisation of best servers areas in GSM networks. In: 11th international conference on knowledge in telecommunication technologies and optics KTTO 2011, Szczyrk, Polan
18. Yang MY (2001) Extending the Kohonen self-organizing map networks for clustering analysis. Comput Stat Data Anal 38:161–180

Normalization of Vietnamese Tweets on Twitter

Vu H. Nguyen, Hien T. Nguyen and Vaclav Snasel

Abstract We study a task of noisy text normalization focusing on Vietnamese tweets. This task aims to improve the performance of applications mining or analyzing semantics of social media contents as well as other social network analysis applications. Since tweets on Twitter are noisy, irregular, short and consist of acronym, spelling errors, processing those tweets is more challenging than that of news or formal texts. In this paper, we proposed a method that aims to normalize Vietnamese tweets by detecting non-standard words as well as spelling errors and correcting them. The method combines a language model with dictionaries and Vietnamese vocabulary structures. We build a dataset including 1,360 Vietnamese tweets to evaluate the proposed method. Experiment results show that our method achieved encouraging performance with 89 % F1-Score.

Keywords Normalization of noisy texts · Spelling error detection and correction · Twitter

1 Introduction

Nowadays, many popular online social networks (OSNs) such as Twitter or Facebook have become some of the most effective channels for users to communicate and share information. Due to the huge magnitude of online social network users, an enormous amount of content has been continuously created every day. According to

V.H. Nguyen · H.T. Nguyen
Faculty of Information Technology, Ton Duc Thang University, Ho Chi Minh City, Vietnam
e-mail: nguyenhongvu@tdt.edu.vn

H.T. Nguyen
e-mail: hien@tdt.edu.vn

V. Snasel (✉)
Faculty of Electrical Engineering and Computer Science, Department of Computer Science and IT4Innovations, VŠB-Technical University of Ostrava, Ostrava, Czech Republic
e-mail: vaclav.snasel@vsb.cz

A. Abraham et al. (eds.), *Intelligent Data Analysis and Applications*,
Advances in Intelligent Systems and Computing 370,
DOI 10.1007/978-3-319-21206-7_16

the Statistics in 2011, the number of tweets people sent on Twitter per day has been up to 140 million tweets[1]. Unlike news or authored textual web content, the contents on OSNs are short, noisy, irregular, temporal dynamics, and they are usually short. In the case of Twitter, a posting on Twitter is limited in 140 characters. Therefore, a user tends to use acronyms, non-standard words, or social tokens. Moreover, he/she tends to compose tweets and comments quickly, which may cause spelling mistakes or typo.

In this paper, we propose a method to normalize Vietnamese tweets by detecting non-standard words as well as spelling errors and correcting them. The method helps improving the performance of applications mining or analyzing semantics of social media contents as well as other social network analysis applications. There have been many methods proposed for text normalization. Most of them are to normalize texts written in English [1, 2, 7, 9, 10, 19, 24, 25, 27] and some other languages such as Chinese [16, 26, 28] or Arabic [15, 23]. We found several methods proposed to Vietnamese spell checking in literature. However, those methods did not take postings on online social networks into account. Our work presented in this paper aims to build a bridge for filling the gap.

This paper presents the first attempt to build a system for normalizing Vietnamese tweets. We propose a method consists of three steps: (i) the first step is to preprocess tweets, (ii) the second is to detect non-standard words, misspelled words by typing, and (iii) the last step is to normalize and correct those errors. For example, spelling errors in tweets: "Tooi đi hocj" can be normalized as "Tôi đi học" (I go to school) or "Banj teen gif?" can be normalized as "Bạn tên gì?" (What is your name?). Other examples of text messages from Twitter and their corresponding normalized forms are shown in Table 1. In this paper, we also propose a method to improve the similarity coefficient of two words according to Dice coefficient [5]. When applying our improving method, the similarity coefficient has increased significantly.

Our contributions in this paper is three-fold: (1) propose a method to detect and normalize Vietnamese tweets based on dictionaries and Vietnamese vocabulary structures combining with a language model, (2) improve Dice coefficient to measure similarity of two words, and (3) build a dataset including 1,360 Vietnamese tweets to evaluate methods proposed for normalizing short texts on OSNs. The rest of this paper is organized as follows: Section 2 presents related work, Sect. 3 presents our proposed method, Sect. 4 presents experiments and results, and finally, we draw conclusion in Sect. 5.

2 Related Work

Nowaday, there are many researches for the spelling error detection and normalization. Normally, every research focuses handle for a specific language. For examples,

[1]https://blog.twitter.com/2011/numbers.

Table 1 Spelling error tweets and their normalized

Spelling error tweets	Normalized tweets
Trời ddang mưa	Trời đang mưa (It is raining)
Hôm nay, siinh viên DDaijj học Tôn DDuwcss Thắng được nghỉ học	Hôm nay, sinh viên Đại học Tôn Đức Thắng được nghỉ học (Today, student of Ton Duc Thang university was allowed to absent.)
ngày maii là Têt rồi	ngày mai là Tết rồi (tomorrow is our traditional Tet's holiday)

in the domain of English, most earlier work on automatic error correction addressed spelling errors and built models of correct using on native English data ([1, 2, 9]). In [25], to normalize non-standard words they developed a taxonomy of non-standard words then they investigated the application of several general techniques including n-gram language models, decision trees and weighted finite-state transducers to the entire range of non-standard words types. In [7, 10], they used discriminative model to propose a mechanism for error detection and error correction at word-level respectively. In [24], they proposed a graph-based text normalization method that utilizes both contextual and grammatical features of text. A log-linear model was used in [27] to characterized the relationship between standard and non-standard tokens. In [19], they proposed a two character-level methods for the abbreviation modeling aspect of the noisy channel model including a statistical classifier using language-based features to decide whether a character is likely to be removed from a word, and a character-level machine translation model. With Chinese language, the majority of studies used model language processing [16, 26, 28] and [18] used unsupervised model and discriminative reranking. With Arabic language, recent researches used supervised learning [15], character based language model [23]. Considering for Vietnamese language, we had several researches involved analyzing the word, phrase, sentence analysis, handling ambiguity, construction dictionary (VLSP[2] [8, 11]) and most recently, studies using ngram language model [17, 21].

In the field of social network, we just have several researches to handle spelling errors. For example, in [12, 13], they have been detected and handled error based on the morphophonemic similarity. In [3], they detected and handled of non-standard words in online social network by using diverse coefficient method such as Dice, Jaccard, Ochiai. [14] used random walks on a contextual similarity bipartite graph constructed from n-gram sequences on large unlabeled text corpus to normalize social text. In [25], they proposed a novel method for normalizing and morphologically analyzing Japanese noisy text by generating both character-level and word-level normalization candidates and using discriminative methods to formulate a cost function. An approach to normalize the Malay Twitter messages based on corpus-driven analysis was proposed in [22]. And the latest is the research of [4], this research proposed a modular approach for lexical normalization applied to Spanish tweets, the system proposed including modules detection and correction candidates for each out of vocabulary word, rank the candidates to select the best one.

[2]http://vlsp.vietlp.org:8080/demo/.

In this paper, we proposed a mechanism to detect and normalize spelling error for Vietnamese tweet based on dictionary and Vietnamese vocabulary structure combining with a language model. The tweet with spelling errors will be detected based on the vocabulary structures, after error detection phase, the system will first normalize the error tweets based on the structure of vowels, consonants and then use the language model to calculate the degree of similarity. In the calculation of the similarity degree task, we have studied and proposed an improvement model based on the model of Dice [5]. The language model using here is SRILM[3] to generate the 3-gram model. The sentence with the highest degree of similarity was selected as the final data.

3 Proposed Method

3.1 The Theoretical Background

Currently, there are several point of views on what is a Vietnamese word. However, to meet the goals automatic error detection, the authors use the views in thesis of Dinh Dien [6], A Vietnamese word is composed of Vietnamese morphemes. And according to the syllable dictionary of Hoang Phe [20], our team split a word into two basic parts: the consonant and syllable, there may exist words without the consonant.

- **Consonant**: Vietnamese language has 26 consonant: "b", "ch", "c", "d", "đ", "gi", "gh", "g", "h", "kh", "k", "l", "m", "ngh", "ng", "nh", "n", "ph", "q", "r", "s", "th", "tr", "t", "v", "x" and 8 tail consonants: "c", "ch", "n", "nh", "ng", "m", "p", "t", Single vowel: Vietnamese language has 12 single vowels including: "a", "ă","â", "e","ê", "i","o", "ô","ơ", "u","ư", "y".
- **Syllable**: the combining of vowels and final consonant. According to the syllable dictionary of Hoang Phe, Vietnamese language has total 158 syllables. And also according to this dictionary, the vowes do not occur consecutively more than once except "ooc" and "oong" syllable.

3.2 Preprocessing

The original text of a tweet can contain various noisy contents such as emotion symbols (e.g: ❤❤,..), hashtag symbol, link url @username, etc. Those noisy symbols can affect the performance of the system. Therefore, we try to clean up those noisy symbols.

Clean up the repeated character: many tweets have repeated character (e.g: Anh yêuuuuuuuu emmmm nhiềuuuuuuuuuu lắmmmmmmmmmm), after clean up it must become: Anh yêu em nhiều lắm (I love you so much). With Vietnamese language,

[3] http://www.speech.sri.com/projects/srilm/.

we can clean those tweets based on Vietnamese vowels and consonants. Normally, Vietnamese vowels do not appear more than twice and Vietnamese consonants just appear once.

3.3 Spelling Error Detection

To perform spelling error detection, we have synthesized and built a dictionary for all Vietnamese words. This dictionary includes more than 7,300 words. A word will be identified error if it does not appear in the dictionary. After a word was identified error, it will be analyzed to classify error and process.

Normally, Vietnamese includes two kind of errors. The first is the error caused by the typing error and the second is misspelled.

3.3.1 Typing Error

To compose Vietnamese text, there are two popular typing: Telex typing and VNI typing. Each input method will be have the combination of the letters to form the Vietnamese vowel marks. Comparison with Latin characters, Vietnamese characters have some extra vowels: â, ă, ê, ô, ơ; one more consonant: đ; Vietnamese has 5 types of mark: acute accent—"á", grave accent—"à", "hook accent"—"ả" , tilde—"ã" , heavy accents—"ạ". The combination of vowels, marks forming its own identity for Vietnamese language.

Example:

- With Telex typing, we have the combination of character to form Vietnamese vowels: aa: â, aw: ă, ee: ê; oo: ô; ow: ơ; uw: ư and we have one consonant dd: đ. For marks, we have: s: acute; f: grave accent, r: hook accent, x: tilde, j: heavy accents.
- Similarly, we have VNI typing: a6: â, a8: ă, e6: ê, o6: ô, o7: ơ, u7: ư, d9: đ. For marks we have: 1: accent, 2: grave accent, 3: hook accent, 4: tilde, 5: heavy accents.

Because the tweet is very short and the speed of typing, the typings will cause the error. Example:

- With the word "Nguyễn" can type with error "nguyeenx", "nguyênx" or "nguyeenxx" with Telex typing, and "nguye6n4", "nguyên4" or "nguye6n44" with VNI typing.
- With the word "người" can type with error "ngươif", "ngươfi", "nguowfi", "nguowif", "nguofwi", "nguofiw", "nguoifw", "nguoiwf", "nguowff" with Telex typing, and "nguwowi2", "ngươ2i", "nguo72i", "nguo7i2", "nguo27i", "nguo2i7", "nguoi27", "nguoi72" with VNI typing.

To handle this issue, we proposed a set of syllable rules to map each syllable combination with mark. For example, with syllable "an", when combined with mark we have five Vietnamese syllables: "àn", "án", "ản", "ãn", "ạn". Then the system will build the set of rules to map the correct syllable to errors, respectively as follows:

- "án": "asn", "ans", "a1n", "an1"
- "àn": "afn", "anf", "a2n", "an2"
- "ản": "arn", "anr", "a3n", "an3"
- "ãn": "axn", "anx", "a4n", "an4"
- "ạn": "ajn", "anj", "a5n", "an5"

3.3.2 Misspelled

This kind of error is popular in Vietnamese. This kind of error usually occurs due to mistakes in pronunciation. Examples of some misspellings:

- Error due to using the wrong mark: "quyển sách" (book) to "quyễn sách"
- Initial consonant error: "bóng chuyền" (volleyball) to "bóng truyền"
- End consonant errror: "bài hát" (song) to "bài hác"
- Region error: "tìm kiếm" (find) to " tìm kím"

3.4 *Normalization*

For the detected spelling error, the system first uses vocabulary structure, the set of syllable rules to normalize, then the result will be input to the next phase to measure the similarity with the word in the dictionary to find words with the highest similarity degree. In the case the result word still exists in the dictionary, the system will use n-gram to normalize the error word.

3.4.1 Two Words Similarity

In this paper, to measure the similarity of two words, we use the results in the research of Dice [5] with our improvement. To use research of Dice, we must split all characters of word to bigrams. Assume that we have two words "nguyen" và "nguye" , bigram of these two word can be represented by as follows: $bigram_{nguyn}$={ng, gu, uy, yn} và $bigram_{nguyen}$={ng, gu, uy,ye,en}.

Dice Coefficient

Dice coefficient is a statistic approach for comparing the similarity of two samples developed by Lee Raymond Dice [5]. Dice coefficient of two words w_i and w_j according to bigram can calculate by Eq. 1:

$$Dice(w_i, w_j) = \frac{2\times \mid bigram_{wi} \bigcap bigram_{wj} \mid}{\mid bigram_{wi} \mid + \mid bigram_{wj} \mid} \qquad (1)$$

Where:

- $\mid bigram_{wi} \mid$ and $\mid bigram_{wj} \mid$: total bigram of w_i and w_j
- $\mid bigram_{wi} \mid \bigcap \mid bigram_{wj} \mid$: number of bigrams appears in w_i and w_j at the same time.

If two words are the same, Dice coefficient is 1. The higher of Dice coefficient, the higher degree of similarity and vice versa.

Proposed Method to Improve Dice Coefficient

As observed from the experimental data using the Dice coefficient. We found that, the above methods will accurately with misspelled words at the end. With those misspelled words in the characters close to the last character at least we will lose the similarity of the two last gram. Especially, for word that has 3 characters, the degree of similarity is 0. For example: Dice("rất" , "rát") = 0; Dice("gân" , "gần") = 0;

From the above problem, we proposed a method to improve the coefficient of Dice. The improvement of coefficient was performed by combination the first character with the last character of two word to form a new pair of bigram. If this pair is different, the system will use the coefficients as shown in Eq. (1). In contrast, we use Eq. (2) as below:

$$iDice(w_i, w_j) = \frac{2 \times \left(\mid bigram_{wi} \bigcap bigram_{wj} \mid + 1\right)}{\mid bigram_{wi} \mid + \mid bigram_{wj} \mid + 2} \tag{2}$$

Let $\text{fbigram}_{\text{w}}$ be an additional bigram of w. Each fbigram is the pair of the first and the last character of w. We can express the formula to improving the Dice coefficient in Eq. (2) as below:

$$fDice(w_i, w_j) = \begin{cases} Dice(w_i, w_j) : \text{if fbigram}_{\text{wi}} \text{ is different from fbigram}_{\text{wj}} \\ iDice(w_i, w_j) : \text{Otherwise} \end{cases} \tag{3}$$

To illustrate for the improvement of Dice coefficient, suppose we have two words to measure the degree of similarity is "nguyen" and "nguyn" as presented in the previous section, we have $\mid bigram_{wi} \bigcap bigram_{wj} \mid = 3$. Combining the first and the last character of two word we have the new pair of diagram which has the same result "nn" . So, using the improvement of Dice coefficient we have **fDice("nguyen" ," nguyn") = 0.727**. If we use the normal coefficient of Dice we have **Dice("nguyen" ," nguyn") = 0.667**.

Table 2 shows the results of measuring the similarity of two words with the Dice Coefficient and the improvement Dice coefficient methods. With the improvement methods, the similarities are obvious improvement.

3.4.2 Two Sentences Similarity

Suppose we need to measure the similarity of two sentences $S_1 = w_1, w_2, w_3, \ldots, w_n$ and $S_2 = w'_1, w'_2, w'_3, \ldots, w'_n$. We compare the similarity of each pair words accord-

Table 2 The result of measuring the similarity of two words comparing between the normal Dice coefficient and the improvement of Dice coefficient

Error Word	Correct word	Dice	fDice
rat	rất	0	**0.333**
rat	rác	0	0
Nguễn	Nguyễn	0.667	**0.727**
Nguễn	Nguy	0.571	0.571
Tượg	Tượng	0.571	**0.667**
Tượg	Tương	0.286	**0.444**

ing to the improvement Dice coefficient. Then we compute the similarity of two sentences by Eq. (3) belows:

$$Sim(S_1, S_2) = \frac{\Sigma_{i=1}^{n} fDice(w_i, w_i')}{n} \quad (4)$$

Where:

- w_i and w_i': corresponding words of S_1 and S_2.
- n: number of words.

If two sentences are the same, the degree similarity (Sim) of two sentences is 1. The higher of Sim coefficent, the higher degree of similarity and vice versa.

4 Experiments

4.1 3-Gram Language Model and Data Using for it

In this paper, to handle misspelled and spelling error that can not normalize by Vietnamese structure, set of syllabus rules, We use 3-gram language model. This model was built from SRILM with a huge data collected from online newspapers (www.vnexpress.net, http://nld.com.vn/, http://dantri.com.vn/...). Data collected from many fields such as current events, world, law, education, science, business, sports, entertainment ... with a total of over 429,310 articles. The total amount of data about 1,045 MB. The 3-gram model was built from SRILM is about 1,460 MB. To ensure the accuracy of results, all trigrams on model from SRILM was selected if occurrence frequency of it is greater than 5 times and 3-gram model with frequency of occurrence more than 5 times is about 81 MB

Table 3 The results using improvement Dice coefficient combines with 2 set data of 3-gram model

Data set	Total error	Detected error	Correct fixed	Wrong fixed	Precision (%)
1	1,360	1,342	1,072	270	**79.88**
2	1,360	1,342	1,207	135	**89.94**

Table 4 The results using fDice and Dice with 3-gram model built from entire data

Method	Precision (%)	Recall (%)	F-Measure (%)
Dice	84.8	83.68	84.23
fDice	89.94	88.75	89.34

4.2 Experiment Results

To test our system, we use data set which randomize collect from Vietnamese tweets. The data set includes 1,360 tweets completely different.

In order to make comparisons for the impact of the data set in the language model. We ran the test two times with the language model built from two input data sets: The first set includes 130 MB randomized data from 1,045 MB data mention above and the second set includes entire 1,045 MB data. The 3-gram model with frequency of more than 5 times of the first set is about 8 MB. In this case, we use the improvement Dice coefficient to measure the similarity of two sentences (trigrams). The results of this test was show in Table 3. From the Table 3, the results of 3-gram model with data from second set achieved the higher accuracy than the results of 3-gram model with data from first set.

To evaluate the improvement Dice coefficient with normal Dice coefficient. We ran the test with 3-gram model built from entire data set (1,045 MB) using Dice and fDice to measure the similarity of two sentences. We also use three metrics Precision, Recall and Balance F-Measure to evaluate our system.

- Precision (P): number of correct fixed divided by the total of detected error.
- Recall (R): number of of correct fixed divided by the total error.
- Balance F-measure (F1): $F1 = \frac{2*P*R}{p+R}$

Combining with the results above, we have the entire results showed in Table 4.

The results in Table 4 show that our improvement Dice coefficient achieves higher performance than normal Dice coefficient.

5 Conclusion

In this paper, we present the first attempt to normalize Vietnamese tweets on Twitter. Our proposed method combines a language model with dictionaries and Vietnamese vocabulary structures. We also extended original Dice coefficient to improve performance of similarity measure between two words. To evaluate the proposed method, we build a dataset including 1,360 Vietnamese tweets. The experiments results show that our proposed method achieves relative high performance with precision approximating to 90 %, recall over 88.7 % and FMeasure over 89 %. Moreover, our improvement on measure the similarity of two words based on the Dice coefficient outperforms original Dice coefficient. We plan to collect larger datasets and build as well as test the language model with not only 3-gram but also 2-gram and 4-gram so that we can have the comparison of some datasets and models.

Acknowledgments This work was supported by the IT4Innovations Centre of Excellence project (CZ.1.05/1.1.00/02.0070), funded by the European Regional Development Fund and the national budget of the Czech Republic via the Research and Development for Innovations Operational Programme and by Project SP2015/146 "Parallel processing of Big data" 2 of the Student Grand System, VŠB—Technical University of Ostrava.

References

1. Banko M, Brill E (2001) Scaling to very very large corpora for natural language disambiguation. In: Proceedings of the 39th annual meeting on association for computational linguistics, pp 26–33
2. Carlson A, Fette I (2007) Memory-based context-sensitive spelling correction at web scale. In: Proceedings of the sixth international conference on machine learning and applications, pp 166–171
3. Choi D, Kim J et al (2014) A method for normalizing non-standard words in online social network services: a case study on twitter. In: context-aware systems and applications second international conference, ICCASA 2013, pp 359–368
4. Cotelo JM et al (2015) A modular approach for lexical normalization applied to Spanish tweets. Expert Syst Appl 42(10):4743–4754
5. Dice LR (1945) Measures of the amount of ecologic association between species. Ecology 26(3):297–302
6. Dien D (2005) Building an English—Vietnamese Bilingual Corpus. Ph.D. thesis, University of Social Sciences and Humanity of HCM City, Vietnam
7. Duan H et al (2012) A discriminative model for query spelling correction with latent structural svm. In: Proceedings of the 2012 joint conference on empirical methods in natural language processing and computational natural language learning, pp 1511–1521
8. Duy NTN et al (2004) An approach in Vietnamese spell checking. Vietnamese. Bachelor's thesis, University of Science Ho Chi Minh city
9. Golding AR, Roth D (1999) A winnow-based approach to context-sensitive spelling correction. Mach Learn 34(1–3):107–130
10. Habash N, Roth RM (2011) Using deep morphology to improve automatic error detection in arabic handwriting recognition. In: Proceedings of the 49th annual meeting of the association for computational linguistics. Human language technologies, vol 1, pp 875–884
11. Hai ND et al (1999) Syntactic parser in Vietnamese sentences and its application in spell checking. Vietnamese. Bachelor's thesis, University of Science Ho Chi Minh city
12. Han B, Baldwin T (2011) Lexical normalisation of short text messages: Makn sens a# twitter. In: Proceedings of the 49th annual meeting of the association for computational linguistics. Human language technologies, vol 1, pp 368–378
13. Han B et al (2013) Lexical normalization for social media text. ACM Trans Intell Syst Technol 4(1):621–633

14. Hassan H, Menezes A (2013) Social text normalization using contextual graph random walks. In: Proceedings of the 51st annual meeting of the association for computational linguistics, Association for computational linguistics, pp 1577–1586
15. Hassan Y et al (2014) Arabic spelling correction using supervised learning. In: Proceedings of the EMNLP 2014 workshop on Arabic natural language processing, Association for computational linguistics, pp 121–126
16. Huang Q et al (2014) Chinese spelling check system based on tri-gram model. In: Proceedings of the third CIPS-SIGHAN joint conference on Chinese language processing, pp 173–178
17. Huong NTX et al (2015) Using large n-gram for Vietnamese spell checking. In: Proceedings of sixth international conference KSE 2014, Springer International Publishing, pp 617–627
18. Li C, Liu Y (2014) Improving text normalization via unsupervised model and discriminative reranking. In: Proceedings of the ACL 2014 student research workshop, Association for Computational Linguistics, pp 86–93
19. Pennell DL, Liu Y (2014) Normalization of informal text. Comput Speech Lang 28(1):256–277
20. Phe H (2011) Syllable dictionary. Hanoi Encyclopedia Publishers, Dictionary Center
21. Quang N (2012) Language model and word segmentation in Vietnamese spell checking. Vietnamese. Bachelor's thesis, University of Engineering and Technology, Hanoi National University
22. Saloot MA et al (2014) An architecture for Malay tweet normalization. Inf Process Manage 50(5):621–633
23. Shaalan KF et al (2012) Arabic word generation and modelling for spell checking. In: Proceedings of the eight international conference on language resources and evaluation (LREC'12), European Language Resources Associations
24. Sönmez C, Ozgür A (2014) A graph-based approach for contextual text normalization. In: Conference on empirical methods in natural language processing (EMNLP), Association for Computational Linguistics, pp 313–324
25. Sproat R et al (2001) Normalization of non-standard words. Comput Speech Lang 15(3):287–333
26. Wu, S-H et al (2010) Reducing the false alarm rate of chinese character error detection and correction. In: Proceedings of CIPS-SIGHAN joint conference on chinese language processing (CLP 2010), pp 54–61
27. Yang Y, Eisenstein J (2013) A log-linear model for unsupervised text normalization. In: Proceedings of the 2013 conference on empirical methods in natural language processing, Association for Computational Linguistics, pp 61–72
28. Yeh J-F et al (2013) Chinese word spelling correction based on n-gram ranked inverted index list. In: Proceedings of the seventh IGHAN workshop on Chinese language processing (SIGHAN-7), pp 43–48

The Influence of the Rhythm with the Pitch on Melodic Segmentation

Michele Della Ventura

Abstract Music segmentation is an important issue for musical analysis in that it provides an overview of the internal structure of a composition. This paper presents a novel algorithm that discovers musical segments in symbolic musical data. The proposed algorithm considers simultaneously the two fundamental components of music, i.e. melody and rhythm and it uses the principles belonging to the information theory in order to identify melodic cells. The algorithm is tested against a small manually-annotated dataset of musical excerpts and results are analysed; it is shown that the technique is promising.

Keywords Entropy • Information theory • Music information retrieval • Music segmentation

1 Introduction

To analyze a piece of music means to divide it into simpler constituents and study their functions within the given structure [1]. The structure this process refers to may be from time to time a part of a composition or an entire composition.

When the listener perceives a piece of music he automatically groups together similar elements [2, 3]: the listener unconsciously performs a segmentation of the piece of music deriving from the fact that, in a melodic context, the notes build motifs, the motifs form the phrases and the phrases develop into the sections that make up a piece of music [4]. The listener, faced with the flow of information contained in the souds of the piece of music, has in a certain way analyzed the piece itself. The analysis will be different in every single listening session and for every single listener [5] and so it becomes a tool that allows filtering the specific pieces of information [1].

M. Della Ventura (✉)
Music Academy "Studio Musica", Department of Technology, Treviso, Italy
e-mail: michele.dellaventura@tin.it

A. Abraham et al. (eds.), *Intelligent Data Analysis and Applications*,
Advances in Intelligent Systems and Computing 370,
DOI 10.1007/978-3-319-21206-7_17

The analysis of a piece of music is fundamental for the musician, inasmuch as it helps him understand the structure and therefore it gives him the necessary information for his interpretation and execution. This article presents an algorithm for the segmentation of a piece of music in text format based on the concept of *Information*. Starting from a MIDI text file, the algorithm identifies the music segments on the basis of simple cognitive/perceptual principles (as for instance repetitivity) and of typical principles in the composition (as for instance the transposition, the inversion, the retrogradation or the variation).

Repetition is considered a fundamental property of music: "Music-making is, to a large degree, the manipulation of structural elements through the use of repetition and change" [6]. Introducing changes to repeated patterns leads to what is called the variation principle in music [7], an important principle involved in the human segmentation of music as illustrated by Lehrdahl and Jackendoff [8].

The segments identified by the algorithm are therefore classified on a level of importance, on the basis of the information that every single one of them carries: the information is quantified by calculating the entropy of every single segment and the lower this value is, the more the information carried by the segment will be.

This paper is structured as follows. We start by reviewing background and related work in Sect. 2. The information theory is described in Sect. 3. The concept of alphabet is described in Sect. 4. Experiments results are shown in Sect. 5. Finally, Sect. 6 concludes this paper and points out future works.

2 Background and Related Work

As opposed to other areas, music is a research area which has been yet little explored as far as Artificial intelligence is concerned. Several studies have been carried out on computer-aided musical analysis or the processing of already-written music texts.

A first attempt is the work of F. Lonati [9], further developed by S. Guagnini [10], who, starting from the typical principles of composition (transposizion, inversion, retrogradation), identifies the music segment of a given piece of music. A major limitation of the algorithm is the fact that the classification of the segments is performed by comparing every fragment found to a specifically-compiled list of rules: rules that may work well for a given historical period but not for a different one.

Meredith et al. [11] present an algorithm for discovering repeated patterns in multidimensional representations of polyphonic music. The proposed algorithm computes all the maximal repeated patterns in a multidimensional data set. The authors maintain that maximal repeated patterns tend to be musically important; however, they acknowledge that the algorithm discovers too many such patterns and that mechanisms for selecting a smaller set of salient patterns is necessary.

A computational model for melodic parallelism that affects the determination of metrical structure is introduced by Temperley and Bartlette [12]. This model calculates the "goodness" of beat intervals in terms of parallelism; this goodness value contributes, via the parallelism rule, to finding a preferred metrical structure. The model calculates "parallelism" values for all the possible pitch interval pairs in a melody (adjacent or further apart); these values depend primarily on whether the intervals are the same (diatonic intervals) or have the same or a different contour. From these values, parallelism scores are computed for beat pairs that reflect the extent to which events in the vicinity of the first beat are "paralleled" in the vicinity of the second beat; these scores are used in the parallelism rule of the metrical structure preference rule system. It should be noted that the proposed model does not explicitly identify patterns; neither does it provide a segmentation of the melodic surface.

Ferrand and Nelson [13] propose a memory-based model for melodic segmentation. Different classes of Markov models are used for acquiring melodic regularities and for determining the probabilities of sequences of symbols. The main assumption is that "segmentation boundaries are likely to occur close to accentuated changes in entropy", that is, points in the melody where the predictability associated with the occurrence of a musical event changes abruptly from low to high or from high to low (these points tend to coincide with the limits of recurring patterns).

Another interesting entry is the work of E. Cambouropoulos [14]: following the assumption that the beginning and ending points of "significant" repeating musical patterns influence the segmentation of a musical surface, the discovered patterns are used as a means to determine probable segmentation points of the melody. "Significant" patterns are defined primarily in terms of frequency of occurrence and length of pattern.

This article will present an algorithm that, inspired by the previous ones has the objective of segmenting a piece of music, identifying and classifiying the music segments on the basis of the entropy value of every one of them. The novelty of the algorithm concerns the alphabet (cfr. Paragraph 3) used for the calculation of the entropy, that considers simultaneously the distance between two consecutive sounds and the time duration passing between the two sounds.

3 Information Theory

The analysis based on the information theory sees music as a linear process supported by its own syntax [15]. However, it is not a syntax formulated on the basis of grammar rules, but rather based on the occurrence probability of every single element of the musical message compared to their preceding element. It follows that a music segment may be considered a message transmitted by a sender to a receiver and therefore subject to the concept of information transmission: there is information transmission when the prevision is disregarded, there is no information transmission when it is confirmed. In other words, the higher the unpredictability degree of the content of the specific message, the higher the amount of information contained in the message [17]. Given a set of objects (or symbols).

$$X = \{\varepsilon_1, \varepsilon_2, \ldots, \varepsilon_c\}$$

having cardinality *C*, i.e. containing *C* elements, the sequence of random variables

$$x(0), x(1), x(2), \ldots, x(\mathrm{n})$$

defines a message or sequence of symbols taken from the alphabet *X*.

In a communication which takes place by way of a given alphabet of symbols, information is associated to every single transmitted symbol [16]. Therefore, information can be defined as *the reduction of uncertainty that would have been obtained* a priori *on the transmitted symbol.*

The first problem of the information theory is how to define, and therefore measure, the amount of information emitted by a source, also known as the entropy of the source.

By imposing two—intuitively desirable—simple conditions, namely that the less the message is expected (i.e. likely), the higher the amount of information carried by a message must be and that the information of a couple of independent messages must be the sum of the respective amounts of information, Shannon defines the information of a specific message x_i having the probability $P(x_i)$ [16] as the non-negative amount

$$I(x_i) = \log_2 \frac{1}{P(x_i)} \tag{1}$$

Given that we want to define the information or entropy of the source, we must take into account the average value, on the whole alphabet, of $I(x_i)$:

$$H(X) = E[I(x_i)] = \sum_{i=1}^{n} I(x_i) \bullet P(x_i) = \sum_{i=1}^{n} P(x_i) \bullet \log_2 \frac{1}{P(x_i)} \tag{2}$$

4 Analysis of the Musical Message

From the definition of the message and of how to quantify the information of the message itself (according to paragraph 3), there follows, in the case of a musical message, the necessity to define first and foremost the alphabet (i.e. the set of symbols) that will be used in order to calculate its entropy.

The musical score is not only a sequence of sounds that make up a (musical) message. The very same sounds have a well-defined duration and they give birth to a sequence of harmonies (Fig. 1).

Fig. 1 Excerpt from the score of "Bolero" by Ravel. The initial notes of the theme, with no indication of rhythm, are represented on the first staff (**a**); the second staff (**b**) contains the same notes with the rhythm assigned by the composer; the third system (**c**) displays the initial notes with the instrumental accompaniment and then the indication of the harmonic sequences of Tonic (T) and Dominant (D) (The sounds of the musical scale acquire a name based on the degree of the scale that they represent. Therefore there will be: Tonic (first degree), Supertonic (second degree), Modal (Third degree), Subdominant (fourth degree), Dominant (fifth degree), Superdominant (sixth degree), Sensitive (seventh degree).)

Hence three possible alphabets can be identified:

(1) **the melodic alphabet**, defined only by the sequence of sounds (Fig. 1a);
(2) **the rhythmic alphabet**, defined only by the rhythmic sequence (Fig. 1b);
(3) **the harmonic alphabet**, defined only by the sequence of harmonies (Fig. 1c).

In the first case, the alphabet is represented by the intervals between the various sounds, i.e. the distance separating a sound from the other. The classification of an interval consists in the denomination (generic indication) and in the qualification (specific indication) [18]. The denomination corresponds to the number of degrees that the interval includes, calculated from the low one to the high one; it may be of a 2nd, a 3rd, 4th, 5th, and so on.; the qualification is deduced from the number of tones and semi-tones that the interval contains; it may be: perfect (G), major (M), minor (m), augmented (A), diminished (d), more than augmented (A+), more than diminished (d+), exceeding (E), deficient (df).

In the second case, the alphabet is represented by the time duration between one sound and another sound. The sound duration will not be expressed in seconds but calculated (automatically by the algorithm) on the basis of the musical sign (be it sound or rest) with the smallest duration existing in the musical piece [19]. The duration of every single sign will therefore be a (integer) number directly proportional to the smallest duration. The rhythmic analysis imposes the necessity to rigorously consider the rests as well as elements of the musical continuum. In order to distinguish between sound and rest, the latter is associated to a duration value directly proportional to the smallest duration, but negative.

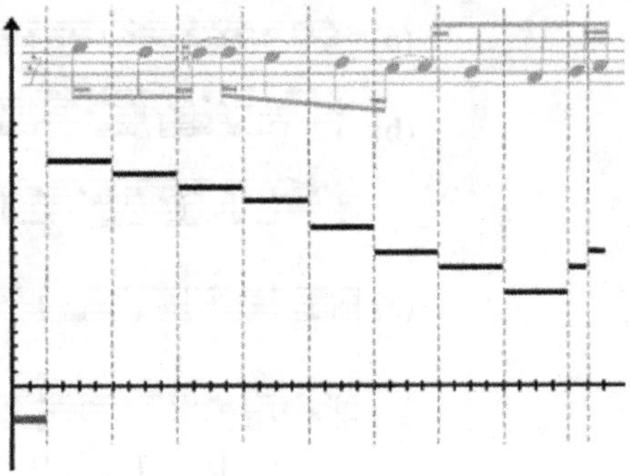

Fig. 2 J.S. Bach's Two Voice Invention in E Major BWV 777 (the first three beats of the first staff)

In the example shown in Fig. 2, the smallest duration sign is represented by the thirty-second note to which the value 1 is associated (automatically): it follows that the sixteenth note shall have the value 2, the eighth note the value 4 and so on.

In the third case, the alphabet is represented by the passage between different harmonies. For the determination of the harmony it is necessary to consider the simultaneity of two or more sounds at a distance of a third interval one from the other. Based on the degree on which the chord develops, it is possible to classify the harmony as a function of the Tonic (T), the Subdominant (S) or Dominant (D). The explanation was limited to these few notions because harmony is a field of particular and complex analysis, in addition to the fact that, as shall be seen below this paragraph, it does not influence the analysis of a musical composition (except for the executive interpretation level that does not concern, in any way, this study). Yet, these definitions deserve a more important observation deriving from the composition grammar and, therefore, from the theory of musical analysis. Musical grammar, nevertheless, provides the composer with a series of tools allowing him to vary, within the same musical piece, an already presented melodic line, by inserting notes which are extraneous to harmony [18]. The sounds of a melodic line, in fact, may belong to the harmonic construction or may be extraneous to it. The former sounds, which fall in the chordal components, are called real, while the latter sounds, which belong to the horizontal dimension, take the name of melodic figurations (passing tones, turns or escape tones). They are complementary additional elements of the basic melodic material that lean directly or indirectly on real notes and also resolve on them.

Based on the above, the score may be simplified for the identification of the harmonies, eliminating from it the melodic figurations (Fig. 3).

The melody assumes, at the same time, another nature, determined not only by the absence of some sounds but also by the change of rhythm of the main sounds, i.e. of the sounds that belong to the harmonic structure. The latter, though, does not change.

The melodic figurations, in the case of analysis of a musical composition, are however an element to be taken into consideration, first of all because they are written by the composer and then because they are meant to define, on the melodic

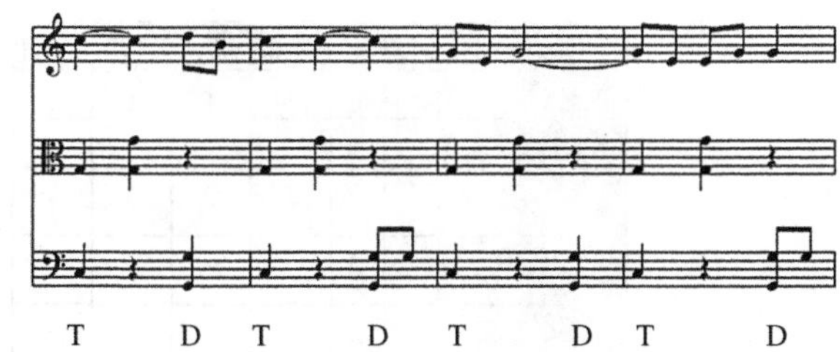

Fig. 3 The melodic line of Fig. 1c without melodic figurations

and/or rhythmic level, the pre-thematic elements of the same composition: elements that characterize the whole musical composition. In this case it's not the melodic figurations that are pre-thematic elements, but they contribute to their definition.

In the case of Ravel's Bolero, one of the pre-thematic elements originated from the melodic figurations is represented by a rhythmic structure of the 𝅘𝅥𝅮𝅘𝅥𝅯 type.

Even if the melodic figurations are eliminated, the internal harmonic sequences of the composition do not change (see Figs. 1 and 2). Based on this important observation, it is possible to define the alphabet for the quantification of the information of the musical message: to this end, the classification of the intervals is contemplated, i.e. the distance separating two consecutive sounds, together with the time duration separating these two sounds.

For every single musical piece, or part of it, a table representing its own alphabet is compiled, considering also for every interval its ascending or descending evolution (Table 1).

Determining an alphabet is not enough to calculate the entropy of a musical segment: the way in which the intervals (according to the Table 1) follow one another within the musical piece must be taken into consideration. To do so the Markov process (or Markov stochastic process) is used: the choice was made to acquire the probability of transition that determines the passage from one state of the system to the next uniquely from the immediately preceding state (Table 2).

Table 1 Example of alphabet

Interval	Direction	Semi-tones	Duration	Number
2° m	Ascending	1	𝅘𝅥𝅯	
2° m	Ascending	1	𝅘𝅥𝅮	
2° m	Ascending	1	𝅘𝅥	
..	..	..	..	
2° m	Descending	−1	𝅘𝅥𝅯	
2° m	Descending	−1	𝅘𝅥𝅮	
2° m	Descending	−1	𝅘𝅥	
..	..	..	..	

Table 2 Example of a matrix of transitions

				2° m	2° m	2° m	2° m
				1	1	1	1
				a	a	a	a
				𝅘𝅥𝅯	𝅘𝅥𝅮	𝅘𝅥	𝅗𝅥
2° m	1	a	𝅘𝅥𝅯				
2° m	1	a	𝅘𝅥𝅮				
2° m	1	a	𝅘𝅥				
2° m	1	a	𝅗𝅥				

5 The Obtained Results

The model of analysis proposed in this article was verified by realizing an algorithm. The algorithm does not provide any limitation with respect to the dimensions of the table representing the alphabet and the matrix of transitions that will be automatically dimensioned in every single analysis on the basis of the characteristics of the analyzed musical piece. This allows conferring generality to the algorithm and specificity to every single analysis (Strength).

The algorithm was tested initially by a set of score extracts for pianoforte, selected so as to belong to different authors of different historical periods: Two Voice Inventions by Bach, sonatas by Mozart, Bagattelles by Beethoven, Moments musicaux by Schubert. The choice of performing a segmentation of score extracts was determined by the will to verify the results, first of all considering as an alphabet the single interval structure and then the interval structure together with the duration (i.e. the method proposed in this article). In both cases, for every composition, the elaboration time, the number of segments identified, the length (in notes) of every segment and the score sections represented by every segment were noted. It could be immediately noticed that the results were clearly different not only in the number of segments and in the length of every single one of them, but, most of all, in the sections of the score they represented.

For illustrative purposes, Table 3 below renders the results of the segmentation of Bach's Two Voice Invention BWV 772.

In the first case, the segment having the most information content identified corresponds to a semi phrase that contains beat 1 and the first sixteenth of beat 2 (Fig. 4). It is composed of two riffs. The first riff ends on the first octave of the third movement the second riff, of a contrasting nature, begins on the note on which the first ends, connection through elision, and it ends with a metrically accented cadence on the sixteenth of beat 2.

In the second case, the first segment corresponds to the first riff in Fig. 4 that characterizes the generating cell of the entire composition. The whole composition is built on this riff (segment), presented in different ways: transposed, inverted, retrograded and even inverted of the retrograde (Fig. 5).

Table 3 This table shows the results of the segmentation considering as alphabet only the intervals; and the intervals and the duration

Starting position segment	Segment length	Entropy value
1	11	0.000064598701357
25	17	0.000000229697945
26	15	0.000006451078562
27	14	0.000030926562129
28	12	0.000058639538319
29	10	0.000691266144383
30	8	0.0019532055046
31	6	0.018549082726523
3	3	0.129519022442096
8	3	0.129519022442096
1	8	0.005310311142706
3	3	0.020956548255987

Fig. 4 Excerpt of the score of Bach's BWV 772 Invention

Fig. 5 Excerpt from J. Brahms's Intermezzo op. 119 n° 3

In a second moment, entire musical compositions were put through segmentation, using exclusively the method proposed in this article.

Initially the same compositions of the first tested were put through analysis and then compositions in polyphonic and orchestral style were added.

The solidity of the algorithm was tested by an analysis of J. Brahms's intermezzo op. 119 n° 3. It is a composition the writing of which is all chord progression: the melody perceived is in the middle of the polyphony, almost concealed.

In this case as well, the results satisfied the expectations: the first segment identified corresponds to the main melody.

6 Conclusions and Future Work

This article introduces the notion of melody, i.e. a sequence of sounds of which every single one has its own duration: melody develops on the rhythm and without it, melody does not exist. A new algorithm has been proposed that discovers melodic segments in symbolic musical data on the basis of *the information* that every single one of them carries. The algorithm has been tested against a handful of musical works and the results have been analysed. The method proposed is new, in that it identifies the important musical segments in a certain composition considering the sounds and the duration of every one of them (the rhythm) joined together, based on a principle intended by the composer: a principle that governs the entire musical composition and defines "the alphabet" of the musical composition, therefore the information concept.

The results are satisfying because they reflect the expectations of a theoretical analysis of a musical composition. Nonetheless, it is appropriate to specify that this type of analysis requires an average elaboration time approximately 15 times longer than the average time employed for an analysis that only takes into consideration the interval structure.

However, further studies are needed in order to clarify the concept: which would also allow approaching styles that are popular nowadays, as the Jazz and the Blues where there is frequent use of swing and of rubatos.

References

1. Bent I, Drabkin W (1990) Music analysis. EDT, Turin
2. Lerdhal F, Jackendoff R (1982) A grammatical parallel between music and language. Plenum Press, New York
3. Carmeci FA (2004) Processi cognitivi nell'ascolto della musica, Psychofenia, vol. VII, no 11
4. de la Motte D (2010) The melody. Astrolabio, Rome
5. Della Ventura M (2011), Analysis of algorithms' implementation for melodical operators in symbolical tectual segmentation and connected evaluation of musical entropy. In: Proceedings 1st Models and Methods in Applied Sciences, Drobeta Turnu Severin, pp 66–73
6. Burns G (1987) A typology of 'hooks' in popular records. Popular Music 6(1):1–20
7. Volk A, de Haas WB, van Kranenburg P (2012) Towards modelling variation in music as foundation for similarity. In: Proceedings ICMPC, pp 1085–1094
8. Lerdahl F, Jackendoff R (1996) A generative theory of tonal music, MIT Press, Cambridge
9. Lonati F (1990) Metodi, algoritmi e loro implementazione per la segmentazione automatica di partiture musicali, tesi di Laurea in Scienze dell'informazione. Università degli studi, Milano
10. Guagnini S (1998) Metodi e strumenti per la segmentazione automatica di partiture musicali rappresentate in NIFF, tesi di Laurea in Scienze dell'informazione. Università degli studi, Milano
11. Meredith D, Lemstrom K, Wiggins GA (2002) Algorithms for discovering repeated patterns in multidimensional representations of polyphonic music. J New Music Res 31:321–345
12. Temperley D, Bartlette C (2002) Parallelism as a factor in metrical structure. Music Percept 20:117–149

13. Ferrand M, Nelson P (2003) Unsupervised learning of melodic segmentation: a memory-based approach. In: Proceedings of the 5th triennial ESCOM conference. Hanover, Germany, pp 141–144
14. Cambouropoulos E (2006) Musical parallelism and melodic Segmentation: a computational approach. Music Percept 23(3):249–269
15. Bent I, Drabkin W (1990) Analisi musicale. EDT, Turin
16. Weaver W, Shannon C (1964) The mathematical theory of information. Illinois Press, Urbana
17. MacKay DJC (2003) Information theory, inference, and learning algorithms. Cambridge Universitry Press, Cambridge
18. Coltro B (1979) Lessons of complementary harmony. Zanibon, Padua
19. Della Ventura M (2012) Rhythm analysis of the "sonorous continuum" and conjoint evaluation of the musical entropy. In: Proceedings latest advances in acoustics and music. Iasi, pp 16–21

13. Ferrand M, Nelson P (2003) Unsupervised learning of melodic segmentation: a memory-based approach. In: Proceedings of the 5th triennial ESCOM conference, Hanover, Germany, pp 141–144
14. Cambouropoulos E (2006) Musical parallelism and melodic segmentation: a computational approach. Music Percept 23(3):249–269
15. Bent I, Drabkin W (1990) Analisi musicale. EDT, Torino
16. Weaver W, Shannon C (1964) The mathematical theory of information. Illinois Press, Urbana
17. MacKay DJC (2003) Information theory, inference, and learning algorithms. Cambridge University Press, Cambridge
18. Coltro B (1979) Lessons of complementary harmony. Zanibon, Padua
19. Della Ventura M (2012) Rhythm analysis of the "sonorous continuum" and conjoint evaluation of the musical entropy. In: Proceedings of latest advances in acoustics and music, Iasi, pp 16–21

PERSONALIZED Source Selection Process: A Social Profile Adaptation Technique

Zakaria Saoud and Samir Kechid

Abstract The previous distributed information retrieval research based on the textual information defined new measures to improve the different process in the distributed information system, but neglected the use of the social information. From this point, we propose an approach which exploits the different social entities, to make a new profile adaptation technique, and to personalize the source selection process in a distributed information retrieval system.

1 Introduction

Centralized Information retrieval is limited by its ability to search through a large quantity of information in the web. Distributed Information Retrieval appeared to remove the restrictions imposed on centralized Information Retrieval. Unlike to the centralized information retrieval, the big potential, in which the distributed information retrieval can achieve it, is the possibility of getting the information from different sources [16]. The broker is the basis and in a distributed information retrieval system, and the mediator between the user and the server. The broker is the responsible of selecting the relevant source according the user's query, and the responsible of merging the returned documents. The source selection process and the merging process represent the most established research area. In this paper, we are interested in the source selection problems. Personalized information retrieval

The erratum of this chapter can be found under DOI 10.1007/978-3-319-21206-7_48

Z. Saoud (✉) · S. Kechid
Computer Sciences Department University of Sciences and Technologies Houari Boumediene, BP 32 EL ALIA Bab Ezzouar, 16111 Algiers, Algeria
e-mail: zakaria.saoud@live.fr

S. Kechid
e-mail: skechid@usthb.dz

A. Abraham et al. (eds.), *Intelligent Data Analysis and Applications*,
Advances in Intelligent Systems and Computing 370,
DOI 10.1007/978-3-319-21206-7_18

aim to filter the results and return the relevant documents which satisfy the users' needs, through a use of users' profiles, which can be constructed using many kinds of personal data, such as search engine history [15], user manually selected interests [14], etc. With the fast web evolution, social networks have become a popular environment for internet users, to communicate, exchange ideas, and share their digital content with other users [6]. Social tagging systems are kind of social networks, which allow users to provide annotations (tags) to different types of resources, such as images, videos, and bookmarks, to give a topical labeling to resources or documents. Several social bookmarking services, such as Flickr[1] and Delicious,[2] are considered an online folksonomy services, and their social tagging data, also known as folksonomies [2]. The set of tags can be used as a source of personal data to build the user profile. In this work we propose new personalized information retrieval approach, which exploit the social annotations, to construct a social profile, and to make a new profile adaptation technique, in order to improve the source selection process. The rest of the paper is organized as follows: Sect. 2 provides an overview of the related work. Section 3 presents our personalization approach, and we describe the experimental setup and results in Sect. 4. Finally, we conclude the paper and list some future work in Sect. 5.

2 Related Work

2.1 Folksonomy

Social tagging systems are a kind of social network, which allows users to add, edit, and share bookmarks of web documents. Users can annotate their bookmarks with a set of keywords, called tags. The collection of a user's tags constitutes their personomy, and the collection of all users' personomy constitutes the folksonomy. A folksonomy is a tuple $\mathbb{F} := (U, T, D, Y)$ where $U = \{u_1, \ldots, u_M\}$ is the set of users, $T = \{t_1, \ldots, t_L\}$ is the set of tags, and $D = \{d_1, \ldots, d_N\}$ is the set of resources or web documents, and $Y \subseteq U \times T \times D$ a set of annotations [2, 7]. In our study, the different web pages represent the elements of D. Each web page is identified by a URL, and each user is identified by a user ID.

2.2 Social Information Retrieval

In social information retrieval, many studies have proposed in the context of search personalization. Most of these studies are based on the folksonomy structure. Zhou et al. [18] analyzed the social tagging system, and proposed a User Recommendation

[1]Flickr—Photo sharing, http://www.flickr.com/.

[2]Delicious—Social bookmarking, http://delicious.com/.

(UserRec) framework for user interest modeling and interest-based user recommendation, in order to provide personalized service for each user, and to boost information sharing among users with similar interests. Michlmayr and Cayzer [19] developed a new technique for construction a user profile from the set of tags, taking into account the structural and the temporal nature of tagging data. The user profile can be used to personalize the search in the web, and to access to user's bookmark collection. Bender et al. [1] exploited the different entities of social networks (users, documents, tags) and social relations between these entities, to apply a query expansion by adding the similar tags to the query keywords, and to make a social expansion to give an advantage to documents tagged by the user's close friends. Noll et al. [2] proposed a graph-based algorithm, called SPEAR (SPamming-resistant Expertise Analysis and Ranking). The aim of this algorithm is ranking the expertise of a user in a collaborative tagging system, to identify the experts and discover the relevant resources through them. Jeon et al. [3] developed an approach called collaborative filtering. They use the links and similarities between the user profiles in the filtering algorithm results, to enlarge the coverage of research using similar profiles. Yang and Chen [20] proposed new technique which exploit the folksonomy structure, using a profile expansion mechanism, in order to improve the effectiveness of personalized recommendation. The proposed technique combines the collaborative filtering algorithm and query expansion mechanism. Vallet et al. [7] presented a personalization model that exploits folksonomy structure. They developed two measures to calculate the relevance between a user profile and a document to re-rank the list of results returned by a search engine.

2.3 Source Selection Approach

Source selection is a decisive step in the metasearching process [9]; it aims to reduce the number of selected sources for a given query, in order to not lose the search time when there are many sources of information, by selecting only the relevant sources to the user's query. Several source selection methods have been developed, and they can be classified into two main categories: manual selection methods, and automatic selection methods. In automatic selection, several approaches have been defined. Callan et al. [17] proposed an approach called CORI (Collection Retrieval Inference network), which consider a collection as a meta-documents, and the selection is made according the similarity between the user query and the source. Kechid and Drias [4, 5] calculated a score for each source. This score combines three measures: 1—The source similarity according to the user interest, 2—The source similarity according to the user query, and 3—The accuracy degree between the source features and the user preferences. Arguello et al. [10] presented a source selection approach that combines multiple sources of evidence to inform the selection decision, and they derive evidence from three different sources: collection documents, the topic of the query, and query click-through data. Si and Callan [8] proposed an approach called UUM (Unified

Utility Maximization Framework for Resource Selection), where they are based on the estimation of the size of the source, to estimate the number of relevant documents can be contained in the source. The estimated number of relevant documents used for the selection of sources. Hong et al. [11] proposed a novel probabilistic model for resource selection process, through combining the evidence of individual sources and the relationship between the sources, to estimates the probability of relevance of information sources.

3 Social Personalization Approach

The previous distributed information retrieval approaches based on the textual information, to define new measures for the personalized search, however they neglect the use of the social information, and they didn't exploit the social information to improve the different process in distributed information retrieval. Hence we decided to define a social profile and exploiting it for personalizing and improving the source selection process in distributed information retrieval.

3.1 User Profile Definition

The user profile is defined by the user's set of tags; as follow:

$$\text{profile}(\text{u}_\text{m}) = \{(t_i, tf_{u_m}(t_i)) | i \in [1..L]\}$$

where:

L: is the number of tags used by the user u_m.
$tf_{u_m}(t_i)$: is the User-based tag frequency, which means how many times the user u_m use the tag t_i. The user profile in our approach is used to select the most suitable sources according to the user profile.

3.2 Document Profile Definition

We use the social annotations (set of tags) to define the document profile, as follows:

$$\text{profile}(\text{d}_\text{n}) = \text{tags}_{\text{d}_\text{n}}$$

where:

$tags_{d_n}$: represents the set of tags used to annotate the document d_n, which is defined as follows:

$$tags_{d_n} = \{(tag_i, tf_{d_n}(tag_i)) | i \in [1..P]\}$$

With:

P: is the number of tags used to annotate the document d_n.
$tf_{d_n}(tag_i)$: is the Document-based tag frequency, which means how many times the document d_n tagged by the tag tag_i.

3.3 User Profile Adaptation

For our profile adaptation, we use the profiles of the query initiator's friends, to find the lists of similar tags for each query term. To find these lists of similar tags, we use the tags expansion measure inspired from the work of Bender et al. [1]. To get the adapted user profile, we make the union between the original user profile and the similar tags lists as follows:

$$profile_{adapted}(u_m, q) = profile(u_m) \cup \bigcup_{t_i \in q} sim(t_i)$$

where:

q: is the current query given by the user u_m.
$sim(t_i)$: refers to the list of tags that are similar to the query term t_i, which belongs to the list of bookmarks of the user (the query initiator) and the user friends. This list is generated by the following proposed formula:

$$sim(t_i) = \{\acute{t} \,|\, TagSim(\acute{t}, t_i) > 0 \text{ and } \acute{t} \in bookmarks(friends(u_m), u_m)\}$$

For example the user "147" send a query which contains the term "learning", and we want to find the similar tags list of this term. To realize that, we browse the list of documents tagged by the close friends of the user "147", in order to collect the similar tags list of the term "learning". To calculate the similarity between two tags $\acute{t}, t_i$, we use the Dice coefficient measure defined as follows:

$$TagSim(\acute{t}, t_i) = \frac{2 \times df_{\acute{t}, t_i}}{df_{\acute{t}} + df_{t_i}}$$

where:

- df_{t,t_i}: is the number of documents which belongs to the list friends(u_m), and that have been tagged by both tags t and t_i.
- df_t, df_{t_i}: are the number of documents which belongs to the list friends(u_m), and that have been tagged with t and t_i, respectively.

3.4 Source Selection Process

Source selection process aims to select the most relevant sources for a given user's query [13]. In our approach we aim to integrate a social profile in the source selection step, and for that we define a score $\text{ScoreSource}_s(u_m, q)$ for each source *S* associated to the user u_m, and based on this score, we select and sort the relevant sources for each user.

The score $\text{ScoreSource}_s(u_m, q)$ is calculated by combination of both measures $\text{SimSource}_s^{\text{Terms}}(q)$ and $\text{SimSource}_s^{\text{Tags}}(u_m, q)$ as follows:

$$\text{ScoreSource}_s(u_m, q) = (1 - \alpha).\text{SimSource}_s^{\text{Terms}}(q) + \alpha.\text{SimSource}_s^{\text{Tags}}(u_m, q)$$

where: $\alpha \in [0, 1]$

The parameter α is used to control the influence of the two measures of similarity $\text{SimSource}_s^{\text{Terms}}(q)$ and $\text{SimSource}_s^{\text{Tags}}(u_m, q)$. For example if α is high, the selected sources will have a high degree of similarity with the adapted social profile of the query initiator u_m and if $\alpha = 0$ these sources will be selected according the degree of similarity with the user's query. The other utility of the parameter α is to normalize the global score $\text{ScoreSource}_s(u_m, q)$, in order to not exceed the interval [0, 1].

$\text{SimSource}_s^{\text{Terms}}(q)$: represent the degree of similarity between the source *s* and the user's query q, according to the set of terms of the source documents. This similarity is calculated as follows:

$$\text{SimSource}_s^{\text{Terms}}(q) = \frac{\sum_{i=1}^{T} t_i * q_i}{\left(\sum_{i=1}^{T} t_i^2\right)^{1/2} \left(\sum_{i=1}^{T} q_i^2\right)^{1/2}}$$

With,

s: is the source; *q* is the user's query;
t_i: is the weight of the term *i* in the source (the set of the first k documents returned by the source);
q_i: is the weight of the term *i* in the query;
T: is the number of terms used in the source. In our approach, the content of a source is represented by the set of its first k returned documents.

$SimSource_s^{Tags}(u_m, q)$: represent the degree of similarity between the source s and the query initiator u_m, According to the set of tags of the source documents. This similarity is calculated using the measure of Noll and Meinel [12] as follows:

$$SimSource_s^{Tags}(u_m, q) = \sum_{j=1}^{K} tf(u_m, d_j)$$

with K is the number of returned documents.
d_j : is the document j of the source s.
$tf(u_m, d_j)$: is the similarity measure of Noll and Meinel, which is defined as follows:

$$tf(u_m, d_j) = \sum_{\substack{i=1 \\ i \in d_j}}^{i=L} tf_{u_m}(t_i)$$

where:
$tf_{u_m}(t_i)$: is the number of times the user u_m has used the tag t_i.

4 Experiments

The evaluation of personalized information retrieval approaches difficult task, because of the absence of personalized relevance judgments. Therefore we have decided to construct a test collection using a social bookmarking dataset and a set of documents downloaded from several search engines to simulate a real distributed environment and to provide the social information.

4.1 Experimental Setup

For our experiments, we used a dataset from the del.ico.us social bookmarking system; this dataset is released in the framework of the 2nd International Workshop on Information Heterogeneity and Fusion in Recommender Systems (HetRec 2011).[3] This dataset contains social networking, bookmarking, and tagging information from sets of 1867 users from Delicious social bookmarking system. It contains 69,226 URLs (resources), 1,867 users, and 53,388 distinct tags.

To evaluate our approach in a distributed environment, we considered the 8 following search engines: GOOGLE, YAHOO, BING, BLEKKO, YANDEX, WOW, ASK, AOL. Each search engine represents a source of information, and for each source we have downloaded the top 30 retrieved documents, and each result

[3]http://grouplens.org/datasets/hetrec-2011/.

list returned by a source is used to create the source description. To examine the benefit of our approach for individual users, we allowed 12 participants to evaluate the returned documents by the selected sources, and each user ran 6 queries. In total, we have tested 72 different queries. These queries are made using the most popular tags in the dataset, in order to increase the proportion of obtaining tagged documents from the search engines to be able to apply our personalization approach, across the most returned documents. To solve the problem of subjectively assessing (relevance judgment) [1], we follow the method of Bender et al. [1], and that by selecting a fictitious profile for each query initiator. In our test, the fictitious profile is extracted from the set of profiles of the social bookmarking dataset del.ico. us, and this profile must contain the greatest sum of tags frequency of the query.

4.2 Experimental Results

4.2.1 Results of Personalization Approach

In this section, we analyze the performance of our personalization approach when only the personalization scores are used to select the different sources of information.

For each query we considered the result in various cases, by varying the parameters α and in the interval [0, 1]. We vary the parameters α to evaluate the influence of the use of the social profile, and to evaluate the importance between the two measures of the source selection score. To evaluate our selection approach without using a merging algorithm, we follow the method of Arguello et al. [10]. This method consider that the source *s* is relevant with a query q, if the source s contain more than t relevant documents which are preset in top T of the full-dataset result. The full-dataset is the set of the different documents returned by the whole sources, which indexed by a function into a centralized indexed list. In our evaluation we select between 1 and 6 sources, and we combine the returned documents by the *k* selected sources into a single list, then we select the top 30 relevant documents according to the users' judgments. We allow each participant to judges the relevance of the 1, 2, 3, 4, 5 and the 6 selected sources. A precision value is computed for each retrieval session according to the following formula:

$$\text{precision} = \frac{\textit{number of relevent documents in the k selected sources}}{\textit{number of returned documents by the k selected sources}}$$

The various cases obtained by varying the parameters $\propto$ in the interval [0, 1] are described as follows:

- case $1: \propto = 0$:
  This case means that the user profile is not used; the relevance of the result is related just to the query.

- case 2: $0 < \propto < 0.5$: $(\alpha = 0.3)$
  In this case, the relevance of the result is related to the user profile and the query, but the query is more significant than the user profile.
- case 3: $\propto = 0.5$:
  In this case, the relevance of the result is related to the user profile and the query, in an equitable way.
- case 4: $1 > \alpha > 0.5$: $(\alpha = 0.7)$
  In this case, the relevance of the result is related to the user profile and the query, but the user profile is more significant than the query.
- case 5: $\propto = 1$:
  This case means that the query is not used; the relevance of the result is related just to the user profile.

In each case we computed an average precision for the whole queries. Table 1 shows the average precision (at $P@\{1, 2, 3, 4, 5, 6\}$) values of the personalization approaches for each case.

From this table we can see that the second and the third case, give better results than the first case, which uses the query and neglects the user profile, and better results than the last case, which uses just the user profile and neglects the query. Thus we can see the advantage and the utility, when we integrate the user profile who presents the social information with the query who presents the text information. The second case gives better results than the fourth case, who gives more importance to the user profile than the query, and better results than the third case, which uses the user profile and the query in an equitable way. Therefore we can deduce that the query is more significant than the user profile for the relevance of their search results. As a result, we can say that the combination of the user profile with the query can improve the source selection process, and gives the best results of the retrieval process when we give more importance to the user query.

Table 1 Average precision values of the personalization approach

k	Case 1	Case 2	Case 3	Case 4	Case 5
1	0.133	0.230	0.230	0.133	0.166
2	0.116	0.200	0.183	0.133	0.133
3	0.155	0.188	0.177	0.177	0.155
4	0.150	0.166	0.175	0.166	0.158
5	0.140	0.160	0.153	0.146	0.146
6	0.127	0.144	0.138	0.133	0.122
Average	0.136	0.181	0.176	0.148	0.146

Table 2 Average precision values of the profile adaptation

k	Without profile adaptation	With profile adaptation
1	0.168	0.179
2	0.151	0.157
3	0.157	0.173
4	0.162	0.169
5	0.143	0.152
6	0.138	0.148
Average	0.153	0.163

4.2.2 Results of Profile Adaptation

In this section, we study the performance of the profile adaptation technique. To realize that we preferred the use of the popular tags to made the queries, and we apply our personalization approach as the previous section. For each query, we have computed the precision values for the 5 and 10 first selected sources. Table 2 shows the averages precisions obtained in each case.

From this table we can remark that the profile adaptation technique can improve the personalized approach results. We mention that the profile adaptation result can vary, depending on the user profile and their friends' profiles. For example if we want to find the similar tag of the term "android" for the user "8691", we obtain the tag "application", but if we expand the same term for the user "6585", we obtain the tag "ipod" because of the difference between the users' profiles, which forms a distinction between the users' interests, and can affect the profile adaptation results.

5 Conclusion and Future Work

Source selection process is a major problem in Distributed Information Retrieval. Our approach consists of using a social profile, to personalize and improve the retrieval process in distributed information retrieval, based on the folksonomy structure. The results obtained shows that the integration of the social profile, in source selection process, improved the relevance of the distributed information retrieval system. In addition, the second evaluation indicates that the profile adaptation technique gave good results for the source selection process. In our future work we plan to integrate the social profile to construct the source descriptions, and to make the search more specific. We would like also to evaluate our approach using a large dataset to obtain reliable results.

References

1. Bender M, Crecelius T, Kacimi M, Michel S, Neumann T, Parreira JX, Schenkel R, Weikum G (2008) Exploiting social relations for query expansion and result ranking. In: ICDE workshops. IEEE Computer Society. pp 501–506
2. Noll MG, Au Yeung CM, Gibbins N, Meinel C, Shadbolt N (2009) Telling experts from spammers: expertise ranking in folksonomies. In: Proceedings of the 32nd international ACM SIGIR conference on research and development in information retrieval. ACM, pp 612–619
3. Jeon H, Kim T, Choi J (2010) Personalized information retrieval by using adaptive user profiling and collaborative filtering. AISS 2(4):134–142
4. Kechid S, Drias H (2009) Personalizing the source selection and the result merging process. Int J Artif Intell Tools 18(2):331–354
5. Kechid S, Drias H (2010) 'Personalised distributed information retrieval-based. Int J Intell Syst Technol Appl 9(1):49–74
6. Chelmis C, Prasanna VK (2013) social link prediction in online social tagging systems. ACM Trans Inf Syst 31(4):20
7. Vallet D, Cantador I, Jose JM (2010) Personalizing web search with folksonomy-based user and document profiles. In: Gurrin C, He Y, Kazai G, Kruschwitz U, Little S, Roelleke T, Rüger SM, van Rijsbergen K (eds) ECIR. Springer, pp 420–431
8. Si L, Callan J (2004) Unified utility maximization framework for resource selection. In: Proceedings of the ninth nternational conference on information and knowledge managementCIKM. Washington, USA, pp 32–41
9. Ipeirotis P, Gravano L (2004) When one sample is not enough : improving text database selection using shrinkage. SIGMOD, pp 767–778
10. Arguello J, Callan J, Diaz F (2009) Classification-based resource selection. In: Proceedings of CIKM, pp 1277–1286
11. Hong D, Si L, Bracke P, Witt M, Juchcinski T (2010) A joint probabilistic classification model for resource selection. In: Proceedings of SIGIR, pp 98–105
12. Noll MG, Meinel C. (2007) Web search personalization via social bookmarking and tagging. In : Proceedings of ISWC 2007, LNCS, vol 4825. Springer, Heidelberg, pp 367–380
13. Markov I, Crestani F (2014) Theoretical, qualitative, and quantitative analyses of small-document approaches to resource selection. ACM Trans Inf Syst 32(2):9:1–9:37
14. Pazzani M, Muramatsu J, Billsus, D (1996) Syskill & webert: Identifying interesting web sites. In: Proceedings of the thirteenth national conference on artificial intelligence, AAAI'96, vol 1. AAAI Press, pp 54–61
15. Paul J, Gowan M (2003) A multiple model approach to personalised information access
16. Steidinger A (2000) Comparison of different collection fusion models in distributed information retrieval
17. Callan J, Lu Z, Croft B (1995) Searching distributed collection with inference networks. In: Eighteenth annual international ACM SIGIR conference on research and development in information retrieval. ACM-SIGIR'95, Seattle, Washington, pp 21–28
18. Zhou TC, Ma H, Lyu MR, King I (2010) UserRec: a user recommendation framework in social tagging systems. In: AAAI
19. Michlmayr E, Cayzer S (2007). Learning user profiles from tagging data and leveraging them for personal (ized) information access. In: Proceedings of the workshop on tagging and metadata for social information organization, 16th international world wide web conference (WWW2007), pp 1–7
20. Yang CS, Chen LC (2014) Personalized recommendation in social media: a profile expansion approach

The School Absenteeism Contributing Factors: Oman as a Case Study

Bader Said Rashid AL-Farsi, Ali Ahmed and Saadat AlHashmi

Abstract Student absenteeism is an acknowledged and pervasive problem in many schools in the sultanate of Oman. It is a major problem in rural areas as a result of lacking the ability to enforce the minimum age requirement in these areas. Practically, the high level of absence has a major impact on the performance of students in the long run. This challenges the researchers of this paper to investigate the root social and demographic factors for such a problem. Abdullah Abin Al-Zubir School is chosen as a case study. To identify such factors, a data mining Decision Support System (DSS) is developed to classify the most probable students to get absent from their schools from the least probable ones. Results show the most contributing factors in the absenteeism problem of Abdullah Abin Al-Zubir Schools students are age: the number of family members, the family income and mothers employment status.

Keywords Data mining · School absenteeism · Social factors · WEKA · Arab world

1 Introduction

Understanding the problem of absenteeism in the schools is a challenging task. Bond defined the absenteeism problem as the unexplained, habitual and persistent absence of students from schools [7]. Absenteeism is a multi-factor problem where reasons

B.S.R. AL-Farsi
University of Liverpool, Liverpool, UK
e-mail: bader.alfarsi@online.liverpool.ac.uk

A. Ahmed (✉)
Faculty of Computers and Information, Cairo University, Giza, Egypt
e-mail: a.ahmed@fci-cu.edu.eg

S. AlHashmi
College of Engineering, Abu Dhabi University, Abu Dhabi, United Arab of Emirates
e-mail: saadat.alhashmi@adu.ac.ae

A. Abraham et al. (eds.), *Intelligent Data Analysis and Applications*,
Advances in Intelligent Systems and Computing 370,
DOI 10.1007/978-3-319-21206-7_19

that led to the absence of students from their schools can be affected by various factors. These reasons range from inadequate community support, inappropriate school or family environment to bad weather conditions, poor health status and transportation difficulties. Students who do not attend their classes miss important information as a result of teacher interaction. This essential learning experience cannot be repeated when teachers teach the course again for absent students, [30, 31].

Students who are absent from school miss the chance to grow socially and academically, which affects their success in the school and community. Those students are at sensitive risk of high school failure [1]. Generally, absenteeism represents the undesired behaviour of students, which may arise from social, psychological and physical factors. It has a major negative effect on the students' academic success [2]. This problem perturbs the dynamic learning environment and negatively influences the whole class environment [25]. Enomoto demonstrated that students who miss their classes are probable to respond wrongly for questions related to materials, which covered without their attendance [10]. As illustrated by Williams, students who are absent from schools suffer socially and academically, where the absenteeism problem implications are felt in and out of the classroom [31]. Barker and Jansen explored that students who are absent from their schools have bad achievement [5]. The Auditor General Victoria recognizes four dimensions for the absenteeism problem; school refusal, truancy, early leaving and school withdrawal. School refusal represents the decline of students to go to their schools even with the presence of punitive and persuasion measures from both parents and schools [3]. McCluskey et al. demonstrated that the repeated unexcused absence of students is a main predictor of unwanted outcomes in teenage years, such as criminal activities, substance abuse, academic failure and gang participation [17]. Teasley introduced that transportation difficulties, family financial concerns, family health, alcohol and drug utilization, conflicting community manners and poor school weather are some of the conditions that related to the repeated absence of students from schools [27]. Ready demonstrated that the attendance is an essential concern in school success for both youth and youth, where better attendance is associated with improved academic achievement for students [23]. Gupta and Lata introduced various solutions for reducing the absenteeism problem in schools [13]. Some of those solutions are: recognizing training needs and conducting workshops, conferences and staff improvement programs for teachers, conducting conferences among school administrators and parents, studying the particular circumstances of schools before taking any plan, depending on the help of both community agencies and parents, involving all school members in programs for decreasing that problem and developing alternative effective programs for students who have problems with the inability of traditional schools to meet students needs.

2 BACKGROUND

2.1 Student Absenteeism in the Arab World

Education is becoming like democracy; everybody has the right to get it. It is a big challenge given the limited resources in many educational institutions especially in the under-developed Arab countries. This might lead to big class size. A bi-product of that is that many students lack the objective, role model, incentive, etc. As a result, a high absenteeism rate is evident in many schools indeed. This is actually because the school climate is not supportive in a way that, as highlighted by Faour, many students do not feel safe physically, socially, and emotionally in schools. Substantial percentages of teachers entered their profession with deficient academic preparation and pre-service training and do not receive adequate and appropriate professional development during service [12]. Student absenteeism is becoming a characteristic of the classroom climate of many Arab countries. Such a low attendance can affect dramatically the learning outcomes of the students indeed bearing in mind that most of the education systems in the Arab world are based on recalling of facts and procedures. When Martin et. al compared that to the international average, results show a serious problem in many Arab countries as shown in Fig. 1 [16].

There are many efforts especially in the Gulf Corporation Council countries to solve the student absenteeism problem. For example, the ministry of education in Kingdom of Saudi Arabia is keen to apply the attendance and absence regulations to absentees who cannot come up with legitimate reasons such as health conditions. Rules enforcement is an issue in most of the Arab countries. For example, in Saudi Arabia, the attendance policy is 80 %. A student can be absent of 1/5 of the semester

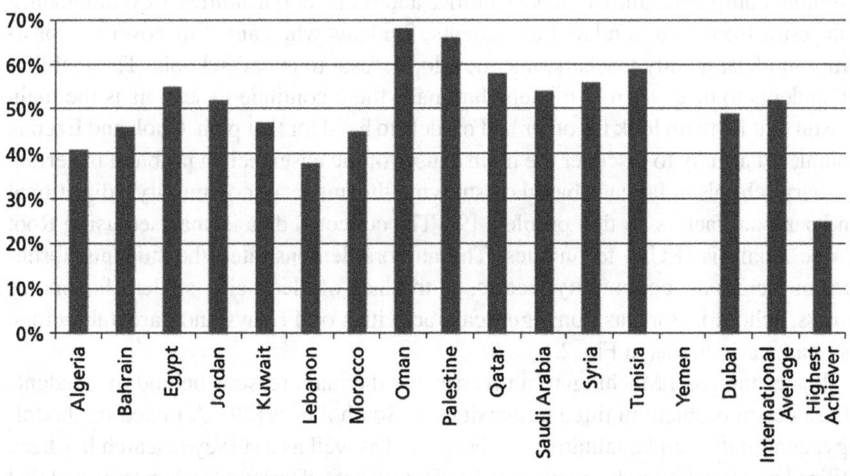

Fig. 1 Percentage of eighth grade students whose mathematics test questions are based on recall of facts and procedures always or almost always [16]

and still be able to pass. The number is quite shocking indeed if we excluded the other absent incidents with legitimate reasons that are not considered in that 20 % policy! The figure may goes down to attending only 75 % of the semester! In fact, the Saudi Arabia ministry of higher education defines the full time education status as, these are full-time students who attend lectures and practical lessons, with a record of attendance of no less than 75 % of all lectures in every course. Regular students are encouraged with a regular stipend paid by the university in return for their full-time devotion to their study. The percentage of regular students is 85 % [18]. One of the results reported by Martin et. al is that the in some Arabic countries there is not a significant effect of absenteeism on the student achievements and learning outcomes especially in the Gulf countries of Kuwait and Qatar [16]. This is basically because of the private tuition culture.

2.2 *Student Absenteeism: The Efforts*

At various schools, teachers are not able to help their students in solving their problems that makes students unwilling to attend classes. Students who cannot get effective counselling service for their psychological concerns are more probable to be absent from their schools. Reimer and Smink explored that the school, family and student factors must be studied together [24]. Authors proposed that the main reasons behind this problem are the incompatible parental discipline, academic weakness perceptions, adverse school perceptions, lack of social skills in class and family conflict. Furthermore, authors proposed that absent students have low consistency in their families and lack of both discipline and parental acceptance. The school environment has a major effect on the low attendance of students due to the lack of essential equipment, attractive possibilities and social opportunities. Low attendance can result from factors related to students. Students who cannot discover responsibility models, mainly teachers, may develop refusal to attend schools. The inability of students to discover role models that make them continue to schools is the main reason that let them look for other bad models to head for that path. Cook and Ezenne conducted a study to discover the main causes of the absenteeism problem in certain primary schools in Jamaica based on studying the impact of community, educational and personal factors on that problem [9]. The collected data is analysed using Root Cause Analysis (RCA) techniques. The authors demonstrated the students absenteeism factors as: community factors, as the lack of electricity, student factors, as illness, school factors, as non significant activities on Fridays and parental factors, as financial issues as in Fig. 2.

Wadesango and Machingambi investigated the main reasons behind the students absenteeism problem in three universities in South Africa [29]. A mixed methodology; quantitative and qualitative has been used as well as a survey research has been utilized as an operation framework for collecting data. The authors demonstrated that the main reasons behind the students absenteeism were: poor relations with lecturers lack of interested objects and having part time jobs. Suhid et al. proposed that there

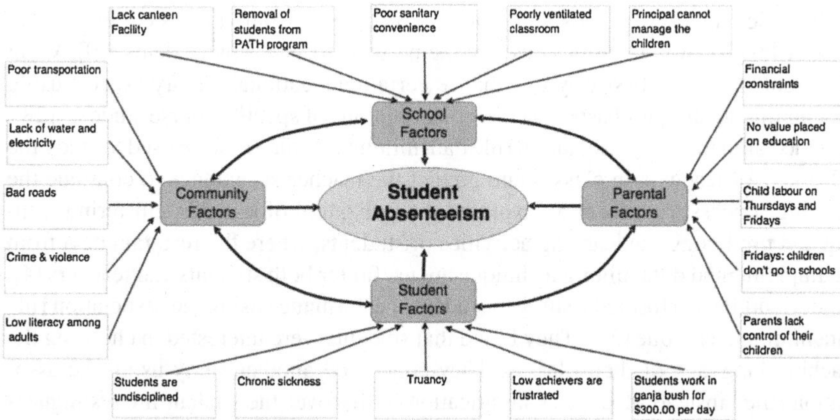

Fig. 2 Absenteeism: the causes [9]

are various reasons for staying students away from schools with no permission [26]. Some of those reasons are the impact of peers, fear of both teachers and being bullied, feeling bored with certain subjects, thinking of fail all the time, no support from parents, school factors and family problems. Habaci et al. demonstrated that families have a major effect on the low attendance of students, where students who have families with difficult time in socio-economic terms are more probable to be absent from their schools [14]. Another problem is the unawareness of parents, where parents who do not care about the attendance of their children to schools are a primary reason for that problem. In practice, there are various parents who do not send their children to schools for several reasons, such as work. Gupta and Lata discovered certain probable reasons for the absenteeism problem at schools, some of those problems are: availability of entertainment opportunities, such as computer centres and movie halls, the gap among students mental capacity and the course opted, lack of subject interest, inadequate teachers skills, adverse learning environment, lack of both allied activities and confidence, poor canteen food, poor school infrastructure facility, lack of expert teachers, performing private coaching for entrance tests, preparation for examination, huge amount of homework, spending much time on TV, transportation problems, poor student socio-economic background, social phobia, using alcohol and drug and providing desired facilities by parents [13].

2.3 Student Absenteeism: Scientific and Technological Endeavours

Erdoan and Timor identified the characteristics of Maltepe University students using the K-means algorithm and examined the relation among the entrance exam results and outcomes [11]. Pandey and Pal proposed a study concerning the performance of students depending on choosing a number of students from various colleges using Bayes Classification in order to expect the performance of new students [20]. Vrani et al. looked for the best way to enhance certain educational quality aspects using data mining techniques based on collecting opinions of specific course students [28]. Both decision tree and association rule data mining techniques were used to study the behaviour of teachers in classes and predict the teacher acceptance to continue the teaching process. Ayesha et al. explored the utilization of k-means clustering technique in predicting the learning activities of students, where the resultant data from the implemented data mining technique are useful for both students and teachers [4]. Pandey and Pal performed a study to students performance using the association rule data mining technique [21]. They found that students were interested in choosing the teaching language of their classes. They conducted another study using the association rules instead of Bayes Classification to discover the student interestingness in certain class teaching language. Bharadwaj and Pal studied the performance of students depending on choosing a group of students from several colleges of computer application using the Bayesian classification technique on 17 attributes [6]. Researchers found that certain factors, such as living location, qualification of mothers, habits, and family income were highly related to the academic performance of students. Parack et al. proposed the application of data mining for the prediction of academic patterns and trends [22]. Nasiri et al. proposed the use of data mining for predicting the academic dismissal in various learning management systems based on testing the C5.0 algorithm [19]. On the other hand, since this algorithm depends on the data distribution, any resultant variation in the data can cause various conclusions. Thus, association rules have been used to improve their work. Bunkar et al. proposed the application of data mining techniques for predicting the performance of students with the use of classification [8]. The main used techniques were Classification And Regression Trees (CART), C4.5 and ID3. Researchers were capable to recognize students who are probable to fail and give them guidance.

3 The Case Study

The question this study tries to answer are, for example, "Will family factors affect students absenteeism at Abdullah Abin Al-Zubir School?", "Will student factors affect students absenteeism at Abdullah Abin Al- Zubir School?", "Will teacher factors affect students absenteeism at Abdullah Abin Al-Zubir School?", "Will school environment factors affect students absenteeism at Abdullah Abin Al-Zubir", "Will electronic factors affect students absenteeism at Abdullah Abin Al-Zubir School?", "Will developed system classify the most probable students to get absent from their", "Will developed system produce reports about the cases of all students?", and "Will developed system help decision makers at Abdullah Abin Al-Zubir School to deal". The methodology of this study depends initially on a survey that was distributed to a random sample of schools students. After that, the collected data are analysed and classified using a WEKA program. Next, a spreadsheet oriented Decision Support

System (DSS) is developed using both the SQL and ASP.NET where its input is the result of the classified data from WEKA and its outputs are reports concerning the status of participating students. The collected data from Abdullah Abin Al-Zubir School students are classified using the decision trees. The main stages are:

1. Preparing the data to be analysed.
2. Preparing the training data, which represent the answers for the distributed questionnaire to be then used for learning the machine.
3. Choosing the appropriate algorithms to work with the data.
4. Applying the decision tree generator for the data to produce the required prediction model.
    (a) Preparing the testing data to predict answers for questions
    (b) Applying the generated model for prediction.

The questions in the questionnaire are divided into two types; multiple choice questions and personal questions. One observation from the collected data is that most of students mothers are housewives. On the other hand, other questions have neutral answers as shown next. This is due to the same culture and standards of the Omani society. Based on the obtained answers from Abdullah Abin Al-Zubir School's students, the main attributes that differ among students, thus they can be used in determining the probability of each student to get absent from his school,are: Age, number of family members, family income and mothers job. Based on the obtained results of questionnaire analysis from 3000 students, the generated association rules for this work are:

1. If the previous grade is in the range 50–60
2. If the number of family member is 9–10 or more than 10 and the family income is 300–600 Riyals, the absence is true.
3. If the age is 11–13 and the mother is not a housewife, the absence is true.
4. If the previous grade is 70–80 or 60–70 and the number of family is 3–5 or 6–8, the absence is false.
5. If the family income is 900–1200 and the previous grade is 50–60, the absence is true.

The training warehouse is a set of data that includes certain information concerning the questionnaire, where the target variable is the absence of students. This variable has two choices based on the generated association rule: true and false. A sample of the training warehouse of this study is shown in Fig. 3. The decision tree generator [15] is used to generate the decision model. Generally, it is not efficient to use all attributes in the developed tree since some rules cannot be real or logical; this is due to the big similarity among students answers to most of the questions, where those questions have been neglected. Thus, there is a need to select certain attributes. As an example, the grade and the number of family member are the most important attributes. Thus, those two attributes can be used with other attributes, such as mothers job. This is shown in the R code snippet shown in Fig. 4.

17-18	Twelfth	50 - 60 %	3- 5	Primary education	Primary education	Good
17-18	Twelfth	50 - 60 %	3- 5	Secondary educatior	Secondary educatior	Good
17-18	Twelfth	50 - 60 %	9- 10	Uneducated	Uneducated	Good
17-18	Twelfth	60 - 70 %	6- 8	Uneducated	Uneducated	Good
17-18	Twelfth	60 - 70 %	3- 5	Primary education	Primary education	ood
17-18	Twelfth	60 - 70 %	6- 8	Uneducated	Uneducated	Good
17-18	Twelfth	60 - 70 %	3- 5	Uneducated	Uneducated	Good
17-18	Twelfth	60 - 70 %	6- 8	Uneducated	Uneducated	Good
17-18	Twelfth	50 - 60 %	3- 5	Uneducated	Uneducated	Good
17-18	Twelfth	50 - 60 %	9- 10	Uneducated	Uneducated	Good
17-18	Twelfth	60 - 70 %	6- 8	Uneducated	Uneducated	Good
17-18	Twelfth	60 - 70 %	9- 10	Primary education	Primary education	Good
17-18	Twelfth	60 - 70 %	6- 8	Uneducated	Uneducated	Good

Fig. 3 Training warehouse sample

```
library(xlsx) : using the library to manage xlsx files
library(rpart) : using library to create decision tree
entrep=read.xlsx(file="C:/dataminingdecision/Prediction.xlsx",1) : read the file that contain the sample of data
ad=rpart.control (minsplit = 1) : this control is the number of lines that make a rule here 1 rule = 1 line
test=read.xlsx(file="C:/dataminingdecision/test.xlsx",1) : read the data of the test
ad.abs<- rpart (pred.classe~S_Previous_Grade+Number.of.family.members+Family.income, entrep, control=ad) : the tree will be generated we have '~' that separates the target variable from the attributes and finally we add the control of tree
```

Fig. 4 Decision tree generation: R code snippet

The obvious result of the case study is that the age, the number of family members, the family income, and the mothers employment status are the main factors contributing to the absenteeism problem. Other actors are insignificant. Factors such as the sex of the student have not been taken into consideration in this study, which we believe might be a significant factor as well. That is why this is one of the future research directions of this study.

4 Conclusion

In this study, the most effective social and demographic factors on the students absenteeism problem at Abdullah Abin Al-Zubir School have been investigated, determined and studied. This was performed based on writing a questionnaire and distributing it to 3000 students about five concerning factors (i.e. family, student, subjects and teacher, school environment, and technology). After that, the collected opinions from students were analysed using both the R and WEKA programs. The analysed data were then used in the designed data mining based spreadsheet oriented Decision Support System (DSS) using the SQL and ASP.NET to classify the most probable students to get absent from their school from the least probable ones and produce reports concerning the cases of students to help decision makers in dealing with those cases expected to get absent from the school before they occur. The analysed questionnaire results demonstrated that there are many questions answered with neutral answers by students. Thus, those questions could not be used in the designed system. Conversely, there were other questions that offer useful data to be used in determining the most probable students to get absent from their school. Those useful data are: age, number of family members, family income and mothers job. This research can be applied in other schools in the region indeed. Practically, there are some limitations concerning the academic and business application of the current work. As described above, only few questions and factors have effects on the absenteeism problem at Abdullah Abin Al-Zubir School, which makes it difficult to find important relations among data when this study is applied to other schools. Factors such as the sex of the student have not been taken into consideration in this study, which we believe might be a significant factor as well. That is why this is one of the future research directions of this study.

References

1. Allensworth EM, Easton JQ (2007) What matters for staying on track and graduating in chicago public high schools. Consortium on Chicago school research, Chicago. Retrieved 17 Dec 2007
2. Altinkurt Y (2008) Reasons for quitting and absenteeism effect of student academic achievement. J Soc Policy Sci 20:129–142
3. Australia: Managing school attendance (2004) Session 2003–04. 108, Auditor General Victoria
4. Ayesha S, Mustafa T, Sattar AR, Khan MI (2010) Data mining model for higher education system. Eur J Sci Res 43(1):24–29
5. Barker D (2000) Using graphs to reduce elementary school absenteeism. Soc Work Educ 22:46–53
6. Bhardwaj BK, Pal S (2011) Data mining: a prediction for performance improvement using classification. Int J Comput Sci Inf Secur 9(4):136–140 arXiv preprint arXiv:1201.3418
7. Bond G (2004) Tackling student absenteeism: research findings and recommendations for school and local communities. Technical report, Report written for Hume/Whittlesea LLEN and Inner Northern LLEN
8. Bunkar K, Singh U, Pandya B, Bunkar R (2012) Data mining: prediction for performance improvement of graduate students using classification. In: 2012 ninth international conference on wireless and optical communications networks (WOCN). IEEE, pp 1–5
9. Cook LD, Ezenne A (2010) Factors influencing students absenteeism in primary schools in jamaica. Editorial Committee 17(1):33–57
10. Enomoto EK (1997) Negotiating the ethics of care and justice. Educ Adm Q 33(3):351–370
11. Erdoğan ŞZ, Timor M (2005) A data mining application in a student database. J Aeronaut Space Technol 2(2):53–57
12. Faour M, Center CME (2012) The Arab world's education report card: school climate and citizenship skills. Carnegie Endowment for International Peace
13. Gupta M, Lata P (2014) Absenteeism in schools: a chronic problem in the present time. Educ Confab 3(1):11–16

14. Habaci I, Küçük S, Erken G, Çekiç O, Bilal L (2013) The most common causes of absenteeism in 6th, 7th and 8th grades in secondary education. Educ Res J 3(3):68–74
15. Maimon O, Rokach L (2005) Data mining and knowledge discovery handbook, vol 2. Springer, Berlin
16. Martin M, Mullis IVS, Foy P (2008) Timss 2007 international mathematics report. Technical report International Association for the Evaluation of Educational Achievement (IEA)
17. McCluskey CP, Bynum TS, Patchin JW (2004) Reducing chronic absenteeism: an assessment of an early truancy initiative. Crime Delinquency 50(2):214–234
18. MOHE: The current status of higher education in the kingdom of Saudi Arabia (2011). http://he.moe.gov.sa/en/Ministry/General-administration-for-Public-relations/BooksList/stat7eng.pdf
19. Nasiri M, Minaei B, Vafaei F (2012) Predicting gpa and academic dismissal in lms using educational data mining: A case mining. In: 2012 third international conference on E-Learning and E-Teaching (ICELET). IEEE, pp. 53–58
20. Pandey UK, Pal S (2011) Data mining: a prediction of performer or underperformer using classification. Int J Comput Sci Inf Technol 2(2):686–690 arXiv preprint arXiv:1104.4163
21. Pandey UK, Pal S (2011) A data mining view on class room teaching language. Int J Comput Sci Issues 8(2):277–282 arXiv preprint arXiv:1104.4164
22. Parack S, Zahid Z, Merchant F (2012) Application of data mining in educational databases for predicting academic trends and patterns. In: 2012 IEEE International Conference on Technology Enhanced Education (ICTEE). IEEE, pp 1–4
23. Ready DD (2010) Socioeconomic disadvantage, school attendance, and early cognitive development the differential effects of school exposure. Sociol Educ 83(4):271–286
24. Reimer M, Smink J (2005) Information about the school dropout issue. Technical report National Dropout Prevention Center
25. Segal C (2008) Classroom behavior. J Hum Res 43(4):783–814
26. Suhid A, Rahman A, Kamal N (2012) Factors causing students absenteeism according to peers. Int J Arts Commer 4(1):342–350
27. Teasley ML (2004) Absenteeism and truancy: risk, protection, and best practice implications for school social workers. Child Schools 26(2):117–128
28. Vranic M, Pintar D, Skocir Z (2007) The use of data mining in education environment. In: 9th International Conference on Telecommunications, ConTel 2007. IEEE, pp 243–250
29. Wadesango N, Machingambi S (2011) Causes and structural effects of student absenteeism: a case study of three south african universities. J Soc Sci 26(2):89–97
30. Weller LD (1996) The next generation of school reform. Qual Prog 29(10):65–74
31. Williams LL (2001) Student absenteeism and truancy: technologies and interventions to reduce and prevent chronic problems among school-age children. Retrieved 12 May 2010

Shipping Schedule-Oriented Production Planning Using Genetic Algorithm

Jana Heckenbergerova and Petr Dolezel

Abstract This paper presents a modified genetic algorithm for the production planning problem solving while distribution of the products to the customers is considered. The aim of the optimization task is to minimize objective function composed of the joint costs of production (manufacturing line rearrangement minimization), transportation (delivery route minimization) and holding (shipping date optimization) while all the customers' orders are satisfied within the defined production horizon. This class of problems is considered as a NP-hard combinatorial problem and it is not possible to solve it analytically for larger input values. Thus, the stochastic procedure is presented in this paper. This procedure, which is based on the genetic algorithm, respects all the components of the objective function and provides acceptable solution in reasonable time. The experiments with real data show, that presented procedure provides the same or (in most cases) better solution than algorithms commonly used in industry.

Keywords Genetic algorithm · Production scheduling · Path planning

1 Introduction

With a spread of manufacturing systems and increasing competition for customers, the optimization of total costs in production and distribution chain of the manufacturing system receives increasing attention. As combinatorial optimization issue, the finding of optimal production schedule is considered to be NP-hard problem [1]. Hence, the attention is set to some heuristics and hybrid search techniques including genetic algorithm [2], ant colony techniques [4], particle swarm algorithm [8] etc.

J. Heckenbergerova (✉) · P. Dolezel
University of Pardubice, Studenska 95,Pardubice, Czech Republic
e-mail: jana.heckenbergerova@upce.cz

P. Dolezel
e-mail: petr.dolezel@upce.cz

A. Abraham et al. (eds.), *Intelligent Data Analysis and Applications*,
Advances in Intelligent Systems and Computing 370,
DOI 10.1007/978-3-319-21206-7_20

Although many works have been dedicated to this problematics, production is an activity of sustained pursuit. Thus, there is plenty of possibilities to improve this area. This particular paper solves the issue of production planning in combination with shipping scheduling and path planning in metal manufacturing production facility situated in the Czech Republic. So far, all mentioned tasks have been treated with separately there. The approach described below brings relevant cost reductions in comparison to current state.

In the beginning of the paper, the particular problem is properly defined. Then, the possible solutions are discussed and finally, the solution using GA is presented and comprehensively tested.

2 Problem Formulation

Production planning is an important part in the production cycle and proper design of the production plan can significantly reduce total production costs.

Generally, production planning means the determination of the required product mix and factory load to satisfy customers needs. In order to develop production plans, one has to optimize customer-independent performance measures such as cycle time and customer-dependent performance measures such as on-time delivery.

For the needs of the facility mentioned above, account material availability, resource availability and some other values are not relevant. Thus, the production planning reduces on the problem defined below. For the need of the proper definition, the relevant terminology is defined in the following paragraphs. First, all the terms are drawn down and then, important quantities (production costs, transportation costs, holding costs) for the objective function are described.

2.1 The Methodology of the Facility

The running of the facility is defined by the following parameters.

Number of types of products	m
Number of products	n
Number of customers	z
Production costs	N_v
Transportation costs	N_d
Holding costs	N_s
Goods vehicle capacity	K_d
Planning horizon	T
Capacity of the manufacturing line	K_s

The aim of the planning is to minimize joint costs of production, transportation and holding—see Fig. 1. In comparison to other approaches commonly used in similar facilities, the transportation costs are directly included in joint costs.

2.2 Production Costs

Production costs included in the optimization task are defined as the costs of reorganisation of the manufacturing line. Other expenses are considered as unavoidable (fixed). In other words, as long as the production schedule includes only one type of products, production costs are set to zero. Once the type of the manufactured product is changed, the manufacturing line has to be reorganised and the production costs rise.

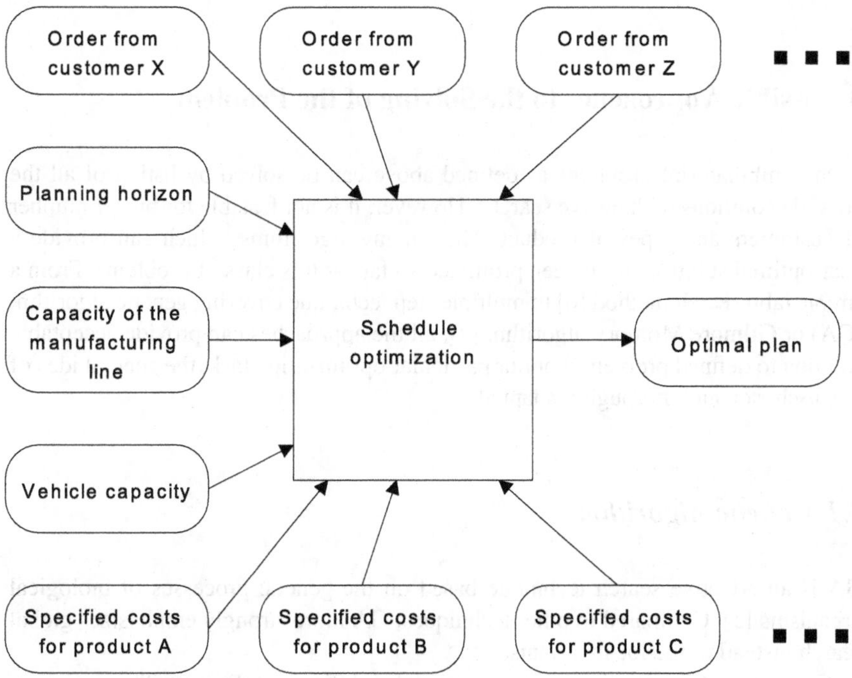

Fig. 1 Problem formulation

2.3 Holding Costs

Holding costs mean the expenses spent on storehouse hiring and are evaluated at the end of every day. If the product is dispatched the same day as it is produced, the holding costs for this product are set to zero. Otherwise, holding costs increase according to the type of stored product.

2.4 Transportation Costs

Transportation costs represent the costs associated with the shipping of the product to the customer. Goods vehicle can load any number of the products up to its capacity K_d and can serve arbitrary number of customers. Transportation costs are then evaluated as the sum of the costs of every goods vehicle ride. All the products have to be dispatched within the planning horizon.

3 Possible Approaches to the Solving of the Problem

Such combinatorial problems as defined above can be solved by listing of all the possible solutions (exhaustive search). However, it is not feasible for bigger number of customers and types of product. Thus, many algorithms, which can provide a near-optimal solution, have been promoted so far for this class of problems. From a simple tabu search method [6] to multiple step technique covering genetic algorithm (GA) or Gilmore Momory algorithm [3], all the approaches can provide acceptable solution to defined problem. For our particular optimization task, the general idea of GA is chosen and thoroughly adapted.

3.1 Genetic Algorithm

GA is an adaptive search technique based on the genetic processes of biological organisms [5]. Compared to other techniques, GA more strongly emphasizes global search instead of the local variants.

GA starts the optimization with several solution candidates. The group of candidates (called population) is transformed then repeatedly using three operators (selection, crossover and mutation) to manipulate the genetic composition of the population. Eventually, the population converges to some near-optimal solution. For more information about GA, see [5].

4 Genetic Algorithm Adaptation to Shipping Schedule-Oriented Production Planning

In this section, the particular solution of the optimization task generally defined in Sect. 2 is presented.

4.1 Phenotype Definition

Each solution candidate maps the production plan and the shipping schedule altogether and is represented by a permutation of genes. The position of the gene among other genes defines the time of production of a particular product. Each gene then carries three values which represent the type of a product, the customer and the goods vehicle to dispatch with, respectively. The situation is illustrated in Fig. 2.

4.2 Fitness Function Evaluation

All the candidates have to be evaluated. In this particular case, fitness function value equals directly to the joint costs of the tested schedule which consists of three elements.

4.2.1 Production Costs

Production costs are defined only by the costs of reorganisation of the manufacturing line. Thus, the data of these costs are required—see an example in Table 1. Then, every time, when the type of currently manufactured product is changed from X to Y, the production costs are increased by the value X → Y.

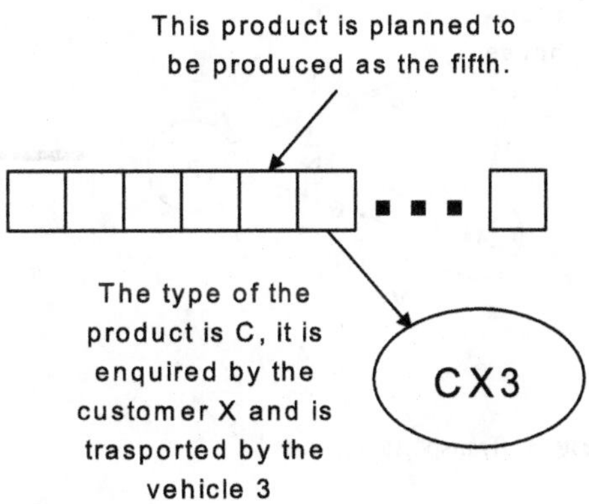
This product is planned to be produced as the fifth.
The type of the product is C, it is enquired by the customer X and is trasported by the vehicle 3
CX3

Fig. 2 Phenotype definition

Table 1 The costs of reorganisation of the manufacturing line

	A	B	C	...
A	0	x	x	...
B	x	0	x	...
C	x	x	0	...
...	...	...	...	...

4.2.2 Holding Costs

Each day of holding of the product in storage instead of shipping it to the customer is charged. The fees can differ according to the type of the product and should be defined as seen in Table 2.

4.2.3 Transportation Costs

Transportation costs are evaluated as a sum of the operational expenses of each goods vehicle used for the deliveries. Every time the vehicle is dispatched, the problem similar to the Travelling Salesman Problem is to be solved. The vehicle begins at the storehouse, it has to attend each customer and then return to the storehouse. In this particular application, Travelling Salesman problem is solved by the Cross Entrophy method [7], which provides optimal solutions in suitable time for reasonable number of customers. Thus, the costs of the routes between each involved node have to be defined—see Fig. 3 and Table 3.

Table 2 Holding costs

Type	A	B	C	...
Fee	x	x	x	...

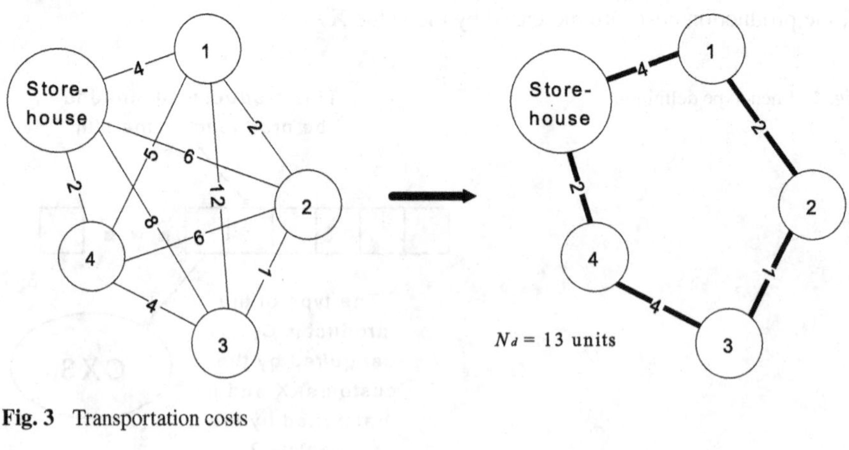

Fig. 3 Transportation costs

Table 3 The route costs between each customer and storehouse

	Storehouse	A	B	C	...
Storehouse	0	x	x	x	...
A	x	0	x	x	...
B	x	x	0	x	...
C	x	x	x	0	...
...	...	...	...	...	...

4.3 Selection

In each iteration, necessary number of individuals has to be selected to breed a new generation. Parents are selected through a fitness value based process. In this case, tournament selection is chosen. Simply, four random individuals are picked and the winner (the one with the best fitness value) is selected for crossover. This procedure is repeated until the proper number of individuals is selected.

4.4 Crossover

Since there is a permutation of products used as a phenotype, ordered crossover with one random crossover point is applied. Thus, a part of one parent is mapped to an opposite part of the other parent. From the replaced part on, the rest is filled up by the remaining genes, where already present genes are omitted and the order is preserved - see an example in Fig. 4. After the crossover, the integrity of each offspring has to be checked especially to detect possible exceeding of the goods vehicle capacity.

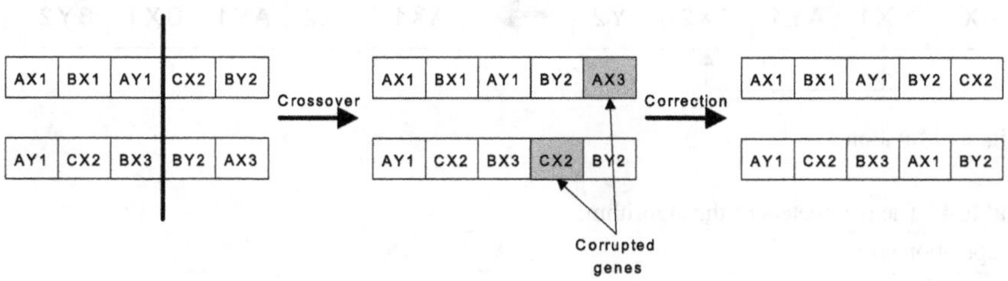

Fig. 4 Crossover

4.5 Mutation

This operator is used to maintain genetic diversity of the population. For permutation phenotypes, the most common version of mutation is the swapping of two random genes—see Fig. 5.

4.6 Elitism

Elitism is used to prevent the population from losing the best individuals. Several best candidates are directly copied into next generation (this operator does not forbid the elite individuals to breed).

4.7 The Whole Algorithm

It is required to define considerable set of parameters before the run of the algorithm—see Table 4. Then whole algorithm runs as seen in Fig. 6.

5 Case Study

The algorithm introduced above was applied to 200 orders handled by the metal manufacturing production facility mentioned at the beginning of the paper in the

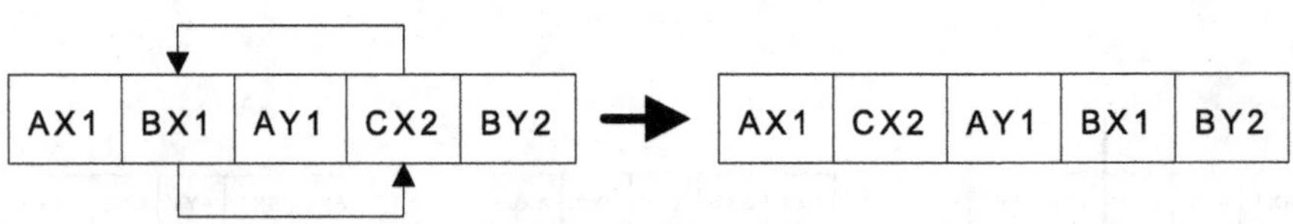

Fig. 5 Mutation

Table 4 The parameters of the algorithm

Population size	X
Number of discarted individuals	V
Number of mutated individuals	M
Number of selected individuals	K
Number of generations	G
Number of generations without improvement	B

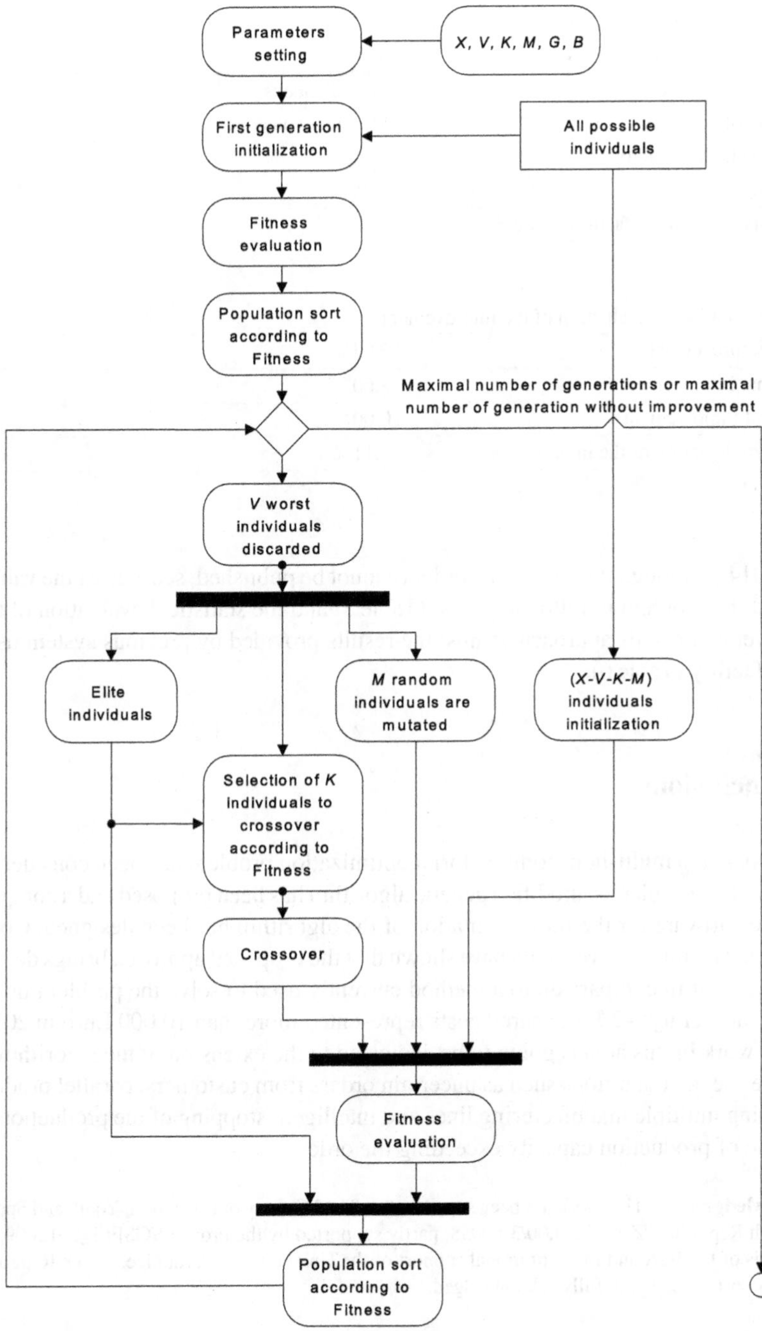

Fig. 6 Diagram of the whole algorithm

Table 5 The parameters of the optimization task

Number of types of products	6
Number of products	Miscellaneous
Number of customers	Miscellaneous
Goods vehicle capacity	3
Planning horizon	5 days
Capacity of the manufacturing line	4

Table 6 Statistical evaluation of the improvement

Mean improvement	2.04 %
Maximal improvement	8.60 %
Minimal improvement	0.00 %
Standard deviation of the improvement	3.11 %

year 2014. Although the data of the orders cannot be published, see at least the values defined the problem in following table (Table 5) and the statistical evaluation of the improvement of this approach against the results provided by previous system used in the facility (Table 6).

6 Conclusions

In this paper, a multi-item combinatorial optimization problem has been considered. To solve this problem, a modified genetic algorithm has been proposed and a comprehensive software for the implementation of the algorithm has been designed. Computational results with real data have shown that the proposed approach brings decent improvement in comparison to a method currently used to solve the problem in the particular facility—2 % of spared costs represented more than 10 000 Euro in 2014. Future work in this area is going to be dedicated to the extension of the algorithm to include special conditions such as uncertain orders from customers, parallel production using multiple manufacturing lines and intelligent stopping of the production in the case of production capacity exceeding the order.

Acknowledgments This work has been supported by The Ministry of Education, Youth and Sports of Czech Republic CZ.1.07/2.3.00/30.0058, partly supported by the project SGSFEI_2015009, by the funds of the IGA and by institutional support of the University of Pardubice, Czech Republic. This support is very gratefully acknowledged.

References

1. Blazewicz J, Ecker K, Pesch E (1996) Scheduling computer and manufacturing processes. Springer, Berlin
2. Costa A (2015) Hybrid genetic optimization for solving the batch-scheduling problem in a pharmaceutical industry. Comput Ind Eng 79:130–147
3. Dawande M, Geismar HN, Hall NG, Sriskandarajah C (2006) Supply chain scheduling: distribution systems. Prod Oper Manage 15(2):243–261
4. Hecker F, Stanke M, Becker T, Hitzmann B (2014) Application of a modified *ga*, *aco* and a random search procedure to solve the production scheduling of a case study bakery. Expert Syst Appl 41(13):5882–5891
5. Holland JH (1992) Adaptation in natural and artificial systems: an introductory analysis with applications to biology., Control and artificial intelligenceMIT Press, Cambridge
6. Peidro D, Daz-Madroero M, Mula J, Navaln A (2014) A tabu search approach for production and sustainable routing planning decisions for inbound logistics in an automotive supply chain. 3:61–68
7. Rubinstein R, Kroese D (2004) The cross-entropy method. Springer, New York
8. Zheng X, Liu Z (2010) The schedule control of engineering project based on particle swarm algorithm. In: 2010 Second International Conference on Communication systems, networks and applications (ICCSNA), vol 1, pp 184–187

Design and Implementation of a *Policy Recommender System* Towards Social Innovation: An Experience With *Hybrid Machine Learning*

Hameed Al-Qaheri and Soumya Banerjee

Abstract Recommender systems are typical software applications that attempt to reduce information overload, while providing recommendations to end users based on their choices and preferences. Conventional structure of recommender system, collaborative filtering and content based filtering are not adequate enough to derive a policy consensus and recommendation from social network based data due to the uncertainty and overlap of data itself. Hence, this paper proposes, a conceptual design and implementation of a recommender system, capable of yielding different social innovation based policies incorporating first level of machine learning on policy data click or access method. The proposed model refers hybrid technology using fuzzy logic led to cumulative prospect theory and machine learning strategies to quantify optimal social innovation based policy recommendation. The repetitive pattern of policy selection and recommendation can also be taken care through learning attribute of the proposed system. The challenge of developing such recomender system is to analyze the pragmatics of contextual multi-optimization behavior of decisions under several social innovation based policy framework.

Keywords Social networks and innovation • Prospect theory • Fuzzy logic • Recommender systems • Machine learning

1 Introduction

The aim of the paper is to develop a recommendation system based on an ensemble of machine learning and fuzzy algorithms to generate an advice, based on the partial knowledge garnered from social networks by tracking the reactions and opinions of

H. Al-Qaheri
Department of Quantitative Methods and Information Systems,
Kuwait University, Safat, Kuwait

S. Banerjee (✉)
Department of Computer Science & Engg, Birla Institute of Technology, Mesra, India
e-mail: dr.soumya@ieee.org

A. Abraham et al. (eds.), *Intelligent Data Analysis and Applications*,
Advances in Intelligent Systems and Computing 370,
DOI 10.1007/978-3-319-21206-7_21

crowd irrespective of their different textures and contexts. The motivation arises primarily from a need to stimulate and support innovative computational platforms for grassroots social innovation. The underlying vision is that individuals and groups can more effectively and sustainably react to societal challenges by acting on the basis of a direct extended awareness of problems and possible solutions. The genesis of the problem has been coined from recent research conducted on multi-criteria based prospect theory [1, 2] to derive a decision from several alternative solutions. Such collective intelligence platforms must include collective decision-making tools and innovation mechanisms to drive individual and community creativity, participation in group decision making and creating situational awareness. This will allow participants on a social network to forge collaborative decisions based on consensual knowledge. The strength of knowledge consensus depends on the aggregation of individual participation. Therefore engagement of all members is crucial to consolidate their opinion and to understand different scenarios. In this paper, the emphasis on the design and development of appropriate machine learning based recomender systems is given to find different lifestyles, passions and opinions. These will be fed to the recommender system firstly to encourage participation and then increase collective situational awareness. Only then can a recommender system successfully enhance collaboration towards more creative solutions to the raised social problems. In this paper, *Cumulative Prospect Theory (CPT)* [3, 4] is a descriptive and general representation of individual preference under risk. It describes utility on positive gains and negative losses where, as in many real world problems like in online social networking, behaviroal choice theory in multi-criteria decision driven paradigm becomes eventually persistent [5]. Collectively, decision making solicits consensus derived from online community and thus, first level of machine learning inspecting the click identification on a social policy set has been initiated. The proposed machine learning is applied after assigned by fuzzy prospect theory and thus algorithm follows a hybrid architecture referring Fuzzy logic and Click through measures on recommendations. The remaining part of the paper has been organized as follows: Sect. 2 elaborates problem statement, followed by contemporary research work on these aspects in Sect. 3. We proposed methodology in Sect. 4. Section 5 details about obtained result and discusses the significant issues about the implementation. Finally, Sect. 6 concludes with present relevance of the proposal and mentions scope of future research in this direction.

2 Problem Statement

Firstly, the paper proposes a collective decision making and recommendation process when there are substantial number of choices are assigned for each decision. The decision making process seeks to achieve an appropriate balance between personalization and aggregation of opinions that are mined from social network users, since both are salient parameters affecting collective decision-making.

The core benefits that can be accrued from such an approach are: incorporating diversity of opinions, increasing acceptability of decisions, widening the coverage of participation and enhancing creativity. Secondly, the proposal will incorporate concrete qualitative metrics to assess activities and processes that are driven by social networks. For example, the *consensus measure* [6] evaluates the degree of agreement between users and is used to segregate participants with different profiles. Likewise, the *proximity measure* of group decision [7] yields the degree of agreement between the experts' individual opinions and the group opinion. Finally, the recommender system will guide the consensus process until a stable final decision crystallizes. These quality-driven guidelines lay the foundation for building a generic framework for a recommendation system, which can be utilized for offering various web based group decisions and services for collective awareness. The paper seeks to develop systematic experimental approaches that can be used to solve social innovation problems, which orient wide range of policy decisions like:

- Crafting immigration policies that have a potential to create a positive long term impact on society at large.
- Finding expert medical opinions to tackle urgent public health situations.
- Forging collective judgment from separate opinions on specific proposals.

The overall aim of the paper is to develop a user-friendly social policy based recommender system allowing the consideration of alternative choices and measuring consensus among participants. Definitely, a natural online referendum is expected where the opinion, sentiment and motif of crowds can be clearly understood.

2.1 Essential Components

The roles of following components in the design the proposed recommender system are essential and mandatory:

- **Cumulative prospect theory** defines the prospect, denoted as a function f : $f = (x_1, p_1; x_2, p_2; \ldots\ldots\ldots x_n, p_n;),$

This theory yields outcome x_i with probability p_i [8]. Specific value function combined with decision weight function determines the prospect value in cumulative prospect theory. Cumulative prospect theory positions the value concerned with specific gains or losses with respect to a pre–defined reference point... The prospect theory here helps to create suitable environment for assigning different choices towards a specific converging result. The flexibility of choosing reference point for the sum of data points, encourages the community to incorporate the theory with multi-criteria based decision making paradigm, where fuzzy logic with different geometric functions available to formulate a decision.

- **Deciding consensus on social innovation policies**

The data set contains several connected conceptual graph structure G, where there will be stronger consensus of opinion, followed by other linguistic modes of operations e.g. moderately, slight moderately (SM), or less strong consensus (LS), slightly less weak etc. (see Fig. 1). These are choices to be imposed on different sets of social innovation policies.

Empirically, we define:

$P^{(i)}$ (t) indicates position p of the social graph node i at time t,

$p^{(i)}(t) \in \{MS^+, SM^-, LS, SW^+, CW^+\}$, therefore, as if both nodes i and j are connected nodes under a social network instance there may a possibility that i and j offers strong opposition affinity, and then finally tie converges to CW^+.

$$\begin{aligned} &|p^{(i)}(t)| > |p^{(j)}(t)| \\ &p^{(i)}(t+1) = p^{(j)}(t+1) = p^{(i)}(t) \end{aligned} \tag{1}$$

$$\begin{aligned} &\text{if } |p^{(i)}(t)| > |p^{(j)}(t)| \text{ and } p^{(i)}(t).p^{(j)}(t) < 0, \text{ then} \\ &p^{(i)}(t+1) = -p^{(j)}(t) \text{ and } p^{(j)}(t+1) = p^{(i)}(t) \end{aligned} \tag{2}$$

$$\text{and if } p^{(i)}(t) = -p^{(i)}(t) \text{ then } p^{(i)}(t+1) = \text{sign}(p^{(j)}(t) \tag{3}$$

- **Social Innovation Policy framework**

Under aegis of prospect and consensus theory, the following framework is constituted:

- Crowd Sourced citizen-led neighbor-funded projects
- Poverty and unemployment
- Anti Bribe movement on social network

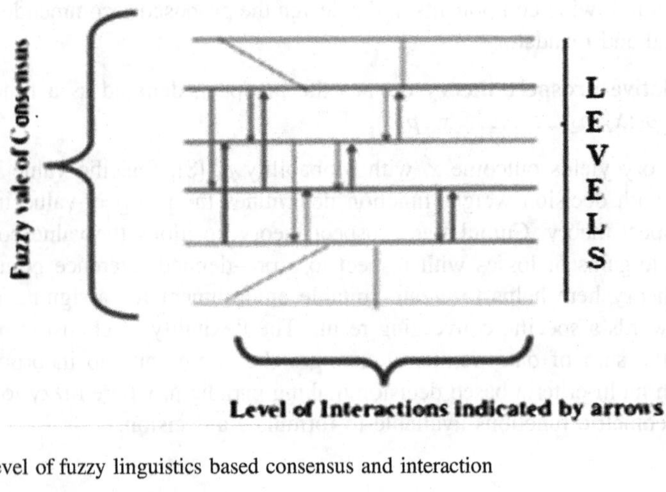

Fig. 1 Level of fuzzy linguistics based consensus and interaction

- Safety of Woman
- Hygiene, sanitation and environment

The above five categories have been chosen to derive consensus from on line social network and media.

2.2 Expected Results

Using an experimental approach, concepts and tools that are developed are verified using real world cases. The following are the expected specific outcomes from this paper at large: (a) The creation of a digitally supported social recommender system on web with features that have the capability to:

- Motivate more and more citizens to participate in the collective decision making process on web.

(b) The participation of common people in the social blogosphere has empowered them to influence public perception. The situation defines the scope of the utility of analytical system, and will be of capable of answering following research questions and specific objective of the paper:

- What are the patterns of choices for a given decision and how/why do they become salient towards the final verdict?
- What is the percentage of success for an open group discussion and to what extent will they will be responsible for achieving final opinion?
- How behavioral choice could affect the multi criteria based decision environment for obtaining consensus?

The automated analysis of blogs and other social media raises several interesting research challenges, which we address below in next section.

3 Similar Works

As the pivotal points of this paper combining fuzzy logic and prospect theory to be deployed in designing a recomender system, therefore we explore certain overlapped research on these areas. Motivated by some contemporary research like from Yingdong He et al. 2014 [3], which investigated fuzzy operators and interaction towards multi-criteria decision making environment, this present model consults certain anomalies from it. Especially, the authors developed intuitionistic fuzzy sets (IFSs) and intuitionistic fuzzy geometric interaction averaging (IFGIA) operators, these operators are flexible to interact with membership and non membership functions of different intuitionistic fuzzy sets. Krohling et al. in 2012 [5] also investigated a case study of oil spill in the sea, where combining prospect theory with Fuzzy logic have been used. In the problem the decision matrix may be affected by symmetric and asymmetric uncertainty. The case study also exhibits poly-behaviroal scenarios while altering the level of the uncertainty to the data. As the proposed model pivots on network centric data sources, therefore interactions from several group members may obtain information from one another simultaneously, it is necessary to understand more than dyadic communication and learning processes to obtain the complexity. Greening et al. in 2015, proposed algebraic topology to detail about the framework during multi-agent interactions [9]. Conventionally, there are number of decision support systems to address the 'large decision table' under intuitionistic fuzzy sets multi-criteria decision-making (MCDM) issues. Hence, risk preferences of decision makers could be evaluated based on the prospect theory and criteria reduction [10] for pure decision problems. The present problem is partially decision metric rather it indicates the recommendations with optimal matching of given criteria. *Wise Social Network Recommender System* [11] has also been proposed to log and aggregate collective intelligence while calculating the trust scores between users. The work was only restricted towards a trusted recomender system.

4 Proposed Model and Methodology

Our proposed innovative model is characterised by the following features

a. Configuring fuzzy logic and cumulative prospect theory within a single module
b. Investigate *Click through Rate* (CTR) on any policy or combination of policies on line.
c. The task of finding/evaluating a subset of labeller (which may comprise of capabilities, motives, knowledge and personality of the participants) that can accurately represent the trend and propagation of social decision.. The system could be able to draw an estimate about this.
d. The approximate correct tendency prediction with minimum number of labellers could be welcome. Hence, therefore unlikely the classifier deployed here is known and task is here to approximate the classifier itself. In addition to the proposed model (usually learning takes place after obtaining a collection of data) has been emphasized the stream of on line data with a simultaneous process of learning having different quality estimates for labellers.

And as such, the proposed model in distinct two phases:

- Data Acquisition block using social posts
- Consensus measure with learning as a parameter

Figure 2 demonstrates the different building blocks of the proposed model including the data collection and process block. It must be noted that while this block diagram outlines the algorithmic strategy it doesn't illustrate the final deliverable module with comprehensive code of learning. Also, the dotted blocks represent embedded modules within the proposed conceptual scheme, e.g. learning,

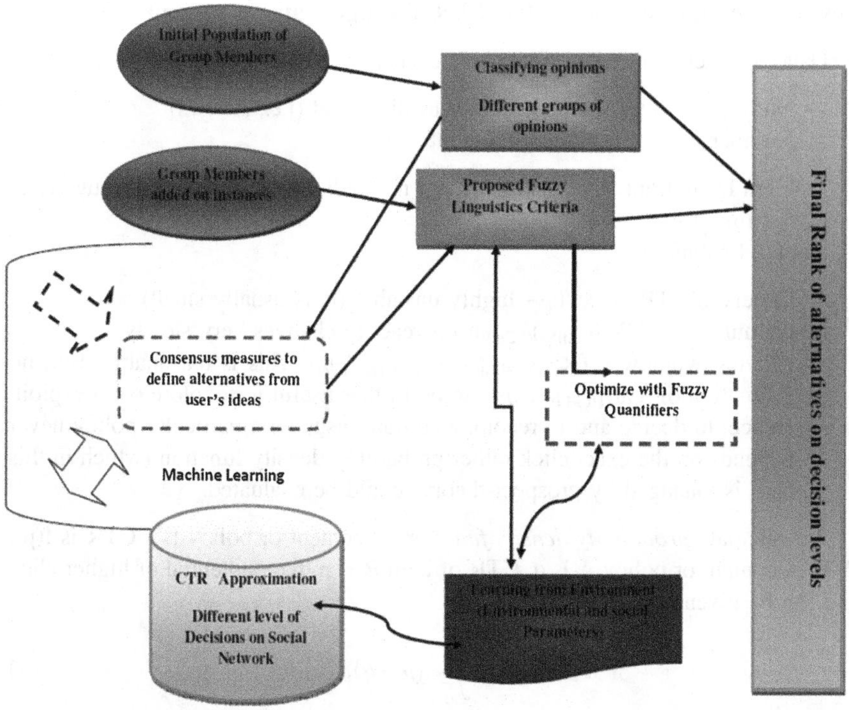

Fig. 2 Block schematic of proposed conceptual model

fuzzy optimized quantifiers and consensus alternatives to be collected from participants through specific instances.

4.1 Learning Phenomena

The learning pattern from environmental and social could be resolved using innovative learning strategy to be followed in final module of the proposed model:

- To trace the pattern and its propagation, proposal uses the action propagation, graph of propagation and log of actions

The proposal used the framework an instance of the simple threshold model, in that case the support data must identify the list or log of active users on the network and how they could be responsible for the next set of instances for the neighbouring users and thus we have to investigate the influence pattern as joint probability but in that case as social expert, the model demands a threshold value against a standard issue raised on social network. The fuzzy linguistics used in the conceptual block is also capable to quantify this threshold value.

How to track the changing CTR (Click through Rate) of content

- Data: for each content, at time t, we observe
  - Number of times the content n_t was displayed (i.e., #views)
  - Number of clicks c_t on the content
- Problem Definition: Given c_1, n_1, …, c_t, n_t, predict the CTR (click-through rate) p_{t+1} at time t + 1
- Potential solutions:
  - Observed CTR at t:$c_t/n_t \rightarrow$ highly unstable (n_t is usually small)
  - Cumulative CTR: $(\Sigma_{all i} c_i)/(\Sigma_{all i} n_i) \rightarrow$ react to changes very slowly
  - Moving window CTR: $(\Sigma_{i \in last\ K} c_i)/(\Sigma_{i \in last\ K} n_i)$. This is reasonable. But, no estimation of $Var[p_{t+1}]$. This can be further useful for explore or to exploit. Hence, to decide and to recommend consensus on a particular policy never depends on the exact click rather probability density function (which in this case is managed by prospect theory) could be evaluated.

We calculate *probability density function* of content or policy B's CTR is f(p), CTR of content or policy A is q, CTR of item B is p, then potential of higher click rate can be given as :

$$\text{Potential} = \int_{p>q} (p-q).f(p)dp \tag{4}$$

This will also help to follow the actual participants entering into or leaving from policy site/blog. Hybrid machine learning justifies the proposed method for choosing recommendation policy as it refers both fuzzy prospect theory framework and click through rate mechanism. After identifying the subsets of different relevant blogs and social media, one can then determine the most influential participants in this space. These are the experts or mavens, whose opinions catch on most rapidly towards a specific social issue. In this proposed model, model itself learns from environment, thereby could be improved by experience. It is important to identify this set of bloggers, since any negative sentiment they express could spread substantially and influence other participants. In addition to authorities, there are influential and dominating, who are not only well connected with the group peers, but also become instrumental for the spread of information in the social media.

Given that we have a network of directed edges indicating the links between posts/blogs, we can apply measures of prestige and importance from Social Network Analysis [12]. In the literature, the authority of a media can be characterized in terms of:

- *Page Rank* which is based on the number and authority of other blogs that link to it
- Influence of a blog can be measured by *Flow Betweenness*. This attribute captures the degree to which the blog contributes to the flow of information between other participants.

There are several alternative approaches to measure a node's rank in a graph like: *Degree Centrality* and *Closeness Centrality.* How do we determine which measure to use? Answering this question becomes non-trivial since the notions of authority and influence vary based on the application and how the pattern of learning could be originated. Suppose, every issues rose on social network may be or may not be stationary, hence an automated social recommender system could be more appreciated, which could infer from scratch input very similar to human being. For the purposes of trapping, we are interested in bloggers whose pattern of exchange could be watched out. Their influence, intention, interaction pattern and subsequently the content blogged by others could be training sample for the proposed learning module to be intituled from uncertain, noisy and non stationary environment [13]. If a blogger is indeed influential, we would expect his ideas to propagate to others. Based on this, we propose a suitable recommendation on a specific agenda, comparing candidates' influence measures on the task of predicting user content generation.

Data collection and Procedure: Data actually originates from live feed data under social network with regional attributes [14]. Although most of the social network organizations do not have the provision of sharing the on line data, therefore special permission could be taken to use conventional real data collection procedure. The present work envisages public source data for the purpose. The high level description and generic steps are as follows:

Step 1: Obtain n values of multiplicative preference relations say R with a set of consequences and positive gain and negative loss of fuzzy membership grades

Step 2: Transfer (n−1) values to other set of consequences say R^+

Step 3: Set the diagonal elements of R^+ are values at predefined threshold 0.5

Step 4: Obtain elements of the lower side triangle

Step 5: Retrieve elements of upper side triangle leaving set R^+

Step 6: Check boundary for all $R^{+\text{elements}}$. If there is element value out of [0 1], linear transformation is mapped to maintain the additive reciprocity of R^+

Step7: Initiate Set $p^{(i)}(t) \in \{MS^+, SM^-, LS, SW^+, CW^+\}$ & test all the relation of $p^{(i)}(t+1) = p^{(j)}(t+1) = p^{(i)}(t)$ or $p^{(i)}(t+1) = -p^{(j)}(t)$ and $p^{(j)}(t+1) = p^{(i)}(t)$

Step 8: Estimate Moving window CTR = $(\Sigma_{i \in \text{last } K} c_i)/(\Sigma_{i \in \text{last } K} n_i)$ and calculate potential of specific content policy based on

$$\int_{p>q} (p-q) \cdot f(p)\, dp$$

Step 9: Follow step 1–6 to obtain preferences of E experts and then take an average for all obtained fuzzy preference relations to get a final R^+./* *Learning pattern**/

Step 10: Obtain relative weights for all alternatives and rank them in order of preferences

The impact of variable behaviroal paradigm of social crowd also has been realized, because the policy framework categorized on five different sections as mentioned earlier. The behaviroal membership grades are defined as: we consider a special type of vector called *Fuzzy Commitment Vector*[1] V_c which span within Vector space V and subspace S at time t. V_c Contains some dimension represented as $V_c = [c_1, c_2, c_3, c_4]$, where the dimensions are as follows:

$c_1 \rightarrow$ Linguistic and Subject relevant consensus
$c_2 \rightarrow$ Linguistic but contradict consensus
$c_3 \rightarrow$ Suggestive consensus
$c_4 \rightarrow$ Not linguistic and no subject relevant consensus

We assign membership grades of 0.7, 0.5, 0.6, 0.3 in fuzzy set for the linguistic terms for *always, often, unspecific* and *never* respectively. Following the different attribute values of participants during consensus measure, the linguistics has been improvised as shown (Table 1):

The attributes in the table are operationally defined as follows:

- **Members:** members are denoted by F_1 to F_{10}.
- **Always :** user who posts opinions something regularly on the community. Value of the attribute is set 1.
- **Often :** user who posts opinions sometimes on the community. Value of the attribute is set 0.75
- **Unspecific:** user who may or may not comment. Value of the attribute is set 0.5.
- **Never :** user who never comments. Value of the attribute is set 0.

Conventional sources of data. There are different sources to collect data from social network; few of them could be mentioned as follows: *FaceBookwalls; Tweets; Delicious bookmarks; You Tube comments; Flickr comments, tags; Yahoo! Answers questions, answers, comments, ratings; Blog articles and Wiki article.* Specifically, *question-answer recommendation* system could be another alternative source of numerical information to be grabbed from social network. The attributes could be: Return value of all answered questions from a category; Question ID, category, question, question description, posting date; Number of answers; Questioner ID, questioner nickname; Chosen answer (best answer), chosen answerer ID, chosen; Pattern of answer; Prediction of inclination of answer; Answerer nickname, chosen answer posting date, chosen answer rating; Return all answers and information from a specific question; Question ID, answer, sources, posting date; Answerer ID, answerer nickname; Return all questions from a specific user. Primarily, this paper adopts most of these attributes for designing analysis framework as per the proposed algorithm. Further, *Comparative Welfare Entitlements Dataset (CWED).*[2] provides systematic data on institutional features of social insurance

[1]tracking emotion and fuzzy linguistics values on analyzing the primary consensus, the standards are adopted from: http://www.wefeelfine.org/api.html.

[2]http://cwed2.org/ http://ec.europa.eu/eurostat/data/database

Table 1 Values different of participant's different activities

Members	Always(A) (1)	Often(O) (0.75)	Unspecific(U) (0.5)	Never(N) (0)	Total feedback (T_F)
F_1	0.3	×	×	0.7	0.3
F_2	×	0.5	×	×	0.375
F_3	×	0.6	×	0.7	0.45
F_4	×	0.6	×	×	0.45
F_5	×	0.7	×	0.6	0.525
F_6	0.6	×	×	0.7	0.6
F_7	0.5	0.6	×	0.7	0.95
F_8	0.6	0.7	×	×	1.125
F_9	0.7	0.6	×	×	1.15
F_{10}	0.7	0.6	0.3	0.5	1.3

programs in 33 countries spanning much of the post-war period. The limitation of data set is to reduce with specific number of given attributes and variables.

4.2 Possible Tools and Measures

The objective of this paper is to design and implement the policy recomender system, therefore it incorporates. *JSON-JavaScript Object Notation* [15]. The proposed model also demonstrates an option to allow and to generate recommendations for assisting experts to change their opinions in order to obtain the highest degree of consensus possible. It is evident that to provide a nearly best recommender, we need to match [16] degrees indicating the agreement among the expert's opinions. Here, proximity measures are used to find out how far the individual opinions are from the group opinion. As the component of basic machine learning drives the proposed model, while putting the recommendation, therefore, a free open source tool known as *Crab*-Recommender Framework in Python has been used. Details can be available at http://muricoca.github.io/crab/, https://github.com/PyBossa/app-epicollect.

5 Results and Discussion

The following plots demonstrate (Figs. 3, 4 and 5) how the social network participants could fluctuate with entry and exit behavior while converging towards a consensus model under different criteria. At some point of time instances, criteria could be different to keep entry towards a post. Therefore, ranking the different yielded decisions could be difficult. Second plot exhibits, (considering the membership grade with different variable criteria) and leads the optimal and near optimal

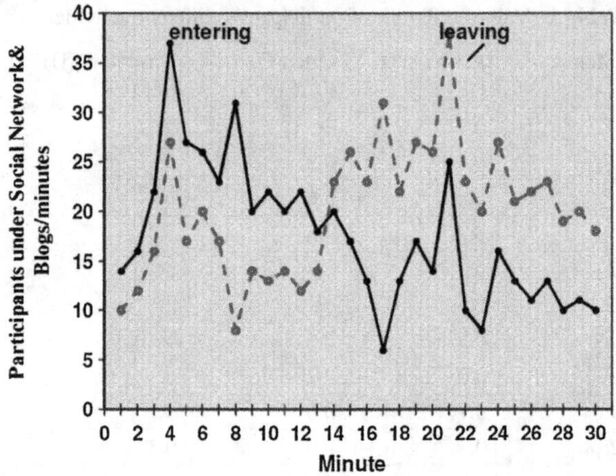

Fig. 3 Blog analytics using Eq. (4) on CTR

recommendation. Figure 3 demonstrates the python simulation from the variable settings (**Refer Appendix**) to quantify the time varying relationship for entering and exit from policy sites. The objective is to learn the pattern for maximum click rate on a particular policy. The more the click rate, the more poplar the policy is.

Based on fuzzy linguistics and prospect theory, the convergence rule for recommending best policy has also been evaluated (Fig. 4) Envisaging the variable of data set and membership grade optimal consensus could be lying nearly $0 > 0.6 > 0.8$. Exact point in terms of recommendation for a policy should also be approximated should also b approximated.

On contradiction, lower click rate also could not yield a smooth curve rather scaled CTR and time scale shows the significant value nearly 0.6 and larger than 0.2 (Refer Fig. 5). More experimentation can signify more precision on the process of recommendation.

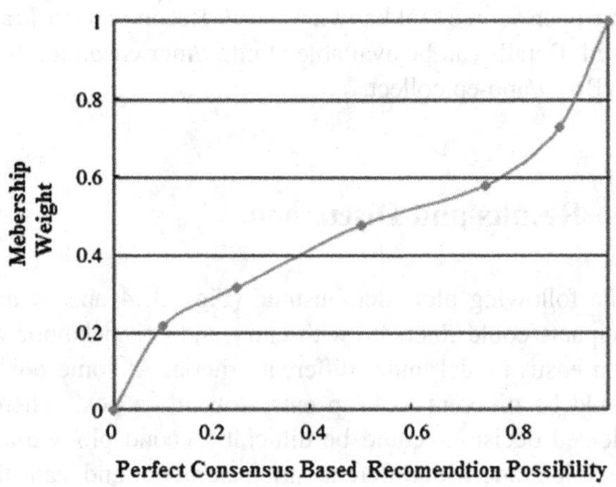

Fig. 4 Convergence of recommendation

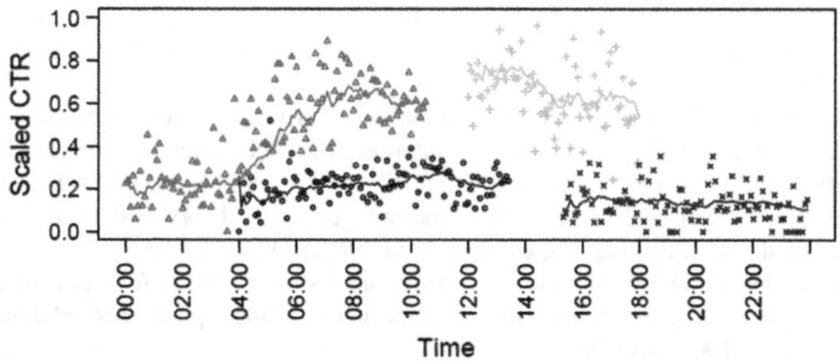

Fig. 5 Low click rate on policies—more temporal softening

6 Conclusion Relevance and Future Scope of Research

This paper suggests a recommender system on the selection of social innovation (primarily crowd sourced citizen-led neighbour-funded projects, poverty and unemployment, anti bribe movement on social network, safety of woman and hygiene, sanitation and environment) and related policy sets. The problem of such recommendation is primarily a problem of multi-objective optimization as seldom there will be number of attributes, number of policies and sub policies with different on-line segments of participants. We use simple machine learning practice *click through principle* hybridized with fuzzy prospect ranges, which is at present based on single objective function. We found that the problem of consensus recommendation could be dependent on feature vectors to visit towards a policy set, reward vectors for referring a particular policy set and final objective function or may the presence of regret function. All these artifacts can put the recommendation efficient with more machine learning applications. The context i.e. the level of dissatisfaction with public support for families with dependents (older relatives or children) was seen to be running high. With the social innovation issues like poverty, unemployment, safety of woman and health care contexts are prominent areas where, machine learning and click identification can be found broadly. Under the aegis of computational social network, the genesis of idea has been mobilized for developing a smart and balanced Europe 2020 [17], which can be socially justified, well analysed and balanced with appropriate consensus *of knowledge.*

References

1. Krohling RA, de Souza TTM (2012) Combining prospect theory and fuzzy numbers to multi-criteria decision making. Expert Syst Appl 39(13):11487–11493
2. He YD, Chen H, Zhou LG, Han B, Zhao QY, Liu JP (2014) Generalized intuitionistic fuzzy geometric interaction operators and their application to decision making. Expert Syst Appl 41 (5):2484–2495

3. He YD, Chen H, Zhou L, Liu J, Tao Z (2014) Intuitionistic fuzzy geometric interaction averaging operators and their application to multi-criteria decision making. Inf Sci 259:142–159
4. Wang J, Sun T (2008) Fuzzy multiple criteria decision making method based on prospect theory. In: Proceeding of the international conference on information management, innovation management and industrial engineering (ICIII '08), vol 1, pp 288–291. Taipei, Taiwan
5. Krohling RA, de Souza TTM (2012) Combining prospect theory and fuzzy numbers to multi-criteria decision making. Expert Syst Appl 39(13):11487–11493
6. García JMT, del Moral MJ, Martínez MA, Herrera-Viedma E (2012) A consensus model for group decision making problems with linguistic interval fuzzy preference relations. Expert Syst Appl 39 pp 10022–10030
7. Perez IJ, Cabrerizo FJ, Herrera-Viedma E (2011) Group decision making problems in a linguistic and dynamic context. Expert Syst Appl 38(3):1675–1688
8. Liu P, Jin F, Zhang X, Su Y, Wang M (2011) Research on the multi-attribute decision-making under risk with interval probability based on prospect theory and the uncertain linguistic variables. Knowl Based Syst 24(4):554–561
9. Greening BR, Pinter-Wollman N, Fefferman NH (2015) Higher-order interactions: understanding the knowledge capacity of social groups using simplicial sets. Curr Zool 61 (1):114–127
10. Liu J, Liu S-F, Liu P, Zhou X-Z, Zhao B (2013) A new decision support model in multi-criteria decision making with intuitionistic fuzzy sets based on risk preferences and criteria reduction. J Oper Res Soc 64;1205–1220
11. Daniel M, Loredana M, Nicolae T (2012) Building a social recommender system by harvesting social relationships and trust scores between users¸ business information system. Workshops Lecture Notes in Business Information Processing, Springer, pp 1–12
12. Terán L, Meier A (2011) Smart participation—a fuzzy-based platform for stimulating Citizens' participation. Int J Inf (IJI) 4(3/4)
13. Blocki J, Blum A, Datta A, Sheffet O (2013) Differentially private data analysis of social networks via restricted sensitivity. arXiv:1208.4586v2
14. Bird et al (2012) Flooding Facebook – the use of social media during the QFs. Aust J Emerg Manage 27(1):27–33
15. Bruns A, Burgess JE, Crawford K, Shaw F (2012) #qldfloods and @QPSMedia: Crisis Communication on Twitter in the 2011 South East Queensland Floods. ARC Centre of Excellence for Creative Industries and Innovation, QUT, Brisbane QLD Australia
16. Alonso S, Herrera-Viedma E, Chiclana F, Herrera F (2010) A web based consensus support system for group decision making problems and incomplete preferences. Inf Sci 180 (23):4477–4495
17. www.2020-horizon.com
18. Cabrerizo FJ, Alonso S, Pérez IJ, Herrera-Viedma E (2008) Modeling decisions for artificial intelligence. In: 5th international conference, MDAI 2008 Sabadell, Spain, 30–31 Oct 2008. Proceedings on consensus measures in fuzzy group decision making lecture notes in computer science, vol 5285

Implicit Authentication System for Smartphones Users Based on Touch Data

Reham Amin, Tarek Gaber and Ghada ElTaweel

Abstract Currently smartphone' users run many crucial applications (such as banking and emails) which contains a very confidential information. To secure this information, the built in sensors equipped with smartphone devices can be utilized. In this paper, based on these sensors, an implicit authentication system for smartphone's users is proposed. A mobile App is developed to collect the data source of users' biometrics and then features (pressure, position, size, and time) are extracted. classifiers were then applied to decide whether a user is the true owner of device or an impostor. The experimental results showed that our implicit authentication system achieved accuracy of 96.5 % which is better than a related work.

1 Introduction

Mobile computing devices (such as Smartphones and tablets) are worldwide commodities that combine phones and desktop computers characteristics [1]. Market analysis predicts that it will be 640 million tablets and 1.5 billion smartphones in use worldwide by the end of 2015 [2]. Inside these devices sensitive information (such as business secrets and even credit card numbers) is stored. Therefore, it's a nightmare for the owner to lose smartphone [1]. Not only theft but also device share with a guest user (e.g. coworkers, partners or family members) is also considered a disaster [3].

To safeguard this information against unintended usage, such as an impostor accessing the bank account [4], user authentication has been proposed. The two common types of authentication are knowledge based and biometric based [2, 5].

R. Amin (✉) · T. Gaber · G. ElTaweel
Faculty of Computers and Informatics, Suez Canal University, Ismailia, Egypt
e-mail: reham_amin@ci.suez.edu.eg

T. Gaber
e-mail: tmgaber@gmail.com

T. Gaber
IT4Innovations, VSB - Techincal University of Ostrava, Ostrava, Czech Republic
e-mail: ghada_eltawel@ci.suez.edu.eg

A. Abraham et al. (eds.), *Intelligent Data Analysis and Applications*,
Advances in Intelligent Systems and Computing 370,
DOI 10.1007/978-3-319-21206-7_22

The knowledge based one depends on a secret such as password, PIN [6] and unlock pattern (i.e. match several points on the screen using one move) [1]. Although this type is simple, cheap and quick enough for frequent logins (e.g. unlock patterns) [6], it is more vulnerable to various attacks such as Smudge attack or shoulder surfing attach.

In the other hand, the biometric one depends on unique human characteristics such as keystroke, face unlock, or finger print. This type is much safe and effective to accommodate some of the above attacks. Biometric can't be forged, stolen or borrowed [7, 8]. However, it's limited on accuracy and usability during unlocked state [2, 9]. In general, biometric authentication is divided into physiological and behavioral biometrics [9]. Physiological biometrics depends on what a user already owns. Such as face, fingerprint, voice, iris and hand geometry. This biometrics can't be stolen or imitated in contrast to the knowledge based one [9]. However, this approach is costly (e.g. requires sensors), and requires user interaction (e.g. frequent logins), not to mention the extra load required to authenticate users on smartphones. [10]. Apart from that, behavioral biometrics depends on how the user behaves. Such as gait, location and keystroke patterns. It provides active authentication by using the built-in smartphone sensors [11].

In this paper, we propose an authentication method which depends on the behavioral biometric of the smartphone's users. This method works in the background while user using the phone's keyboard typing phone numbers. It utilizes the various sensors equipped with the smartphone device, thus there is no need for password/PIN (i.e. avoiding password remembering problem and surf attack). Also there is no need for external hardware like the case of physiological based methods. Not to mention ease of use and user intrusive manner in gathering data among other behavioral biometrics (e.g. Gait recognition).

The main contributions of this paper are threefold. Firstly, this paper presenting a touch behavior based authentication system by using only touch data available in most smartphones without using any external hardware. Secondly, developing a mobile App to collect our own dataset from different type of smartphone's users. Thirdly, proposing an implicit authentication approach using our collected dataset and based on SVM and KNN classifier.

The rest of the paper is organized as follows. Section 2 presents a background about authentication system based on touch data and Sect. 3 discusses the related work. Section 4 introduces the proposed system, Sect. 6 concludes and gives some open points for further search.

2 Tapping Background

There are many reasons motivate this work. First of all, data provided by the touchscreen sensors of mobile devices is considerably richer data than that available from personal computer hardware keyboards. The capabilities of such screens could be utilized as input devices of keystroke biometric which is considered as means of authentication on touchscreen devices. Secondly, this biometrics is unique to an individual and difficult to imitate [12]. Thirdly, even if imposter sees what user input, he couldn't reproduce the user's behavior through shoulder surfing or smudge attacks [1]. This is because of the non-visual cues for tapping behavior. Last but not the least, such mechanism require no extra hardware and done in user intrusive manner as person typed information [10].

2.1 Tapping Types

Types of touch operations include stroke, slide, pinch and handwriting [3, 10].

- Keystroke(Tap): is a finger press on some point of the screen to click item, text, type PIN for example. This type differs when inputting different words.
- Slide: is a finger move (i.e. curve) on the screen to navigate mails, photos, messages or contacts. This type differs on any of 4 directions.
- Pinch: is a two-finger gesture on the screen to read EBook, zoom in/out photo or webpage. This type differs based on case: open or close.
- Handwriting: is a free form gesture for entering characters. This type differs on different letters.

In practice, people interaction with mobile is not limited to these principle gestures, and they may use double touch, open pinch or long press to deal with phone. Such gestures are achieved for daily usage of a fraction less than 5 % [5]. As a result, we neglect other gestures in this paper and focus on the tap gesture.

2.2 Main Modules of an Authentication System Based on Touch Data

To authenticate a user based on touch operations, a user model has to be built for identifying him/her. Building such a model for a legitimate user requires a training phase in which touch data of a labeled user is collected. Then a feature extraction module define what features should be extracted from touch data. Finally, classifier should recognize users based on these features. Again to decide if a user is owner or impostor during authentication phase, feature extraction and classification modules are required. So the main key steps needed to be addressed are what and how. What features to extract (feature extraction's mission during enrollment phase) and how features can be used to recognize user (classifier's mission during verification phase). These modules are shown at Fig. 1.

3 Related Work

There are a number of efforts done to support solutions for implicit authentication for mobile's users. This section discusses a number of these solutions. Latent Gesture [4] collected a suite of behavioral features associated with a user interaction (i.e. touch

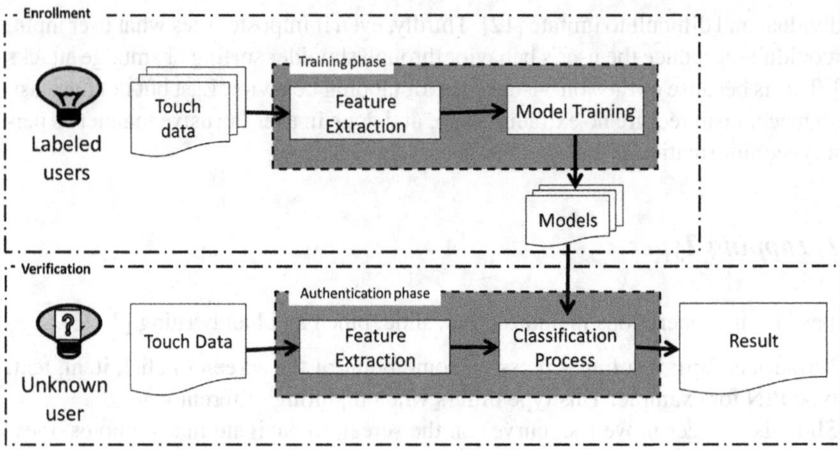

Fig. 1 User authentication typical system [10]

pressure, locations) on common user interface. This method used the common user interface to gather tapping behavior without any user interaction or any external hardware. It also makes usage of Support Vector Machine (SVM) and Random Forests classifiers to achieve the owner's identification which was 96.87 % TPR. However, almost of time the imposter uses this common user interface. This consumes much energy to analyze users interaction continuously.

Jain et al. [12] compared the Equal Error Rates(EER) obtained from the touch screen sensor with the rates of keystrokes in hardware keyboard. A developed keyboard application replaces the system keyboard to capture features in any application that uses a keyboard. The features are then stored in a SQLite database and classified by a one-class SVM. This method achieved EER of 10.5 % for keystroke data, 3.5 % for touch data and 2.8 % for all data (touch and keystroke). However, Additional sensors on the devices, such as gyroscopic and rotational sensors, are ignored in this study. These sensors could differentiate accurately touch data from the keyboard timing data.

Alariki et al. [2] has suggested a framework to be implemented later for authentication using touch biometrics. They review methods and features used for this field. Then, they proposed an approach in which a user has to enter gesture in any direction to build user's model to be matched later through a score computation. However, no implementation is done to show accuracy achieved by the proposed framework.

Table 1 summarizes the related work based on: # of input : needed training samples to build a model for owner, Classifier : classifier used to classify user and give a decision, Performance: performance achieved from the experiment, and Users: Number of users participated in the experiment.

Table 1 Tapping paper summary

Ref.	# of Inputs	Classifier	Performance
[4]	9 per user	Naive Bayes, Random Forests	TPR of 96.87 %
[13]	25 samples per PIN	Z score	EER of 3.65 %
[12]	A single passcode	1 class SVM	EER: 10.5 % Touch, 3.5 % All Data, 2.8 % keystroke
[1]	15009 samples	SVDE libSVM with RBF	EER of 0.5 %
[2]	6 trials per direction	SVM classifier	No implementation
[3]	min 15 samples	SVM classifier	FAR: 18 % for tap, 22 % for fling, 8 % for scroll
[5]	Thousands of actions	SVM classifier	EER of 20 %
[10]	Different inputs	SVM with RBF	EER of 10 %
[14]	Roughly 10 times	Protractor recognition algorithm	EER of 3.34 to 13.16 %

4 Proposed Solution

Our proposed solution consists of three phases: *Data collection, feature extraction and user's classification*. In the data collection phase, we have developed our own mobile App and used it for gathering touch data from different users, students and employees. In the feature extraction phase, from the data collected, the features of the pressure, position, size, and time of pressing were extracted for each user. The classifiers, e.g. SVM, were used in the classification phase to differentiate between the Mobile's owner and the impostor.

4.1 Data Collection

Touch data was gathered from the entered subscriber's code of Egyptian mobile operators (i.e., 0106 for Vodafone, 0128 for Mobinil, 0114 for Etisalat) and from the national landline telecommunication with 0643 code for a local phone code. These codes are chosen because it spans the keyboard: 6 for the leftmost side, 4 for the rightmost, 8 for the downside, 1 and 2 for the upside. While a device's owner uses his smartphone to contact any unsaved number which started with the subscriber's code. A mobile app was developed using Android OS. It was then run on a smartphone of Samsung Galagxy Note N7000.

The touch data was collected by two different modes: Captcha and free entry. For the Captcha mode as seen in Fig. 2, the developed touchscreen number board prompts

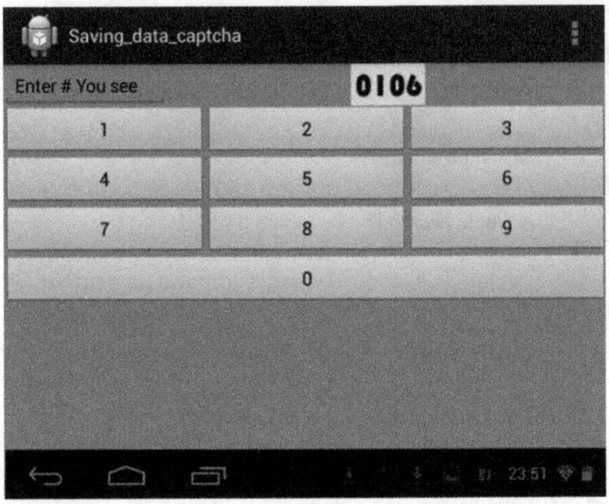

Fig. 2 Screenshot of data collection App: captcha mode

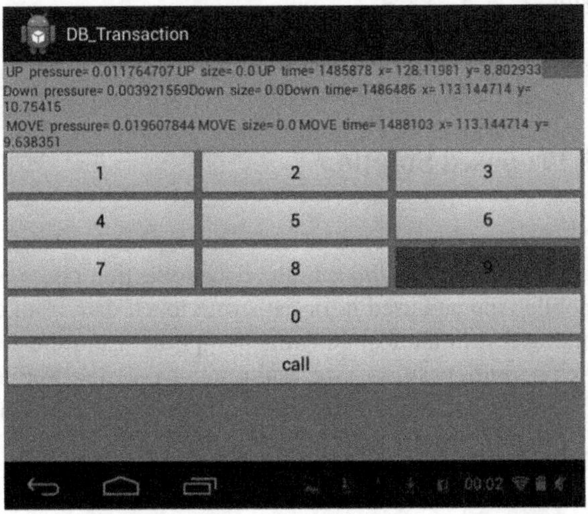

Fig. 3 Screenshot of data acquistion App for free entry

the captcha at the top right of the screen where a user can type in the text field. In our system, Captcha could be any subscription code of the Egyptian mobile operators (e.g. 0106 for vodafone, 0128 for Mobinil, 0114 for Etisalat and 0643 for local land-line phone code). For the **Free Entry mode** , a user could enter any free sequence of numbers, e.g. the full mobile number to complete the code entered through Capcha. Figure 3 shows the layout of free entry Android application for the data collection.

Participants Twelve participants were recruited for collecting touch data. Participants were with age range(20–50) and jobs ranging from faculty students to corporate employees. Also they were mixed between male and female. More details about the participants can be seen in Table 2.

Table 2 Type of participants

Participants	Age	Gender	Job	Used hand
4	22	Female	Undergraduate students	Right handed
2	20	Male	Undergraduate students	Right handed
1	20	Female	Undergraduate students	Right handed
1	23	Female	Employees	Left handed
1	25	Female	Employees	Right handed
1	50	Female	Employees	Right handed
1	28	Male	Employees	Right handed
1	50	Male	Employees	Right handed
12	**Total users**			

All participants followed the detailed procedure described below. They were free to use the mobile with any hand of their hand. One was left-handed and the others was right-hand for typing.

Data Collection Procedure The data collection procedure is described as follows. First of all, each participant has to sign in/up with any chosen username and password. During this process, ID is assigned to each participant to allow anonymous data collection. Secondly, he/she was asked to enter a network operator code for Captcha mode and other numbers for *Free Mode*. In this step, (1) the session lasted for about 20 min where participants sat on a chair and were reminded to enter Captcha code and any chosen number, (2) the phone was held in portrait orientation, (3) the participants were asked to touch screen naturally as they usually do while using their smartphone, and each participant typed 7 taps on the numbers keyboard: 4 in the same order as requested by Captcha and 3 in random order.

Data was submitted with the call key appeared in Fig. 3. Data are stored in Android DB inside mobile device and then exported into Excel file. The number of attempts was unlimited to get as much samples as possible for building the user model. However, the data, from 7 taps from each user's data, was used during the training and testing the system.

4.2 Feature Extraction

From the database collected in the above phase, as shown in Table 3, features of size, pressure, time, and position are recorded when touching any key during the raw touch events(Up, Down and Move). The *Up* action happens when a pressed gesture has finished while the *Down* action takes place when a pressed gesture has started. The *Move* action occurs when a change has happened during a press gesture (between ACTION_DOWN and ACTION_UP) while the *Time* action is the time taken during the pressing action. Table 3 shows the data which gathered from *one single user* using the data acquisition tool when touching some number.

Table 3 Tapping raw data sample

Tap	Pressure			Size			Time in milliseconds			Position					
	UP	DOWN	MOVE	UP	DOWN	MOVE	UP	DOWN	MOVE	UP		DOWN		MOVE	
										X	Y	X	Y	X	Y
1	0.5	0.5	0	0.234	0.234	0	67905343	67905242	0	50	32	50	32	50	32
1	0.45	0.45	0.45	0.2	0.17	0.2	68492904	68488901	68492854	49	42	49	68	49	42
1	0.65	0.65	0.55	0.134	0.134	0.1	69187800	69187686	69187201	94	86	94	86	213	75
2	0.3	0.83	0.3	0.034	0.167	0.04	69373956	69371134	69373906	188	63	169	52	188	63
2	0.99	0.99	0.99	0.2	0.2	0.2	69511770	69511673	69511684	214	57	214	57	214	57
2	1.15	1.15	1.15	0.3	0.5	0.3	69636591	69636421	69636541	237	82	237	82	237	82

These features are collected while tapping "1" and "2" on the developed soft keyboard. In this example, each tap "1" and "2" are repeated for three times. These values are used to build a feature vector for each user. Thus a feature vector for each tap on the screen consists of the shown 15 features for each user. These features can be described as followed:

1. Pressure: the finger's pressure when touching up/down on a soft key. This includes 3 features: pressure at touchup, pressure at touchdown and pressure at touchmove.
2. Size/Orientation: length and orientation of major and minor axes of finger-press. This includes 3 features: size up, size down and size move.
3. Tap timing: times of holding and releasing the soft key. This is also available on hardware keyboards. This includes 3 features: time up, time down, time move.
4. Position: the position where finger touches a soft key with x-y coordinates. This includes 6 features: xposition up, yposition up, xposition down, yposition down, xposition move and yposition move.

4.3 User Classification

The aim of the classification phase is to authenticate the legitimate owner of the smartphone. For this purpose, two classifiers, SVM (Support Vector Machine) with its four kernel functions [15], and the KNN (k Nearest Neighbor) [16] were used.

For training each classifier, supervised learning methods were used to build a classifying model on groups of patterns belonging to both the owner and other user's groups. During the training process, each sample was provided with its known class (user). This model was then used in the authentication phase for an unknown user. By this way, it is possible to authenticate the legitimate owner of the smartphone using his/her touch data.

5 Results and Analysis

A number of experiments were conducting to evaluate the proposed system, i.e. to decide if a user is the legitimate owner of a smartphone or an impostor. The KNN and the SVM with its kernel functions (linear, polynomial, quadratic,RBF, an MLP) were applied to the extract features The results obtained from the classifiers are shown at Table 4.

Our proposed system was evaluated with two well-known methods: error rates and accuracy (i.e. the number of true identification from the number of all identification attempts). For the error rates, the False Acceptance Rate (FAR) and the False Rejection Rate (FRR) were used [9, 10]. The FAR means the probability of accepting an impostor falsely, while the FRR means the probability of rejecting a rightful owner falsely. The FER and FAR are calculated according to Eqs. (1) and (2) respectively.

Table 4 Classifier performance

Classifier	SVM					KNN
	Quadratic	RBF	Linear	MLP	Polynomial	Euclidean distance
FRR	0.0853	0.1067	0.0587	0.7333	0.1040	0.0053
FAR	0.1040	0.0267	0.0907	0.4601	0.2213	0.1387
Correction Rate	83.20 %	89.33 %	89.87 %	48.55 %	67.73 %	96.80 %

$$FRR = \frac{N_{FR}}{N_{IA}} \tag{1}$$

where N_{FR} is the number of false rejections and N_{IA} is the number of identification attempts

$$FAR = \frac{N_{FA}}{N_{IA}} \tag{2}$$

where N_{FA} is the number of false acceptances and N_{IA} is the number of identification attempts

For a good authentication system, FAR and FRR rates should be as small as possible. As it can be seen from Table 4, the smallest value of FRR(0.0053) was achieved by KNN with the Euclidean Distance while the smallest value of FAR (0.0267) was achieved by the SVM with RBF kernel. From these results, the following remarks can be drawn. Firstly, the KNN classifier with the Euclidean Distance was the best by achieving the highest correction rate and the lowest FRR. Secondly, these features are able to distinguish stroke behavior among users (discriminating users). Last but not least, the seven taping of different numbers provided by collected for a participant is encouraging as it the results that even users although users with few times touching the soft keyboard (i.e. only 7 taps) can still be a rich source of data to distinguish among own and impostor.

6 Conclusion and Future Work

This paper presented a proposed system for authenticating smartphone's users based on their behavior while touching their mobile screen. We built our own dataset by developing a mobile App and recruiting a number of participates from different background and ages. Feature are then extracted from the collected data and then SVM with its 4 kernel functions and KNN classifiers are used to classify the legitimate owner of a mobile and an impostor. Our proposed system was evaluated using FRR and FAR error rates. It was found that the smallest value of FRR(0.0053) was achieved by KNN with the Eculidean Distance while the smallest value of FAR (0.0267) was achieved by the SVM with RBF kernel. Also based on the accuracy rate, it was found that our system is comparable with related work achieving rate at 96.8 %. For the future work, we plan to collect more data by increasing the number of participates and then try other classifiers, e.g. Random Linear Oracle.

Acknowledgments This paper has been elaborated in the framework of the project New creative teams in priorities of scientific research, reg. no. CZ.1.07/2.3.00/30.0055, supported by Operational Programme Education for Competitiveness and co-financed by the European Social Fund and the state budget of the Czech Republic and supported by the IT4Innovations Centre of Excellence project (CZ.1.05/1.1.00/ 02.0070), funded by the European Regional Development Fund and the national budget of the Czech Republic via the Research and Development for nnovations Operational Programme.

References

1. Muhammad S, Alex XL, Arjmand S (2013) Secure unlocking of mobile touch screen devices by simple gestures: you can see it but you can not do it. In: Proceedings of the 19th annual international conference on mobile computing & #38; networking. MobiCom '13, New York, NY, USA, ACM (2013), pp 39–50
2. Ala AA, Azizah AM (2014) Touch gesture authentication framework for touch screen mobile devices. J Theor Appl Inf Technol 62(2)
3. Cheng B, Lan Z, Xiang-Yang L (2013) Silentsense: silent user identification via touch and movement behavioral biometrics. In: Proceedings of the 19th annual international conference on mobile computing & networking. MobiCom '13, New York, NY, USA, ACM, pp 187–190
4. Premkumar S, Samuel C, Duen HPC, Hongyuan Z (2014) Latentgesture: active user authentication through background touch analysis. In: Proceedings of the second international symposium of Chinese CHI. Chinese CHI '14, New York, NY, USA, ACM, pp 110–113
5. Cheng B, Lan Z, Xiang-Yang L (2013) Silentsense: silent user identification via dynamics of touch and movement behavioral biometrics. arXiv preprint arXiv:1309.0073
6. Vibha KR (2011) Integration of biometric authentication procedure in customer oriented payment system in trusted mobile devices. Int J Inf Technol Conver Serv 1(6):15–25
7. Kresimir D, Mislav G (2004) A survey of biometric recognition methods. In: 46th International symposium on electronics in Marine. Proceedings Elmar 2004. IEEE pp 184–193
8. Tharwat A, Ibrahim A, Ali H (2012) Personal identification using ear images based on fast and accurate principal component analysis. In: 8th international conference on informatics and systems (INFOS), IEEE MM-56
9. Reham A, Tarek G, Ghada E, Aboul Ella H (2014) Biometric and traditional mobile authentication techniques: Overviews and open issues. In: Hassanien AE, Kim TH, Kacprzyk J, Awad AI (eds) Bio-inspiring cyber security and cloud services: trends and innovations, vol 70. Intelligent Systems Reference LibrarySpringer, Berlin Heidelberg, pp 423–446
10. Hui X, Yangfan Z, Michael RL (2014) Towards continuous and passive authentication via touch biometrics: An experimental study on smartphones. In: Symposium on usable privacy and security (SOUPS 2014), Menlo Park, CA, USENIX Association, pp 187–198
11. Hugo G, Sebastian U, Christopher W, Konrad R (2014) Continuous authentication on mobile devices by analysis of typing motion behavior. In: Sicherheit, vol P-228, Gesellschaft fr Informatik, Bonn, pp 1–12
12. Lohit J, John VM, Michael JC, Charles CT (2014) Passcode keystroke biometric performance on smartphone touchscreens is superior to that on hardware keyboards. Int J Res Comput Appl Inf Technol 2:29–33
13. Nan Z, Kun B, Hai H, Haining W (2014) You are how you touch: user verification on smartphones via tapping behaviors. In: IEEE 22nd international conference on network protocols (ICNP), IEEE, pp 221–232

14. Michael S, Gradeigh C, Yulong Y, Shridatt S, Arttu M, Janne L, Antti O, Teemu R (2014) User-generated free-form gestures for authentication: security and memorability. In; Proceedings of the 12th annual international conference on mobile systems, applications, and services, pp 176–189
15. Tharwat A, Gaber T, Hassanien AE, Hassanien HA, Tolba MF (2014) Cattle identification using muzzle print images based on texture features approach. In: Abraham A, Kömer P, Snášel V (eds) Proceedings of the fourth international conference on innovations in bio-inspired computing and application IBICA 2014, vol 303. AISCSpringer, Heidelberg, pp 217–227
16. Tharwat A, Gaber T, Hassanien AE, Shahin M, Refaat B (2015) Sift-based arabic sign language recognition system. In: Afro-european conference for industrial advancement, vol 334. Springer International Publishing, pp 359–370

Part II
Secure Multimedia Applications

Development of Iris Security System Using Adaptive Quality-Based Template Fusion

M.M. Eid, M.A. Mohamed and M.A. Abou-El-Soud

Abstract Recently, the interface of computer technologies and biology has an enormous impact on society. Human recognition research projects promise a new life to various security consulting. Iris recognition is considered to be one of the exceedingly reliable authentication systems. To account for iris data variations, nearly of all iris systems store multiple templates per each user. Approaching to overcome the storage space and computation overheads, this paper proposes an intelligent fusion technique, algorithms, and suggestions. The quality of the input image has been checked firstly to ensure that "qualified iris samples" will be only treated. The proposed system, with the aid of the image selection stage and template quality test, has the advantage of being adaptive and simple, but can come at the expense of reject extremely inadequate data for any user. Complete the eye shape using the convex hull is the main key for image selection module. Shape-based thresholding is integrated with morphological features to avoid the dark iris problems, elliptical pupil and iris shapes. For best recognition rate, the optimum values of the 1D log Gabor filter parameters and different sizes of the iris code are recorded. From experimental results, an HD value of 0.4 can be chosen as a suitable separation point, and the optimum code size was found to be 20 × 480. An experimental work reveals a reduction in database size by nearly a 78 % and an increase of verification speed of about 85.25 % is achieved while 14.75 % of computation time in the shifting process is only required. Comparing with existing algorithms, the proposed algorithm gives an accuracy of 96.52 and 99.72 % for TRR and TAR, respectively for closed loop tested dataset. In addition, with a number of experimentations, the proposed algorithm gives the lowest FAR and FRR of 0.0019 and 0.00287 % respectively and EER of 0.0024 %.

M.M. Eid
Higher Institute of Engineering, Delta University, Mansoura, Egypt
e-mail: marwa.3eeed@gmail.com

M.A. Mohamed (✉) · M.A. Abou-El-Soud
Faculty of Engineering, Mansoura University, Mansoura, Egypt
e-mail: mazim12@yahoo.com

M.A. Abou-El-Soud
e-mail: mohyldin@yahoo.com

A. Abraham et al. (eds.), *Intelligent Data Analysis and Applications*,
Advances in Intelligent Systems and Computing 370,
DOI 10.1007/978-3-319-21206-7_23

Keywords Convex hull • Circular hough transform (CHT) • Discrete wavelet transform (DWT) • Equal error rate (EER) • Histogram concavity

1 Introduction

Recently, with an expanding emphasis in the iris recognition field, numerous researchers substantially concern with accomplishing more flexible, faster, and reliable system. Various drawbacks of using a single still image can be avoided via the high-resolution, extremely expensive, acquisition mechanism with speedy frame rate. Conventional iris systems routinely store diverse templates for a separate individual user [1–3]. For the sake of reducing the search time and computational complexity, it is desirable to classify irises in an adequate and consistent framework. Ma et al. [4] suggested analyzing several images and keeping the best-quality image. Three templates of a given iris had been employed and they recorded the average of triple scores as the final matching scores [1–6]. Du et al. [7] rendered every subject with three images, and they perpetuated the minimum proportion of their association measures. Du et al. [7] noted that applying several recorded images contrary to a unique template improved the rank-one recognition rate.

Furthermore, privacy protection for the individuals is extremely critical issues in all authentication systems. Each stored iris template will be conserved for a long period of time, thus and so it may be liable to any breakthroughs and extremely exposed to any cracking or attacking. To handle these problems, an exceptional base template is created which could be rescinded and reissued via blending the different frames when the old ones are lost or stolen as in [4]. In this paper, dynamic iris template fusion method could generate a distinct number of base templates based on the original iris patterns whenever the existing one is stolen or hacked.

Computing the convex hull of the complemented eye shape and comparing the percentage of pupil size and occlusion for the available patterns are the principal keys for an image selection stage. The foremost problem of the proposed method is binarization arises in the case of persons having dark iris [7, 8]. Shape-based optimum thresholding technique has been implemented to avoid all the dark iris obstacles and perform isolation process easier and extensively accurate. Toward templates integrating, a fused template is created based on DWT multiresolution fusion. Each qualified quality image could be combined with other still images to produce an improved one. The choice between using two, three, or seven base templates for each user is adaptable according to the quality of available data. The generative iris template can handle missing iris data in a certain template, by including its counterpart from another available iris sample. A quality-based features fusion is presented that gives higher weight to the evidence from only reliable templates. In this paper, with the aid of the proposed image selection strategy and template quality test, a dynamic image fusion technique based on wavelet transform will be presented.

2 The Proposed Iris Recognition System

The iris is an overt body that is available for remote noninvasive assessment. Mainly in the enrollment phase, the system stores the created template code of iris features in the system database. In addition, in the identification phase the system compares the created template code of iris features with all the templates stored in the system to identify the individuals.

2.1 Image Selection Stage

Iris-Imaging Issues: Poor quality patterns are the principal causes for matching errors and the implementations vulnerability [9–12]. Chinese Academy of Sciences Institute of Automation (CASIA) iris image database version 1.0 [8], particularly, a dataset of 100 persons every person has a database of 7 images, i.e., 700 images.

Image Selection Stages in Enrollment Phase: The proposed algorithm firstly checks the quality of the input images by comparing the percentage of the pupil size and the occlusion for each separate image. Beginning with 30 % allowed noise occlusion of image size. If no input, samples are proper, the occlusion threshold auto-increase 5 % commission storing more data; images that surpass the certain threshold will be particularly processed according to the following steps:

Step-1 Enroll iris samples I, system count number of available samples '$N_{samples}$', number of bits 'N_{bits}', and initialize failed images, N_f, and good images N_g.

Step-2 Enrolled images are converted to black and white images according to its suitable threshold, subtract background, complete and specify neighborhoods of only the eye shape mask all unexpected factors affecting the location of the iris, complete the convex hull of the complemented eye shape as:

$$\left\{ \sum_{i=1}^{|S|} \alpha_i x_i \middle| (\forall_i : \alpha_i \geq 0)^\wedge \sum_{i=1}^{|S|} \alpha_i = 1 \right\} \quad (1)$$

where the convex hull of a finite point set $S \in \Re^n$ is the set of all convex combinations of its points, it forms a convex polygon when $n = 2$, or more generally a convex polytope in $\Re^n$.

Step-3 Begin with allowed eyelids and eyelashes noise threshold 'T_{occ}', the number of connected pixels formed pupil area 'P.S', and accepted pupil threshold 'T_{pup}' as:

Case-a: T_{occ} = 30 % of image size (N_{bits}), T_{pup} = 5 % of image size (N_{bits}), and 'P.S' = 0 [started-point].

Case-b: Adapt T_{occ} = T_{occ} + 5 % of image size (N_{bits}), T_{pup} = T_{pup} + 5 % of image size (N_{bits}), and adapt 'P.S' if calculated [adapted-point].

Step 4 Check the percentage of occlusions for each image 'I_{occ}', with T_{occ}:

Case-a: If (I_{occ} < = T_{occ}) then compute the percentage of pupil size [then step-5].

Case-b: If (I_{occ} > T_{occ}) then count N_f, and check T_{occ}, and T_{pup} (If (T_{occ} > 55 % N_{bits} & T_{pup} > 30 % N_{bits}) [terminate, if not [then case-c]).

Case-c: Check N_f, If (N_f > 40 % $N_{samples}$) [go to case-b in step-3 with predefined 'P.S' and I_{occ} for each image as no repetition for the same image]), and if not [then step-5 to check 'P.S']).

Step-5 A specific morphological operations such as opening and erosion are applied to the binarized images to extract pupil area as in [1]. Compute the size of pupil 'P.S' for each image, to manner step repetition; store it for this certain sample. Then, check 'P.S' between available images, less 'P.S' is better, according to the following strategy:

Case-a: If (P.S <= Tpup) then store good quality sample [go to iris localization stage & count NG].

Case-b: If (P.S > Tpup) then [go to case-b in step-3]), results are shown in Fig. 1 and Table 1.

Image Selection Stages in Identification Phase: The proposed module could detect and discard the faulty images, which don't possess enough information to identify persons as in the following steps:

Step 1 Enroll iris samples 'I', define sample size (number of bits) 'N_{bits}'.

Step 2 Compute 'I_{occ}' as [step-2 in enrollment phase] and check I_{occ} as:

Case-a: If (I_{occ} < 55 % Nbits), then [go to case-b in this step] and if not, i.e., (I_{occ} >= 55 % identification])

Case-b: Compute the percentage of pupil size 'PS' as [step-5 in enrollment phase].

Step 3 Check the percentage of pupil size for each image 'P.S' as:

Case-a: If (P.S < 30 % N_{bits}) then localize good quality sample [go to iris localization stage].

Case-b: If (P.S. >= 30 % N_{bits}) then [Terminate, as the input is not suitable]).

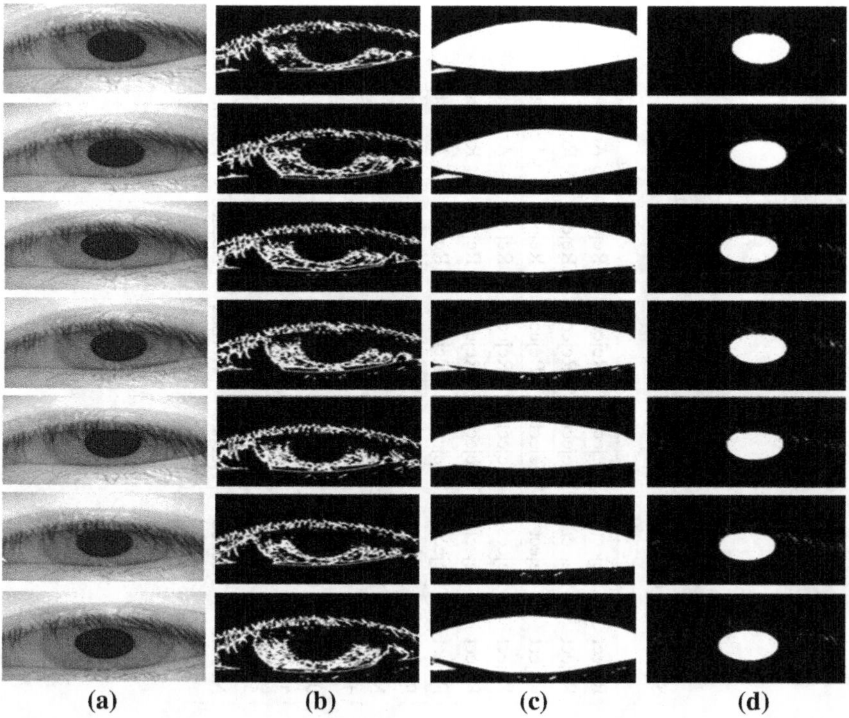

Fig. 1 Samples of the proposed image selection: **a** input samples, **b** binarized images, **c** extracted eye shapes, and **d** the extracted pupils

2.2 Iris Localization

Daugman approach based on integro-differential operator is the best known tested algorithm through the history of iris localization [14]. Furthermore, Wildes [15] in its well-known algorithm establishes the iris boundaries via creating a binary edge map using gradient-based edge detection. Several attempts have been proposed in a robust iris localization based on morphological features via previous experiments [16, 17, 19]. Immediately, noise occlusion is the foremost cause of the localization failures, samples of rejected unfortunate instances.

2.3 Iris Normalization Process

From the 'doughnut' iris region, the homogenous rubber sheet model devised by Daugman produces a 2D array in polar coordinates. In order to prevent non-iris region data from corrupting the normalized representation data points, another 2D array was created for marking noise occluded [18].

Table 1 Sample of image selection result

Input samples	Noise occlusion (I_{occ} %)	Pupil size (I_{pup} %)	System response					
Case_1_1	54.7891	7.693	Reject	Reject	Reject	Reject	Reject	Accept
Case_1_2	55.1741	7.073	Reject	Reject	Reject	Reject	Reject	Reject
Case_1_3	54.8348	7.099	Reject	Reject	Reject	Reject	Reject	Accept
Case_2_1	53.7076	7.35	Reject	Reject	Reject	Reject	Reject	Accept
Case_2_2	56.9665	7	Reject	Reject	Reject	Reject	Reject	Reject
Case_2_3	54.3002	7.118	Reject	Reject	Reject	Reject	Reject	Accept
Case_2_4	48.5949	7.805	Reject	Reject	Reject	Reject	Accept	Accept
Adaptation process for the input samples	Image counter		Number of Images each trial					
	Rejected images N_f		7	7	7	7	6	2
	Accepted images N_g		0	0	0	0	1	5
	Occlusion threshold T_{occ}		30	35	40	45	50	55
	Pupil size threshold T_p		5	10	15	20	25	30
	Adaptation Request ($N_f > 50\ \%\ N_s$)		Needed	Needed	Needed	Needed	Needed	Aborted and Stored

2.4 Template Quality Test

However, all noise factors are seemed to be isolated, parts of the unwrapped images might be included with occlusion. Towards an equitable evaluation of the generated code, it is significant to appraise the percentage of iris texture. Moreover, to diminish the system complexity and the elapsed time, template quality examination will be easily accomplished via the statistical analysis earlier to the fusion stage.

Template Quality Test in Enrollment Phase: Presently, the challenge is detecting false-segmented image and enable the system to ignore them if it allowed. With the aid of this module, the percentage of occlusions and iris texture could be computed and checked using quality measures. The quality measures will be adapted according to the quality of input data as in the following strategy:

Step-2 Compute 'I_{occ}' as [step-2 in enrollment phase] and check I_{occ} as:

Case-a: If (I_{occ} < 55 % N_{bits}), then [go to case-b in this step] and if not, i.e., (I_{occ} >= 55 % identification])

Case-b: Compute the percentage of pupil size 'P.S' as [step-5 in enrollment phase].

Step-1 Enroll 'N_g' result from image selection stage, unwrapped iris template and the corresponding noise mask 'M', define mask size (number of bits) 'M_s', and start the failed templates, 'T_f' = 0, and good templates 'T_g' = 0.

Step-2 Compute the percentage of mask noise occlusion 'M_{Occ}' via: Firstly, get the local standard deviation in a 3 × 3 around each pixel in unwrapped iris mask. Secondly, binarize the result image, and then compute the percentage of noise occlusion according to the number of zero elements in the result mask.

Step-3 Begin with allowed eyelids and eyelashes noise threshold 'MT_{occ}'

Case-a: MT_{occ} = 25 % of mask size (M_s) [started-point].

Case-b: Adapt $MT_{occ} = MT_{occ}$ + 5 % of mask size (M_s)[adapted-point].

Step-4 Check the percentage of occlusions for each mask 'M_{Occ}', with MT_{occ}:

Case-a: If (M_{occ} < = MT_{occ}) then [go to template fusion stage (store the template and corresponding mask as it can be qualified as a base template) & count T_G].

Case-b: If (M_{Occ} > MTocc) then check T_f, and MTocc, (If (T_f > 90 % Ng & MT_{occ} > 50 %) then [terminate, as the input templates are not suitable for enrollment]), and if not then [go to case-b in step-3]).

Template Quality Test in Identification/Verification Phase: The amount of available iris region can affect the recognition performance. The quality of the result depends on the presence of eyelashes, glare and light spots on the eyelids borders. In practice, these factors caused defects in identification and verification results, which do not lead to a perfect result for all iris cases. It is important to localize the richness of iris texture in tested template to take the decision correctly. In this unit, an occlusion score is developed to estimating the percentage of significant pixels of the input image and is checked as follows:

Step-1 Enroll 'N_g' result from image selection stage, unwrapped iris template and the corresponding noise mask 'M', define mask size (number of bits) 'M_s'.

Step-2 Compute the percentage of mask noise occlusion 'M_{Occ}' as [step-2 in enrollment phase] and check M_{occ} as: if ($M_{oc}c$ < 50 % M_s), then [go to feature encoding stage; template is qualified] and if not, i.e., (I_{occ} >= 50 % $M_{s)}$ [Terminate, as the input template is not suitable for verification/identification]).

Moreover, the strategies for both phases and experimental tests showed that the local entropy with the standard deviation map could provide results that are more reliable. Recently, the field researchers have attempted to address this problem with varying degrees of accomplishment. Matej et al. [19] proposed a global quality index operating in the cosine transform (DCT) field. Through comparing the proposed quality check measures with the Matej et al. [19] quality index based on DCT, the same result in reordering the samples according to its quality is easily obtained; Fig. 2, e.g. of template quality test, rejecting the last three templates according to percentage of occlusion as shown in Fig. 3 and Table 2.

2.5 Multi-sample Dynamic Fusion

The image fusion procedure could combine various input images or their features into a single representation without information losses [2, 20]. The fusion procedure

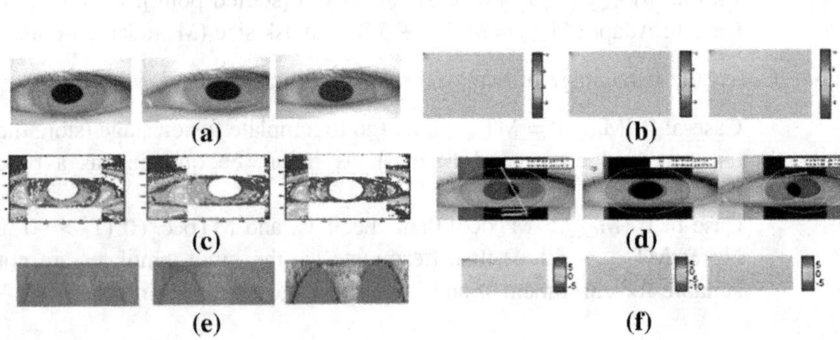

Fig. 2 Deformation of the iris area: **a** different samples from CASIA database, **b** the logarithm normalized DCT images, **c** localized iris after noise masking in spatial domain, **d** matching iris areas between each two different pairs in spatial domain, **e** corresponding polar templates, and **f** logarithm normalized DCT for each unwrapped image; the coefficients with higher values are shown in red

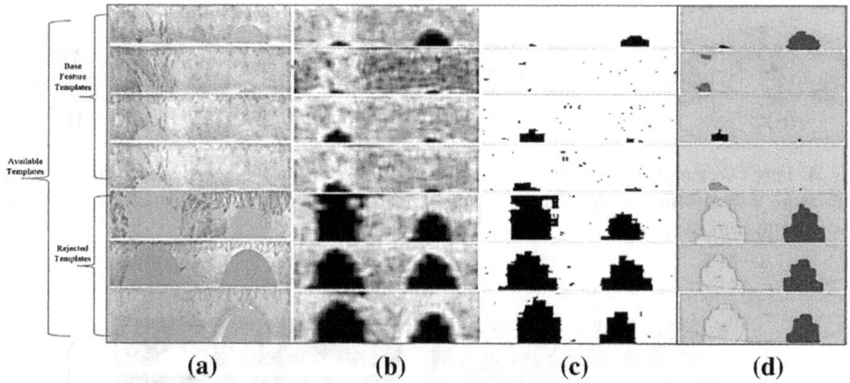

Fig. 3 Template quality test for dynamic fusion stage **a** 7-generated iris templates, **b** corresponding local entropy, **c** noise map, and **d** detected noise percentage

Table 2 Sample of template quality test result

Feature measure	Entropy	Standard deviation	Percentage of noise (%)	Order	System response
Template-1	5.038	8.3270	2.9722	Second	Accepted
Template-2	5.571	12.155	0.3750	First (Best)	Accepted
Template-3	5.166	9.3113	3.5208	The fourth	Accepted
Template-4	5.228	9.3113	3.1250	The third	Accepted
Template-5	3.979	6.1199	31.9375	The last	Rejected
Template-6	4.538	11.069	31.4167	The sixth	Rejected
Template-7	4.634	10.124	28.5208	The fifth	Rejected

based on DWT generates a new composite multiresolution representation utilizing a certain fusion rule [19]. Although numerous scenarios and results of iris fusion could be applied in the spatial and polar representation at level 2 using db2 using various fusion rules, the iris area will be deformed in whole cases as shown in Fig. 4. Consequently, iris templates fusion should be performed in the polar domain by taking the maximum values for both approximations and details for base templates and minimum values for masks; Fig. 5.

2.6 Iris Encoding and Matching

Convolving the final template with 1D Log-Gabor wavelets; the output of filtering is then phase quantized to four levels using the Daugman method [17], with each filter producing two bits of data for each phasor. Hamming Distance (HD) could be utilized as a matching metric for iris templates comparing. False accept rate (FAR),

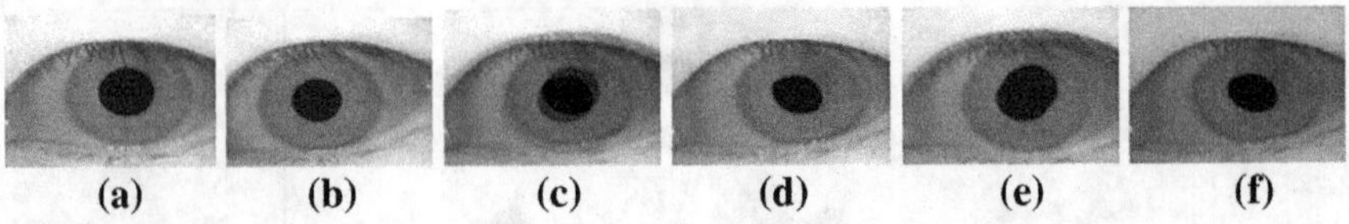

Fig. 4 Fuse two sample: **a** first image, **b** first image, **c** fusion using average, **d** second fusion using minimum, **e** third fusion using average of approximations and minimum of details, and **f** fourth fusion of maximum

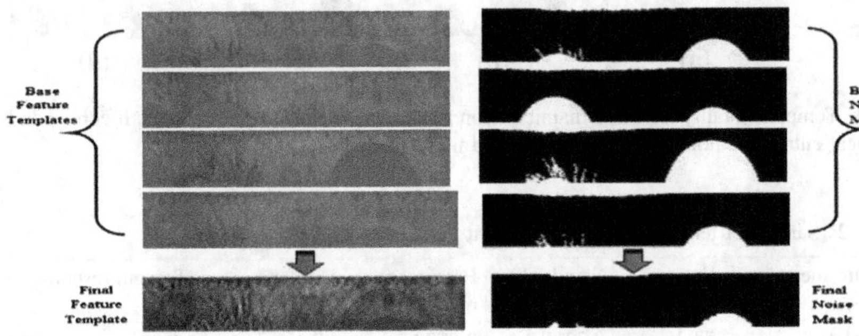

Fig. 5 Sample of dynamic fusion process applied for the 4-qualified iris templates

false reject rate (FRR), equal error rate (EER), true accept rate (TAR) and true reject rate (TRR) [14, 15] could be the evaluation parameters for the system experiments through encouraging in terms of reducing EER furthermore increasing TRR.

3 Results and Discussions

Clearly, the separation point will influence the error rates, since a lower separation HD will decrease FAR while increasing FRR, and vice versa. The optimized parameters will be taken in fusion stage. From experimental results a HD value of 0.4 can be chosen as a separation point, so that if any two templates generate a HD value greater than 0.4, they are deemed to be generated from different irises. If two templates generate a HD value, lower than 0.4 then the two are deemed to be from the same iris as shown in Fig. 6. The optimum code size for the data set was found to be 20 × 480, which is a radial resolution of 20 pixels, and an angular resolution of 240 pixels for best performance rate. Code size with 20 × 240 also offers good recognition rate with less memory and time requirements as shown in Fig. 7.

There is a tradeoff between the number of templates, and the storage and computational overheads introduced by multiple templates. From 700 original eye images from CASIA database, image quality modules auto-reject 22 bad images from different subjects. This phase results in 678 base feature templates and 678

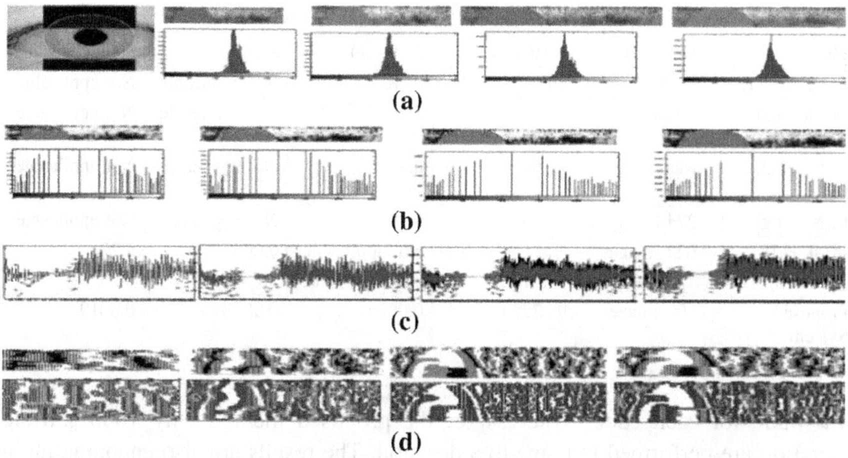

Fig. 6 Normalization: **a** polar templates [(20 × 60), (20 × 120), (20 × 240), and (80 × 180)] respectively and the corresponding histogram, **b** enhanced templates using histogram equalization and the corresponding histogram, **c** a box plot of each template data, **d** the encoded iris pattern template respectively and its corresponding contour plot

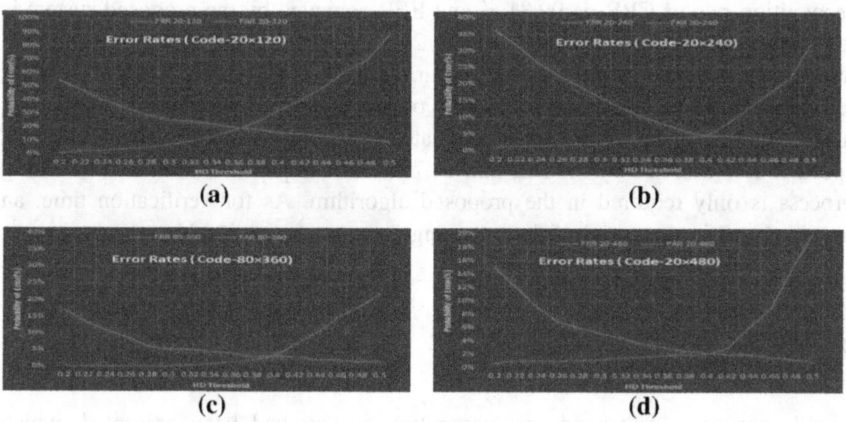

Fig. 7 Error rates: **a** 20 × 120, **b** 20 × 240, **c** 20 × 480, and (80 × 360) bit resolution respectively at different HD threshold

extracted base masks. The available base templates are fused into a final template and an associate mask weight resulting in a 100 final mask and a 100-weight mask for 100 persons considered. Each template is compared with all the enrolled final templates and a HD is estimated, result in 490,000 comparisons, whereas for the proposed method only 4000 comparisons is needed. According to Libor Masek algorithm [17] 16 shifts are performed for each template comparison. Thus, for the 700 base templates one would perform 11,200-shift operation to find the minimum

Table 3 Comparison iris recognition methodologies with the proposed method

Methods	Data size	TAR (%)	EER (%)	FAR (%)	FRR (%)
Daugman [13]	4258 images	100	0.08.	Not applicable	Not applicable
Wildes and Asmuth [15]	60 images	99.90	1.76	Not applicable	Not applicable
Boles et al. [24]	Real images	92.64	8.13	Not applicable	Not applicable
Li Ma et al. [4]	2245 images	99.60	0.29	Not applicable	Not applicable
Mask [17]	624 images	Not applicable	Not applicable	0.005	0.238
Jong et al. [21]	756 images	98.21	Not applicable	5.26	0.25
Proposed System	700 images	99.7234	0.0024	0.00287	0.0019

HD value for each code. Whereas for the proposed method only 1600 shifting operation are performed to come to a decision. The results are also encouraging in terms of accuracy, efficiency and reduced computational complexity and the time elapsed for the system. Results are compared with published result of methods [15–17, 23].

Table 3 provides a comparative study of the most famous presented techniques. The proposed method shows encouraging results with reduced EER of 0.0288 %, recognition rate of CRR, is 99.81 % and EER achieved by the proposed system is less than methods proposed by [4, 15, 17, 24]. Yet, a great deal of computation time and storage capacity saving is achieved in verification phase. The storage capacity requirements of the proposed method is only 8 KB and conventional approaches needs 56 KB for each iris-stored templates. The improvement in computational speed is also asserted by the fact that 14.75 % of computation time in the shifting process is only required in the proposed algorithm. As for verification time, an improvement in the speed of the matching process was noticeable.

4 Conclusion

In this paper, an enhanced iris recognition system had been presented, which combines Gabor filters and statistical operators to characterize iris texture and a quality check module to reject inadequate quality images. Attention was drawn towards the current trend of the use of image fusion techniques, especially in the qualified iris data. An improved dynamic fusion scheme was introduced based on feature multiresolution fusion level. With the support of adapting occlusion threshold in quality test modules, different number of base templates could be taken into account. Therefore, only fitting iris images possessing sufficient texture information and correctly segmented could be just processed. Moreover, this method could simply discard very improper data for any user. From experimental results analysis, the proposed approach has capable of handling various problems such as invariance to noisy instances, occlusion, and the deformed elliptical pupil and iris shapes. Comparing with existing algorithms, the proposed algorithm gives the average accuracy of 96.52 and 99.72 % for TRR and TAR respectively for only closed loop tested dataset. A Hamming distance value of 0.4 had been chosen as a separation point gave on the average the lowest FAR and FRR of 0.0019 and 0.00287 % respectively and EER of 0.0024 %.

References

1. Garg A (2013) Comparative survey of various iris recognition algorithms. Int J Comput Commun Inf Technol 1(1):47–59
2. Campos S, Salas R, Allende H, Castro C (2009) Multimodal algorithm for iris recognition with local topological descriptors. In: Progress in pattern recognition, image analysis, computer vision, and applications, pp 766–773
3. Bowyer K, Hollingsworth K, Flynn P (2008) Image understanding for iris biometrics: a survey. Comput Vis Image Underst 110(2):281–307
4. Ma L, Tan T, Wang Y (2003) Personal identification based on iris texture analysis. IEEE Trans Pattern Anal Mach Intell 25(12):1519–1533
5. Ross A, Jain AK (2003) Information fusion in biometrics. Pattern Recogn Lett 24 (13):2115–2125
6. Wildes RP, Asmuth J, Green G, Hsu S, Kolczynski R, Matey J, McBride S (1994) A system for automated iris recognition. In: Proceedings IEEE workshop on applications of computer vision, Sarasota, pp 121–128
7. Du Y, Ives RW, Etter DM, Welch T (2006) Use of one-dimensional iris signatures to rank iris pattern similarities. Opt Eng 45(3):1–10
8. Institute of Automation, Chinese Academy of Sciences. CASIA iris image database (2004). http://www.sinobiometrics.com. Accessed 24 April 2015
9. Daugman J (1993) High confidence visual recognition of persons by a test of statistical independence. IEEE Trans Pattern Anal Mach Intell 15(11):1148–1161
10. Zuo J, Schmid NA (2013) Adaptive quality-based performance prediction and boosting for iris authentication: methodology and its illustration. IEEE Trans Inf Forensics Secur 8 (6):1051–1060
11. Mahlouji M, Noruzi A (2012) Human iris segmentation for iris recognition in unconstrained environments. Int J Comput Sci Issues 9(1):149–155
12. Belcher C, Du Y (2008) A selective feature information approach for iris image quality measure. IEEE Trans Inf Forensics Secur 3(3):572–577
13. Daugman J (2004) How iris recognition works. IEEE Trans Circ Syst Video Technol 14 (1):21–30
14. Daugman J (2003) The importance of being random: statistical principles of iris recognition pattern recognition society 36(2):279–291
15. Wildes RP, Asmuth JC (1996) A machine-vision system for iris recognition. Mach Vis Appl 1 (9):1–8
16. Rosenfeld A, Torre PDL (1983) Histogram concavity analysis as an aid in threshold selection. IEEE Trans Syst Man Cybern 13:231–235
17. Masek L (2003) Recognition of human iris patterns for biometric identification. Technical report, The School of Computer Science and Software Engineering, The University of Western Australia
18. Russ J (1998) The image processing handbook. CRC Press LLC, Boca Raton
19. Matej K, Pers J, Matej P, Kovaci S (2006) A bayes-spectral-entropy-based measure of camera focus using a DCT. Pattern Recogn Lett 27(13):1431–1439

20. Sahu DK, Parsai MP (2012) Different image fusion techniques–a critical review. Int J Mod Eng Res 2(5):4298–4301
21. Vatsa M, Singh R, Noore A (2008) Improving iris recognition performance using segmentation, quality enhancement, match score fusion, and indexing. IEEE Trans Syst Man Cybern Part B Cybern 38(4):1021–1035
22. Du Y (2006) Using 2D log-Gabor spatial filters for iris recognition. In: Proceedings of SPIE 6202: Biometric Technology for Human Identification III, 62020, pp 1–8
23. Ko JG, Gil YH, Yoo JH, Chung KI (2007) Method of iris recognition using cumulative-sum-based change point analysis and apparatus using the same. US Patent 20,070,014,438
24. Boles W, Boashash B (1998) A human identification technique using images of the iris and wavelet transform. IEEE Trans Signal Process 46(4):1185–1188

3D Model Reconstruction Method Using Data Fusion

E.A. Othman, Abdelhameed Ibrahim and M.A. Mohamed

Abstract Recently, significant efforts have been done in order to get more accurate 3D models which can be used in several applications. This paper proposes a 3D reconstruction method to improve the 3D modeling results. The proposed method uses two images for each scene taken from different viewpoints in the reconstruction process. Using different algorithms for feature extraction and matching, multiple 3D models can be obtained. We use different fusion techniques, such as addition, simple average, Principal Component Analysis (PCA) and Discrete Wavelet Transform (DWT), for the reconstructed models to produce a more accurate 3D models. The most desirable information and characteristics of each input 3D model are preserved. We focus on modeling from reality rather than computer graphics. The effectiveness of the proposed framework is illustrated in experiments.

Keywords Data fusion · 3D modeling · Feature extraction · 3D reconstruction

1 Introduction

The reconstruction of the 3D modeling is one of the most important research areas in computer vision. Because of the development of computer hardware and software, 3D models have been used in many applications such as creating movies, computer games, face recognition [1, 2], industry applications [3]. 3D models can be obtained using different methods [4, 5]. A common method is based on two images of a scene acquired by a fully calibrated camera or a regular hand-held camera, named left and right views. Extracting features and then matched find correspondences between the images to restore the 3D structure.

E.A. Othman
Delta Academy of Science for Engineering and Technology, Mansoura, Egypt

A. Ibrahim (✉) · M.A. Mohamed
Faculty of Engineering, Mansoura University, Mansoura, Egypt
e-mail: afai79@mans.edu.eg

A. Abraham et al. (eds.), *Intelligent Data Analysis and Applications*,
Advances in Intelligent Systems and Computing 370,
DOI 10.1007/978-3-319-21206-7_24

There are various algorithms that can be used for extracting features and matching processes such as Scale Invariant Feature Transform (SIFT) [6, 7], Kanade Lucas Tomasi (KLT) [8], and Speeded up Robust Features (SURF) [9]. SIFT is one of the most popular algorithms to extract stable feature points with robustness against occlusions. Spatial intensity information can be extracted by KLT which can track features and directs the search to the best match position. SURF is fast and more robust against different image transformations.

Data fusion combines data from different data sources together [10–12]. The objective of applying fusion techniques in this paper is to produce a fused 3D model that provides a more accurate 3D models since fusing multiple 3D models together produces a more efficient representation of the original models. The fusion process can take place at different levels of pixel, feature, and decision levels. Combination of raw data from multiple source images into a single image is pixel level fusion. Fusion at feature level requires extraction of features from source data before features are merged together while combines results of multiple algorithms to yield a final fused decision is called decision level fusion [13].

In this paper, we propose an improved 3D modeling method using different fusion techniques. For feature extraction and matching, three different algorithms, SIFT, KLT and SURF, are tested. Thus, three different 3D models will be obtained as a result of using different algorithms for feature extraction and matching. Fusion techniques that are used between the resulted 3D models are addition, simple average, Principal Component Analysis (PCA) and Discrete Wavelet Transform (DWT). We focus on modeling from reality rather than computer graphics. The effectiveness of the proposed framework is illustrated in experiments.

2 Related Work

There are various researches which used two images of an object as human visual system for 3D object reconstruction over the last few years. Most of these methods depend on smoothing or regularizing the depth estimation for the reconstruction. One of the most successful methods is the Conditional Random Field (CRF) framework [14, 15]. A probabilistic model was used to associate adjacent pixels to encourage them to have similar depth values. A Second order Markov Random Field (MRF) was proposed to avoid the front-parallel bias [16–18]. Authors in [19] proposed a new framework for local feature-based object recognition, which is based on the local features and their 3D information with stereo cameras. A number of 3D reconstruction techniques based on data fusion are proposed [20–22]. Recently, an improved 3D modeling technique based on multistage feature extraction and matching was used to improve accuracy [23].

3 Feature Extraction and Matching Algorithms

Feature extraction and matching algorithms are considered as important stage for reconstructing the 3D model. In this paper, three different algorithms will be tested, Scale Invariant Feature Transform (SIFT), Kanade-Lucas-Tomasi (KLT) and Speeded up Robust Features (SURF).

3.1 *Scale Invariant Feature Transform (SIFT)*

The SIFT algorithm has four main steps of scale space extreme detection, key-point localization, orientation assignment, and key-point descriptor [7]. The scale space of an image is computed based on the variable-scale Gaussian and the input image [6]. The stable key-point locations are detected using the difference of Gaussian functions which can be computed from the difference of the two nearby scales. Every candidate key-point is fitted to a detailed model and low contrast points are discarded, in addition to the poorly localized edge responses, in the key-point localization process. Based on local image gradient direction, orientations are assigned to each key-point location. Finally, the local image gradients are measured at the selected scale in the region around each key-point in the key-point descriptor.

3.2 *Kanade-Lucas-Tomasi (KLT)*

The two main steps in the KLT framework are feature point detection and feature point tracking [8]. In the feature point detection step, for each pixel in the first image, a structure matrix can be computed using a small rectangular region centered at a given point. The eigenvalues of the structure matrix can determine if the rectangular region is completely homogeneous, contains an edge, or indicates a corner. Larger values mean stronger corners. In the feature point tracking, for each feature point in the first image, the corresponding vector in the second image is calculated. This can measure the image intensity deviation between a neighborhood of the feature point position in the first image and its potential position in the second image and should be zero in the ideal case.

3.3 *Speeded Up Robust Features (SURF)*

The SURF Feature extraction method [9] consists of two main steps, named orientation assignment and description. For the orientation assignment, Haar-wavelet responses are calculated in a circular neighborhood around the interest point. All responses within a sliding orientation window are then summed. Then, Haar wavelet responses in the horizontal and vertical directions are calculated for the description. These responses, over each sub-region, are summed up separately forming a first set of entries to the feature vector. Finally, the vector is normalized into a unit length to achieve invariance to contrast.

4 Fusion Techniques

There are many types of fusion techniques. In this paper, four fusion techniques have be used to produce the 3D models, named addition, simple average, Principal Component Analysis (PCA), and Discrete Wavelet Transform (DWT).

Fusion Using Addition Technique: The pixel value in the fused 3D model is simply computed by adding the value of each pixel in the first 3D model and the corresponding pixel value of in the second 3D model.

Fusion Using Simple Average Technique: In this technique, the average value of each pixel in the first 3D model and the value of the corresponding pixel in the second 3D model are assigned to the pixel of the fused 3D model [11].

Fusion Using PCA Technique: Principal component analysis (PCA) is a mathematical procedure; its objective is to transform a number of correlated variables into a number of uncorrelated variables called principal components [24]. Fusion using PCA technique is shown in Fig. 1. This technique consists of the following steps:

1. Images to be fused (I_1 and $_I2$) are organized into column vectors, this result in a matrix "Z".
2. The empirical mean "*M*" along each column is computed.
3. The empirical mean "*M*" is subtracted from each column of the data matrix "Z" to obtain a new matrix "*X*".
4. Find the covariance matrix "*C*" of "*X*" as $C = XX^T$

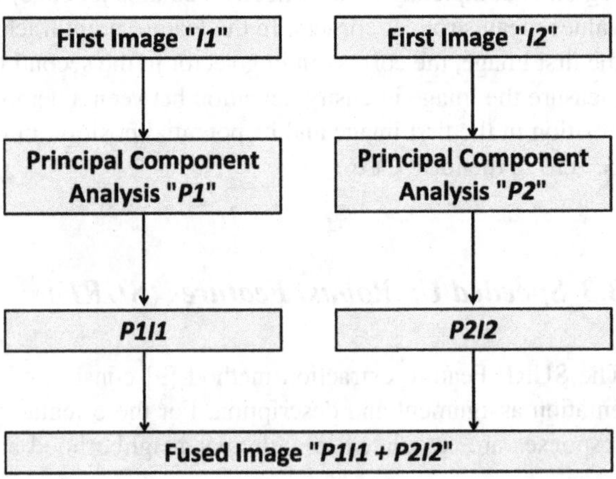

Fig. 1 Fusion using PCA technique

5. The eigenvectors "V" and eigenvalue "D" of "C" are computed.
6. P_1 and P_2 are calculated as; $P_1 = \frac{V(1)}{\sum V}$, $P_2 = \frac{V(2)}{\sum V}$. Finally, the fused image is obtained as; $I_f = P_1 L_1 + P_2 L_1$

Fusion Using DWT Technique: Wavelets are functions with zero average value. They have finite energy and they are suited for analysis of transient signals. Wavelets can be described by using two functions; Scaling function and Wavelet function. A wavelet series expansion is similar in form to the well-known Fourier series expansion, in which it maps a function of a continuous variable into a sequence of coefficients [12]. The process of computing these coefficients is referred to as DWT analysis. On the other hand, inverse DWT (IDWT) is used to reconstruct the fused image with these coefficients [13]. DWT-based fusion technique is illustrated in Fig. 2. Fusion using DWT technique can be achieved according to the following steps:

1. The images are decomposed into approximation and details coefficients are combined according to a certain fusion rule.
2. Fusion rule can take the maximum, minimum, mean of the approximations, or the details structures. The first rule takes the mean of the approximations and the mean of the details coefficients (DWT(mean&mean)). In the second rule, the maximum value of the approximations coefficients and the maximum value of the details coefficients are taken in account (DWT(max.&max.)). The third rule takes the minimum value of the approximations coefficients and the maximum value of the details coefficients (DWT(min.&max.)).
3. After applying fusion rule, new coefficients will be obtained.
4. The fused image can be obtained by taking the inverse discrete wavelets transform (IDWT).

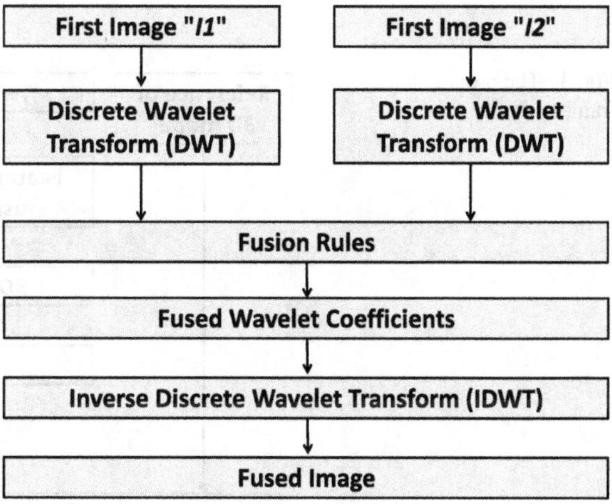

Fig. 2 Fusion using DWT technique

5 Proposed Framework

Figure 3 shows the overall structure of the proposed framework. First, two images of an object taken from different views are used. Features are then extracted from the images and matched to find correspondences. Three different algorithms (SIFT, KLT, and SURF) are used for features extraction and matching. 3D model is reconstructed depending on the depth information. The required depth information is obtained by calculating the horizontal offset of the pairs of matched up points. This offset is then multiplied by a gain to scale the depth with respect to X and Y directions. Three different 3D models are obtained based on the features extraction and matching algorithms. In order to obtain a 3D model that provides the most detailed and reliable information possible, different data fusion techniques (Addition, Simple Average, PCA, and DWT) is applied between each pair of the resulted 3D models. Fused 3D models are compared with a reference 3D model of the object by using different performance metrics, such as Mean Square Error (MSE), Peak Signal-to-Noise Ratio (PSNR), Normalized Cross Correlation (NCC), Average Difference (AD) and Structural Content (SC)).

6 Performance Metrics

To measure the performance of the proposed framework, five different methods of Mean Square Error (MSE), Peak Signal-to-Noise Ratio (PSNR), Normalized Cross Correlation (NCC), Average Difference (AD), and Structural Content (SC) methods are used [25]. In this analysis, m and n represent the number of pixels in row and column directions, respectively. Samples of the 3D models are denoted by $x(m, n)$, while $x'(m, n)$ denotes samples of the reference 3D model.

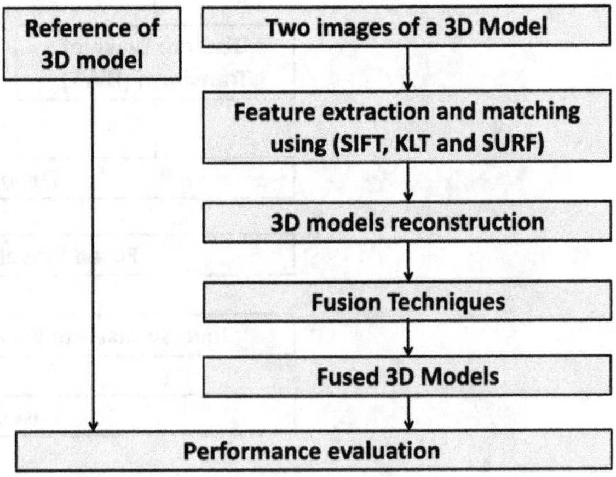

Fig. 3 The proposed framework

The simplest method for image quality measurement is the MSE. The large value of MSE means that image is in poor quality. MSE can be calculated as $MSN = \frac{1}{MN} \sum_{m=1}^{M} \sum_{n=1}^{N} (x(m,n) - x'(m,n))^2$. The small value of PSNR means that image is in poor quality. PSNR is defined as $PSNR = 10 \log 255^2 / MSE$. The small value of NCC means that image is in poor quality. NCC is defined as $NCC = \sum_{m=1}^{M} \sum_{n=1}^{N} x_{m,n}.x'_{m,n} / \sum_{m=1}^{M} \sum_{n=1}^{N} x^2_{m,n}$. The large value of AD means that image is in poor quality. AD is computed as $AD = \sum_{m=1}^{M} \sum_{n=1}^{N} (x_{m,n} - x'_{m,n})/MN$. The large value of SC means that image is in poor quality. SC is obtained as $SC = \sum_{m=1}^{M} \sum_{n=1}^{N} x^2_{m,n} / \sum_{m=1}^{M} \sum_{n=1}^{N} x'^2_{m,n}$.

7 Experimental Results

In this section, the performance of the proposed method is tested. Two RGB images as shown in Fig. 4 with a resolution of 150 × 200 × 3 is used. A reference 3D model [26] for performance evaluation is shown in Fig. 5. The three algorithms of (SIFT, KLT, and SURF) for feature extraction and matching are used to obtain three different 3D models from the left and right images as shown in Fig. 6. Note that, the 3D model obtained from using SURF algorithm shown in Fig. 6 is not good, thus it will affect the fusion results. The resulted 3D models after applying fusion techniques between 3D models obtained from using SIFT and KLT for feature extraction and matching are shown in Fig. 7. The proposed method is compared with the original method in which a 3D model was reconstructed from two images using KLT algorithm.

The performance metrics between the fused 3D models and the reference 3D model experimentally show the follow. The 3D model produced from applying fusion using average technique has the lowest MSE. The highest value of PSNR is obtained by applying fusion using average technique. The 3D model resulted from

Fig. 4 Two test images; *left view*, *right view*

Fig. 5 Reference 3D model

(a) (b) (c)

Fig. 6 Three different 3D models; **a** Using SIFT, **b** Using KLT [27], and **c** Using SURF

applying fusion using PCA technique has the highest value of NCC The 3D model with the lowest value of AD is produced from applying fusion using DWT (min.& max.) technique. The lowest value of SC is achieved by applying fusion using DWT (min.&max.) technique. The 3D model obtained from using SURF algorithm for feature extraction and matching consumes the shortest time.

From the previous results, we can notice the follow. Using SIFT algorithm for feature extraction and matching achieves better results than using KLT and SURF algorithms. 3D models obtained by applying different fusion techniques between the resulted 3D models are more accurate than 3D models obtained from using traditional techniques. The best 3D model can be produced by using PCA-based fusion technique between the 3D models obtained from using (SIFT and KLT) algorithms for feature extraction and matching. The time required to get the best 3D model

Fig. 7 3D models from applying fusion techniques between SIFT and KLT outputs using; **a** addition, **b** average, **c** PCA, **d** DWT (mean&mean), **e** DWT (max.&max.), and (f) DWT (min.&max.)

Table 1 Comparison between PCA-based model with the reference 3D model

Performance metric	PCA-based model
MSE	0.0207
PSNR	64.9616
NCC	0.8799
AD	0.0212
SC	0.4406

is (78.11) second. Table 1, shows the values of (MSE, PSNR, NCC, AD and SC) resulted from comparing the PCA-based model with the reference 3D model of the used object.

8 Conclusions

This paper proposed a 3D modeling technique based on data fusion of different 3D models. Applying fusion techniques between the resulted 3D models produced more accurate 3D model that retains the most desirable information and characteristics of each input 3D model. Experimental results of different fusion techniques were illustrated to explore the effectiveness of the proposed method. We conclude that the best 3D model is achieved when applying fusion using PCA technique between the 3D models resulted from using SIFT and KLT algorithms. As a future work, we will test the proposed framework using different feature extraction and matching algorithms, and using different data fusion techniques.

References

1. Ng SC, Ismail N, Yusof JM, Kassim MSM, Raham KA (2009) IV-FMC: an automated vision based part modeling and reconstruction system for flexible manufacturing cells. In: IEEE international conference on robotics and biomimetics, pp 1383–1387
2. Paysan P, Knothe R, Amberg B, Romdhani S, Vetter T (2009) A 3D face model for pose and illumination invariant face recognition. In: IEEE international conference on advanced video and signal based surveillance, pp 296–301
3. Al-Kindi GA, Shirinzadeh B (2007) An evaluation of surface roughness parameters measurement using vision-based data. Int J Mach Manuf 47(34):697–708
4. Moons T, Gool LV, Vergauwen M (2008) 3D reconstruction from multiple images part 1: principles. Comput Graph Vis 4(4):287–398
5. Hartley RI, Sturm P (1997) Triangulation. Comput Vis Image Underst 68(2):146–157
6. Lowe D (2004) Distinctive image features from scale-invariant keypoints. Int J Comput Vis 60(2):91–110
7. Lowe DG (1997) Object recognition from local scale-invariant features. In: Proceedings of international conference on computer vision, vol 2, pp 1150–1157
8. Shi J, Tomasi C (1994) Good features to track. In: Proceedings of the IEEE conference on computer vision and pattern recognition, pp 593–600
9. Kandil H, Atwan A (2012) A comparative study between SIFT-particle and SURF-particle video tracking algorithms. Int J Sign Proc Image Proc Pattern Recogn 5(3):111–122
10. Godse D, Bormane D (2011) Wavelet based image fusion using pixel based maximum selection rule. Int J Engi Sci Technol 3(7):5572–5577
11. Sahu D, Parsai MP (2012) Different image fusion techniques a critical review. Int J Mod Eng Res 2(5):4298–4301
12. Zheng Y, Essock E, Hansen B (2004) An advanced image fusion algorithm based on wavelet transform incorporation with PCA and morphological processing. In: Proceedings of SPIE 5298, image processing: algorithms and systems III, pp 177–187
13. Maruthi R, Sankarasubramanian K (2007) Multi focus image fusion based on the information level in the region of the images. J Theor Appl Inf Technol 3(4):80–85
14. Campbell NDF, Vogiatzis G, Hernández C, Cipolla R (2008) Using multiple hypotheses to improve depth-maps for multi-view stereo. In: Forsyth D, Torr P, Zisserman A (eds) ECCV 2008, part I, vol 5302. LNCS, Springer, Heidelberg, pp 766–779
15. Hernandez C, Vogiatzis G (2010) Shape from photographs: a multiview stereo pipeline. Comput Vis Stud Comput Intell 285:281–311
16. Kohli P, Ladicky L, Torr PHS (2009) Robust higher order potentials for enforcing label consistency. Int J Comput Vis 82(3):302–324
17. Woodford O, Torr PHS, Reid I, Fitzgibbon A (2009) Global stereo reconstruction under second order smoothness priors. IEEE Trans Pattern Anal Mach Intell 31(12):2115–2128
18. Komodakis N, Paragios N (2009) Beyond pairwise energies: efficient optimization for higher-order MRFS. In: IEEE conference on computer vision and pattern recognition, CVPR 2009, pp 2985–2992
19. Yoon K, Shin M (2010) Recognizing 3D objects with 3D information from stereo vision. In: International conference on pattern recognition, pp 4020–4023

20. Yemez Y, Wetherilt CJ (2007) A volumetric fusion technique for surface reconstruction from silhouettes and range data. Comput Vis Image Underst 105(1):30–41
21. Aliakbarpour H, Dias J (2011) Multi-resolution virtual plane based 3D reconstruction using internal-visual data fusion. In: International conference on computer vision theory and applications, pp 112–118
22. Izadi S, Kim D (2011) KinectFusion: real-time 3D reconstruction and interaction using a moving depth camera. In: Proceedings of 24th annual ACM symposium on user interface software and technology, pp 559–568
23. Mohamed MA, Ibrahim A, Othman EA (2014) Improved 3D modeling using multistage feature extraction and matching. In: International conference on computer engineering and systems (ICCES), pp 396–400
24. Naidu VPS, Raol JR (2008) Pixel-level Image Fusion using wavelets and principal component analysis. Defence Sci J 58(3):338–352
25. Vora VS, Suthar AC, Makwana YN, Davda SJ (2010) Analysis of compressed image quality assessments. Int J Adv Eng Appl 10:225–229
26. Graphics and Media Lab (GML) (2015). http://graphics.cs.msu.ru/en/research/projects/3dreconstruction/sfm. Accessed 24 April 2015

Safe Vehicle Driving Using Android Based Smartphones

Bassant M. El-Den, M.A. Mohamed and A.I. AbdelFattah

Abstract Expert sys, driving with your eyes closed can kill you. The human body needs to sleep like his needs to food and drink, this paper will discuss the sleep phenomenon which occurs during driving a car which leads to many accidents and irregularities occurring. The proposed system save the driver life by monitoring the eye movement using a mobile supplied with camera and capturing images of the face and repeat this process every 3 s to note if their inattention or drowsiness that may occur while driving or not. The application uses the Android environment in simulation of the system. In the case of proven snoozing or sleepiness of the driver, the system translates the reflexes of the person and give response to firing the car alarm or to stop the car by using brakes, Sensor ELM327_V1_5 Bluetooth interface OBD2 will automatically send and receive information from mobile and sent to the control unit in the car to take appropriate action. One of the biggest concerns when implementing the system is to achieve respond quickly and effectively to save the life of the driver. The simulation was conducted with a Mitsubishi Lancer car and Samsung Galaxy_S4 phone to simulate the system.

Keywords Eye tracking · Smartphones · Android · Open CV · ELM327 bluetooth interface OBD2 · Template matching method

1 Introduction

Sleep while driving, people lost their lives during their driven a car Just nap; leads to a disaster. This paper is an attempt to treat the consequences of sleeping while driven. Sleep great blessing from God, yes almighty, a complex physiological

B.M. El-Den
Faculty of Engineering, Delta University, Gamassa, Egypt

M.A. Mohamed (✉) · A.I. AbdelFattah
Faculty of Engineering, Mansoura University, Mansoura, Egypt
e-mail: mazim12@yahoo.com

A. Abraham et al. (eds.), *Intelligent Data Analysis and Applications*,
Advances in Intelligent Systems and Computing 370,
DOI 10.1007/978-3-319-21206-7_25

process of going through human day in order to regain his body and regain his waking mind and focus. Although this process occurs every day, ignoring or not realizing their importance which will reflect negatively on human health and life affairs. The sleep is caused by the lack of a natural number hours of sleep leads to increase drowsiness and lack of concentration and the slow reaction of the beams sudden which is understood concept.

On the serious complications of sleep disorders, increased motor vehicle accidents caused by sleep drivers. The reasons for increased sleepiness that may lead to the driver sleeping behind the wheel, not hours to get enough sleep at night is a common cause relation to lifestyle the person and the circumstances of his work.

The lack of sleep has become a global problem resulting from the changes that accompanied the modern civil. The number of hours of sleep needed by the human varies from one person to another; every man knows the number of hours they need to be active. The United States that 51 % of drivers continue drive their car, even when feelings of deep sleepy driver million and admitted that they slept through driving a car which resulted in accidents [1, 2].

In 2006 Lexus introduced the first driver monitoring system based on LS460 sensor, providing a warning if the driver takes his eye off the road and/or it detects any sign of sleeping. Official reports indicates there is a tremendous increase in the numbers of traffic accidents all over the world are due to diminished driver's vigilance level. Due to this reason, developing system to monitor and alert drivers during any abnormal circumstances is essential for accident prevention. Therefore, the target of this paper is to minimize the issues of rising sleep accidents during driving. In this paper, we adopted development some of novel, proposed and/or modified methods to detect any driver suffering from drowsiness through the number of blinking. On the other hand, there are some researches that attempt to measure brain electrical activity during driving or put a web-camera at the bottom of the car to detect car deviation and other attempts; researchers aware about seriousness of sleeping during driving [3].

Android is an open source operating system featuring both a platform for software development and system to run Android Smartphones. Inherently, Android provides programmers/developers to enhance/update/develop Smartphones applications. Android allows the development of applications for the next generation of Smartphones and no longer only confined to making a phone call, but became support the latest trends of technology and of equipment such as a camera, a reader, multimedia communication system, Wi-Fi, GPS, etc. [4]. The proposed system depends on a camera attached to the front of the driver to interact with surrounding events. Other systems when monitor the driver's drowsiness they appear immediately on the front screen a picture of a cup of coffee and underneath the words say, "O driver it's time to a break". Finally, always remember and be very careful in the event of feeling drowsy while driving, stop in safe place and take a rest.

1.1 Eye Tracking: An Overview

Eye tracking is a fundamental task in the field of computer vision research, which has broad applications in human-computer interaction (HCI). In the few coming years, the increase of sophisticated accessibility of eye tracking technologies has a great deal of interest in security systems. Measuring of the point of gaze; where the person is looking, or the movement of an eye relative to the person's head claims to eye tracking. The device used for measuring eye positions and eye movement is called eye tracker; which is used in: (i) visual system research; (ii) cognitive linguistics, and (iii) psychology and in product design. There are many different methods for monitoring eye movement: (i) the most popular method by using video images to determine eye position using feature extraction and (ii) search coil is other method based on the Electrooculogram, Fig. 1 provides different methods of eye tracking [5].

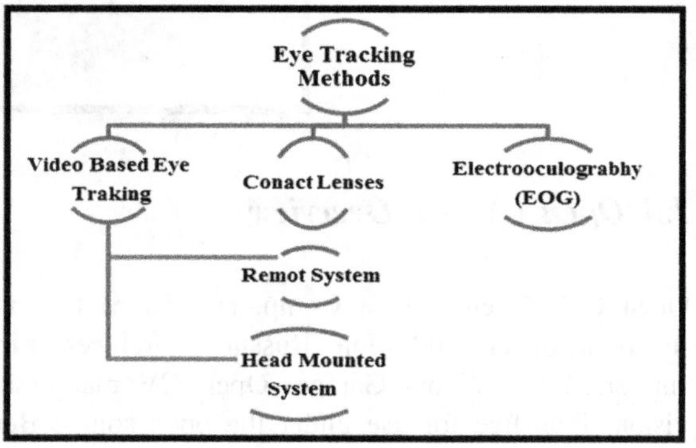

Fig. 1 Structure of eye tracker methods

Nowadays, there is a trend technology for standard camera components such as a webcam or mobile cam to do the same object of what an eye tracker can do. Standard webcam technology applied to determine the direction of interest of website users also in the indicative study of visual attention. The precise not accurate as standard eye tracking, however provides a general cost-effective solution to non-scientific studies for online websites and other visuals provided on a computer. On the other hand, mobile cameras technology applied in some Smartphones for eye presence detection. Primarily a power-saving feature, in contrast to eye tracking technology with illumination, which limited by the ambient light in the environment.

1.2 Android: An Overview

Smartphones are not only cell phones, they may also be considered as portable computing platform. Smartphones have different forms of operating system that controls the phone's functions: (i) Android; (ii) IOS; (iii) Windows, etc. Android is one of the emerging leading operating systems for Smartphone's as an open source platform. Smartphone's have adopted this platform and Fig. 2 shows the top smartphone operation system in 2014 in the whole market [6].

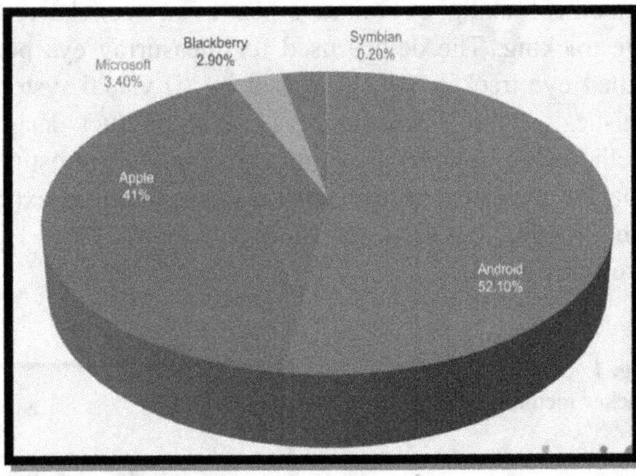

Fig. 2 Top smartphone operation system in end of 2014

1.3 Open CV: An Overview

Open CV (Open Source Computer Vision) tool is a library of programming functions, developed by Intel Russia research center in Nizhny Novgorod, and now supported by Willow Garage. Open CV mainly aimed at real-time computer vision. It is free for use under the open source Berkeley Software Distribution (BSD) license. The library is cross-platform [7].

1.4 Goals and Challenge

The goals of this work in general aims to design android based software system to get over the following challenges and problems: (i) the incorporation of eye tracking algorithms into smartphone devices; (ii) Integration of eye-tracking cameras into automobiles., (iii) provide the vehicle with android system to assess in real-time the visual behavior of the driver and thus avoid accidents., and (iv) Give a warning to the driver to tell him/her with a violation or Car irregularities if it occurs. The Challenging of this work in general (i) Eye tracking have major problems such as head motions, closure of the eye lid, blink frequency, existence of glasses, lighting., and (ii) implementing real time performance. (iii) All these problems may cause an eye tracking system to fail in detecting eye position and affect system performance [8].

This remaining of this paper is organized as follows: (2) describes the eye tracking concept and methods; (3) Sect. 3 presents the system model. Section 4 presents the description of development environment. At last, Sects. 5 and 6 offers the result and conclusions.

2 Eye Tracking

Human face image analysis, detection and recognition have become one of the important research topics in the field of computer vision and pattern classification. Human faces recognition out of still images or image sequences is an active developing research field. New technology begins to emerge in last few years for the field of computer engineering this gives reason to develop other methods of HCI such as eye-gaze or through voice. Different researches on human-computer interaction or other types of research based on eye gaze have been done. Apart from it, eye detection is a crucial aspect in useful topics ranging from face recognition and face detection to human computer interface design, driver behavior analysis, and compression techniques development. The gaze can be determined, by locating the position of the eyes. In the last decade a large number of works have been published on this subject. The main purpose of this work is to present an eye tracking method that is suitable with a standard camera, in a real world condition [9]. In general, fast eye tracking and detection tools handle inputs from phone camera using Open CV's. In this paper, we will use template matching technique for finding small parts of an image which match a template image [10]. As previously mentioned, since template-based matching may potentially require sampling of a huge number of points, it is possible to reduce the number of sampling points by reducing the resolution of the search and template images by the same factor and performing the operation on the resultant downsized images, providing a search window with sample data points within the search image so the template does not need to search every viable data point, or a combination of both.

3 The System Model

The driver's operations as well as vehicles behavior can be realized using a set of parameters: (i) steering wheel movement; (ii) accelerator/brake patterns; (iii) vehicle speed; (iv) lateral acceleration, and (v) lateral displacement. In general, these parameters are non-inserting tools of detecting drowsiness, but are limited to driver conditions and vehicle type. This periodically involves requesting the driver to send a response to the proposed system to running alertness; the disadvantage of this system is that it will consume sometimes to the driver [11].

3.1 Monitoring Physiological Characteristics

The human physiological phenomena technique can be performed in two ways: (i) Measuring the variation in physiological signals, such as heart rate or brain waves or; (ii) recording physical changes such as leaning, eye blinking or the open/closed states of the eyes or sagging posture, of the driver's head. The first technique, since sensing electrodes would have to be connected directly onto the driver's body, and hence be annoying and distracting the driver so it's not realistic. Furthermore, long time driving would cause in perspiration on the sensors; reducing their ability to monitor precisely as shown in Fig. 3 the camera module and IR sensors that is to be used for measuring the changes in physiological signals brain waves. The second technique [12] which discussed and implemented in this paper is well suited for real world driving conditions since it can be non-inserting in the driver body and by using smartphone cameras to detect variation Fig. 4 show an overview of the proposed system.

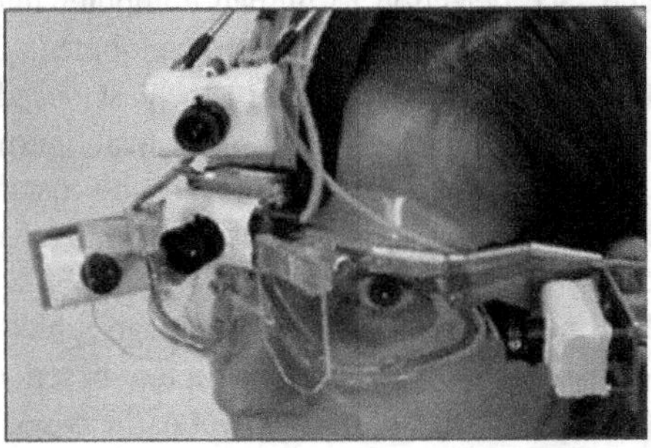

Fig. 3 Module for measuring the changes in physiological signals brain waves

3.2 ELM327-OBD2

On-Board Diagnostics (OBD) is used to refer to vehicle's self-diagnostic and reporting capability. The vehicle owner or repair technician has the ability by OBD systems to access to the status information of different vehicle sub-systems. The OBD provides a suitable amount of diagnostic information which varied widely. The previous versions of OBD would simply illuminate a malfunction indicator light or "idiot light" when damage was detected but would not provide any information about the damage nature. Standardized digital communications port used in modern OBD implementations to provide real-time data in addition to a standardized series of diagnostic trouble codes (DTCs), which allow one to fast identify and remedy damages within the vehicle. OBD systems to vehicle sub-systems, sensor information, allow access to a variety of data about the engine. Access to data from the engine control unit (ECU) is provided by OBD-II as well as it offers a valuable source of information when troubleshooting problems inside a

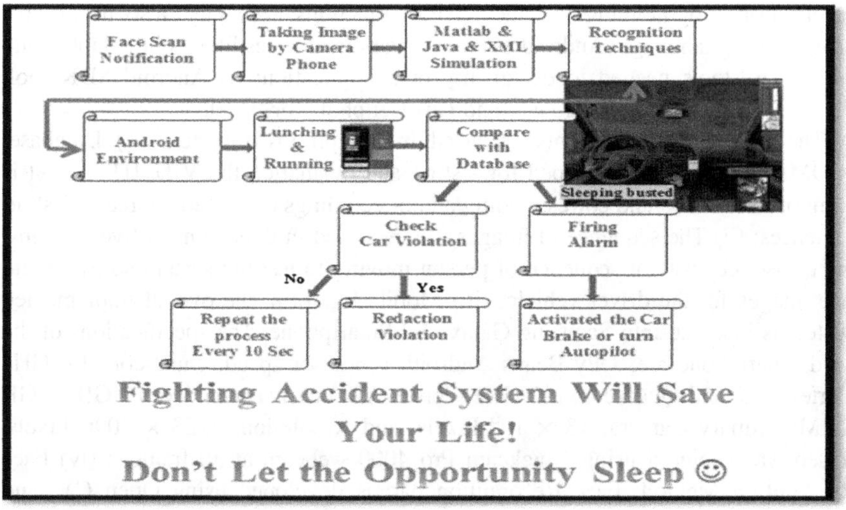

Fig. 4 An overview of the proposed system

vehicle. For requesting various diagnostic data The SAE-J1979 standard defines such method and a list of standard parameters that might be available from the ECU. The different parameters that are available are addressed by PIDs "parameter identification numbers" which are defined in J1979. The list of basic PIDs, their definitions, and the formula to convert raw OBD-II output to significant diagnostic units. In J1979 and they are allowed to include proprietary PIDs that are not listed so manufacturers are not required to implement all PIDs listed. The PID request and data retrieval system gives access to real time performance data as well as flagged DTCs. For OBD-II Codes which is generic list suggested by the SAE the table of OBD-II Codes illustration in [13]. The OBD-II code enhances often by individual manufacturers by adding additional proprietary DTCs. ELM327 is the latest PC-based scan tool technology. Supports OBD-II protocols and comes with a compatible program. OBD-II Software for ELM327 is free also allows receives information from the vehicle computer and display it in cellphone. Output protocol: RS232. Different versions have the ability and equipment. Wired ones are available at authorized service and repair shop. Who are different in size, Wi-Fi, Bluetooth, USB connection, which has its version properties?

4 Development Environment

Recently, mobiles devices use Android, is software stack middleware and key applications, as an operating system. Developing applications on Android platform we will use Software Development Kit (SDK) tool, which will provide provides the

Application Programming Interface (APIs) functions necessary; also using Java programming language. Android operate in various devices like mobile, laptop and tablet. The tools needed to design my own application: (i) Android SDK tool; (ii) Java Programming Language, and (iii) Cellphone or tablet.

The following components are used in the proposed system; (i) Database: VIDMIT database is developed for testing and evaluation the VIDMIT dataset is comprised of video and corresponding audio recordings of 43 people, reciting short sentences; (ii) The sequence of images was recorded in three consecutive sessions; the dataset consists of sequence of person moving to test the system software and live images for the driver vehicle; (iii) Mobile Platform: the overall implemented system is installed into Samsung Galaxy-S4 Smartphone. The specifications of the used Smartphone are: Jelly Beans Android; Processor speed: Quad-core 1.6 GHz Cortex-A15 and Quad-core 1.2 GHz Cortex-A7; Internal memory: 16 GB; 2 GB RAM; Primary Camera: 13×10^6 Pixels, and Resolution: 4128×3096 Pixels. Video was captured using Quickcam Pro 4000 webcam at 30 frame/s; (iv) Face Recognition Method: Face Recognition part is developed using Open CV, and (ix) OBD-II Sensor; Fig. 5: Engine speed; Calculated load value; Antifreeze temperature; Fuel system status; Vehicle speed,; Short-distance fuel consumption; Long distance fuel consumption; Intake manifold; pressure; Timing Advance; Air intake temperature; Air flow rate; Oxygen sensor voltage, and Fuel pressure OBD-II scanner has developed a variety of software for kits.

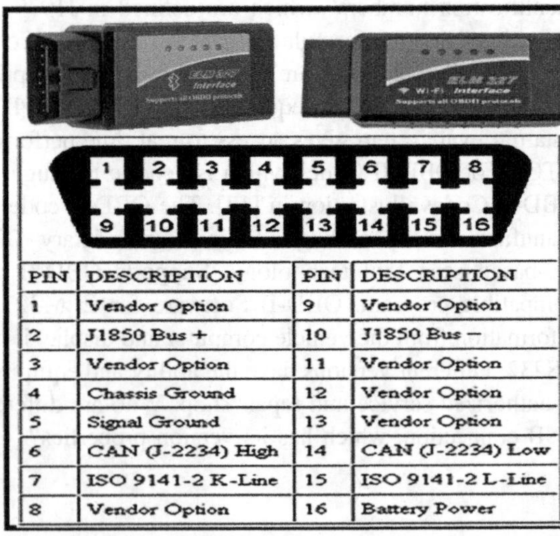

PIN	DESCRIPTION	PIN	DESCRIPTION
1	Vendor Option	9	Vendor Option
2	J1850 Bus +	10	J1850 Bus −
3	Vendor Option	11	Vendor Option
4	Chassis Ground	12	Vendor Option
5	Signal Ground	13	Vendor Option
6	CAN (J-2234) High	14	CAN (J-2234) Low
7	ISO 9141-2 K-Line	15	ISO 9141-2 L-Line
8	Vendor Option	16	Battery Power

Fig. 5 ELM327 OBD II connector and pinout

5 Results and Discussions

The implemented Android application to capture face images and performing face detection and eye tracking was successfully created. Furthermore, the adopted eye tracking method was executed on Android, Figs. 6, 7, 8, 9, 10 and 11. Integrating eye tracking code with the Android platform encountered many technical difficulties; the separate components of face detection and eye tracking worked successfully dependently; both of them integrated together on the Android platform. Eye tracking while driving a car is a difficult issue for studying the eye movement VIDMIT database and live images while driven have been filmed with a phone camera as shown in Fig. 4 the drivers had their eye-movement recorded while approaching a bend of a narrow road. The sequence of images has been taken from the original film to show eye fixations per image for better comprehension. Each of these stills corresponds approximately to 0.5 s in real time. The series of images shows an example of eye fixations for a typical novice and an experienced driver [14]. All images were processed using various utilities from the VIDMIT database or live images. Figure 4 shows the block diagram for the proposed system. After capturing images and detecting the eyes blinking and testing the eye blinking and tracking method; we installed it on Samsung Galaxy S4 model of DROID phone. Compare it with the stored images in mobile memory there is two results: (i) if the comparison gives sleeping busted alarm working and link between mobile and motor give signal to activate alarm and another sign for the brake to stop the vehicle gradually and send a signal in with GPS positioning and get help [15, 16] and (ii) if there is not sleeping busted keep taken images every 10 s. An Android app that captured eye images was successfully created from the ground up as showed in Figs. 6, 7, 8, 9 and 10. Furthermore, the eye detection algorithm was executed on

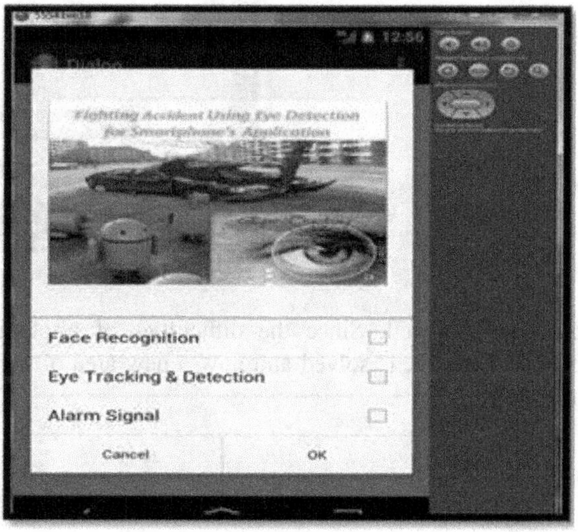

Fig. 6 Main menu for the safety drive system application

Fig. 7 Face recognition progress

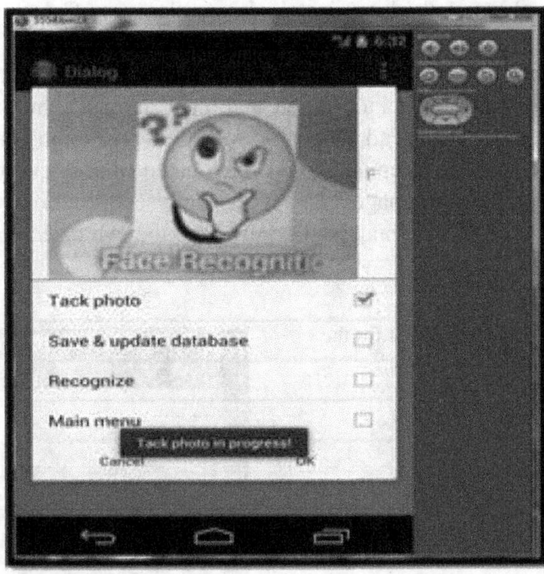

Fig. 8 The process of organize the owner system in sub-menu

Android, as well. Since the difficulties of implementing eye detection on an Android device, is solved and now a new area of using it as a security feature for the device.

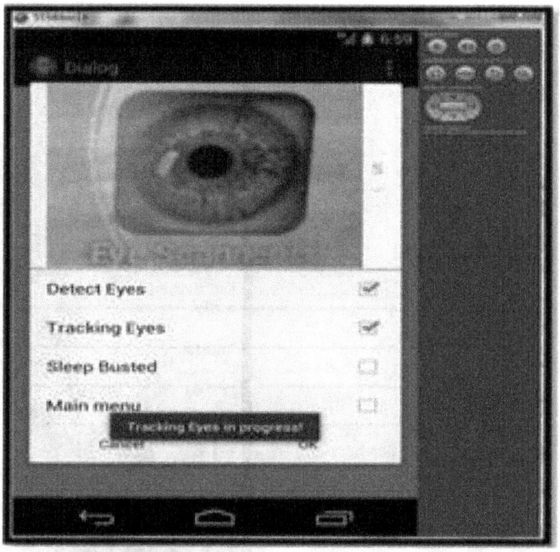

Fig. 9 Eye detection and tracking sub

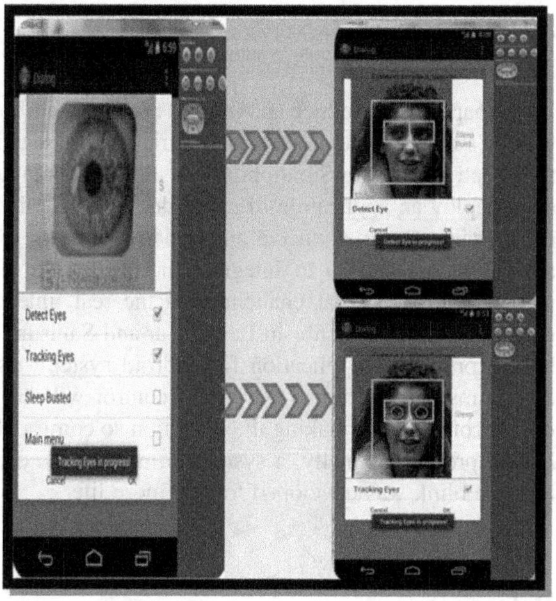

Fig. 10 Eye detection and tracking progress

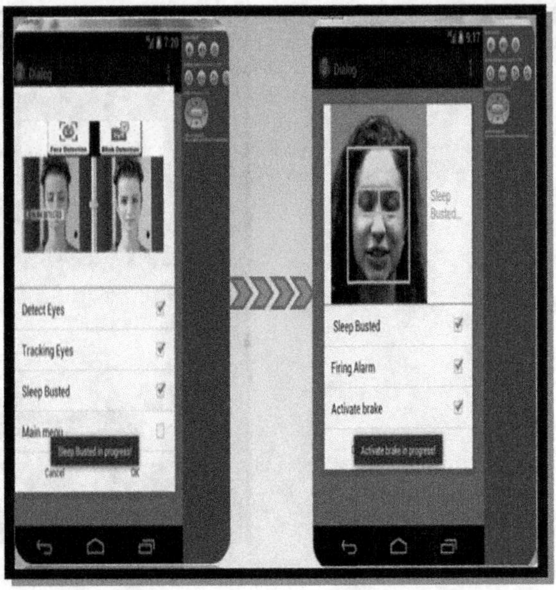

Fig. 11 Sleep busted progress

6 Conclusion

In this paper we introduce an Android based system for safety and authentication of vehicle's drivers. The discussed is a trial for aggregating modifications to improve the effectiveness of a Smartphones in the field of authentication and safety, as well as to deploy an application of smart driver to help in preventing or at least reduce drive accident. An extensive and exhaustive work for implementing eye tracking algorithms as well as to integrate these algorithms with Android based Smartphones; with the great challenges in the real time domain. The simulation was conducted with a Mitsubishi Lancer car and Samsung Galaxy S4 Smartphone. The human protection application for android system has been developed using Java algorithms. In the future eye based control will be the common of all types of device control, thus making the operation so comfortable and much easier with less human presence. Finally, a system to monitor the driver fatigue by detecting face and eye blink was developed to safe more life.

References

1. Global Status Report on Road Safety (2009). World Health Organization (WHO), Geneva
2. Rau P (2005) Drowsy driver detection and warning system for commercial vehicle drivers: field operational test design, analysis, and progress. National Highway Traffic Safety Administration, Washington
3. http://www.newcarnet.co.uk/Lexus_news.html?id=5787. Accessed 27 Feb 2015

4. Owen K (2011) An executive summary of research in android and integrated development environments, April 2011
5. Boening G, Bartl K, Dera T, Bardins S,. Schneider S, Brandt T (2006) Mobile eye tracking as a basis for real-time control of a gaze driven head-mounted video camera. In: Proceedings of the eye tracking research and applications symposium, p 56
6. http://marketshare.hitslink.com/operating-system-marketshare.aspx?qprid=8&qpcustomd=1&qptimeframe=M. Accessed 29 Feb 2015
7. Itseez leads the development of the renowned computer vision library OpenCV (2015). http://itseez.com/Accessed 05 March 2015
8. Ai N, Lu Y, Deogun J (2008) The smart phones of tomorrow. ACM SIGBED review. Special issue on the RTSS forum on deeply embedded real-time computing, vol 5, issue 1, no 16
9. Viola P, Jones MJ (2004) Robust real-time face detection. Int J Comput Vis 57(2):137–154
10. Brunelli R (2009) Template matching techniques in computer vision: theory and practice, Wiley, Chichester. ISBN:978-0-470-51706-2, (http://eu.wiley.com/WileyCDA/WileyTitle/productCd-0470517069.html) TM book
11. Kulkarni S, Agrawal P (2008) Smartphone driven healthcare system for rural communities in developing countries. In: Proceedings of the 2nd international workshop on systems and networking support for health care and assisted living environments, no 8
12. Bhaskar TN, Keat FT, Ranganath S, Venkatesh YV (2003) Blink detection and eye tracking for eye localization. In: Proceedings of the conference on convergent technologies for Asia-Pacific region (TENCON 2003), pp 821–824, Bangalore, India, 15–17 Oct 2003
13. http://en.wikipedia.org/wiki/OBD-II_PIDs. Accessed 05 March 2015
14. Ballagas R, Borchers J, Rohs M, Sheridan JG (2006) The smart phone: ubiquitous input device. IEEE Pervasive Comput 5(1):70–77
15. Loscri V, Tropea M, Marano S (2006) Voice and video telephony services in smartphone. EURASIP J Wirel Commun Netw 2006(2):16–24
16. http://cswww.essex.ac.uk/mv/allfaces/faces94.html. Accessed 05 March 2015

QoS Optimization for Multimedia Streaming Using Hybrid RRM and DASs Over LTE Networks

Hegazi Ibrahim and M.A. Mohamed

Abstract According to the nature of streaming technologies, it is very sensitive to packet loss and has strict minimum speed requirements. Network QoS metrics is proposed in order to evaluate streaming video and audio data over wireless networks under the effect of two techniques RRM and DASs. The objective of this paper is to improve QoS for streaming audio and video content over LTE networks for all users at the minimum delay and higher throughput rates. In this paper, we investigate the performance over LTE networks by the combination of these techniques to enhance the overall system performance. The Simulation results show that LTE makes a great enhancement for multimedia streaming systems under the effect of combination between both techniques DASs and RRM where it can lead to a significant and QoS enhancement by minimizing packet losses and maximization for packet delivery with total enhancement for network performance by 80 % for voice and 70 % for video streaming packets.

Keywords Long term evolution (LTE) · Orthogonal frequency division multiplex (OFDM) radio resource management (RRM) · Distributed antenna systems (DASs) · Quality of service (QoS) · Quality of experience (QoE)

1 Introduction

The increase in multimedia content on the Internet has created a renewed interest in quality QoS. In this paper we overview the most relevant challenges to perform QoE assessment in IP networks and highlight the particular considerations necessary when compared to alternative mechanisms, already deployed, such as QoS. To assist with the handling of such challenges we first discuss the different approaches to Quality of Experience assessment along with the most QoS metrics, and then we

H. Ibrahim (✉) · M.A. Mohamed
Faculty of Engineering, Mansoura University, Mansoura, Egypt
e-mail: mazim12@yahoo.com; Hegazibrahim@gmail.co

A. Abraham et al. (eds.), *Intelligent Data Analysis and Applications*,
Advances in Intelligent Systems and Computing 370,
DOI 10.1007/978-3-319-21206-7_26

discuss how they are used to provide objective results about user satisfaction [1, 2]. Today, video communication over mobile broadband is challenging due to limitations in bandwidth and difficulties in maintaining high sensitivity, the best quality, and higher bandwidth for streaming multimedia applications. In the meantime, mobile video traffic is growing at an immense rate due to significant consumer demand, with the projected share of video constituting more than two-thirds of the total mobile traffic by 2015 [3]. With the increased video traffic in state-of-the-art cellular networks, it is imperative to enhance the QoS of video transmissions. On the other hand, the video quality of service is gaining significant interest as a method to quantify the multimedia experience of mobile users [4], LTE motivating the needs to enhance video service capabilities of future cellular and mobile systems. Now, it is very important to understand both the potential and limitations of these networks for delivering video content, which will contain video streaming and uploading in the uplink direction and not only for old systems for video broadcasting [5].

Singh et al. [6] explained techniques for Video Capacity and QoE Enhancements methods over LTE networks. This paper presents a QoE-based evaluation methodology to assess the LTE system video capacity in terms of the number of unicast video consumers that can be simultaneously supported for a given target QoE. The impact of QoE-based outage criteria is also investigated for the downlink video capacity; also this study proposes a QoE-aware RRM framework which allows the network operator to further enhance the video capacity. This study had shown that there is a significant potential to optimize video capacity through QoE awareness both at the application level and radio access network (RAN) level.

Liu et al. [7] explained QoS-driven and Fair Downlink Scheduling for Video Streaming over LTE Networks with Deadline and Hard Handoff in which one major challenge of this system is to design an effective schedule that can guarantee QoS to users under various mobility scenarios, including the hard HO procedure. They develop a QoS-driven downlink scheduling scheme for video streaming that considers both QoS metrics of video packet deadline. Their Simulation results confirm the efficiency of the proposed scheme.

Yaacoub and Dawy [8] introduced Fair Optimization of Video Streaming Quality of Experience in LTE Networks Using DAS and RRM to capture the overall performance of RRM algorithms in terms of video quality perceived by the end users and they aimed to ensure a fair QoE for all users in the network, also to investigate both the uplink (UL) and downlink (DL) directions, and it consider the use of DASs to make an enhancement for the overall performance. The overall performance of different RRM methods in terms of QoE metrics is studied with and without DAS deployments.

The simulated system in this paper is introduced to get over many issues of the previously presented algorithms. The proposed system has been developed to address (i) delay reduction: this technology is suffered from long delay values during transmission so the access of LTE reduces this delay and gives a small delay ranges; (ii) throughput: during the use of this multimedia systems large amount of data transmission will be lost so the quality of access is very bad so with LTE can increase the number of packets successfully delivered to the destination; (iii) improving network behavior: by using LTE to access video streaming transmission user can access this technology anytime anywhere without delay by the combination of two techniques RRM and DAS with excellent video buffering and resolution during SVC video coding which considered as the best from all video codec's types and (iv) increasing speed of access to that system: this element represents the point of strength where we are using LTE networks that has several features enabling us to improve the quality of service and increase the speed in display and video playback without any noise and impairments whatever the number of subscriber stations (SS) or any loading. Also this system shows that by using RRM, DAS and the combination of both techniques; we can develop this technology to reduce packet loss ratio during transmission during the available bandwidth.

The novelty in this work is in proposing metrics for assessing the QoS performance of the LTE network, taking into account four QoS metrics that affects directly on network performance. Furthermore, we study the impact of RRM and DAS mechanisms on optimizing the network QoS performance and ensuring fairness towards the various users in the network at the effect of combination of both techniques on data flow and our contributions is to give a point of view of how to reduce delay and losses in network up to the optimized level.

This remaining of this paper has been organized as follows: (2) Streaming Multimedia Overview; (3) LTE Overview; (4) Media Streaming Protocols and Basic Problems; (5) DASs and RRM Techniques; (6) QoS Metrics; (7) Development of Multimedia streaming over LTE Networks; (8) System Design; (9) Simulation Results and Discussion; and (10) Conclusion and Future Work.

2 General Overview

2.1 Architecture of Video and Audio Streaming

Architecture for video streaming where raw video and audio data are pre-compressed by video compression and audio compression algorithms and then saved in storage devices is presented in [9]. Upon the client's request, a streaming server retrieves compressed video/audio data from storage devices and then the application-layer QoS control module adapts the video/audio bit-streams according to the network status and QoS requirements [9].

2.2 Services and Applications

2.2.1 Voice over IP (VOIP)

VOIP is a technology for delivery of voice communication over IP networks such as the Internet. Voice traffic channels require low bandwidth, but to support quality live communication, packets should be transmitted with minimum latency and jitter.

Packets associated with voice traffic must be given very high priority and assigned to a guaranteed bandwidth channel to ensure the packet delivery within an acceptable delay limit. Even within voice traffic, differentiated, high-priority service must be provided for important calls such as emergency (911) calls (i.e., based on destination) and critical communication among emergency service personnel (i.e., based on source and/or destination) [10].

2.2.2 Video Streaming

Real-time, user-generated and on-demand video streaming applications are a huge factor requiring QoS in carrier networks. For quality voice and video streaming, the network has to satisfy high bandwidth and stringent latency requirements. Streaming can be person-to-person, real-time video sharing or content-to-person, and each has different QoS needs. For example, real-time video sharing has far more stringent latency requirements than content-to-person streaming. Real-time video sharing also requires that bandwidth on both uplink and downlink bandwidths are high. The network should be able to provide the QoS based on the streaming application [10].

3 LTE Networks

The Third Generation Partnership Project (3GPP) LTE system is the latest generation of wireless cellular technology expected to deliver higher data rates. To access video services in mobile systems, providing a fair QoS to video users is a key objective for LTE system design [11].

3.1 LTE Physical Layer

The LTE physical layer implements a number of technologies to deliver on requirements for high data rates and spectral efficiency. The design of the LTE physical layer (PHY) is heavily influenced by the requirements for high peak transmission rate about 100 Mbps for the downlink and 50 Mbps for the uplink, spectral efficiency, and multiple channel bandwidths (1.25–20 MHz). To fulfill these requirements, OFDM was selected as the basis for the PHY layer. Developments in electronics and signal processing since that time has made OFDM a mature technology widely used in other access systems like 802.11 (Wi-Fi) and 802.16 Worldwide Interoperable Microwave Access (WiMAX) and broadcast systems (Digital Audio/Video Broadcast—DAB/DVB).OFDMA with MIMO allows the downlink to provide as high as 100 Mbps in link throughput while

SC-FDMA on the uplink reduces design complexity for the user terminals by reducing PAPR [12].

4 Streaming Protocols and Basic Problems

This section briefly describes the network protocols for media streaming. In addition, we highlight some of the current popular specifications and standards for video streaming and audio, including 3GPP.

4.1 Streaming Protocols

This section briefly highlights network protocols for video streaming over LTE networks [13].

4.1.1 Session-Initiation Protocol [SIP]

SIP is an end-to-end signaling protocol designed for sessions involving voice and audio over IP networks. Applications such as video-conferencing, instant messaging and online games all use SIP to manage the active session state. SIP runs over any of the common transport layer protocols such as TCP, UDP, and SCTP, and since it is a text-based protocol, it incorporates many of the same elements found in HTTP and SMTP. While SIP is used to provide session setup and configuration, it doesn't do any of the actual heavy-lifting involved in data transmission. SIP usually relies on other protocols such as RTP for the actual transfer of data.

4.1.2 Real-Time Transport Control Protocol [RTCP]

RTCP is an application layer protocol which runs on top of IP networks. It is used in conjunction with the RTCP to deliver streaming multimedia across a network. RTP is used as the actual medium for which the data is transmitted; whereas RTCP is a control protocol used to provide asynchronous QoS metrics. RTP is typically provided by UDP, which favors speed over reliability. This is important for streaming multimedia because TCP packets can often reach a bottleneck where one packet needs to be retransmitted before other packets can be received.

4.1.3 Real-Time Transport Protocol for Multimedia Streaming

The real-time transport streaming protocol is similar to SIP in that it is an endpoint protocol designed to facilitate streaming media over a network. RTSP employs several commands that are useful for streamed data, such as PAUSE, PLAY, and RECORD and so on. RTSP typically uses TCP to hold connection, similar to HTTP.

Unlike protocols such as HTTP, however, RTSP keeps track of state, which is used to keep track of concurrent sessions.

4.2 Basic Problems in Streaming Technologies

There are a number of basic problems faced multimedia streaming. In this paper, we focus on the case of video/audio streaming over LTE [13]. Video/audio streaming over the LTE networks is reliable as it provides guarantees on different factors as system bandwidth, delay jitter and loss rate. Specifically, these characteristics are unknown and dynamic. Therefore, a key goal of video and audio streaming is to design a system to reliably deliver high-quality video over the wireless networks.

4.2.1 Bandwidth

Multimedia (video and audio) streaming required higher bandwidth. If the sender transmits faster than the available bandwidth then packet conflicts occurs, packets are lost, and there is a drop in the overall video quality. If the sender transmits slower than the available bandwidth then the receiver produces sub-optimal video quality. It is very important to solve the bandwidth problem to estimate the available bandwidth and then match the transmitted video bit rate to the available bandwidth.

4.2.2 Delay Jitter

The end-to-end delay that a packet experiences may fluctuate from each other which referred to the delay jitter. Delay jitter for receiver is a problem because the receiver must receive/decode/display frames at a constant rate.

5 DASs and RRM Techniques

5.1 Distributed Antenna Systems [DASs]

A DAS Network is used as a method of distributing RF signals from a central hub to specific areas with poor coverage or inadequate capacity and aggregating the return signals at the hub for interconnection with the larger telecommunications network. Two primary components are featured in each DAS Network solution: (i) a number of remote communications nodes in DAS, each containing at least one antenna for the transmission and reception of a wireless service provider's RF signals. Depending on the particular DAS Network architecture as shown in Fig. 2, DAS communication nodes may include equipment with the antennas, amplifiers, remote radio, converters and power supplies and (ii) a high capacity signal transport medium connecting each DAS Node back to a central communications hub site [14, 15].

5.2 Radio Resource Management [RRM]

The radio interface of LTE is based on the OFDM technology. Radio resources appear as one common shared channel. The scheduler located in e-NB controls and assigns of time-frequency blocks to UEs in an orthogonal manner The smallest unit in the resource grid is called "Resource Element" (RE) it corresponds to one subcarrier during one symbol duration 7 symbols (with the normal CP) makes one slot with the length of 0.5 m and 12 subcarriers in one slot is called "Resource Block" (RB) and Slot can include 6 symbols in case of extended Cyclic Prefix (CP) and 2 consecutive time slots make a sub-frame 10 sub-frames create a type-1 frame. [17].

6 QoS Performance Metrics

6.1 End-to-from End Delay

End-to-End Delay (measured in a sec) is the average time taken by a data packet to arrive in the destination. Only the data packets that successfully delivered to destinations that counted [17]. The end-to-end delay should be less than or equal to 150 m for one-way communication.

6.2 Packet Delay Variation (PDV)

PDV (measured in a sec) measures the variation in the delay of unidirectional, consecutive packets (packet 1 and 2, 2 and 3 etc.) which flow between two hosts through an IP. Low IPDV is very important for applications in which requiring timely delivery of packets such as multimedia applications, VoIP, video etc. Packet delay variation should be less than or equal 45 ms.

6.3 Throughput

Average throughput (measured in packet/sec) is a measure of number of packets successfully delivered in a network. Throughput values should be high or else it affects every service class defined in LTE. Throughput for variable bit rate (VBR) traffic loading is dynamic in nature and it is a function of the scene complexity and associated audio content. Variable bit rate (VBR) traffic loads are typically quoted as peak throughput ranges from 221 to 5321 packet/s.

6.4 Mean Opinion Score (MOS)

MOS provides a numerical measure of the quality of voice and video telecommunications. MOS value ranging from 1 to 5 where 1 is the worst quality and 5 is the best quality.

7 Development of Multimedia Streaming Over LTE Network

7.1 RRM Over LTE Networks

We study the impact of RRM technique on optimizing the network QoS performance and ensuring fairness towards the various users in the network through observing the network performance by accessing RRM through LTE networks and show that how the delay ratings can be reduced and average throughput can be maximized with lower rates for packets lost so these targets can't be achieved without the important role of the second technique (DASs).

7.2 DASs Accessed in LTE

We consider the use of DAS to enhance the performance such that the deployment of DASs considered as an important method to improve performance in cellular systems as DASs are used to increase the coverage and capacity of wireless networks in a cost effective way. Accessing DASs through LTE networks adds more advantages on overall network performance such that lower delay values, maximization for average throughput and minimization of packets lost can be achieved.

7.3 Combination of Both Scheduling Techniques RRM and DAS

The combination of DAS and RRM techniques can lead to QoS enhancements for all the users in the network. When RRM and distributed antennas are deployed, proportional fair scheduling was able to maximize the network performance so it is led to excellent results with the entire network QoS metrics.

8 System Design

8.1 Network Topology and Simulated Scenarios

Video and Voice streaming over wireless networks are challenging due to the high bandwidth required and the delay-sensitive nature of video than most other types of application. Different scenarios have been designed in our simulation with different video codes to measure the performance of some parameters such as throughput, jitter, packet loss and packet end-to-end delay as well as to meet the goal to analyze the quality of service of streaming video and voice packets over LTE networks. The simulation was performed to evaluate the performance of multimedia streaming over the LTE networks and to study the effect of combing both of scheduling techniques RRM and DAS over LTE networks on streaming video and voice systems. OPNET Modeler used to facilitate the utilization of inbuilt models of commercially available network elements with reasonably accurate emulation of various real life network topologies as shown in Fig. 1.

Different six scenarios; as shown in Fig. 1; (i) video streaming: scenario-1: RAM; scenario-2: DAS, and scenario-3: hybrid (RAM and DAS), and (ii) audio streaming: scenario-4: RAM; scenario-5: DAS, and scenario-6: hybrid, have been designed in our simulation, and their simulated results were compared to analyze the effect on overall performance. This study focuses on analyzing QoS parameters through OPNET Modeler that effect on multimedia streaming over LTE networks. The goal of this simulation was to test the deployment, and to analysis the

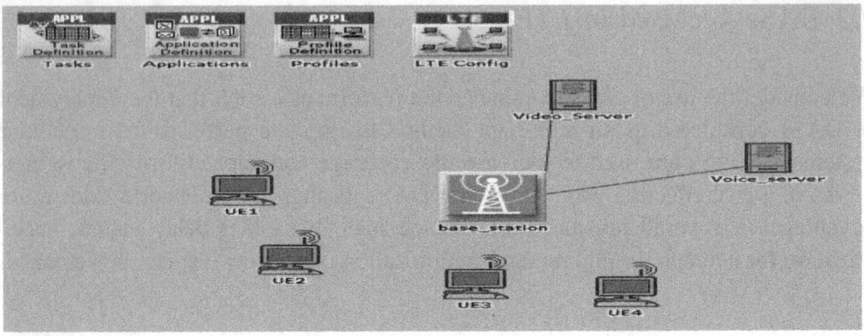

Fig. 1 Network topology

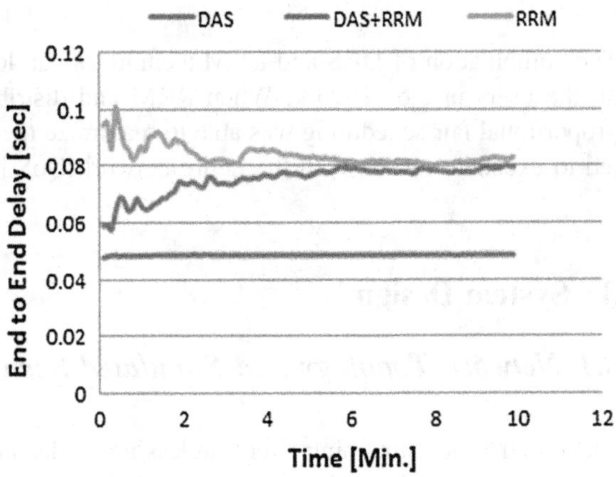

Fig. 2 End-to-end delay (sec) for video scenarios over LTE networks

performance metrics for multimedia streaming data over LTE networks. The simulation results of our model are averaged across the 10 min. Average values of end-to-end delay, PDV, and throughput are used for the analysis in all the Figures. The following results based on sufficient strength to elaborate on our system for simulating streaming traffics by RRM; DASs; and the combination between them. At the first view, the proposed system provided a stable and monotonic delay response during the entire simulation time.

8.2 Simulation Results

The proposed scenarios from "1" to "3" are configured at a fixed modulation and maximum power transmission in watt. This configuration shows a minimized value for system delay, maximize for average throughput, maximization of packet delivery ratio and minimization in packet loss ratio at the combination of both of two scheduling techniques RRM and DAS that can leads to the optimized performance instead of accessing network by one of these techniques as shown in Figs. 2, 3, 4 and 5. The proposed scenarios from "4" to "6" are configured at a fixed modulation and maximum power transmission in watt. This configuration shows a minimized value for system delay, maximize for average throughput, maximization of packet delivery ratio and minimization in packet loss ratio at the combination of both of two scheduling techniques RRM and DAS that can leads to the optimized performance instead of accessing network by one of these techniques as shown in Figs. 6, 7, 8 and 9.

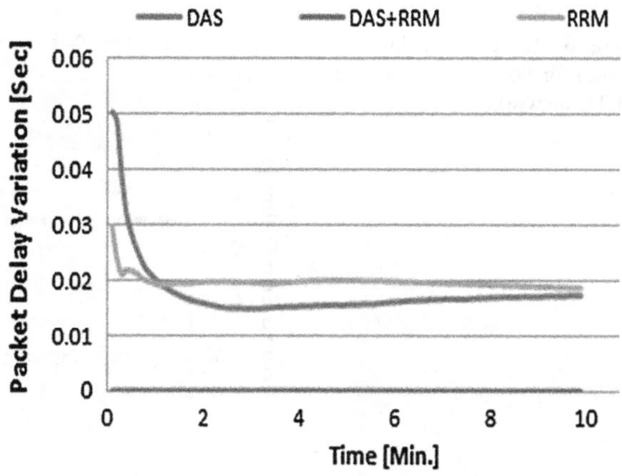

Fig. 3 Packet delay variation (sec) for video scenarios over LTE networks

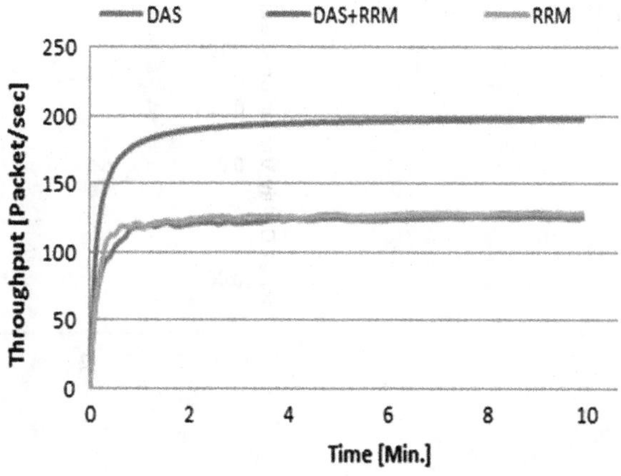

Fig. 4 Throughput (packet/sec) for voice scenarios over LTE networks

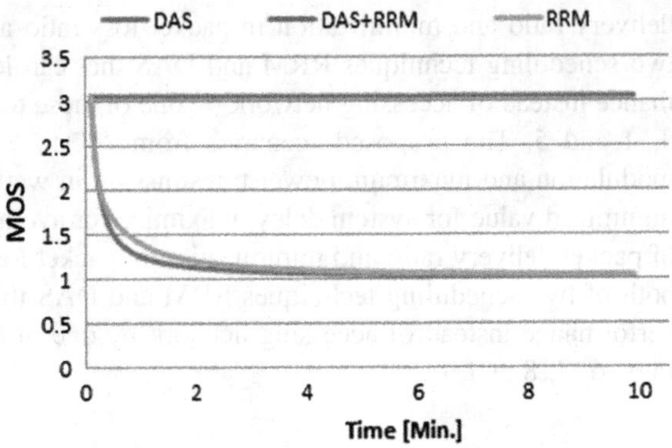

Fig. 5 Mean Opinion Score (MOS) for voice scenarios over LTE networks

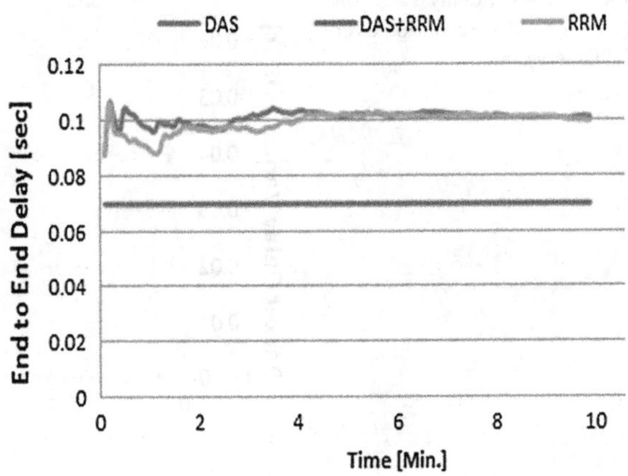

Fig. 6 End-to-end delay (sec) for voice scenarios over LTE networks

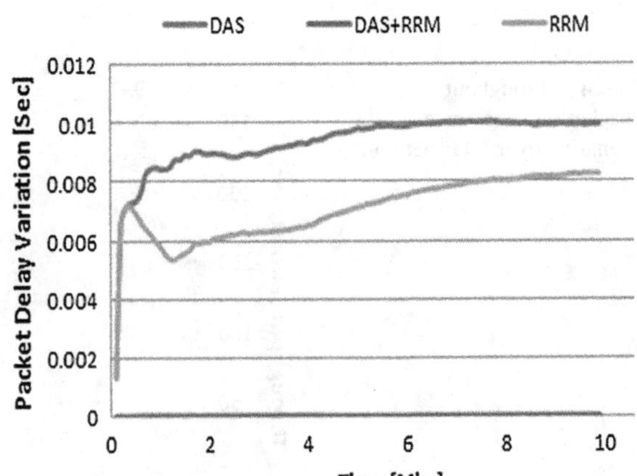

Fig. 7 Packet delay variation (sec) for voice scenarios over LTE networks

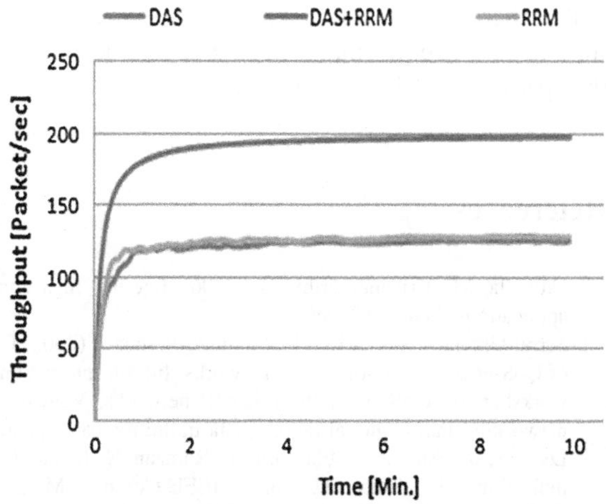

Fig. 8 Throughput (packet/sec) for voice scenarios over LTE networks

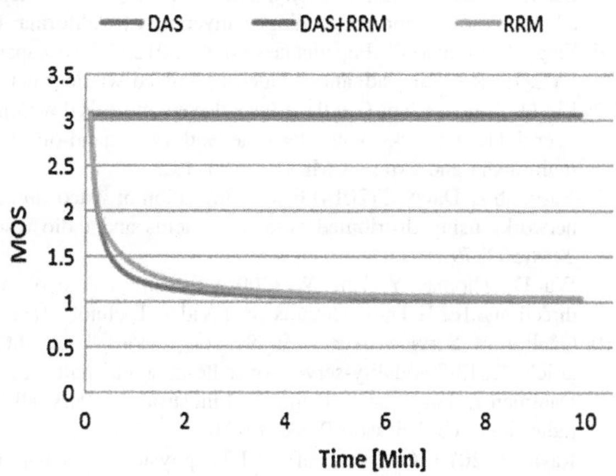

Fig. 9 Mean Opinion Score (MOS) for voice scenarios over LTE networks

9 Conclusion

LTE is able to support different types of services including HD video streaming, VoIP, and Video on demand.3GPP-LTEsystem is the latest generation of wireless cellular technology expected to deliver higher data rates and meet the burgeoning data demand. The simulation results indicate that LTE can minimize packet delays and packet delay variation meets the requirements of multimedia streaming content. This paper presents OPNET simulated networks to show the effect of scheduling techniques such as RRM, DAS and the combination of both techniques over LTE networks on multimedia streaming technologies by supporting streaming protocols at different types of techniques to satisfy a good empirical values for multimedia streaming over LTE networks. A high level of QoS is required for multimedia, and will be therefore be evaluated in the analysis. Future work includes a suitable model for mapping QoS for multimedia streaming technologies such as Voice and Video data packets over LTE networks.

References

1. Crovella M, Krishnamurthy B (2006) Internet measurement. infrastructure, traffic, and applications. Wiley, Chicester
2. Serral-Gracià R, Barlet-Ros P, Domingo-Pascual J (2008) Coping with distributed monitoring of QoS-enabled heterogeneous networks. In: 4th International telecommunication networking workshop on QoS in multiservice IP networks, Venice, Italy,(2008) Cisco, "Cisco visual networking index: global mobile data traffic forecast update, 2010–2015, pp 142–147
3. Lee J-S, de Simone F, Ebrahimi T, Ramzan N, Izquierdo E (2012) Quality assessment of multi-dimensional video scalability. IEEE Commun Mag 50(4):38–46
4. So-In C, Jain R, Al Tamimi AK (2010) A scheduler for unsolicited grant service (UGS) in IEEE 802.16e Mobile WiMAX networks. IEEE Syst J 4(4):487–494
5. Pande A, Ramamurthi V, Mohapatra P (2010) Quality-oriented video delivery over LTE using adaptive modulation and coding. University of California, USA, pp 1–5
6. Singh S, Oyman O, Papathanassiou A (2012) Video capacity and QoE enhancements over LTE. In: Realizing advanced video optimized wireless networks, pp 7071–7076
7. Liu Q, Zou Z, Chen C (2012) QoS-driven and fair downlink scheduling for video streaming over LTE networks with deadline and hard hand-off. In: IEEE International conference multimedia and expo (ICME), pp 183–192
8. Yaacoub E, Dawy Z (2014) Fair optimization of video streaming quality of experience in LTE networks using distributed antenna systems and radio resource management. J Appl Math 2014(562079):1–12
9. Wu D, Thomas Y, Zhu W (2001) Streaming Video over the Internet: approaches and directions. IEEE Trans Circuits Syst Video Technol 11(3):282–300
10. Quality of Service over LTE Networks—Part 1 of 3http://www.wirelessdesignmag.com/articles/2013/09/quality-service-over-lte-networks-part-1-3. Accessed 24 April 2015
11. Dahlman E, Furuskär A, Jading Y, Lindström M, Parkvall S (2008) Key Features of the LTE radio interface. Ericsson Rev 2:77–80
12. Rayal F (2010) Overview of the LTE physical layer part I. In: Design less automotive tele system, pp 1–2
13. Video streaming through Ustream. http://www.slideshare.net/shresthasaagar/ustream-38710002, pp 19–20. 24 April 2015
14. Hejazi S, Stapleton S (2013) Traffic monitoring in LTE distributed antenna system. J Sel Areas Telecommun 19–25
15. Ozan O, ElGamal H (2011)Cooperative encoding with partially known non-causal interference. In: IEEE transaction on information theory, pp 5682–5694
16. Radio Resource Management in LTE Systems, Lecture 8, Aalto University Course, S-72.3260 Radio resource management methods course
17. Packet Delivery Ratio Packet Lost End to End Delay. http://harrismare.net/2011/07/14/packet-delivery-ratio-packet-lost-end-to-end-delay/. Accessed 24 April 2015

Enhanced Algorithm for Rate Adaption for IEEE 802.11 Networks

Salma S. Mohamed, Waleed M. Bahgat and M.A. Mohamed

Abstract The IEEE 802.11 standards provide multi-rates capabilities; therefore, network devices adapt their transmission rate dynamically under varying conditions to achieve high network performance. Rate adaption is a critical component of network performance. High mobility environment and large transmission distance are challenging problems. These problems cause dramatically degradation in wireless network performance. In this paper we will present a solution to overcome these problems based on RRAA. There are two main contributions of the paper: introducing/updating performance evaluation study conducted in IEEE 802.11 networks for most existing rate adaptation algorithms; the simulation results showed that the RRAA has the best performance except for long transmission distance and mixed mobility channel condition and (ii) introducing a modified RRAA algorithm to address the weakness of RRAA. Modified RRAA uses hybrid technique between fixed and adaptive rates. Moreover, it uses adaptive packet size during transmission to achieve maximum throughput of channel. The extensive experiments showed that, Modified RRAA outperforms pure RRAA solutions; with throughput improvement up to 63.27 % in scenarios containing mixed mobility at long transmission distance between nodes.

Keywords Robust rate adaptation algorithm (RRAA) • Adaptive automatic rate fallback (AARF) • Collision rate adaption algorithm (CARA)

S.S. Mohamed (✉) · M.A. Mohamed
Faculty of Engineering, Mansoura University, Mansoura, Egypt
e-mail: mazim12@gmail.com

W.M. Bahgat
Faculty of Computer and Information, Mansoura University, Mansoura, Egypt

A. Abraham et al. (eds.), *Intelligent Data Analysis and Applications*,
Advances in Intelligent Systems and Computing 370,
DOI 10.1007/978-3-319-21206-7_27

1 Introduction

Wireless local area networks (WLAN) first entered the market in the late 1990s after the IEEE 802.11 working group was established to work on the first standard [1]. They finished their work and published the standard in 1997. Since the original standard was published, there have been several extended versions of the same base technology. The legacy 802.11 standard was designed using two bit-rates at the physical layer: 1 and 2 Mbps, respectively. This was the basic design principle of the first rate adaptation algorithm; auto rate fallback (ARF); this algorithm was designed to select the best performance rate out of the two rates under a changing wireless environment. The current 802.11 specifications mandate multiple transmission rates at the physical layer that use different modulation and coding schemes, e.g. 802.11b supports "4" transmission rates (1–11 Mbps); 802.11a offers "8" rates (6–54 Mbps), and 802.11g supports "12" rates (1–54 Mbps) [2]. As such, one of the key components of an 802.11 system is the rate adaptation mechanism, which adapts the data rate used by a wireless sender to the wireless channel conditions. If the sender uses a rate that is too high, many of the packets will be dropped due to bit errors, however if he/she uses a rate that is too low, the wireless channel is not fully utilized. The effectiveness of a rate adaptation scheme depends on how fast it can respond to the variation of wireless channel.

In addition, in a multi-user environment where frame collisions are inevitable due to the contention nature of the 802.11 distributed coordination function (DCF), the effectiveness of a rate adaptation scheme also depends greatly on how the collisions may be detected and handled properly [3]. The implementation of rate adaptation algorithms is not specified by the 802.11 standard. This intentional omission nourishes the studies on this active area; where a variety of rate adaptation algorithms have been proposed. Common challenge of rate adaptation algorithms is how to match unknown channel condition optimally such that the network throughput is maximized [4].

According to the methods of estimating channel conditions, rate adaptation algorithms can be divided into two major categories: open-loop schemes; in which a station makes the rate adaptation decision solely based on its local acknowledgment (ACK) information, are generally standard-compliant and very simple to implement, and closed-loop schemes; which rely on the interaction between transmitter and receiver, and the rate adaptation is dictated by the receiver, which diagnoses the cause of a loss and appropriately adjust the data rate. Most of them are designed to work in realistic scenarios without Request to Send (RTS) and/or Clear to Send (CTS) [5]. The research on rate adaptation is challenging and far from being completed, especially in recent extremely complicated communication environments, e.g. more and more mobile equipment's need to access the networks, which will lead to more, congested wireless bands and more collisions. Furthermore, different applications have different node motion patterns, which will make the link highly dynamic and unpredictable. As a result of the performed evaluation of many rate adaptation algorithms, it was found that all the rate adaptation methods under evaluation are unqualified for the future application scenarios, in the sense that they cannot promptly and accurately differentiate collision induced packet losses from channel error induced packet losses. In addition, they are unable to respond quickly to changes of channel condition in highly dynamic environments. The adaptive RTS/CTS mechanisms cannot eliminate the impact of collision on rate selection. We believe that there is room for improvement, especially with the help of information gained from the nodes themselves [5].

This paper will present a proposed rate adaption algorithm, which is an evolution of automatic rate control technique, RRAA [6] and fixed rate selection technique that appeared in the literature. Details of both approaches will be reviewed shortly. It will then become obvious that the proposed algorithm is a culmination of the best attributes of both RRAA and fixed rate while removing their respective weaknesses. The presented modification based on simulation based evaluation of the performance of six rate adaption algorithms. This paper aims to achieve two objectives: (i) study the performance of six rate adaptation algorithms, namely AARF, Onoe, and Minstrel from open-loop approaches, RRAA, CARA, and RBAR from closed-loop approaches. The evaluation is conducted in the following two simulation scenarios: static and mobility-concerned. The static-concerned scenarios and mobility scenarios are used to evaluate the performance of rate adaptation algorithms in the presence of varying degrees of nodes mobility. Mixed mobility pattern also used in evaluation i.e. (scenario contains mobile and fixed nodes), and (ii) Secondly, present a modification for RRAA to improve its performance in highly changeable environment which contain mixed pattern of mobility node within large area. The modification helps to improve the average throughput of the networks. The remaining of this paper is organized as follows: (2) summarizes related rate adaptation algorithms; (3) discusses the performance evaluation of current rate adaptation algorithms; (4) provides a modified RRAA algorithm and (5) gives general conclusions.

2 Related Rate Adaption Algorithms

Rate adaption algorithms may be placed in: (i) MAC-layer; (ii) physical layer, and/or (iii) both layers. The WiFi node should decide an appropriate transmission rate at a certain time, according to one or more related network attributes such as signal-to-noise ratio (SNR); received signal strength indicator (RSSI); transmission error rate, … etc. In general, there are two approaches to decide transmission rate: fixed bit-rate: A fixed rate setting prevents the device from retransmitting at a lower rate after a failed transmission [7, 8], and adaptive bit-rate selection algorithm: In order to maintain the best transmission rate under time-varying conditions WiFi nodes adapts the transmission rate. An adaptive bit-rate selection algorithm needs to evaluate the current radio-channel condition accurately, estimate an appropriate bit-rate, and then adjust the transmission rate accordingly. Although the idea is simple, it is not easy to figure out the current radio-channel condition accurately.

There are also several challenges in adapting the transmission rate at a certain time: (i) it is difficult to estimate the best possible bit-rate to maximize the transmission throughput, and (ii) it is also difficult to decide when the bit-rate needs to be changed. In addressing these issues, a number of adaptive bit-rate selection algorithms have been proposed [9].

2.1 Open-Loop Approach Rate Adaptation Schemes

Open-loop approaches use either consecutive packet transmission successes or failures, or packet delivery ratios in a time window, to sequentially increase or decrease bit rates. AARF [10], SampleRate [11] and Minstrel [12] are collision-ignored and bit rates are reducing once packet loss occurs. Frame-based ones by nature are not responsive to variations of link status, because they require multiple frame transmissions to converge to a meaningful estimated value of link status. In addition, they usually adopt the fixed threshold mechanism (e.g., 10 consecutive successes or 1 failure) to adjust the bit rate, which cannot work well in different environments.

2.2 Closed-Loop Approach Rate Adaptation Schemes

Closed-loop approaches use a timely collected SNR value to select an appropriate bit rate through looking up a predefined SNR-rate table. Receiver based auto-rate (RBAR) [12], Opportunistic auto-rate media access protocol (OAR) [13], CARA [14], and RRAA [2] use adaptive RTS/CTS exchanges belong to this category. Compared with frame-based ones, the SNR-based ones react more quickly to the changes of link status. However, it is difficult to obtain the accurate SNR values, and capture the exact SNR and bit error rate (BER) relationship in different propagation environments, especially in mobile environments [15]. Both the statistics-based and the SNR-based approaches have their advantages and disadvantages. The statistics-based approach gives robust performance and inherently maximizes the throughput in the long term. However, the main drawback is its slow response to changing link conditions, which can be a source of problems for real-time applications. The SNR-based rate control can respond very fast, but due to the uncertain and fluctuating relation between SNR information and BER of the link, it lacks stability and reliability [16].

2.3 Related Evaluations of Rate Adaption Algorithms

There are little works comparing the performance of rate adaptation algorithms based on simulations. Most of previous works are either conducted in a real environment or in a controlled emulator-based environment. Furthermore, many of these studies focus on static networks, single access Point infrastructure network or use very simple topology, such as only a few mobile nodes for mobile scenario evaluation. In the following sections related evacuation and works are discussed and compared with our evaluation: (i) *Bicket Evaluation* [18]; (ii) *Judd* et al. *Evaluation* [19]; (iii) *Joseph and Edward Evaluation* [20]; (iv) *Wong* et al. *Evaluation* [6]; (v) *Vutukuru* et al. *Evaluation* [21]; (vi) *Shaoen and Biaz Evaluation* [22], and (vii) *Mohamed* et al. *Evaluation* [23].

3 Performance Evaluation of Current Rate Adaption Algorithms

In this section, we conducted comprehensive experiments for studying the performance of various bit-rate selection algorithms in Ad Hoc networks having the characteristic of rapidly changed wireless channel condition due to mobility of nodes and losses in channel. The performance of three open-loop algorithms: (i) AARF; (ii) Onoe, and (iii) Minstrel, as well as three closed-loop algorithms: (i) CARA; (ii) RBAR, and (iii) RRAA, will be presented. The performance will be evaluated under four different scenarios: (i) the effect of distance between nodes; (ii) the effect of the number of nodes; (iii) the speed of nodes, and (iv) the effect of the packet size.

3.1 Evaluation View Point

Most of the existing evaluations are conducted either through test-bed experiments or controlled emulations, which use a channel emulator to generate various channel conditions. It is hard to configure many network parameters for both of these two methods. As a result, it is impossible to repeat the results of evaluations. In this thesis, ns-3 [24] network simulator used to conduct the evaluations. Our simulations considering the following two factors: static and mobility state condition of nodes. Also all previous evaluation works depends on simple topology and data traffics, there is no work of previous evaluation rely on Ad Hoc networks and they use simple parameter to measure performance in there evaluation. Experiments were carried out with IEEE 802.11a MAC on simulator evaluate the performance of six rate adaption algorithms. Three schemes were chosen form Open-Loop based Approaches: AARF, Minstrel, and Onoe. From Closed-Loop Approaches: CARA,

RRAA and RBAR. We choose these algorithms based in compatibility with IEEE 802.11 protocol and implemented in wireless drive.

3.2 Performance Metrics and Measured Parameters

To compare the performance of the six Algorithms we must define the metrics and parameters to be used in the experiment. In most of the WLANS performance testing for channel selection, roaming, mobility and interoperability with other networks the most common test parameters are the system delay, jitter, error ratios and throughput. For the performance testing of rate adaptation algorithms the most suitable test parameter will be the throughput of the system which gives a fair representation of the physical IEEE.802.11 rate selected under different channel conditions [12]. Ad hoc networks are dynamic and distributed entities with no centralized controller where nodes need to adapt their transmission parameter depending on the channel status and network dynamics e.g. variation in node density, traffic, and mobility. So we choose Ad Hoc to measure the performance of algorithms under uncontrolled and variation channel conditions, the values of the network parameters configured in simulations are summarized: (i) Corei5 Dell Laptop with Ubuntu version 12.10.LTS operating system; (ii) WLAN standard: IEEE.802.11g; (iii) Channel loss model: log distance propagation; (iv) MAC Type: Ad Hoc WiFi; (v) Data Rate: 8 Mbps; (vi) Packet Size: 1024 Bytes; (vii) Number of Packets: 1200 Packets; (viii) Number of Nodes: 16; (ix) Mobility Model: 2D random direction; (x) Mobility Speed: Variable speed: Pause Variable (10.0–0.2) m/s; (xi) Topology Size: grid of rectangular range: [0 200 m, 0 200 m], and (xii) Run Time: 200 s.

3.3 Experiment Scenarios

The performance evaluation of the rate adaptation algorithms, through simulation scenarios described in the following sections. The Impact of distance, speed, number of nodes and packet size evaluated using 8 Mbps UDP data traffic flows between nodes as shown in Fig. 3 in all the following scenarios.

(i) *Impact of Transmission Distance:* In this experiment, the nodes send constant UDP traffic to its neighbor for ranges of transmissions distances: 30, 60, 90 and 120 m. Nodes start to move far away from its neighbor in these ranges of transmissions distances with random speed and mobility states. The simulation results, shown in Fig. 1 explain the following: (i) As expected, with the increase of transmission distances average throughput generally decreases; (ii) Most of algorithms nearly have steady throughput up to 60 m transmission distance expect RRAA, but it descends severely upon arrival at the point 90 m as maximum allowable transmission distance; (iii) RRAA throughput descends severely upon

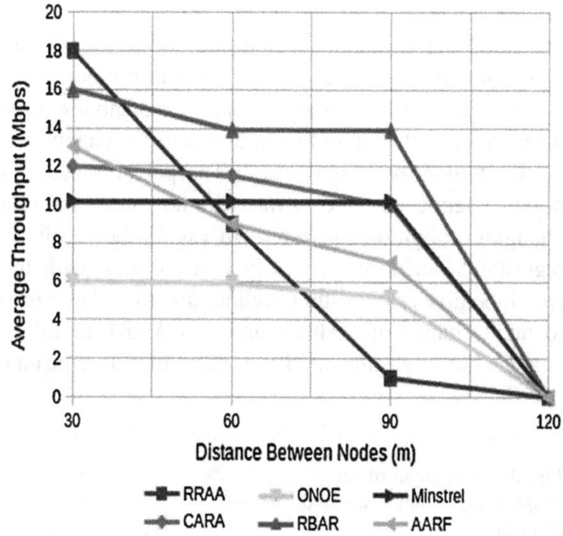

Fig. 1 Simulation results of throughput of rate adaption algorithms versus various distances between nodes

arrival at the point 30 m due to mixed mobility pattern and large transmission and arrives to zero throughputs at point 60 m, and (iv) IEEE.802.11g only support range from (35–120) m as transmission distance, so all algorithms give zero throughput at the end of range expect Minstrel give a few Km throughput.

(ii) *Impact of Number of Nodes:* The impact of number of nodes in the network measured for each algorithm as shown in the Fig. 2 the results explain the following: (i) As the number of node increase, RRAA get highest throughput and

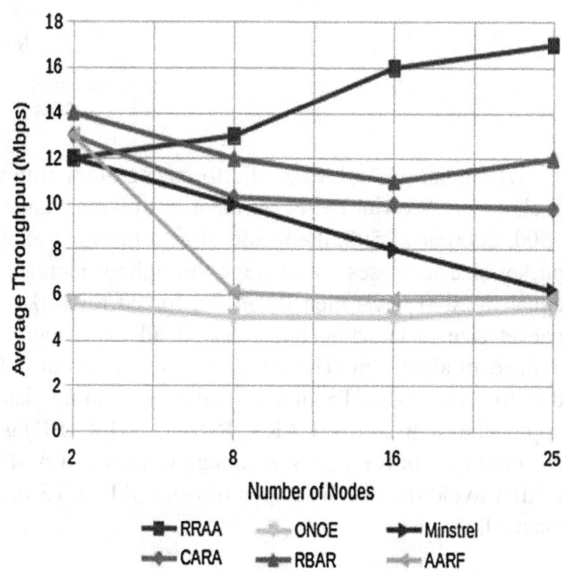

Fig. 2 Simulation results of throughput of rate adaption algorithms with vs. number of nodes

remain constant after 25 clients are attach to the network; (ii) The experiment measured the collision in data due to increasing the number of nodes effects in algorithms performance, and (iii) All closed loop approaches algorithms handle the collision due to increasing the number of nodes using RTS/CTS protocols, it can differentiate between collision and error in channel.

(iii) *Impact of Nodes Speed:* The performance of the algorithms with different nodes speed: 2, 4, 8 and 12 m/s with no pause of node movement to get fast channel changing condition measured. In Fig. 3 the results explain the following: (i) All algorithms performance nearly is a constant with increasing of the nodes speed; (ii) Onoe gives smallest throughput due to its large response time, so it is insensitive to rapid change in channel, and (iii) AARF double it's up rate threshold at high changeable environment. Its performance is constant with the increase of nodes speed.

Fig. 3 Throughput of rate adaption algorithms vs. speed of nodes

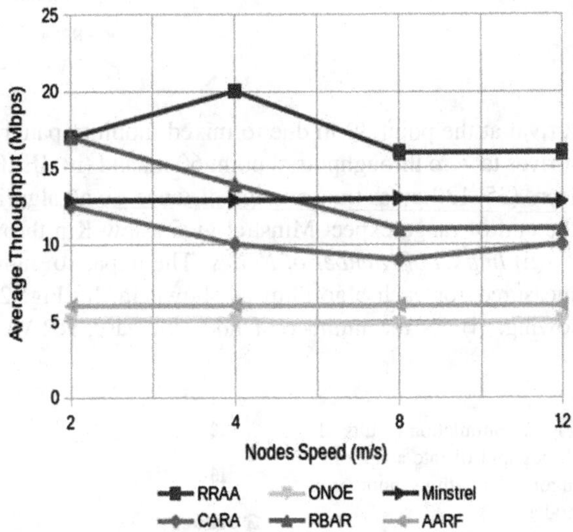

(iv) *Impact of Packet size:* Performance of algorithms measured at different type packet size. We run the experiment at different ranges of packet size: 500, 1000, 1500, 2000 and 2500, the results shown in Fig. 4 explain the following: (i) As the packet size increases the average throughput increase also until reach packet size equal to 2000 bytes then RTS/CTS forced to work; (ii) The RTS/CTS is fired at packet size equal 2000 bytes and at all experiment we leave RTS/CTS as it is defined in algorithm. The throughput decreases at 2500 bytes with all algorithms due to uses RTS/CTS at each time of sending data, and (iii) All closed loop approaches algorithms (CARA, RRAA, and RABR) get more throughputs because its smart uses of RTS/CTS, Although CARA and AARF have similar technique but CARA avoid the disadvantaged of using RTS/CTS in utilization and throughput of channel.

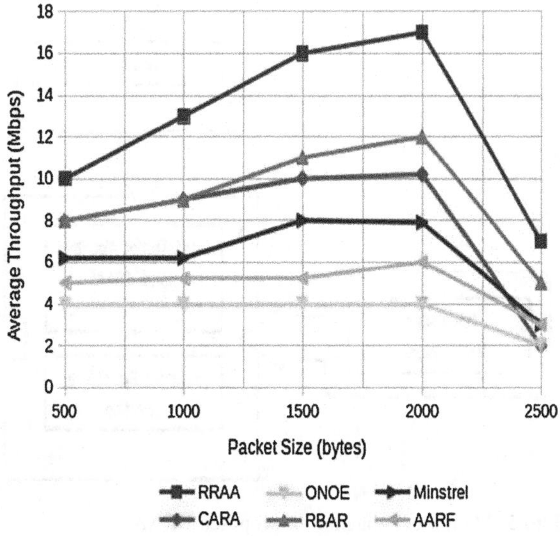

Fig. 4 Throughput of rate adaption algorithms vs. packet size

4 Proposed Rate Adaption Algorithm

4.1 Standard RRAA

RRAA algorithm tries to maximize the aggregate throughput in the presence of various channel dynamics. The design of RRAA is based on two ideas: issues short-term loss ratio to assess the channel and opportunistically adapt the runtime transmission rate to dynamic channel variations and leverages the RTS option in an adaptive manner to filter out collision losses with small overhead, Fig. 5. The basic idea is to leverage per frame RTS option in IEEE 802.11 standards, and selectively turn on RTS/CTS exchange to suppress collision losses. While RTS is well known as an effective means to handle hidden terminals, the main design challenge is to decide when/how long RTS should be turned on/off, Fig. 6. RRAA addresses these two issues by applying two checks in integrating loss estimation, rate change algorithm and A-RTS. While loss estimation, rate change algorithm and A-RTS address channel fluctuations and hidden terminals respectively, one might think that a simple combination of both algorithms would suffice.

4.2 Modified RRAA Algorithm

Hybrid adaptive rate algorithm based on switching between two types of rate selection approaches (fixed/adaptive) and, node state (mobile/static) to solve the long transmission distance and mixed mobility pattern. Switching between fixed and adaptive algorithms give high probability to send data over channel with low

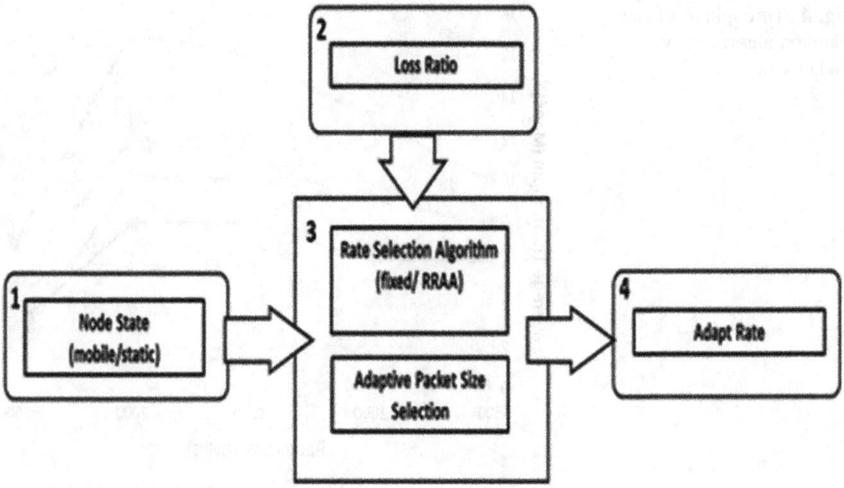

Fig. 5 Main block diagram of proposed RRAA

packet loss, and fixed rate in static case in mobility or long pause motion give more stable channel. Loss ratio with the node state i.e. mobile or static used in dissension of switching between using fixed or adaptive rate in data transmission as shown in Fig. 8. The Proposed algorithms consist of the following blocks: (i) Sensor in mobile node which send a hint to main algorithms; (ii) Network card calculate the loss ratio and also send it to block 3; (iii) In block 3 both adaptive and fixed rate defined, the packet size can change according to loss ratio so the channel utilization will be under control, and (iv) Based in information from 1 and 2, block no.4 adapt the most optimum rate. The proposed algorithm added in existing C++ RRAA code implementation in ns-3 library [6]. The ns-3 simulation tool is used to run the same evaluation scenarios in [24]. The flow chart of proposed algorithm is shown in Fig. 9. Start transmission with maximum transmission rate available in standard. First, calculated loss ratio every 100 m s the loss ratio and the state of node. Then, if static node states i.e. (zero velocity) use minimum rate in standard in next data transmission. Afterwards, if loss ratio less than 10 % then increase the packet size to 2000 Bytes. Finally, if node mobile with speed greater than zero then adaptive rate selection RRAA use to select the transmission rate of next sending date.

4.3 Performance of Proposed Algorithm Analysis

The performance of proposed RRAA and standard RRAA algorithms evaluated and compare. Impact of various size of packets, distance of transmission, and number of nodes evaluated. The experiment focused on measuring throughput over time and calculating the average percentage of improved throughput. Mobility pattern of

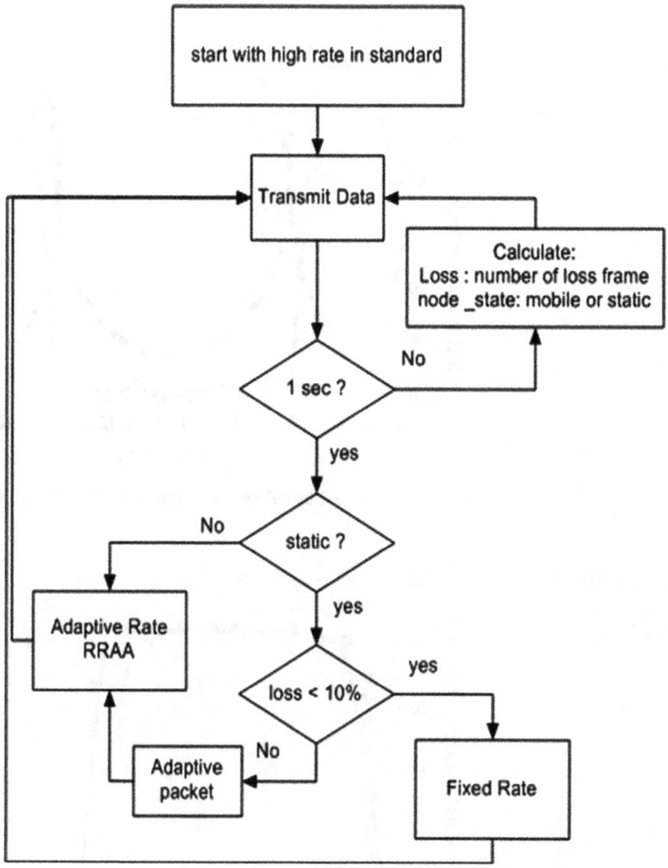

Fig. 6 Proposed RRAA flow chart

nodes in this experiment contains two patterns. Nodes randomly move in free 2D direction (200 × 200) m and take long pause time as a static time movement. In first experiments, nodes allowed to move with different mobile state, node start from center of experiment area and move forward to the edge of the area. Nodes also change its mobile pattern i.e. switching between static and mobile state, randomly. The instantaneous throughput value captured in Fig. 7: (i) proposed RRAA outperform on RRAA in region A, B and C. Nodes change its state from region A (static), region B (mobile) and C (static); (ii) proposed RRAA outperform fixed rate in region B i.e. static case which it fails completely in high mobility movement node environment; (iii) proposed RRAA achieved 63.23 % average throughput than RRAA; (iv) some fixed bit-rates, such as 48 and 54 Mbps, also performed well, and even performed better than some of adaptive bit-rate selection algorithms, and (v) result indicates that higher fixed bit-rates in some cases are more effective than adaptive bit-rate selection algorithms some networks and channel condition because

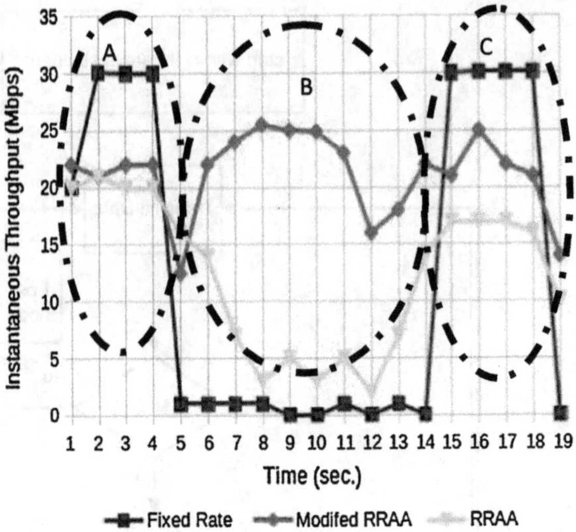

Fig. 7 Instantaneous throughput of mobile nodes in the networks

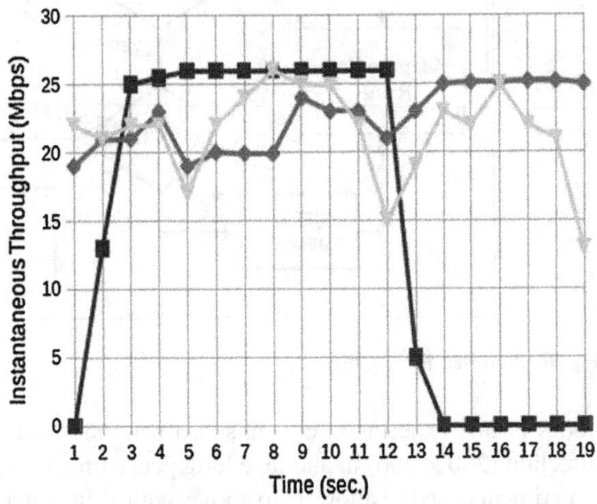

Fig. 8 Proposed RRAA vs. RRAA and fixed rate under node speed = 2 m/s

of similarity/simplicity of the algorithm. Figures 8 and 9 show the instantaneous throughput at a speed of 2 and 10 m/s, respectively. It can be concluded from the figures that: (i) increasing the speed cause degradation in fixed rate performance, because channel conditions change rapidly with the movement of node. Therefore, using the same rate with different condition degrades the performance; (ii) the performance degrades severely when the time reaches 5 s (that represent distance = 40 m as shown in the evaluation results), where RRAA cannot handle fast mobility with long transmission distances. RRAA fail completely at these scenarios, and (iii) proposed RRAA gives a stable throughput at two scenarios with fast mobile nodes and long transmission area.

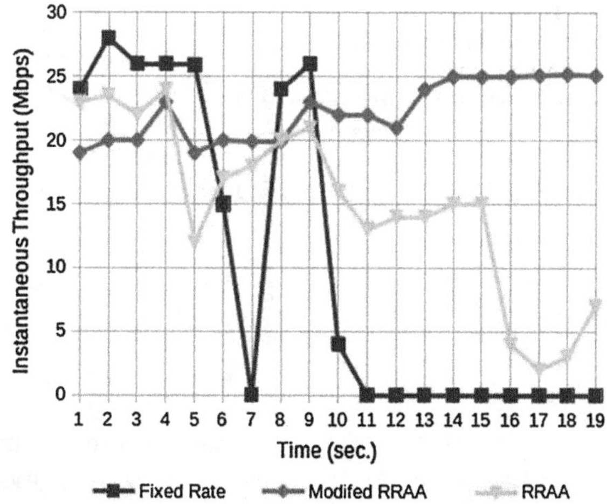

Fig. 9 Proposed RRAA vs. RRAA and fixed rate under node speed = 10 m/s

(i) *Impact of Nodes Speed*: Figure 10 shows the average throughput performance of Proposed RRAA versus RRAA at range of speed (2–8) m/s. The results show: (i) At the beginning of curves i.e. (2–4 m/s) due to lower speed of nodes so channel change more slowly, fixed rate work here more effective and improve the performance; (ii) When the node speed reaches 4 m/s, nodes start to move fast and rate selection process algorithms need to switches between two algorithms to handle the node state, and (iii) It is important to mention that most of rate algorithms cannot differentiate between the values of speed that node move with it, and this seen in results when the throughput becomes constant with increasing the speed i.e. more than 6 m/s.

(ii) *Impact of Packet Size:* Fig. 11 shows the average throughput performance of Proposed RRAA versus RRAA at range of packet size (500–2500) bytes. The result

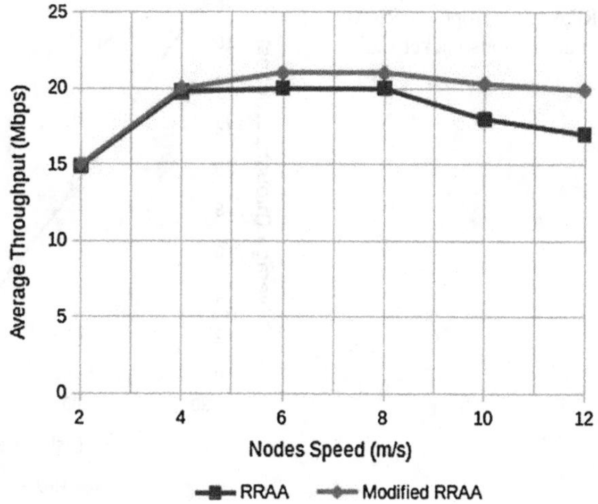

Fig. 10 Average throughput for RRAA vs. proposed RRAA under nodes speed [2–8]

Fig. 11 Average performance throughput for RRAA vs. proposed RRAA under various packet size

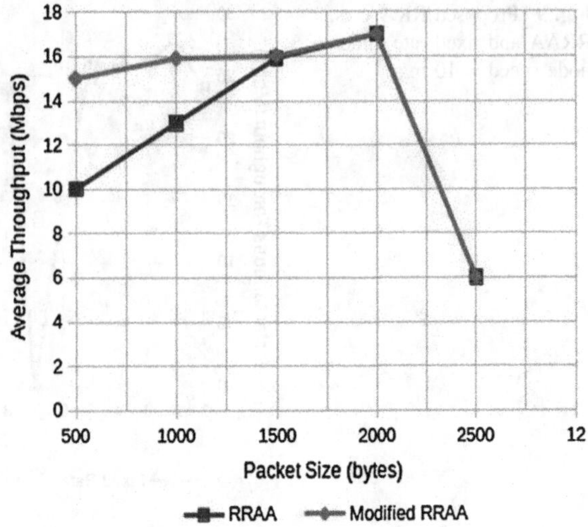

shows the following: (i) The Proposed RRAA improves the performance especially at small packet size and achieves similar performance in the period (1500–2000) bytes due to firing RTS/CTS protocol at each sending frame and (ii) Both Proposed RRAA and RRAA forced to use RTS/CTS without Adaptive RTS at packet size equal to 2000 so throughput degraded under uses RTS/CTS with each transmitted frame.

(iii) *Impact of transmission distance:* Last scenario, the results are shown in Fig. 12. The figure show the average performance of proposed RRAA versus the RRAA under transmission distance. The results show the following: (i) Proposed RRAA gain more throughput than pure RRAA at the range of transmission

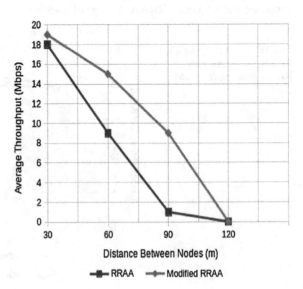

Fig. 12 Average performance throughput for RRAA vs. proposed RRAA under various packet size

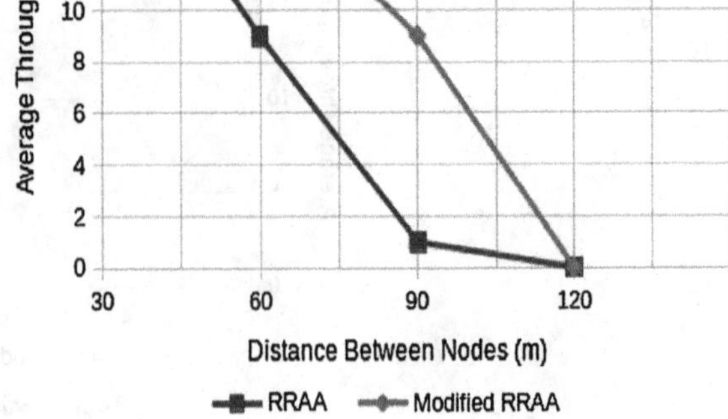

distance from (30–120) m and (ii) The results also show that Proposed RRAA improves the transmission distance after 90 m, which original RRAA fail completely to cover it under mixed mobility pattern of nodes.

5 Conclusions

Six algorithms were selected for performance evaluation; AARF, Onoe and Minstrel algorithms selected from open-loop category, and CARA, RRAA and RABR algorithms selected from closed-loop. The performance evaluation study is conducted in IEEE.802.11 based Ad Hoc networks. An extensive comparative simulation study was performed on the six algorithms. The simulation was conducted using ns-3 simulator. The simulation results showed that standard RRAA has the best throughput performance compared to the other five algorithms. Unfortunately, standard RRAA still have some deficiencies, e.g., the performance needs to be improved to work well under many simulation scenarios. In addition, there are remarkable performance degradation in case of mixed mobility pattern with long transmission distance. Based on the previous results, a modified RRAA was presented to improve the performance of the network. The main concept in modified RRAA is based on using nodes state (static or mobile) to decided which rate adaption technique should be used i.e. (fixed or adaptive rate selection). Our simulation results proved that there is a significant improvement in the performance of modified RRAA compared to RRAA. The average throughput performance of modified RRAA is increased by 63.23 % than standard RRAA in transmission distances scenarios. Furthermore, the average throughput performance of modified RRAA is increased by 14.63 % than standard RRAA in case of improvement in high mixed mobility pattern environment.

References

1. Negus K, Petrick A (2009) History of wireless local area networks (WLANs) in the unlicensed bands. Info 11(5):36–56
2. Crow B, Widjaja I, Kim J, Prescott T (1997) IEEE 802.11 wireless local area networks. IEEE Commun Mag 35(9):116–126
3. IEEE Std 802.11-1999, Part 11: Wireless LAN medium access control (MAC) and physical layer (PHY) specifications, Std., Aug 1999
4. Prado D, Choi S (2003) Link adaptation strategy for IEEE 802.11 WLAN via received signal strength measurement. IEEE Int Conf Comm 2:1108–1113
5. Kim S, Choi S, Qiao D, Kim J (2007) Enhanced rate adaptation schemes with collision awareness. In: Networking 2007. Ad hoc and sensor networks, wireless networks, next generation internet, pp 1179–1182. International federation of information processing (IFIP). Springer, Berlin

6. Starsky W, Hao H, Lu S, Bharghavan V (2006) Robust rate adaptation for 802.11 wireless networks. In: Proceedings of the 12th annual international conference on mobile computing and networking, pp 146–157. ACM
7. Huang T, Chen H, Zhang Z, Cui L (2012) EasiRA: A hybrid rate adaptation scheme for 802.11 mobile wireless access networks. In: Wireless communications & networking conference, pp 1520–1525
8. Thomas H (2013) A measurement-based joint power and rate controller for IEEE 802.11 networks. PhD diss., Oxford Brookes University
9. Xia Q, Hamdi M (2005) Smart sender: a practical rate adaptation algorithm for multirate IEEE 802.11 WLANs. IEEE Trans Wirel Comm 7(5):1764–1775
10. Bit-rate selection algorithms. https://madwifi-project.org/wiki/UserDocs/RateControl. Accessed 24 April 2015
11. Xia D, Hart J, Fu Q (2012) On the performance of rate control algorithm minstrel. In: Proceedings of the 23rd IEEE international symposium on personal, indoor and mobile radio communications, Sydney, Australia
12. Holland G, Vaidya N, Bahl P (2001) A rate-adaptive MAC protocol for multi-hop wireless networks. In: The 7th annual international conference on mobile computing and networking, pp 236–251. ACM
13. Sadeghi B et al (2005) OAR: an opportunistic auto-rate media access protocol for ad hoc networks. Wirel Netw 11(1–2):39–53
14. Kim J, Kim S, Choi S, Qiao D (2006) CARA: collision-aware rate adaptation for IEEE 802.11 WLANs. In: The 25th IEEE international conference on computer communications, IEEE-INFOCOM, pp 1–11
15. Giyeong S (2011) Experimental performance evaluation of bit-rate selection algorithms in multi-vehicular networks
16. Haratcherev L, Taal J, Langendoen K, Lagendijk R, Sips H (2005) Automatic IEEE 802.11 rate control for streaming applications. Wirel Commun Mobile Comput 5(4):421–437
17. Xia Q, Hamdi M, Chan T (2006) Practical rate adaptation for IEEE 802.11 WLANs. In: Global telecommunications conference, GLOBECOM. IEEE
18. Bicket J (2005) Bit-rate selection in wireless networks. PhD diss., MIT
19. Judd G, Wang X, Steenkiste P (2008) Efficient channel ware rate adaptation in dynamic environments. In: 6th international conference on mobile systems, applications, and services, Breckenridge, CO, USA, pp 118–131
20. Joseph C, Edward K (2008) Modulation rate adaptation in urban and vehicular environments: cross-layer implementation and experimental evaluation. In: 14th ACM international conference on MobiCom, pp 315–326
21. Vutukuru M, Balakrishnan H, Jamieson K (2009) Cross-layer wireless bit rate adaptation. ACM Comput Commun Rev 39(4):3–14
22. Saad B, Wu S (2008) Rate adaptation algorithms for IEEE 802.11 networks: a survey and comparison. In: Computers and communications, ISCC, pp 130–136
23. Mohamed MA, Bahget WM, Mohamed SS (2014) A performance evaluation for rate adaptation algorithms in IEEE 802.11 wireless networks. Int J Comput Appl 99(4):54–59
24. Network simulator-ns-3. http://www.nsnam.org/. Accessed 24 April 2015

IPv6 Adoption in the Kingdom of Saudi Arabia

Ahmed Abed, Ali Ahmed and Saadat AlHashmi

Abstract This paper reports a case study on the factors that affect IPv6 adoption at Al-Mouwasat Hospital in Al-Dammam, Kingdom of Saudi Arabia (KSA). IPv6 was blamed for the high network overload after enabling it on the hospital computer network workstations and devices. This resulted in disabling IPv6 on the network as well as operating systems. There is a need to investigate such a case to identify what factors significantly contributed to that decision (i.e. rolling back IPv6). This includes evaluating the actual effect of IPv6 on the network to realize the factors that may hinder IPv6 adoption. The study aims at developing an approach for adopting IPv6 in a dual-stack set-up with IPv4 in order to increase the adoption rate of IPv6 in other networks.

Keywords IPv4 · IPv6 · Affecting factors · Adoption · Kingdom of Saudi Arabia

1 Introduction

Internet Protocol[1] (IP) addressing is the cornerstone for allowing various devices to communicate without misidentifying themselves. Wu explained that the rapid increase in the Internet users especially with the evolution of mobile phones cannot be satisfied by IPv4 any more [14]. Schwankert stated that IPv4 addresses are being exhausted and the solution for this is IPv6 [11]. Xianhui and Dylan explained

[1] This is where the first author is working as a Network Engineer.

A. Abed
University of Liverpool, Liverpool, UK
e-mail: ahmed.abed@online.liverpool.ac.uk

A. Ahmed (✉)
Faculty of Computers and Information, Cairo University, Cairo, Egypt
e-mail: a.ahmed@fci-cu.edu.eg

S. AlHashmi
College of Engineering, Abu Dhabi University, Abu Dhabi, United Arab of Emirates
e-mail: saadat.alhashmi@adu.ac.ae

A. Abraham et al. (eds.), *Intelligent Data Analysis and Applications*,
Advances in Intelligent Systems and Computing 370,
DOI 10.1007/978-3-319-21206-7_28

various aspects that push for IPv6 deployment such as the limitation of IPv4 scalability, NAT breaking the end-to-end network model, VoIP applications, and the IPv6 features of easier network management [15]. Interestingly, on the 20th of May, 2014, ICANN allocated the remaining blocks of IPv4 which means IPv4 reached a very critical level of running out of addresses. It is even stated that a quick adoption of IPv6 has become necessity now [1]. With the increased demand of IP addresses in the 1990s, the Internet Engineering Task Force (IETF) has introduced Classless Inter-Domain Routing (CIDR) and private address space while starting a plan for a new version of the Internet Protocol called IP version 6. This was mainly because of the addressing limitations of IPv4. IPv6 have been around for many years but the adoption rate in the KSA is slow. In the KSA, A National IPv6 task force was created with representatives from key Service Providers in 2008[2] to accelerate the adoption rate. Generally, there are many potential factors that are hindering the adoption of IPv6 such as the network readiness and compatibility, the transition costs, and the performance impact. In the KSA, those factors may have different significance in IPv6 adoption. For example, in an Oil-rich country, cost might not be a problem. There is a need to evaluate these factors to study their significance. Security is a big concern in IPv4. Indeed the protocol was not designed with security in mind, network congestion caused by the broadcast feature, packet loss hindering good performance of VOIP and video streaming. Sharma explained that IPv4 suffers from various attacks such as Reconnaissance, Unauthorized access, Header manipulation and fragmentation, Layer 3 and 4 spoofing, Address Resolution Protocol (ARP) and DHCP attacks, Broadcast smurf attacks and Viruses and worms [12]. Rooney explained that IPv6 improves upon IPv4 in terms of security, mobility, multicast, and auto-configuration [10]. However, the main difference is the large increase in the size of the address.

2 IPv6 Adoption: Opportunities and Challenges

In order to determine the current status of IPv6 adoption, Kim et al. analysed the traffic over 14 months including both the World IPv6 day in 2011 and the World IPv6 Launch in 2012 from the viewpoint of a huge Internet eXchange Point (IXP) in Europe [7]. The methodology used was via a developed analysis tool that can identify the type of transition method whether it is 6in4 or Teredo based on the IP version, port numbers, and protocol number. The conclusion is that about 0.5 % of all the Internet is based on IPv6. This is surprising indeed. out of the 0.5 %, the amount of native IPv6 traffic exceeds the one via IPv6-over-IPv4 methods. Google statistics of IPv6 presented in Fig. 1 show that by 1/1/2014, a total of 2.8 % of the users accessing Google run IPv6 with 2.78 % using native IPv6 while the remaining 0.02 accessing via 6to4/Teredo [5]. However, the graph shows a continuous increase. For example, on 30/9/2014 around 4.57 % of the total IPv6 with 4.56 % comes from native IPv6.

[2]http://www.ipv6.org.sa/.

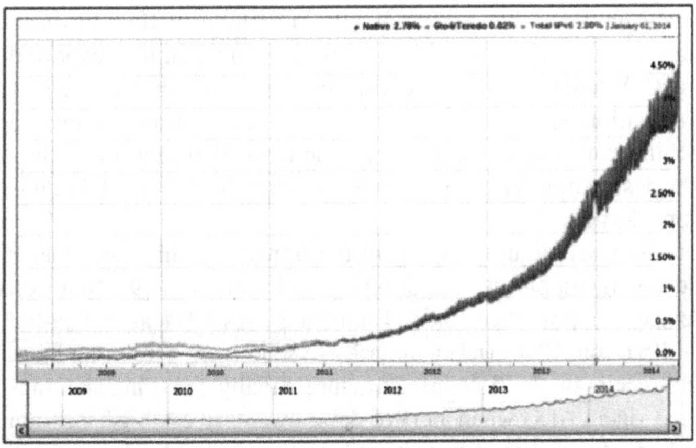

Fig. 1 Users accessing Google over IPv6 [5]

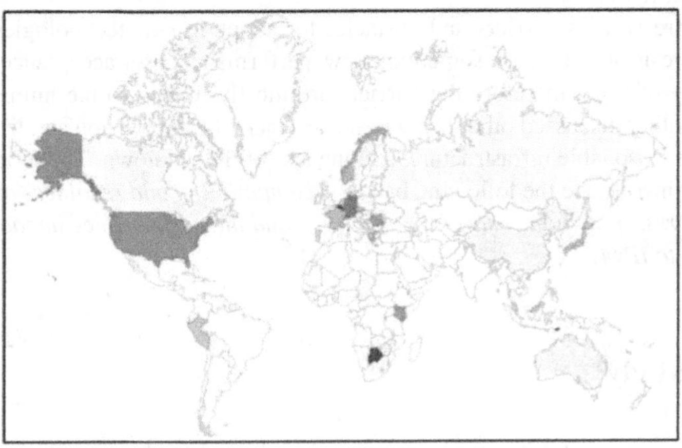

Fig. 2 Map of IPv6 Users' adoption by location [4]

Google provides a map of the world showing the IPv6 adoption by location as in Fig. 2. The green regions have already widely adopted IPv6 while the red ones have not yet adopted IPv6 widely. It seems that the most areas adopting IPv6 are Americas, and Europe. Interestingly, the highest country with IPv6 adoption according to the map is Belgium with 28.46 %, followed by 11.9 % in Germany, 11.01 % in Switzerland, 9.5 % in the US, and about 0 % in the Middle East region including the Gulf Cooperation Council (GCC) countries.

The growth of IPv6 adoption and deployment is steady although it is not as expected [3]. Interestingly, Yadav et al. explained main reasons for the slow growth of IPv6 [6]. According to the survey, the transition costs count for 40 % followed by compatibility issues of 27 % and finally security concerns with 20 %. The challenges of IPv6 include upgrading requirements, resilience of people to change, lack of testing, and unknown business ROI. The migration to IPv6 has security concerns because of the various transitioning mechanisms that may expose the system to attackers [8]. While there are some Internet nodes that run dual-stack of IPv4 and IPv6, two transitioning methods exist which are: translation and tunnelling [14]. The idea behind translation is to transform the IPv4/IPv6 packet into the same type matching the destination. This is somehow similar to NAT in IPv4, but it is not practical at all on a large scale.

There are four techniques helping in the transition which are dual-stack, tunnelling, addressing simplicity, and proxying and translation [9]. New software and network devices support dual-stack. Tunnelling uses IPv4 as a data-link layer to encapsulate IPv6 into IPv4 packets in order to send them across the Internet which is still counting a lot on the IPv4 infrastructure. Finally, proxying and translation are accomplished via an ALG when an IPv6 device wants to establish a communication with an IPv4 one. Stockebrand recommends to always start by dual-stacking the network devices when planning to adopt IPv6 and also filter the old devices that can only run on IPv4 [13].

There are various barriers and obstacles for adopting new technologies including costs, resistance to learn something new, performance, user acceptance, and the availability of time, therefore the barriers are mostly related to the human factor. There are other discussed other barriers such as lack of future vision and leadership, lack of the reasonable infrastructure, and the fear of the unknown. The scope of this study is to investigate the following barriers: *compatibility and readiness of the current networks, transition costs of the adoption, and the performance impact of IPv6 compared to IPv4.*

3 Case Study

Al-Mouwasat Hospital has many branches but this research focuses on the main branch located in Al-Dammam city, KSA. The hospital has an employee size of around 1500 employees (i.e. medical and non-medical). The hospital mainly consists of two buildings known as "old building" and "tower building". The computer network consists of one core switch in the old building and another main switch in the tower building connected directly back to the core switch via fibre optics. From each of these two switches, direct connections are going to many edge switches that connect the workstations and servers. The switches are Cisco Catalyst with both the core one in old building and the main one in tower building being Cisco 4507R while there are layer 3 and layer 2 edge switches such as 3560 and 2960, respectively.

The story behind this study is there was a need to have multiple new workstations in each clinic in the tower building. This workstation increase is about 25 % of the current running workstations. By the time the workstations were running, a huge increase in the network traffic observed resulting in the core switch to reach 99 % of the CPU utilization. As a result, too many packets were dropped causing the Oasis Hospital Management Information System (HMIS) of the hospital to malfunction. Furthermore, the workstations became unable to normally obtain IP addresses dynamically. At that time, the IT department decided to disable many network services/protocols such as IPv6, LLMNR, and NetBIOS. As a result, the core switch CPU utilisation went back to normal readings. The blame was on IPv6 causing the increased traffic. This has been considered as a failure to adopt IPv6. The researchers of this paper doubt that IPv6 was the reason for such dramatic increase in network traffic. The performance impact of IPv6 needs to be studied in order to realize whether it was really causing that low performance and the high load on the network. An investigation of the network compatibility to run IPv6 should be studied as well, and any costs involved should be addressed in order to validate whether transition costs of IPv6 are significant on IPv6 adoption. Therefore, the scenario to be evaluated is briefly to ensure that the workstations are receiving IPv6 addresses with the current IPv4, and then to evaluate and check the aforementioned factors for their validity.

3.1 Data Collection and Analysis

A survey was conducted and the main parties were involved in the earlier decision of abandoning IPv6 were given the chance to comment. Questions 2, 4, and 9 were taken from NRO as it provides a global IPv6 deployment survey that has been used by the researchers [2]. The questionnaire questions are as follows:

1. Why the hospital has not yet considered IPv6 allocation/assignment? (choose all that apply)
   (a) Cannot meet the requirements
   (b) Cannot afford the risk of transition from my IPv4 base
   (c) Cannot afford the expense
   (d) ISP does not support IPv6
   (e) Lack of configuration management tools of IPv6
   (f) Our infrastructure does not support IPv6
   (g) Could not convince business decision makers
   (h) Do not see the business need now
   (i) Other (please specify)
2. Under which set-up do you plan
   (a) Dual-Stack (IPv4 and IPv6 on the same hardware)
   (b) Only IPv6
   (c) Separate infrastructure for IPv4 and IPv6
3. Do your current main network devices (core switch, edge switches, routers, etc.) support IPv6?
   (a) Yes
   (b) No

(c) Not Sure

4. Do your current main computers, servers, network printers, and other client devices support IPv6?
   (a) Yes
   (b) No
   (c) Not Sure

5. What would you do if additional costs are required to adopt IPv6?
   (a) Only financial costs are acceptable
   (b) I am willing to adopt IPv6 regardless of any costs involved
   (c) Only time-related costs are acceptable
   (d) I will not accept any additional costs needed for IPv6 adoption

6. What performance impact do you expect by running IPv6 on your network?
   (a) Good
   (b) Bad
   (c) Not Sure

7. Rank in order the biggest hurdle(s) that caused your IPv6 adoption to fail thus pushed your organization to totally disable IPv6 afterwards. (1 is the biggest)
   (a) Availability of knowledgeable staff
   (b) Business case to non-technical decision makers
   (c) Costs (required financial investment / time of staff)
   (d) Information security
   (e) Performance impact
   (f) Vendor support (Network readiness and compatibility)

The investigation of the case study shows that there are various factors that affect IPv6 adoption but the most technical ones are: transition costs and performance impact. Still, the other factors are not far behind. While it seems that there is a good-level of confidence among the staff when it comes to understanding IPv6 and being sure that the end-user devices are supporting IPv6, the outcome of investigating the factors will determine whether that is true or not. Interestingly, the agreement over running IPv6 over a dual-stack set-up supports the earlier literature review in that the dual-stack set-up is the most welcomed initial approach when going for IPv6 adoption. Indeed, the main reason for failing to adopt IPv6 was failing to maintain the dual-stack set-up by totally disabling IPv6. The lack of knowledgeable staff caused this and pushed for such a decision which should not have been rushed without a thorough analysis of the variables in order to isolate the problem as much as possible.

Table 1 Sample of IPv6 addressing plan

Location	Building	VLAN number (Hex)	VLAN IPv6 address
Al-Dammam (1)	Old building (1)	23 (17)	2001:DB8:9999:1117::0/64
Al-Dammam (1)	Old building (1)	101 (65)	2001:DB8:9999:1165::0/64
Al-Dammam (1)	Tower building (2)	3 (03)	2001:DB8:9999:1203::0/64
Al-Dammam (1)	Tower building (2)	5 (05)	2001:DB8:9999:1205::0/64

3.2 Implementation and Adoption of IPv6

While there is not IPv6 allocation provided by the ISP, the network address to be used will be a global address in order to keep the network ready for any IPv6 allocation given in the future by the ISP. The network address to be used is 2001:DB8:9999::0 /48. For this study, the IPv6 addressing mode shown in Tables 1 and 2 is used.

For the end-user devices such as servers, computers, network printers, or any other devices that favour static IPv6 addresses, an additional exclusion of up to 45 devices is made which means addresses between hexadecimal 6 and 32 are all reserved in each VLAN. Accordingly, Table 3 shows the network range and the address exclusion of each IPv6 scope for dynamic addressing in the DHCPv6.

3.2.1 Network Readiness and Compatibility

The network readiness and compatibility is the first factor to be investigated in this case study. Therefore, checking all the existing network devices that are involved in the IPv6 adoption is required. Table 4 lists all the Cisco switches models in the hospital and the IOS versions used. Interestingly, all models support IPv6; therefore the hardware part is successfully passed.

On the software side of the switches or more specifically the IOS, it also supports IPv6 even on the oldest 3560 IOS version among them all which is IPBASE 12.2 (25) SEB4. However, the IPv6 capabilities on the 3560 switches were not active because of the SDM being used on them. The current SDM is "desktop default", therefore it is required to switch the SDM of the switches to another one which supports dual-stacking of IPv4 and IPv6 such as "desktop IPv4 and IPv6 default", and then all the IPv6 capabilities available on that certain IOS will be active and so do any commands for configuring the switch for IPv6. The available commands are different depending on the current IOS version. While the very old IPBASE 12.2 (25) SEB4 supports IPv6, it is not possible to assign an IPv6 interface IP for the VLAN, unlike the newer firmware IP BASE 12.2 (55) SE9 which offers a lot of IPv6 commands.

With the previous findings, it is obvious that the network readiness and compatibility factor is no longer a valid one because both Operating System (i.e. Windows 7 and Windows Server 2003) and network devices that support IPv6 are outdated (i.e. since 2005). It is unlikely that companies are still running network devices or

Table 2 Sample of IPv6 interfaces addressing

VLAN IPv6 address	Hospital core switch	Tower main switch	Edge switches
2001:DB8:9999:1117::0/64	2001:DB8:9999:1117::1	2001:DB8:9999:1117::2	2001:DB8:9999:1117::3 2001:DB8:9999:1117::4 2001:DB8:9999:1117::5
2001:DB8:9999:1203::0/64	2001:DB8:9999:1203::1	2001:DB8:9999:1203::2	2001:DB8:9999:1203::3 2001:DB8:9999:1203::4 2001:DB8:9999:1203::5

Table 3 Sample of IPv6 scopes in DHCPv6

VLAN IPv6 address	Network range	Exclusion
2001:DB8:9999:1116::0/64	2001:DB8:9999:1116::33 to 2001:DB8:9999:1116:FFFF:FFFF:FFFF:FFFF	2001:DB8:9999:1116::1 to 2001:DB8:9999:1116::32
2001:DB8:9999:1203::0/64	2001:DB8:9999:1203::33 to 2001:DB8:9999:1203:FFFF:FFFF:FFFF:FFFF	2001:DB8:9999:1203::1 to 2001:DB8:9999:1203::32

Table 4 Switches models and IOS versions

Model	IOS versions
4507R	BASIC L3 12.2 (25) EWA4, and IPBASE K9 15.0 (2) SG7
3560	IPBASE 12.2 (25) SEB4, 12.2 (25) SEE3, 12.2 (35) SE5, 12.2 (50) SE5, and 12.2 (55) SE9
2960	LANBASE K9 12.2 (44) SE6

operating systems that are older than that because it is recommended to upgrade the devices and systems once they are out of their support life-cycle.

3.2.2 Performance Impact

The results of the performance impact factor were significant because there were fears and doubts about the performance impact of IPv6 on the network according to the literature review and the survey done.

The suggested inside-out approach when adopting IPv6 has given us the chance to keep monitoring the performance impact of IPv6 on the clients side, the edge switch, and the core switch in order to find how each one of these sides may affect the performance of the network when adopting IPv6. Table 5 shows that the initial baseline of the core switch CPU utilization was showing an average of 30–40 % with IPv6 being totally disabled on the whole network and computers. After enabling IPv6 on the computers, the average CPU utilization has not changed which was encouraging. Afterwards, enabling the IPv6 on the edge switches did not affect the percentage badly at all but kept the same range of 30–40 % of the CPU utilization on the core switch. The last step was very crucial because enabling IPv6 on the core switch means finally the IPv6 traffic will flow across the network and the computers will start receiving IPv6 addresses as well as the IPv4 ones and both of these should be registered on the DHCPv6 and the DHCP, respectively. The outcome was impressive with more than 65 computers receiving an IPv6 address thus running the dual-stack

Table 5 Performance impact of IPv6

Stage	Average core switch CPU utilization
Initial baseline	30–40 % but mostly in the 30 % range
After Step 1	30–40 % with more frequent average in the 40 % range
After Step 2	30–40 % similar to previous stage
After Step 3	30–40 % but mostly in the 40 % range

set-up successfully while the CPU utilization of the core switch was still maintained on the same range of 30 to 40 % despite the slight increase.

Accordingly, this removes any doubts that IPv6 was the reason for causing the extra CPU utilization earlier and in fact eliminates any worries about adopting IPv6 in a dual-stack network set-up because it was very smooth and did not cause any bad performance impact although the computers were successfully able to receive both IPv4 and IPv6 address and communicate successfully. Proving that, the actual reason for the dramatic CPU utilization is insignificant to this study.

3.2.3 Transition Costs

The success of the first two factors reflects on the transition costs. We hardly found any compatibility issues related to the first factor that has a significant financial transition costs which may hinder IPv6 adoption. Although some of the hardware and software used in this case study is considered old and outdated, the existing support of IPv6 omitted any need for hardware or software upgrades that are very costly. Any hardware or licenses upgrades required involve financial costs; on the other hand, any software and firmware upgrades or new configurations are likely to involve time-related costs only. Therefore the transition costs depend on the case and network set-up. With the performance impact being very minimal and IPv6 running very smoothly and causing no issues on the core switch performance, there were no further costs recorded from the second factor apart from the time taken to carefully study IPv6 and adopt it which is something that is required when dealing with any new technology by putting a good plan in order to follow thus manage the time as much as possible.

4 Conclusion

The aim of this case study is to investigate whether IPv6 was the reason for the high load and thus dramatic CPU utilization of the core switch and to analyse and evaluate the factors significantly affect IPv6 adoption, which are the network readiness and compatibility, the performance impact, and the transition costs. The result of the study reveals:

1. The factor of the network readiness and compatibility for running IPv6 should not be a hurdle to adopt IPv6.
2. There were doubts against the poor performance impact of IPv6, however, the test results proved that IPv6 does not cause a severe and serious bad performance impact even when running on dual-stack set-up with IPv4.
3. The test results showed that the transition costs are very minimal and they are mostly time-related costs not financial ones.

4. The suggested IPv6 addressing plan for running a dual-stack set-up proved to be useful and practical for providing easy-to-remember IPv6 addresses thus easier management and troubleshooting without the burden of not being able to memorize the IPv6 addresses and subnets.

Acknowledgments The authors of this research would like to thank Al-Mouwasat Hospital in Al-Dammam, Kingdom of Saudi Arabia for their support throughout this research by authorising the authors to run the experiments on their network.

References

1. Announcements I (2014) Remaining ipv4 addresses to be redistributed to regional internet registries|address redistribution signals that ipv4 is nearing total exhaustion. Technical report, ICANN, May 2014. https://www.icann.org/news/announcement-2-2014-05-20-en
2. Botterman M (2013) Ipv6 deployment survey. Techical report, NRO (2013). https://www.nro.net/wp-content/uploads/Global-IPv6-Deployment-Survey-2013-final.pdf
3. Colitti L, Gunderson SH, Kline E, Refice T (2010) Evaluating IPv6 adoption in the internet. In: Krishnamurthy A, Plattner B (eds) PAM 2010. LNCS, vol 6032. Springer, Heidelberg, pp 141–150
4. Google (2014) Per-country ipv6 adoption. http://www.google.com/intl/en/ipv6/statistics.html#tab=per-country-ipv6-adoption
5. Google (2014) Statistics ipv6 adoption. http://www.google.com/intl/en/ipv6/statistics.html#tab=ipv6-adoption
6. Kaul A (2012) YPAHSA Ipv6 protocol adoption in the u.s.: Why is it so slow? Techical report, Telecommunication System Laboratory, University of Colorado Boulder
7. Kim J, Sarrar N, Feldmann A (2014) Watching the ipv6 takeoff from an ixp's viewpoint. CoRR abs/1402.3982. http://arxiv.org/abs/1402.3982
8. Leavitt N (2011) Ipv6: any closer to adoption? Computer 44(9):14–16
9. Olabenjo Babatunde OAD (2014) A comparative review of internet protocol version 4 (ipv4) and internet protocol version 6 (ipv6). Int J Comput Trends Technol (IJCTT) 13(1):11–13
10. Rooney T (2010) IP address management principles and practice. Wiley-IEEE Press, Hoboken
11. Schwankert S (2008) Sound the alarm for ipv6. NetworkWorld Asia 4(4):8
12. Sharma V (2010) Ipv6 and ipv4 security challenge analysis and best-practice scenario. Int J Adv Netw Appl 1(4):258
13. Stockebrand B (2007) IPv6 in practice: a Unixer's guide to the next generation internet. Springer, Berlin
14. Wu P, Cui Y, Wu J, Liu J, Metz C (2013) Transition from ipv4 to ipv6: a state-of-the-art survey. Commun Surv Tutor IEEE 15(3):1407–1424
15. Che X, Lewis D (2010) Ipv6: current deployment and migration status. Int J Res Rev Comput Sci (IJRRCS) 1(2):22–29

Software Engineering for Security as a Non-functional Requirement

Noha Ragab, Ali Ahmed and Saadat AlHashmi

Abstract Interactions between software engineering and security requirements or engineering can be carried out in almost all software process phases, i.e. requirements analysis, design, implementation, verification, and deployment. In current information era, modern societies and emergent business increasingly rely on technology and communication. It becomes inevitable that every software system developed today must defend itself from malicious adversaries. Organizations used to adhere to well-defined and proved models for software development, but these models were originally proposed for functional requirements. The non-functional requirements don't receive the same level of concern. Non-functional requirements address how the system should behave. It is understood that adding non-functional requirements after the system (functional) requirements and design are done is both difficult, expensive, and sometimes impossible. This paper comprehensively studies how security as a non-functional requirement is incorporated in the software development life-cycle. It shows how to unify the security policies with the development models earlier in development life cycle. The work accommodate any non-functional requirement but the case study is centred on security.

Keywords Functional requirement · Non-functional requirement · Security · RBAC · XACML

N. Ragab
University of Hertfordshire, Hertfordshire, UK
e-mail: noha.ragab@gmail.com

A. Ahmed (✉)
Faculty of Computers and Information, Cairo University, Cairo, Egypt
e-mail: a.ahmed@fci-cu.edu.eg

S. AlHashmi
College of Engineering, Abu Dhabi University, Abu Dhabi, United Arab of Emirates
e-mail: saadat.alhashmi@adu.ac.ae

A. Abraham et al. (eds.), *Intelligent Data Analysis and Applications*,
Advances in Intelligent Systems and Computing 370,
DOI 10.1007/978-3-319-21206-7_29

1 Introduction

In current information era, modern societies and emergent business increasingly rely on technology and communication. It becomes inevitable that "every software system developed today must defend itself from malicious adversaries" [2]. Software security is the way of developing software such that it can function correctly and continuously under malicious attacks, usually, comes as an afterthought to the software process [8]. A software system is typically specified both by its functional requirements (what the system should do), and non-functional requirements (how the system should behave) [4, 13]. Functional requirements are intuitively emphasized in a software development process, but the non-functional requirements don't receive the same level of concern [2]. It is understood that adding non-functional requirements after the system (functional) requirements and design are done is both difficult, expensive, and sometimes impossible [2]. Currently, it is becoming more and more demanding for software which satisfy more than its functional requirements [1, 13]. Fortunately, There is a recent research trend towards unifying security requirements with software engineering, particularly, unifying security and system models [2]. It is worth noting that security as a functional requirement is not the primary focus for this research.

Since the software evolution in the 1960s, software process models started to be applied for the development of large software systems. A plethora of process models have been introduced to facilitate the overall development of different types of software projects, ranging from commercial websites to safety-critical systems. Different process models provide for the particular planning, organizing, coordinating, staffing, budgeting and directing of a wide spectrum of applications. These models could be categorized into generic models or paradigms. "These generic models are not definitive descriptions of software processes. Rather they are abstractions of the process that can be used to explain different approaches to software development" [12]. For example, the linear sequential model has been designed for straight-line development; The most well-known linear sequential model is the waterfall model, which has been around since 1970 by Royce [11]. Feedback is essential in those phases, which is collected after each phase. The feedback of one phase affects the next phase. This helps when requirements change and is vital should the final test lead to fatal errors. However, nowadays, the waterfall model refers to software models that are not flexible and non-iterative although the intention for the model was to be modified into an iterative model. Thus the waterfall paradigm has overwhelmed serious problems of requirements volatility. Outstanding problems in requirement changes and system design that arise thereafter can only be resolved through maintenance with a rework of the system, because the process does not allow for backwards revising and iterations. Models other than the conventional waterfall model include the enriched waterfall model, and the 'V' life-cycle model.

The iterative development model, was propose to solve the current problems of the linear sequential model; it is based on the recognition that in real systems. It is not really achievable that requirements is fully understood or never going to change. In the iterative development, requirement elicitation is done once for the whole system at the beginning of the process, then a rough product is created in first iteration, review it, then looping through same iterations for improvement until finished. The software development phases are fitted into each iteration. Iterative and incremental developments are essential parts of Extreme Programming (XP) and generally the various agile software development frameworks. Many models fall under the Agile umbrella such as: Test Driven Development (TDD), Scrum, Crystal, Extreme Programming, Adaptive Software Development, Feature Driven Development, Dynamic Systems Development Method (DSDM), Unified Process (AUP), etc. Extreme Programming (XP) is an Agile method that is intended to improve software quality, team productivity as well as the ability to adopt according to changing requirements. This is basically achievable by an iterative and incremental process, with short iterations and continues feedback from customers that helps to develop plans for further iterations, and to eventually reach up to the full function desired system. The good thing about XP, from our point of view, is that it combines the best practices or feature from many other process model and encapsulates it in a very light, fast, and simple way. Not only XP affects the product development, but also the developing team and the customer. It facilitates the coordination between team members and management, as well we the team members themselves.

2 Engineering the Software Non-functional Requirements

In many cases, the functionality of a system may not be useful or even usable without the complementary non-functional characteristics. The work on non-functional requirement regarding, definition, description and ways of representation is all recent, as the market is always demanding for better software quality. The Non-Functional Requirements (NFRs) have only been described by informal natural language for long time. As a result, the outcome product may fail short of accuracy, integrity, and reliability, as well as making the software design and development more difficult. Franch and Botella introduced an interesting model for software architecture based on the component/ connector scheme, where components present their non-functional specifications and their non-functional behaviour, and connectors describe which non-functional aspects are relevant to components connections [4]. Both components (active entities) and connectors (interaction definition between components) has specification and implementation, each of which has functional and non-functional parts. Then the authors introduce on the language NoFun on which they rely on to represent non-functional requirements in terms of components/connectors attribute values. The main aim of the authors in [3] is to propose a modelling method that helps the integration with software architecture design methods (such as UML). They introduced a quiet simple approach based on XML to describe and represent NFR, and hence integrate this easily with other functional requirements design models being XML a general data exchange language. The method is challenging an simple although it's not provide a complete solution to dealing with NFR and needs further work and refinements.

With the evolution of network and internet applications, which became very prominent in the last few years, the market is more demanding for software that does not only satisfy its expected functionalities but also concerns about other non-functional aspects such as: reliability, accuracy, maintainability, and security. The non-functional requirements have to be handled from the beginning of the software development life-cycle and hence to be dealt with throughout that cycle. There are some proposals to incorporate the NFRs in software development life-cycle. Examples of those proposals include the Process to Support Software Security (PSSS), and one of its sub-processes, model security threat have been introduced in [10]. PSSS is designed for iterative projects, and is considered a heavy process, the interesting about PSSS that is offers 37 positive security activities to be adapted according to corporate software development preferences. Jurjens proposed the UMLsec as an extension to the standard UML to allow the presentation of security relevant information within the diagram in system specifications in order to aid the development of security-critical systems [9]. Another modelling language that based on UML is Secure UML that is a modelling language for the model-driven development of secure, distributed system. it's designed for integrating the specification of access control into application models, based on the role based access control (RBAC). The authors in [6] proposed a framework represents and analyses the security requirements. They emphasised the definition of security requirement and the importance of the context in which the software is running. They also proposed "a structure for the satisfaction arguments for validating whether the system can satisfy the security requirements. The Systematic Software Development Process for Non-Functional Requirements (SSDP" [7] is one of the interesting proposals for engineering the non-functional requirements. The proposal answers the important question of "where in the development process should non-functional requirements be dealt with?" The proposed follows six steps to accommodate NFRs. Those steps are: Elaborating Non-Functional Requirements, Developing Non-Functional Requirement Models, Estimating Non-Functional Requirements, Assessing Non-Functional Requirements, Specifying Non-Functional Requirements in a Formal Method, and Analysing Non-Functional Requirements. The model is not tailored for a certain NFR, however the authors have given examples of Security, Availability, and Performance. The authors argues that it could be used for any other NFRs though. As we highlighted before, the model is interesting as it introduces a complete process to consider the NFRs intuitively. In addition, the model also helps in integrating functional and non-functional requirements. This is why, the SSDP proposal is adopted in the proposed case study.

3 The Case Study

This case study demonstrates a smart (i.e. pervasive) hospital that utilizes both wired and wireless communication mediums to provide medical services to patients. The clinicians are utilizing, for example, their smart devices to view patient records, giving prescriptions, off-shore consultation, etc. The nursing staff are able also to log into the system by their smart devices to provide supporting services everywhere and at any time. Patients may be using body sensors that may be forming a body sensor network (BSN). It is worth noting that the system uses different authorisations for the users as a way to protect sensitive information in the patient record. As the adopted model (i.e. SSDP) suggests, the 6 steps applied to the case study are detailed in the following paragraphs.

The first Step is titled, "Elaborating NFRs". As this step suggests refining and classifying the resources to be secured, the patient Information System (PIS) system is the main focus for this step. The patient sensitive information of the patient record is identified as a possible asset. In addition, the communication channel needs to be secure such that no information is leaked to unintended entities while transmitted. For example, the following information/assets are identified to be of a primary concern: Patient's family medical history, Patient's Identifying Information (PII), Patient's family identifying Information (PFII), WIFI communication especially for those BSN related, Sensors (i.e. location, Gas, Fire, etc.), Hospital modality device (e.g. x-Ray), Hardware (e.g. PCs, printers, plotters, smart devices, etc.), and Software (i.e. operating systems, applications, etc.). The actors (i.e. roles) in the system are: Administrator, Doctor/Clinician, Nurse, Patient, Database, and Policy Manager. The Role-Based Access Control Model (RBAC) is adopted as the authorization engine. Thus, every subject in the system will have a role and those roles will be associated with rights/permissions. To be able to carry out a certain task, all rights should be possessed beforehand. The access control policy will be written in XACML policy language.

The second step, Developing NFR models, starts with some resource classification regarding the ownership. From our point of view, this is already accomplished in the previous step where the access rights versus roles have been identified and elaborated. To build the trust model, one should determine the trust relationships amongst the actors and then check whether an actor that offers a service is authorised to have. In fact, building that trust model is not an easy task due to the lack of proposals in the literature [5]. The work in [5] shows an interesting trust model that is applied to electronic health care system. The same foundations are adopted in this research. The trust model of this study assumes that sensors (i.e. body sensors, location sensors, movement sensors, etc.) are part of our trust computation base (TCB). The threat model has to capture the threat profile of a system. Threat modelling is useful for architects as they help providing accurate security comprehension. One of the well-known tools in threat modelling are the misuse/abuse cases and the attack trees. The abuse cases are adopted in this study as a tool to model the threat of the proposed case study. The abuse case is an extension of the standard use cases. While

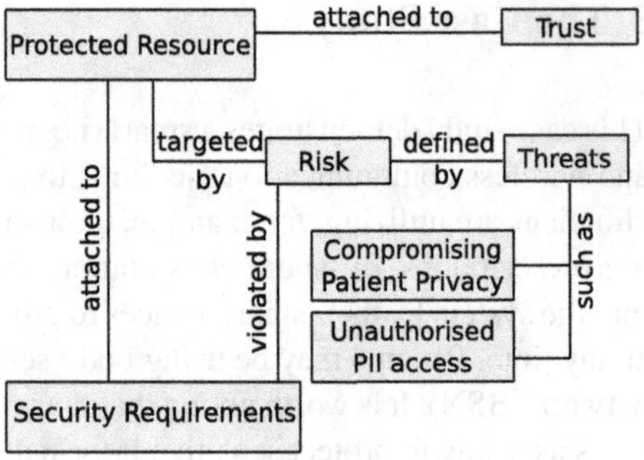

Fig. 1 Conceptual system threat model

the standard use cases show the scenario of actions/services an actor can do, the abuse cases show the actions/services a malicious actor can not do. Figure 1 shows our conceptual risk model and potential threats associated with illegitimate resource access.

In general, attacks to patient's records could be categorized a: Compromise Client, Compromise Server,and Compromise Network. The threats/vulnerabilities an attacker could exploit to compromise the system are threefold: Compromise Client, Compromise Server, and Compromise Network. Compromising the client can be done via Shoulder surfing, Use unattended logged-on client, Obtain valid user-name/password (i.e. Social engineering, Network interception, Key-logging, Phishing emails, etc.), Infect with malware, Steal client if portable (e.g. smart phones), or Destroy client (i.e. Denial of Service). Compromising the server could be done by Gaining remote access (i.e. Use default user-name/password (e.g. guest), Use exploit(s), etc.), Gain local access by either Gain physical access or Obtain administrator user-name/password, and Server Denial of Service attack. Compromising the network is done by Eavesdropping traffic, Modifying or injecting traffic, and making network unavailable by cutting network cables or destroying the wireless access points. As a proof of concept, some use cases from Fig. 2 will be investigated and the corresponding misuse cases will be shown. The misuse case of the patient registration is shown in Fig. 3. The login misuse case is shown in Fig. 4. For the doctor appointment, there is no need to address the misuse case here thanks to the patient log in service, which is called before the doctor appointment service accommodating. The staff registration misuse case is not considered here assuming that all entities are trusted. It is worth noting here that a collusion may take place between multiple entities faking the identity of the user (i.e. the one to register). However, this is beyond the scope of this research.

The fundamental element in step 3 (i.e. Estimating NFRs) is to establish a proper control framework for the flow of information specially the sensitive one. The design of the detection framework is shown in Fig. 5. The proposal can actually control the flow of information from the patient information stores to those legitimate subjects trying to gain access. The proposal works as follows: A subject may request access to a certain data item. The request is sent encompassing: Subject ID, Subject creden-

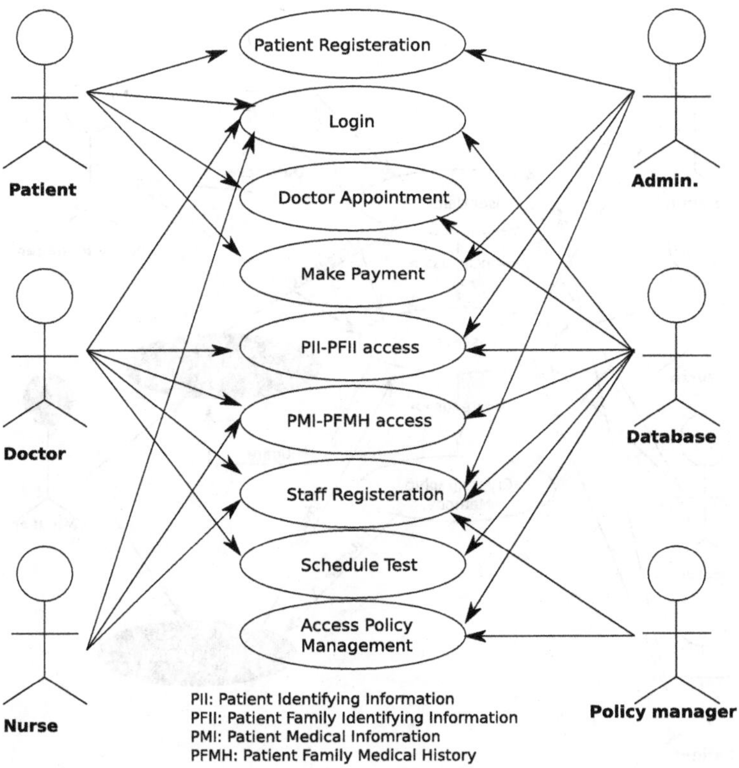

Fig. 2 The hospital overall UseCase model

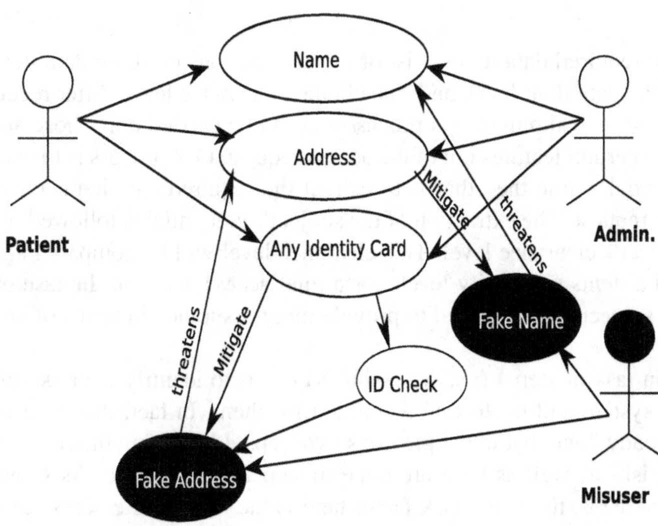

Fig. 3 The patient registration misuse case

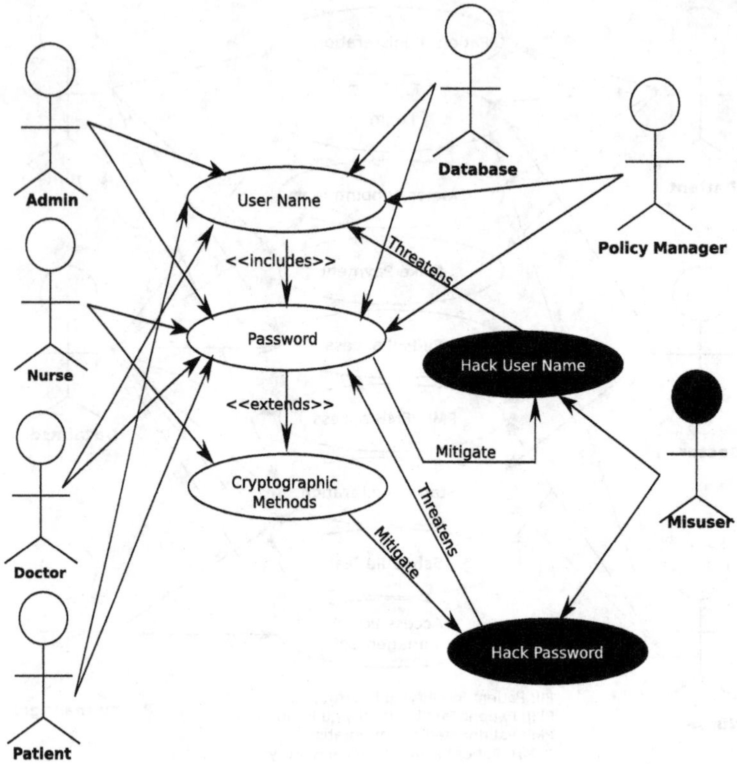

Fig. 4 The login misuse case

tials, List of required data items, List of required actions on those data items (what s/he needs the data item for?), and The claimed clearance level. After receiving the access request, two simultaneous processes could be carried out. Those are related to extracting certain features out of the access request. One process is to extract subject's credential, while the other is to extract the required data items contained in the original request. The extraction of the subject's credential is followed by retrieving the subject's clearance level. This clearance level will be compared against the required data items sensitivity levels for a final access decision. In case of a deny access, the subject may be asked to provide more assurance in terms of credentials (i.e. PKI).

The main task of step 4 (i.e. Assessing NFRs) is to identify the risk factors that impact the system and try to find a solution for them. In fact, this is repetition as we have already done that in the previous two steps. Identifying the risk factors and model the risks as well as trust are done in step 2 and 3 above. As shown in the previous two steps, the major risk factor here is the illegitimate access to sensitive resource objects. PII is of a high sensitivity level and all access requesters should attain certain privileges before PII is released. Based on that, the RBAC model is

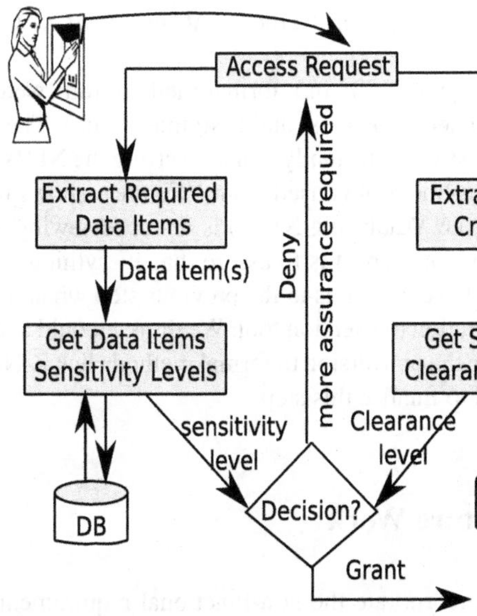

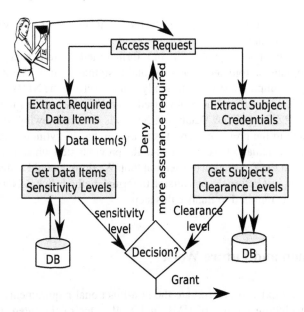

Fig. 5 The authorisation workflow

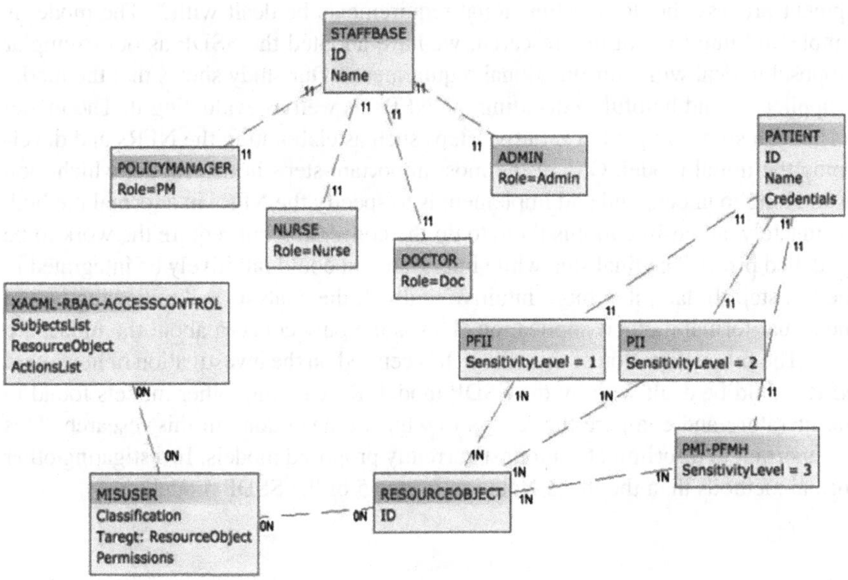

Fig. 6 The ERZ NFRs modelling

used to control such access. More precisely, the RBAC profile of XACML is used to provide such functionality.

Step 5 is all about specifying NFRs in a formal method, thus we choose the Z-notation since many automated tools exist and it simpler to use. The Z-notation is used to describe computing systems formally. For converting the NFRs as well as our entities to Z-notation, an on-line tool is used from.[1] Figure 6 shows the Z-notation modelling in ERZ. Step 6 (i.e. Analysing NFRs) is about reviewing and analysing the NFRs stated in the previous step. It is to assure that everything regarding NFR works fine and we actually have done this in the previous step when using the UML model along with the Z-Notation conversion tool. We think it could actually be done within step 5 and just before the conversion to formal methods lick Z-Notation. Thus, no further work is required to finalise this step.

4 Conclusion and Future Work

Many techniques exist to incorporate the non-functional requirements in software engineering including the use of UML and XML. One of the interesting models found is the SSDP. (SSDP) answers the important question of "where in the development process should non-functional requirements be dealt with?" The model is simple and intuitive. In this research, we have adopted the SSDP as our exemplar proposal to deal with non-functional requirements. Our study shows that the model is applicable and helpful in deciding the NFRs as well as evaluating it. The model starts with some simple but intuitive steps such as elaborating the NFRs and developing the initial model. One of the most important steps in this model which took us too long to understand and implement is to specify the NFRs in a formal method. Fortunately, an on-line tool is there to do the conversion but require the work to be modelled priori. The final step while important but could intuitively be integrated in the last step. In fact, it is more intuitive to do all the analysis and reviewing before the actual formal method application. This is the only concern about the model we have. The future direction of this research is centred on the investigation of how other NFRs could be dealt with by the SSDP model. Researching other models found in the literature and compare the results to what we have done in this research. This is important to discriminate amongst currently proposed models. Investigating other formal methods than the the Z-Notation in step 5 of the SSDP model.

[1]http://erz.comli.com/er.php?action=new.

References

1. Cysneiros L, do Prado Leite J (2004) Nonfunctional requirements: from elicitation to conceptual models. IEEE Trans Softw Eng 30(5):328–350
2. Devanbu PT, Stubblebine S (2000) Software engineering for security: a roadmap. In: The future of software engineering. ACM Press, pp 227–239
3. Fei Y, Xiaodong Z (2007) An xml-based software non-functional requirements modeling method. In: ICEMI '07. 8th International conference on electronic measurement and instruments, pp 2–375 -2-380, 16–18 July 2007
4. Franch X, Botella P (1998) Putting non-functional requirements into software architecture. In: 9th International workshop on software specification and design, pp 60–67
5. Giorgini P, Massacci F, Mylopoulos J (2003) Requirement engineering meets security: a case study on modelling secure electronic transactions by visa and mastercard. In: ER, pp 263–276
6. Haley C, Laney R, Moffett J, Nuseibeh B (2008) Security requirements engineering: a framework for representation and analysis. IEEE Trans Softw Eng 34(1):133–153
7. Jung HT, Lee GH (2010) A systematic software development process for non-functional requirements. In: 2010 International conference on information and communication technology convergence (ICTC), pp 431–436
8. Jürjens J (2001) Towards development of secure systems using UMLsec. In: Hussmann H (ed) FASE 2001. LNCS, vol 2029. Springer, Heidelberg, p 187
9. Jürjens J (2002) UMLsec: extending UML for secure systems development. In: Jézéquel J-M, Hussmann H, Cook S (eds) UML 2002. LNCS, vol 2460. Springer, Heidelberg, p 412
10. Nunes F, Belchior A, Albuquerque A (2010) Security engineering approach to support software security. In: 2010 6th World congress on services (SERVICES-1), pp 48–55
11. Royce WW (1970) Managing the development of large software systems: concepts and techniques. In: ICSE '87: Proceedings of the 9th international conference on software engineering. IEEE Computer Society Press, Los Alamitos, CA, USA, pp 328–338
12. Sommerville I (2011) Software engineering, vol 9/E. Addison-Wesley, New York
13. Umar M, Khan N (2011) Analyzing non-functional requirements (nfrs) for software development. In: 2011 IEEE 2nd International conference on software engineering and service science (ICSESS), pp 675–678

Part III
Security and Privacy

Localization for Jamming Attack in Wireless Sensor Networks

Jing Zhang, Li Xu, Qun Shen and Xiaorong Ji

Abstract Wireless Sensor Network is easily suffered from various kinds of attacks, it is essential to localize the jamming attacker in wireless sensor networks, for the implementation and deployment of security mechanisms. In this paper α-hull is applied to calculate the Minimum Circumscribed Circle (MCC) of a set of points. All vertices of the α-hull will be on the same circle, if $1/\alpha$ is equal to the radius of points' MCC. Based on this nature, an effective and accurate method for MCC detection is established, it is mainly through finding the least squares circle of the set of points and iteratively approaching the MCC with recursive subdivision. We provide the analysis of correctness and time complexity of our algorithm through theoretical proof, and the comparison between our algorithm and others through practical simulation. Experimental results show this algorithm is reliable and it is able to achieve higher accuracy than other comparable localization algorithms in most cases. Such as the changes of the attack range with the network node density changes.

Keywords Wireless sensor networks · Jamming attack · Localization · Minimum circumscribed circle

The authors wish to thank National Natural Science Foundation of China (Grant NO: 61072080). Fujian Normal University Innovative Research Team (No. IRTL1207). The Natural Science Foundation of Fujian Province (NO: 2013J01222, J01223). The Education Department of Fujian Province science and technology project (JA13215, 2014H0008).

J. Zhang (✉) · X. Ji
School of Information Science and Engineering,
Fujian University of Technology, Fuzhou, China
e-mail: jing165455@126.com

L. Xu
School of Mathematics and Computer Science,
Fujian Normal University, Fuzhou, China

Q. Shen
College of Computer and Information Science,
Fujian Agriculture and Forestry University, Fuzhou, China

A. Abraham et al. (eds.), *Intelligent Data Analysis and Applications*,
Advances in Intelligent Systems and Computing 370,
DOI 10.1007/978-3-319-21206-7_30

1 Introduction

Wireless sensor networks are a continuation of the evolution of networks toward larger-scale, distributed computing, they are made up of mostly small sensor with limited-resources and capabilities [1]. As they become increasingly pervasive, ensuring the dependability of wireless networks deployments will become an issue of critical importance. One serious class of threats that will affect the availability of wireless networks are radio interference, or JA (radio frequency Jamming Attack). In order to destroying wireless communications, the attackers sending radio frequency. There are a variety of ways to achieve this, such as full-channel barrage jamming, spot barrage jamming and so on. Full-channel barrage jamming belongs to high-power complete cover jamming, the jamming can cover the whole frequency band of target area completely, and block the transmission of all almost normal signals efficiently, but the jamming effect is related to distance, it may be poor if beyond a certain distance. Spot barrage jamming is only intended to a certain period of interference bands, the attacker needs to break the target's communication frequency-band, and then sends proportional jamming information without any influence on the attacker himself. When the network appears in JA, the sensor performance will be obviously decreased, not even running normally. The traditional method, for the reply of this kind of attack, is to use complex and costly physical layer technology. Currently, it has been focused on research on all kinds of methods of prevention strategies, such as anti-jamming technology based on wormhole, time slot channel, channel surfing and so on [3].

For the next step of implementation and deployment of security mechanisms, it is rather essential to find the location of Jammer. Once JA is detected, we must find and locate the attacker as soon as possible. However, it is not easy to locate the attacker. First of all, since the energy of sensor itself is limited, if the location methods of Jammer bring too much communication overhead, it will not be feasible; In addition, the majority of location algorithms, protocols need to depend on special equipment, so the location is not easy to achieve. It is imperative to find an effective Jammer location method.

The main contributions of this paper are as follows.

(1) α-hull is applied to calculate the Minimum Circumscribed Circle (MCC) of a set of points. All vertices of the α-hull will be on the same circle, if $1/\alpha$ is equal to the radius of points' MCC.

(2) An effective and accurate method for MCC detection is established to locate the Jammer, it is mainly through finding the least squares circle of the set of points and iteratively approaching the MCC with recursive subdivision.

(3) We provide the analysis of correctness of our algorithm through theoretical proof, and the comparison between our algorithm and others through practical simulation.

The rest of this paper is organized as follows: Sect. 2 presents the related work. In Sect. 3, we present Algorithm of α-MCC. Simulation results are demonstrated in Sect. 4. The conclusion is in Sect. 5.

2 Related Work

It is crucial matter of the location information in wireless sensor network applications, the data almost doesn't make sense without location information [1]. Because of the uncontrollability of the sensors in the deployment, such as scattering by plane, most of the nodes in the network location can not be predetermined, but the applications of the wireless sensor network require the nodes location information, thus informed of the exact location of the sources of information. in addition, use of the node location information, routing algorithm can be designed to improve the efficiency of routing. Therefore, the node localization problem has become an important research direction in wireless sensor networks.

Node location can be obtained directly by installation of a GPS receiver, because the global positioning system (GPS) technology is mature. But there are some difficulties because of its price, size, power consumption, signal shielding and many other factors. Currently, most of the positioning technologies are use a small number of known locations of the nodes in the network to obtain the location of unknown nodes by location algorithm [2]. In order to respond to the threat posed by the jamming attack, it is crucial to understand a variety of jamming attack model. How to map an interference region studied in the literature [3], several ways to detect jamming attacks has given in it.

Positioning algorithm can be divided into two categories according to the positioning mechanism: the Range-based and the Range-free positioning technologies. The Range-based localization algorithms require specific hardware to measure the distance or angle information between nodes, then use trilateration, triangulation or maximum likelihood estimation method to calculate the node position. This algorithm can accurately measure the location information of the nodes, but hardware overhead to pay. In the jamming attack environment, interference area of communication has been all or partially damaged. therefore, this method is not suitable for jammer positioning. The latter simply by network connectivity, signals and other information, precise enough to locate in the case of no distance, angle and other information, there is a great advantage in the calculation of costs, overhead, etc. centroid localization algorithm [4], positioning algorithm based on convex programming [5] algorithm based on Distance Vector Routing (DV-Hop) [6, 7], algorithm based on Approximate Point-In-Triangulation test(APIT) [8] and so on are all typical of the current algorithm by Range-free positioning technologies. A virtual force it-elative localization algorithm (VFIL) proposed by Liu H [10]. Bahrepour M [11] studied the fire detection, the establishment of a model based on the fusion detection of two layers, first of all nodes judged by using the measured data, then the cluster head integration the detection results of each nodes in the network, final calculation of the detection results. The weighted subtractions negative add on positive multi-source location algorithm (WSNAP) was applied to localize the multiple sources [12].

Findings from the existing literature [4], for the wireless sensor networks which distribution in the form of a random shed, built to self-organization. The nodes generally have the following distribution: Uniform distribution, Gaussian distribution,

Rayleigh distribution and so on, which the Gaussian distribution model is the most common. Convenient for the research problem, assume that all nodes dispenser with a flat system, and select the radio communication model. Centroid algorithm only based on the nodes connectivity, beacon nodes broadcasts a beacon signal to the neighbor nodes from time to time, the signal contains its own ID and location information. The unknown node can determine its own position when the beacon signal it receives from different beacon nodes exceeds a preset threshold, or receive a certain period of time, the location is the centroid of the polygon of these beacon nodes. Assume that the coordinates of the n beacon nodes were $(x_1, y_1)...(x_n, y_n)$, according to the above assumptions, issues were discussed in the two-dimensional space, the unknown node coordinate is calculated by the formula (1).

$$(x, y) = (\frac{\sum_{k=1}^{N} x_k}{N}, \frac{\sum_{k=1}^{N} y_k}{N}). \tag{1}$$

The test showed that the unknown nodes coordinates with the centroid, there are about 90 % positioning accuracy are lower than 1/3 of the beacon node spacing [4], due to the nodes density, nodes distribution model, signal transmission of environmental noise, nodes performance impact.

Niculescu et al. proposed the DV-hop algorithm [6, 7], the distance between the unknown nodes and the anchor nodes, represent by the product of average hop distance and hops between them. Perform trilateration positioning when the unknown node has calculated the distances between its own and more then three anchor nodes. Assume that the coordinate of the unknown node is $A(x, y)$, and the coordinates of the anchor nodes are $L_1(x_1, y_1)...L_k(x_k, y_k)$, the distances between the unknown node and the anchor nodes are $r_1, r_2, ..., r_k$, then it can establishment of linear equations. In this paper, the Algorithm for Minimum Circumscribed Circle Detection Based on α-hull (α-MCC), in order to locate the jammer.

3 Algorithm of α-MCC

There are many ways of detecting jamming attacks [3, 13], this paper will not go into it, but direct use of its conclusion, to mapping an interference region. It will send a message to all its neighbors, let they bypass, to prevent the information is captured, when sentencing a node in the interference range. The attack graph is initialized by the interference node's neighbors that who has received the jamming information, this nodes are defined as the edge nodes, gradually increase the interference nodes, then the attack graph is finally jointed by edge nodes.

Definition 1 α**-hull** [14]: For the real numbers α and a finite set of points P in a plane, the intersection of all the α- disks that inclusive P defined as the α-hull of P. All the points on the border of the α-hull defined as the α -hull vertex of the set P.

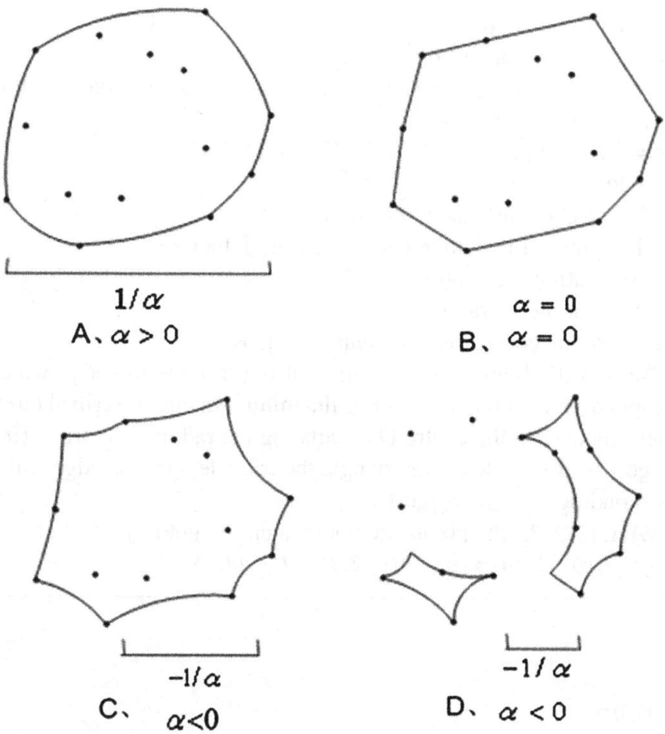

Fig. 1 α-Hull of a set of points

For $\alpha < 0$, the α-hull of *P* is the complement of the union of all disks of radius $-1/\alpha$ that do not contain any point of *P*, for $\alpha = 0$, the α-hull is just the convex hull and for $\alpha > 0$, the α-hull of *P* is the intersection of all disks of radius $1/\alpha$ that contain *P*. Figure 1 illustrates α-hull of *P* for three values of α.

All vertices of the α-hull will be on the same circle, if $1/\alpha$ is equal to the radius of points' minimum circumscribed circle (MCC). And then based on the number of the vertices, the radius of the minimum circumscribed circle can be calculated.

Algorithm1. Algorithm of α-MCC
Input: $P = \{p_1, p_2, p_3, \ldots p_n\}$, which is a set of points which are the disturbed cells.
Output: the coordinates of centre *O* and *r*.
01: choose three points which are non-collinear.
02: calculating the centre *O* of the circle that the three points are all in it.
03: Calculating the distances between the points of *P* and the centre *O*.
04: find the farthest distance *rb* and the closest range *ra*, let $r = rb$.
05: if ($ra \geq rb$), then do
06: *ra* is the radius of the minimum circumscribed circle,
07: calculating the centre *O*,

08: calculating the radius r,
09: return the coordinates of centre O and r.
10: else find the α-hull ($1/\alpha = r$) of the set P, the P_1 set up by the α-hull vertexes;
11: if ($P_1 \subset \phi$) then do
12: $r = (ra + rb)/2$; $ra = (ra + rb)/2$; goto 5
12: else goto 13.
13: if ($\forall A(x, y) \in P_1$, and the points of a circle) then do
14: if the centre of the circle inside the α-hull, then
15: calculating the centre O,
16: calculating the radius r,
17: return the coordinates of centre O and r.
18: else find the biggest obtuse angle of the triangle the edge which obtuse angle corresponding to is the diameter of the minimum circumscribed circle
19 then calculating the centre O calculating the radius r, return r. (It is define that the biggest obtuse angle of the triangle the triangle with the edge which obtuse angle corresponding to is the biggest.)
20: if ($\forall A(x, y) \in P_1$, the points are not on a circle) goto 21.
21: $r = (ra + rb)/2$; $ra = (ra + rb)/2$, $P = P_1$ goto 5.

---

4 Evaluation

In the algorithm of α-MCC, it is use the method of α-hull to solving the position of the jammer. So this problem is transformed to find the minimum circumscribed circle so that it can cover all the interference nodes. The center can be estimated to be the location of the jammer. The following theorems are to prove its correctness.

Theorem 1 *For a finite set of points P in a plane, there is only one minimum circumscribed circle, r0 is the radius of the minimum circumscribed circle, then all vertices of the α-hull ($1/\alpha = r0$) will be on the same circle.*

Proof If assume there is a point p in set P which is in the minimum circumscribed circle. We can do with $r0$ as the radius of the circle and p is in the circle, and this circle inclusive of any points that is in set P, because of $r > r0$, therefore, there must be a circle with r as the radius which contains all the points in the set P, and also p is in the circle. According to the concept of α-hull, so this point must be the α-hull ($1/\alpha = r$) vertex of the set P;

Anther, assumption that the α-hull($1/\alpha = r$) of the set P is existing, so there are at least one point which is the α-hull vertex. According to the concept of α-hull, there is at least one circle C cross the point p with r as the radius which contains all the points in the set P, like that $r \geq ro$. This is the original assumptions of which $0 < r < r0$ contradict, so the assumption does not hold. This completes the proof.

Last, for any points p which is the α-hull ($1/\alpha = r0$) vertex of the set , there are at least one circle C with $r0$ as the radius which contains all the points in the set P.

circle C is the minimum circumscribed circle. So point p is in the circle. Empathy may permit, all of the α-hull ($1/\alpha = r0$) vertices of the set P are on the minimum circumscribed circle. This completes the proof. ■

By the above theorem reasoning, for the already determined attack graph, the interference node set with the minimum circumscribed circle radius is $r0$ in the graph, for $\forall r$ which is positive real number, the number of the α-hull ($1/\alpha = r$) vertices of the set P, n and r related, and satisfy the laws as follows:

(1) For $r < r0$, $n = 0$;

(2) For $r = r0$, all vertices of the α-hull ($1/\alpha = r$) are on the same circle;

(3) For $r > r0$, n gradually decreases with the decrease of the r;

(4) For $r = \infty$, n is the number of vertices of the convex hull of the set P.

It can be seen from the above rules, For a finite set of points P in a plane, it can according the number of the α-hull ($1/\alpha = r$) vertices to judge the relation of the r and the minimum circumscribed circle radius $r0$. By constantly projected r, when all vertices of the α-hull ($1/\alpha = r$) are on the same circle, vertices are located points on the minimum circumscribed circle, then calculate its radius. The correctness of the algorithm has demonstrated.

As shown in Fig. 2, a size of 1000 * 1000 (unit length) rectangular area of wireless sensor networks has been simulated by VC++6.0, the upper left corner coordinate is (0, 0), lower right corner coordinates is (1000, 1000). Sensor nodes are randomly and uniformly distributed in this area. Assuming that the jammer is on the center of the simulated regional, and its coordinate is (500, 500), the length of the 375 units for the radio interference radius. The nodes in the black circular internal region in Fig. 2 are considered to be the disturbed nodes.

The incremental algorithm and the algorithm of α-MCC which proposed in this paper are used for the simulation experiments. 1000 sensor nodes are placed within

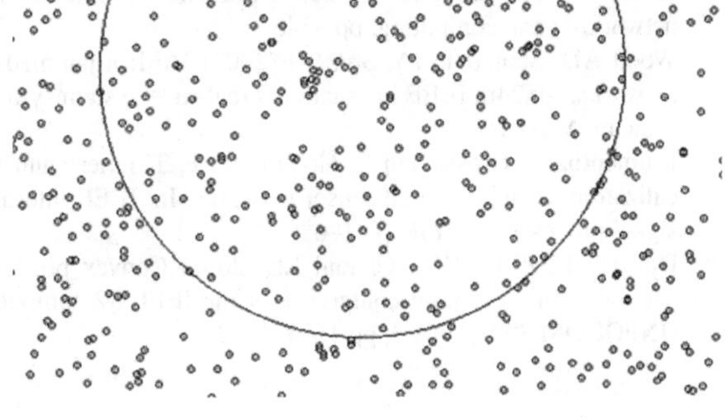

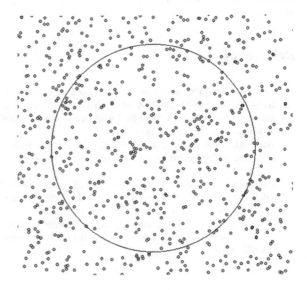

Fig. 2 Simulation network diagram

Fig. 3 Compare the radius of the two algorithms

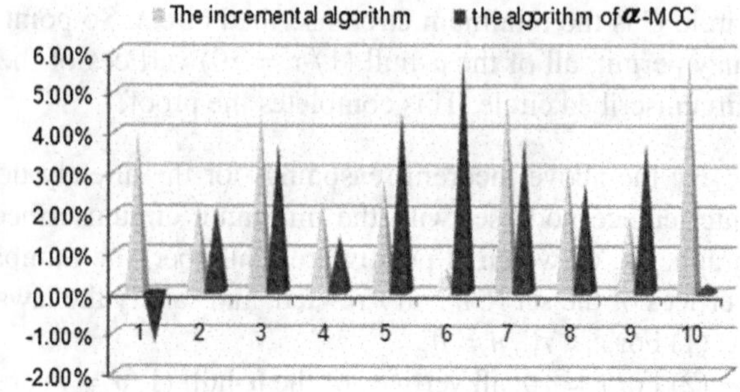

the region, 10 times of comparisons are given. As shown in the Fig. 3, the radius errors of the two algorithms are relatively small, the error can be able to control in less than 6 %, that is within an acceptable range.

5 Conclusion

To achieve positioning RF attacks, this paper bring the concept of α-hull, which is in the computation geometry applied to α-MCC (Algorithm for Minimum Circumscribed Circle Detection Based on α-hull). According to α-hull concept and their properties, it is proved that all vertices of the -hull will be on the same circle, if $1/\alpha$ is equal to the radius of points' MCC. With the number of α-hull vertices as reference, realize the attacking radius search of the attackers and the solution of the circle, as to realize the RF positioning. The simulation results show that, the algorithm changes as the measures changes, such as the density of the network nodes, the attack range and so on.

References

1. Li JZ, Gao H (2008) Survey on sensor network research. J Comput Res Dev 45(1):1–15
2. Hady SA, Stephan O (2009) A 3d-localization and terrain modeling technique for wireless sensor networks. In: 2nd ACM Int'l workshop on foundations of wireless ad hoc and sensor networking and computing, pp 3746
3. Wood AD, Stankovic JA, Son SH (2003) JAM: a jammed-area mapping service for sensor networks. In: 24th IEEE international real-time system symposium, pp 286–297. IEEE Press, Cancun, Mexico
4. Blumenthal J, Grossmann R, Golatowski F, Timmermann D (2007) Weighted centroid laocalization in zigbee-based sensor networks. In: IEEE international symposium on intelligent signal processing, WISP, pp 1–6
5. Doherty L, Pister KSJ, Ghaoui LE (2001) Convex position estimation in wireless sensor networks. In: 20th joint conference of the IEEE 32 computer and communications societies (INFOCOM 2001), vol 3, pp 1655–1663

6. Nicolescud D, Nath B (2001) Ad-hoc positioning systems (APS). In: 2001 IEEE global telecommunications conference (IEEE GLOBECOM 01), vol 5, pp 2926–2931
7. Niculescu D, Nath B (2003) DV based positioning in ad hoc net-works. J Telecommun Syst 22(1 /4):267–280
8. He T, Huang C, Blum BM et al (2003) Range-free localization schemes for scale sensor networks. In: International conference on mobile computing and networking (MOBICOM'03), pp 81–95 (2003)
9. Savarese C, Rabay J, Langendoen K (2002) Robust positioning algorithms for distributed adhoc wireless sensor networks. In: USENIX technical annual conference, Monterey, CA, USA, pp 317–327
10. Liu H, Xu W, Chen Y et al (2009) Localizing jammers in wireless networks. In: IEEE percom international workshop on pervasive wireless networking (IEEE PWN) (Held in conjuction with IEEE PerCom), pp 401–407
11. Bahrepour M, Meratnia N, Havinga PJM (2009) Sensor fusion-based event detection in wireless sensor networks. In: 6th annual international MobiQuitous, Toronto, Canada
12. Cheng L, Wu C, Zhang Y (2011) Multi-source localization with binary sensor networks. J Commun 32(10):158–165
13. Chen Y, Francisco J, Trappe W, Martin RP (2006) A practical approach to landmark deployment for indoor localization. In: Third annual IEEE communications society conference on sensor, mesh and ad hoc communications and networks (SECON)
14. Zhang J, Xu L, Zhang S (2012) α-hulls based localization for jamming attack in wireless sensor networks. J Comput Appl 32(02):461–464
15. Zhang Y, Chen Q (2007) Algorithm for minimum circumscribed circle detection based in computational geometry technique. J Eng Graph 3:97–101

(1) Exception Ex: Exis the most representative point in number field representing qualitative concept A. In another word, it is the most typical sample point towards this quantitative concept. The value of Ex is the central value in the discourse domain, the value most representable for the qualitative concept. Regularly, Ex is the value of x in the core of the cloud. Ex exactly subject to the qualitative concept. Ex reflects information's central value about the related qualitative knowledge.
(2) Entropy En: En reflects the nondeterminacy of qualitative concept A. On one hand, En represents the range of size for the cloud drop group acceptable by linguistic value A. In another word, En is the degree of fuzzy. En is the metric for qualitative concept both this and that.. On the other hand, En reflects the randomness exists in the representative qualitative concept of cloud drop. What's more, the associativity between fuzzy and randomness is revealed by the value of En. En could be also used to represent the granularity of a qualitative concept. In most cases, macro concepts comes with big En values, with the randomness and fuzzy degree gets bigger, making quantitative more difficult.
(3) Hyper Entropy He: He represents the coherency of the non-determination for all the points stand for concepts in data space. In another word, it is the entropy of the entropy. He reflects the degree of dispersion for cloud drops. High level dispersion comes with big En value and a high level stochastic characteristic of membership degree. In another word, High level dispersion makes the cloud thicker. There's none sense exists for a single cloud drop. Characteristics of all cloud drops together make sense in reality.

Gaussian (Normal) cloud is the most fundamental while important cloud model for the reason that normal distribution is proved universal in almost every branch of natural science [7]. The curve for one dimensional normal cloud is a type of normal curve, shown as Fig. 1. In Fig. 1, we set Ex = 2, En = 1, He = 0.15, the number of cloud drop is N = 1000

The overall shape of the cloud reflects the important characteristics of qualitative concepts. There's little influence on the overall cloud model even if some of the cloud drops are changed. Thus we can use the overall cloud as the watermark information and use the 3 parameters Ex En and He as keys. By applying the information to the one dimensional cloud generator, we can generate N cloud drops needed [8]. We can use inverse cloud generator to transfer the valid precise value exists in the extracted cloud watermark sequence to appropriate qualitative linguistic values in the process of watermark extraction. By this way, we transfer qualitative concepts into quantitative values.

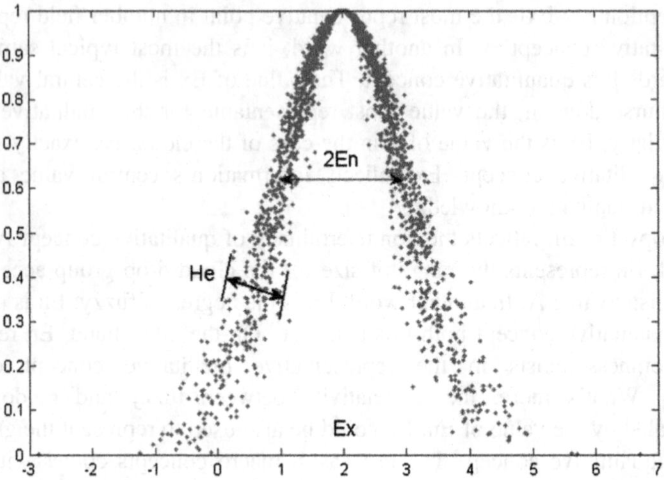

Fig. 1 One-dimensional normal cloud sequence

3 One Dimensional Normal Cloud Watermarking Scheme

3.1 Generation of Cloud Watermarking Sequence

By applying one dimensional forward cloud generator we get cloud watermarking sequence, the algorithm applied could be described as *Forward – Cloud* (Ex, En, He, N), shown as below:

Input: Three character values representing qualitative concepts C and the number of cloud drops N, those are: Ex, En, He and N
Output: N quantitative values for cloud drops and the certainty degree value of each cloud drop representing concept C.
Procedure:

Step 1: Generate a normal stochastic number x_ii, with exception equals to Ex and variance equals to *En*.

Step 2: Generate a normal stochastic number En' with exception equals to En and variance equals to *He*

Step 3: Calculate $y_i = \exp\left[-\frac{(x_i - Ex)^2}{2(Ex')^2}\right]$

Step 4: Let (x_i, y_i) be a cloud drop, it is a concrete realization in quantity for the linguistic value this cloud representing. Here x is a value the qualitative concept corresponded in the discourse domain. y_i is the metric for the degree of this linguistic value belongs.

Step 5: Repeat Step 1 to Step 4, until got the required amount of cloud drops.

Cloud generated using this algorithm has the natural characteristic of uneven in thickness. The above mentioned three numeric value is enough to describe the shape of the overall cloud. The waist, the top and the bottom of the cloud don't need to be defined preciously [10].

3.2 Embedding Watermark Sequence

To reduce the influence of the algorithm on the watermark and to better embodies the reality cloud modeling used in watermarking area. We adopt MSB watermarking algorithm in frequency domain to embed watermarking sequence got, here's the total procedure.

Step 1: Apply DCT transform on the carrier image.

Step 2: Generate a watermark sequence with concrete characteristics

Step 3: Embed the watermarking sequence into carrier image's larger amplitude sectors after DCT transformation, in order to improve the robustness for the embedded watermark.

Step 4: Using inverse DCT transformation and then got the watermarked carrier image.

3.3 Extracting Cloud Watermark

Using the embedded carrier images to extract the watermark is the process using MSB extraction algorithm and the inverse cloud generator to get it. Inverse cloud generator is a generator that summarize cloud model's overall characteristic based on a given number of cloud drops. The summarized representative characteristics are Ex, En and He. The procedure of inverse cloud generator is shown as below:

Input: N quantitative values for exactly N cloud drops and the certainty degree (x_i, y_i) of each cloud drop representing the concept.
Output: expectation Ex of qualitative concept C represented by this N cloud drops, entropy En and Hyper Entropy He.
Procedure:

Step 1: Use x_i to calculate the sample average value for this group, $\bar{X} = \frac{1}{n}\sum_{i=1}^{n} x_i$, and to calculate the first order central moment $\frac{1}{n}\sum_{i=1}^{n} |x_i - \bar{X}|$, also with the sample's variance $S^2 = \frac{1}{n-1}\sum_{i=1}^{n} (x_i - \bar{X})^2$

Step 2: From step 1 we get $Ex = \bar{X}$

Step 3: From sample's average value we get the entropy: $En = \sqrt{\frac{\pi}{2}} \times \frac{1}{n} \sum_{i=1}^{n} |x_i - \bar{X}|$

Step 4: From variance got from step 1 and entropy got from step 3 we can calculate hyper entropy $He = \sqrt{S^2 - En^2}$

Make the qualitative judgment if the cloud extracted is the embedded one based on the extracted values that cloud watermark sequence presents is a key problem that cloud watermark technology needs to resolve. Through the definition of cloud in this paper, we can infer that determine the membership degree for a single point ignoring the overall shape and coherence characteristic is meaningless. Also it is not possible to judge the overall membership degree through a single point. For the reason that every cloud could be described using three numeric characteristic values to represent the overall shape, we can use the three numeric characteristic parameters to study the judging similarity problem among different clouds. In this way, we can treat the clouds that satisfy several mathematical parameter conditions as the cloud having a particular linguistic value representing a same qualitative concept.

3.4 Similarity Cloud Judgment

There's deviation exists in the process transferring determined quantitative value to none-determined points set and the process inverse due to model transfer problem. Here we get the deviation based on experiments shown as Table 1. We use Euclidian Distance in Table 1. Here the deviations are Ex En and He before and after transforming.

From Table 1 we can infer that the number of cloud drops has some influence on the deviations exist in cloud model. Thus we need similarity metric analyze to determine the similarity between the two cloud compared. Similarity metric analyses cloud be measured based on distances between cloud drops. The algorithm to measure the similarity is shown as below:

Input : cloud drop number: N, cloud 1: Ex_1, En_1, He_1; cloud 2: Ex_2, En_2, He_2, the threshold u.
Output: Is cloud similar to cloud 2?
Procedure:

Step 1: Cloud 1 generates n cloud drops with their horizontal ordinate reserved: $\mathrm{Drop}_1(\mathrm{I}) = \mathrm{G}(\mathrm{Ex}_1, \mathrm{En}_1, \mathrm{He}_1, \mathrm{n})$;
Step 2: Cloud 2 generates n cloud drops with their horizontal ordinate reserved: $\mathrm{Drop}_2(\mathrm{I}) = \mathrm{G}(\mathrm{Ex}_2, \mathrm{En}_2, \mathrm{He}_2, \mathrm{n})$

Table 1 Influence of count to deviations

Ex	En	He	Count	Error
2	1	0.15	1000	0.0497
2	1	0.15	1500	0.0398
2	1	0.15	2000	0.0460
2	1	0.15	3000	0.0666

Step 3: Sort descending for cloud drop r1 and r2: $Drop_2(I) = G(Ex_2, En_2, He_2, n)$

Step 4: Choose cloud drops fall into $[Ex - 3En, Ex + 3En]$, then got the number of cloud drops from cloud 1 fallen into this interval $Drop'_1$, that of cloud 2 is $Drop'_2$

Step 5: Assume that $n_1 \leq n_2$, the combination of $Drop_2^j$ for $Drop_2'$ is then $C_{n_2}^{n_1}$, $j \in [1, 2, 3, \ldots, C_{n_2}^{n_1}]$

Step 6: Calculate the square of the difference between $Drop_1^j$ and $Drop_2^j$ in turn, that is D(j);

Step 7: $S = SQRT\left(\sum D(j)/C_{n_2}^{n_1}\right)/m$, if $S < u$ then we can make the belief that the wo cloud are similar.

4 Experimental Results

We use 256*256 standard cameraman gray scale image (Fig. 2) as the carrier image to perform our simulation experiment. We set $Ex = 2, En = 1, He = 0.15$ and the number of cloud drop N = 1000 in our adopted one dimensional normal cloud watermark. The image embedded is Fig. 3. We compare the embedded cloud watermark (Fig. 1) and the extracted one (Fig. 4) by applying the cloud similarity measurement algorithm and got the calculation result $S = 9.367 * 10^{-7}$ less than the given threshold value $\sigma = 0.003$. Thus the two clouds are similarity cloud before and after extraction.

Fig. 2 Carrier image

Fig. 3 Image watermarked

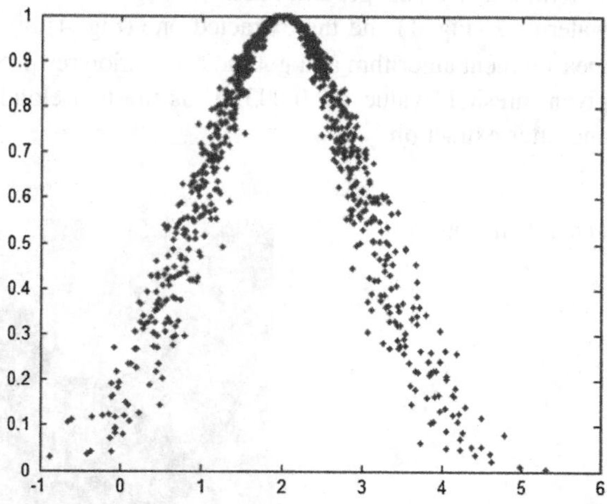

Fig. 4 Cloud watermarking sequence extracted

5 Conclusions

In this paper we use one dimensional normal cloud model as a watermark sequence and apply MSB algorithm to embed the information. Then we extract the information embedded and judge the similarity between cloud models got and embedded properly. Experimental results proved the effectiveness of cloud model and its relative theory used in digital watermark area. Because of the meaningless for a single cloud drop and the meaningful characteristic for overall cloud drops, there's high level resistance for the watermark in attacks such as shearing. It deserves further research in digital watermarking area to judge overall characteristics ignore single cloud drop.

Acknowledgement The authors would like to thank for the support from the project NSFC (National Natural Science Foundation of China) with the Grant number 61202456 and the HIT Innovation Fund with Grant number HIT. NSRIF. 2015087.

References

1. Cox IJ, Miller ML, Jeffrey A (2007) Bloom digital watermarking, 2nd edn. Morgan Kaufmann, Burlington
2. Watson AB (1993) DCT quantization matrices optimized for individual images. Human Vision, Visual Processing, and Digital Display IV. SPIE Press, San Jose, pp 202–216
3. Li D (2000) Uncertainty in knowledge representation. Eng Sci 2(10):73–79
4. Li D, Shi X, Ward P et al Soft inference mechanism based on cloud models. In: Proceedings of the 1st international workshop on logic programming and soft computing, pp 38–63
5. Li D, Han J, Shi X (1998) Knowledge representation and discovery based on linguistic atoms. Knowl Based Syst 10:431–440
6. Li D, Cheng DW, Shi X et al (1998) Uncertainty reasoning based on cloud models in controllers. Comput Math Appl 35(3):99–123
7. Li D, Hwang K (2010) Trusted cloud computing with secure resources and data coloring. IEEE Internet Comput 5:14–22
8. Liu Y-C, Ma Y-T, Zhang H-S, Li D-Y, Chen G-S (2011) A method for trust management in cloud computing: data coloring by cloud watermarking. Int J Autom Comput 8(3):280–285
9. Zhang Y, Zhao D-N, Li D (2004) Similar cloud and its measurement analysis method. Inf Control 33(2):129–132
10. Li D, Liu C (2004) Study on the universality of the normal cloud model. China Eng Sci 6(8):28–34

A Dual Watermarking Scheme by Using Compressive Sensing and Subsampling

Jeng-Shyang Pan, Jia-Jiao Duan and Wei Li

Abstract This paper proposes a dual watermarking scheme based on subsampling and Compressive Sensing Theory. In this scheme, one robust watermark is embedded into the DCT domain of two sub-images and another watermark is embedded into the CS domain. Bit Correction Rate (BCR) between original secret message and extracted message are used to calculate the accuracy of this method. Extensive experimental results demonstrate the validity of the proposed scheme and high security of security information.

Keywords Dual watermarking · Compressive sensing (CS)

1 Introduction

Since Tirkel [1] proposed the concept of digital watermarking and the possible applications in the paper entitled "Electronic watermark" in 1993, digital watermarking algorithms and its application have attracted more and more attention.

As different watermarks have different properties, they can be used in different application scenarios. But there is a problem that the watermark with single

---

J.-S. Pan (✉)
College of Information Science and Engineering, Fujian University of Technology, Fuzhou 350118, Fujian Province, China
e-mail: jengshyangpan@gmail.com

J.-S. Pan · J.-J. Duan
IIIRC, School of Computer Science and Technology, Harbin of Technology, Shenzhen Graduate School, Shenzhen 518055, China
e-mail: enpainduan@163.com

W. Li
School of Computer Science and Engineering, University of New South Wales, Sydney, Australia
e-mail: weil0819@gmail.com

A. Abraham et al. (eds.), *Intelligent Data Analysis and Applications*,
Advances in Intelligent Systems and Computing 370,
DOI 10.1007/978-3-319-21206-7_32

function often can not meet the needs of practical application. To achieve multiple goals, protecting the copyright of one product with several owners in different phases [2], we must embed more than one watermark into the same host image. And that is multiple watermarking. In fact most multiple-watermarking algorithms generate from dual-watermarking algorithms, which embed two watermarks into host image. Shen H described that there are three dual watermarking strategies in [2]. And we propose a novel dual watermarking scheme inspired by strategy 3 [2].

In recent years, compressive sensing (CS) [5, 11] has earned more and more attention for its sparsity and incoherence properties. Through the appropriate reconstruction algorithms, we can use a small amount of sparse measurements to restore the original signal [5, 11]. Some researchers focus on CS with the application of information hiding. In 2014, Pan et al. [6] proposed a novel image steganography algorithm based on subsampling and compressive sensing in the CS transform domain with high security.

In this paper, we propose a novel dual watermarking scheme by using subsampling technique and compressive sensing theory, whose results show the validity and high transparency and robustness. The rest of the paper is organized as follows. In Sect. 2, we briefly describe the basic concept of subsampling and compressive sensing. Section 3 will contains a detailed introduction of the proposed dual watermarking algorithm. The experimental results and analysis will be presented in Sect. 4. Finally, some conclusions will be given in Sect. 5.

2 Related Knowledge

2.1 Subsampling

Subsampling is a kind of way to decompose the original image into 4^n sub-images which have similar perceptual effect. Here n is an integer number. For example, if we want to decompose a N*N image **I** into four sub-images $\mathbf{I_1}$, $\mathbf{I_2}$, $\mathbf{I_3}$ and $\mathbf{I_4}$. We can use the equations below to complete the process.

$$\begin{aligned} I_1(row, col) &= I(2*row, 2*col) \\ I_2(row, col) &= I(2*row + 1, 2*col) \\ I_3(row, col) &= I(2*row, 2*col + 1) \\ I_4(row, col) &= I(2*row + 1, 2*col + 1) \end{aligned} \quad (1)$$

Here $\mathbf{I_i}$ (row,col)(i = 1, 2, 3, 4; row,col = 0, 1, 2…, INT[N/2]) are pixels in the sub-images and **I**(row,col) is pixel in the original image. It is clearly that the pixels in the same position of the sub-images are next to each other in the original image, so the value of them are very close.

2.2 Compressive Sensing

Compressed sensing include three procedures: sparse representation of signal, measurement sampling and reconstruction of signal.

Sparse representation of signal. The main idea of CS is to reconstruct the original signal from much fewer signals or samples than traditional methods. And these signals should be sparse signal at first. Suppose there is an original signal **X**, $\mathbf{\Psi} = \{\psi_n | n = 1, 2, 3\ldots N\}$ is an orthonormal basis which has N vectors in it. If **X** can be represented by a linear combination of m vectors in **Ψ**.

$$X = \Psi \mathbf{a} = \sum_{m=1}^{N} a_m \psi_m \tag{2}$$

Where the signal **X** is a sparse signal. **Ψ** is a sparse basis **a** is the sparse decomposition coefficient for **X** at the basis **Ψ**. the value of **a** is:

$$\mathbf{a} = \Psi^T X \tag{3}$$

Here $\mathbf{\Psi}^T$ is the Transpose for matrix **Ψ**. For images, typical choices of include the Discrete Cosine Transform (DCT) and Discrete Wavelet Transform (DWT).

Measurement sampling. Almost none of common natural signals have the sparsity feature. So the original signal should suffer a linear projection procedure.

For Signal **X** that is not sparse, **Ψ** is the sparse basis, and **Φ** is a M*N matrix and M<<N. The linear projection for **X** on **Φ** basis is:

$$y = \Phi X = \Phi \Psi a = \Theta a \tag{4}$$

where **Φ** is the measurement matrix in compressed sampling, **Θ** is the sensing matrix and y is a M*1 column vector consisting of linear measurements. After measurement sampling processing, the original signal is compressed into a M*1 vector **y**.

Reconstruction of signal. The reconstruction of signal is to recover the original signal in under-sampling situation. E. Candès [7] has proved that if you use the priori knowledge to the sparsity or compressibility of signals, and other requirements for **y** and **Φ**, accurate reconstruction can be get by a high probability through solve the l_0 norm.

$$\tilde{x}_1 = \arg\min \|x\|_0 \; subject\, to.\, y = \Phi X \tag{5}$$

$\tilde{x}_1$ is the reconstructed signal. $\|x\|_0$ is l_0 norm which indicates the number of elements not equal to zero in **X**.

3 Proposed Scheme

This paper proposes a novel dual watermarking scheme based on subsampling and compressed sensing. At first the host image **I** will be decomposed into four sub-images by the subsampling method. And then the sub-image $\mathbf{I_2}$ and $\mathbf{I_3}$ will be used to embed a robust watermark in the DCT domain. The other two sub-images $\mathbf{I_1}$ and $\mathbf{I_4}$ will be used to embed another watermark in the compressed sensing domain. After that we could get four new sub-images:$\mathbf{I'_1}$, $\mathbf{I'_2}$, $\mathbf{I'_3}$ and $\mathbf{I'_4}$. At last we composed these sub-images to a watermarked image **I'**. Through the corresponding extracting method, we could extract the dual watermarks from the watermarked image.

3.1 Data Embedding

This part introduce the process of the embedding of dual watermarks to the host image. Given the host image **I** and its size is N*N. Use the subsampling method to sub-sample **I** into four sub-images $\mathbf{I_1}$, $\mathbf{I_2}$, $\mathbf{I_3}$ and $\mathbf{I_4}$.

Here we choose the DCT as sparse basis. So we can transform the four sub-images to DCT domain and get four groups of DCT coefficients $\mathbf{X_i}$, where i = 1, 2, 3, 4. And the size of $\mathbf{X_i}$ is N*N/4. Each pixel at the same position of each group can be represented by $x_{ti}(m,n)$ where $0 < m \leq N/2$, $0 < n \leq N/2$. Because each sub-images is very similar to each other, the difference of the DCT coefficients for each group is very small.

We select $\mathbf{I_2}$ and $\mathbf{I_3}$ to embed the first robust watermark in DCT domain. After DCT transform operation for $\mathbf{I_2}$ and $\mathbf{I_3}$,we get the DCT coefficient matrices $\mathbf{X_2}$ and $\mathbf{X_3}$ respectively. And then we embed the robust watermark $\mathbf{W_1}$ into them and the size of $\mathbf{W_1}$ is M*M. Suppose x_{t2} and x_{t3} are two coefficients from $\mathbf{X_2}$ and $\mathbf{X_3}$ and they are at the same position of $\mathbf{X_2}$ and $\mathbf{X_3}$. So we can get $x_{t2} \approx x_{t3}$. We insert the watermark information by changing the relationship between x_{t2} and x_{t3}. And w_t is the watermark pixel value to be embedded. The embedding method is as follows:

When $\omega_{t1} = 1$:

$$\begin{cases} x'_{t2} = x_{\text{avg}} \\ x'_{t3} = x_{\text{avg}} \end{cases} \quad \text{mod}(T,2) = 0 \tag{6}$$

$$\begin{cases} x'_{t2} = T^*\alpha + sign(x_{\text{avg}})^*T \\ x'_{t3} = T^*\alpha - sign(x_{\text{avg}})^*T \end{cases} \quad \text{mod}(T,2) = 1 \tag{7}$$

When $\omega_{t1} = 0$:

$$\begin{cases} x'_{t2} = x_{avg} \\ x'_{t3} = x_{avg} \end{cases} \quad mod(T, 2) = 1 \tag{8}$$

$$\begin{cases} x'_{t2} = T * \alpha + sign(x_{avg}) * T \\ x'_{t3} = T * \alpha - sign(x_{avg}) * T \end{cases} \quad mod(T, 2) = 0 \tag{9}$$

In the above equations, $x_{avg} = (x_{t2} + x_{t3})/2$, α is a factor that can be set. And $T = x_{avg}/\alpha$. Sign(x_{avg}) means the sign of x_{avg}. After all the pixels in the watermark $\mathbf{W_1}$ is embedded, we use inverse DCT transform to recover the sub-images that is named $\mathbf{I'_2}$ and $\mathbf{I'_3}$.

And then we embed the watermark $\mathbf{W_2}$ to the compressed sensing domain of $\mathbf{I_1}$ and $\mathbf{I_4}$. Here we choose DCT as sparse basis. Transform $\mathbf{I_1}$ and $\mathbf{I_4}$ to DCT domain and get the DCT coefficient matrices $\mathbf{X_1}$ and $\mathbf{X_4}$, which size is n*n and n = N/2.

Choose a m*n^2 matrix $\boldsymbol{\Phi}$ to do compressed sensing measurement for $\mathbf{X_1}$ and $\mathbf{X_4}$. Then we get two m*1 measurements vector $\mathbf{y_1}$ and $\mathbf{y_4}$. We change the relations between the elements at the same position of $\mathbf{y_1}$ and $\mathbf{y_4}$ to embed the watermark information. Suppose y_{t1} and y_{t4} are elements in $\mathbf{y_1}$ and $\mathbf{y_4}$, and ω_{t2} is a value in $\mathbf{W_2}$ to be embedded. The embedding method is as below:

$$y'_4 = \begin{cases} y_1 + \beta \times \omega_{t2} & if \ |y_1 - y_4| < thd \\ y_4 & other \end{cases} \tag{10}$$

Here thd is a threshold that can be set. And β is a scaling factor. From the Eq. (10), we can know when the difference between y_1 and y_4 are relatively large, no watermark information will be embedded and the value of y_4 will retain the same. And the watermark information will be embedded when the difference is relatively small. Whether the difference is large or small is decided by the value of thd. After above operations, we can get the new measurements vector $\mathbf{Y'_1}$ and $\mathbf{Y'_4}$.

After watermark embedding, we use the measurements $\mathbf{Y'_1}$ and $\mathbf{Y'_4}$ to recover the two sub-images by Total Variation (TV) reconstruction algorithm [15].

At last, we recompose the image using the sub-images $\mathbf{I'_i}$ (i = 1, 2, 3, 4) and what we get is the stego-image $\mathbf{I'}$.

3.2 Data Extracting

This part introduce the process of the extraction of hiding information in the stego-image.

At first, the watermarked image $\mathbf{I'}$ will be decomposed into four sub-images $\mathbf{I'_1}$, $\mathbf{I'_2}$, $\mathbf{I'_3}$ and $\mathbf{I'_4}$ as the above subsampling operation.

For the sub-images $\mathbf{I'}_2$ and $\mathbf{I'}_3$, we apply the DCT transform onto them and get the DCT coefficient matrices $\mathbf{X'}_2$ and $\mathbf{X'}_3$.

And then we extract the first watermark from $\mathbf{X'}_2$ and $\mathbf{X'}_3$. Suppose that the corresponding elements in $\mathbf{X'}_2$ and $\mathbf{X'}_3$ are x'_{t2} and x'_{t3}, the extracted watermark is $\mathbf{W'}_1$ and one pixel in $\mathbf{W'}_1$ is extracted from x'_{t2} and x'_{t3}. We names this pixel value ω'_{t1}. If T is an even number, the value of ω'_{t1} can be obtained by below equations:

$$\begin{cases} \omega'_{t1} = 0 & |x_{avg} - T*\alpha| < \alpha/2 \\ \omega'_{t1} = 1 & else \end{cases} \tag{11}$$

Here, $x_{avg} = (x'_3 + x'_2)/2$. $T = x_{avg}/\alpha$. And the value of α is previously set already. If T is an odd number, the value of ω'_{t1} is the oppsite to Eq. (11).

In this way, we can obtain all the hiding information in $\mathbf{X'}_2$ and $\mathbf{X'}_3$. These values compose the watermark $\mathbf{W'}_1$.

For the sub-images $\mathbf{I'}_1$ and $\mathbf{I'}_4$. After we get their DCT coefficient matrices $\mathbf{X'}_1$ and $\mathbf{X'}_4$, we use the same matrix $\mathbf{\Phi}$ as above to do compressed sensing measurement for them. And the measurement vector is $\mathbf{Y'}_1$ and $\mathbf{Y'}_4$. Suppose the watermark embedded is $\mathbf{W'}_2$ and one of the value of $\mathbf{W'}_2$ is ω'_{t2}. And then we can obtain ω'_{t2} as the Eq. (12) shows:

Here, y'_{t1} and y'_{t4} are corresponding elements at the same position of $\mathbf{Y'}_1$ and $\mathbf{Y'}_4$. The thd1 and thd2 are thresholds that can be set. In this way, all the values in $\mathbf{W'}_2$ can be computed and we will get the second watermark information.

$$\omega'_{t2} = \begin{cases} 0 & if\ |y'_{t1} - y'_{t4}| < thd1; \\ 1 & if\ thd1 < |y'_{t_1} - y'_{t_4}| < thd2; \\ nothing & otherwise. \end{cases} \tag{12}$$

4 Experimental Results

The experiment in this paper is completed on Matlab 2011a. The gray-level image ‘Lena’, ‘Baboon’ and ‘Peppers’ are used as host images. And the size of each image is 512*512. The first watermark is a 64*64 binary image and the second watermark is a sequence of random 0, 1 bit.

Figure 1 is an example of sub-sampling. We can see from this figure, the original Baboon image is sub-sampled into for sub-images. And the sub-images are very similar to the original image and they are very similar with each other. That is the foundation for our algorithm.

After embedding the two watermarks, we get the watermarked images. In Figs. 2 and 3 we can see the comparison between the original images and the watermarked images. We will find that the difference between them is too small to be find by your naked eyes. That means the transparency of this algorithm is very good and it has a rather high security to a certain extent.

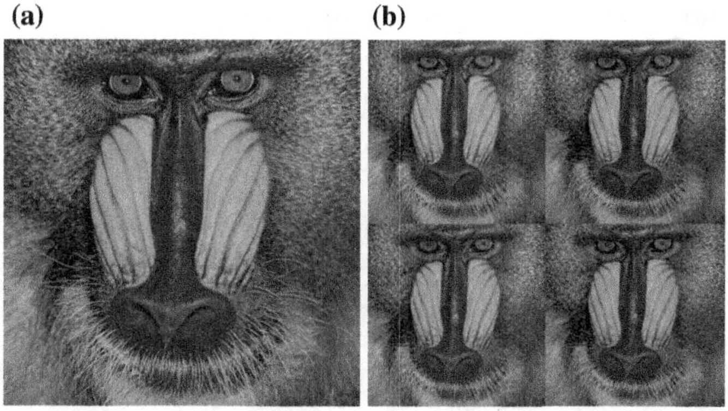

Fig. 1 A sub-sampling example. a Original image. b Sub-sampling image

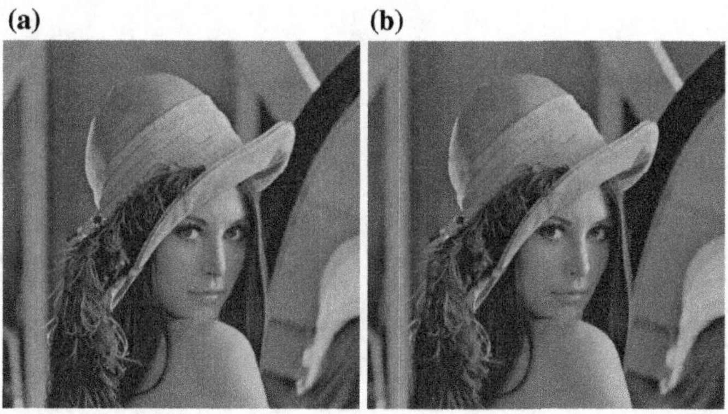

Fig. 2 512 × 512 Lena host-image and watermarked-image by proposed method. a Host image. b Watermarked image, PSNR = 34.63 dB

We use PSNR (Peak Signal to Noise Ratio) value to evaluate the transparency of the algorithm. In general, when the PSNR value are bigger than 30 dB, it will be difficult to find the difference of two pictures on people's eyes. The proposed algorithm PSNR values for Lena and Pepper are 34.63 and 34.90 dB. And through many experiments for different cover images and watermarks, we find that all the PSNR values are more than 30 dB and near 35 dB. This means that this algorithm are pretty good in transparency.

At last, we conduct the proposed extracting algorithm and we get the extracted watermarks. Because the first watermark is a picture and the second is a text. And we show the first as a picture and for the second we just compute the BCR (Byte Correct Rate) value for it. The BCR value is between 0 and 1. If the watermark is completely extracted, the BCR value is equal to 1. And we will notice that the watermarks in Lena and Peppers are correctly extracted and the BCR value for them is 1. And the watermarks in Baboon has a little change to the original one. But it is

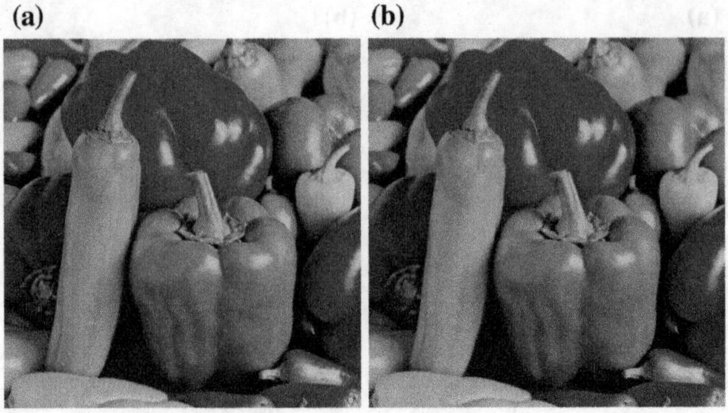

Fig. 3 512 × 512 Pepper host-image and watermarked-image by proposed method. **a** Original image. **b** Watermarked Image, PSNR = 34.90 dB

Table 1 Extracted watermark information

Image	Lena	Peppers	Baboon
Extracted watermark 1			
BCR for watermark 1	1	1	0.9850
BCR for watermark 2	1	0.9943	0.9950

very close to the original one and the BCR values for it are 0.9850 and 0.9950. This data in all experiments indicates that almost all the embedded information is correctly extracted. So this proposed method is pretty good in use (see Table 1).

5 Conclusion

This paper proposes a novel dual watermarking scheme based on subsampling and compressed sensing. By subsampling, the host image can be decomposed into similar sub-images and that makes it easy to choose 2 sub-images to embed information. Through embedding the watermark in DCT domain and compressive sensing domain, the information of watermarks are distributed on each pixel and that makes the watermarked image more security. The experiment shows that this algorithm has a pretty good transparency and robust as well.

References

1. Tirkel A, Rankin G, Van Schyndel R et al (1993) Electronic watermark. In: Proceedings DICTA, vol 12, issue 5, pp 666–672
2. Shen H, Chen B (2012) From single watermark to dual watermark: a new approach for image watermarking. J Comput Electr Eng 38(5):1310–1324
3. Cox I, Miller M, Bloom J (2001) Digital watermarking: principles and practice. Morga Kaufman, San Francisco
4. Candes EJ, Wakin MB (2008) An introduction to compressive sampling. IEEE Signal Process Mag 25(2):21–30. doi:10.1109/MSP.2007.914731
5. Donoho DL (2006) Compressed sensing. IEEE Trans Inf Theory 52(4):1289–1306. doi:10.1109/TIT.2006.871582
6. Pan JS, Li W, Yang CS, Yan LJ (2014) Image steganography based on subsampling and compressive sensing. Springer Science+Business Media, New York
7. Candès E, Romberg J, Tao T (2006) Robust uncertainty principles: exact signal reconstruction from highly incomplete frequency information. IEEE Trans Inf Theory 52(2):489–509
8. Zhou YX, Jin W (2011) A novel zero-watermarking scheme based on DWT-SVD. IEEE Trans Inf Theory 57:978–982
9. Wei F (2014) Research of digital image blind watermarking algorithm based on comperssive sensing. Anhui University, Hefei
10. Chen GF, Guo SX, Li Y (2012) Digital image watermarking based on compressive sensing. Mod Electr Technol 35(13):98–104
11. Candes EJ, Romberg J, Tao T (2006) Robust uncertainty principles: exact signal reconstruction from highly incomplete frequency information. IEEE Trans Inf Theory 52 (2):489–509
12. Valenzise G, Tagliasacchi M, Tubaro S, Cancelli G, Barni M (2009) A compressive sensing based watermarking scheme for sparseimage tampering identification. In: 16th IEEE International Conference on Image Processing (ICIP). IEEE, pp 1265–1268
13. Rachlin Y, Baron D (2008) The secrecy of compressed sensing measurements. In: 46th Annual Allerton conference on communication, control, and computing. IEEE, pp 813–817
14. Shannon CE (1949) Communication theory of secrecy systems. Bell Syst Tech J 28 (4):656–715
15. Chambolle A (2004) An algorithm for total variation minimization and applications. J Math Imaging Vis 20(1–2):89–97
16. Li W, Lin CC, Pan JS (2015) Novel image authentication scheme with fine image quality for BTC-based compressed images. Multimedia Tools Appl 34:1–23
17. Pan JS, Li W, Lin CC (2014) Novel reversible data hiding scheme for AMBTC-compressed images by reference matrix. In: Multidisciplinary social networks research. Springer, Heidelberg, pp 427–436

Part IV
Signal Processing and Applications

A Unified Framework of Single Image Haze Removal under Different Weather Conditions

Yi-Fan Li, Qiang Su, Jeng-Shyang Pan, Jun-Bao Li, Wei Cui and Huan-Yu Liu

Abstract Outdoor images are usually affected by haze, fog and smoke which are such phenomena due to atmospheric scatting, so that the degraded images suffer from the loss of contrast and color fidelity. In this paper, we summed up a variety of haze removal techniques which are grouped into two categories: physics based and non physics based, and processed some hazy images under different weather conditions. Then, we make a comparison among these results, and analyze their advantages and shortcomings, estimate the appropriate haze removed technique for each weather condition. Finally, we lay further discussions on technical challenges and future development.

Keywords Single image · Haze removal · Image enhancement

1 Introduction

Outdoor images and some works about computer vision are often influenced by bad weather. In any outside scenes, the light reflected from object is influenced from air scattering more or less before they reach camera. Under different weather conditions, the particles in the air and the degrees that the images are depleted are different. In the aerology, with the thickness of haze, the fog is often divided into three classes by the size of visibility, namely, dense fog, fog and mist. At present, the technology of image

Y.-F. Li · Q. Su (✉) · J.-S. Pan · W. Cui
Innovative Information Industry Research Center, Shenzhen Graduate School, Harbin Institute of Technology, Shenzhen 518055, China
e-mail: hitsuquiang@163.com

J.-S. Pan
College of Information Science and Engineering, Fujian University of Technology, Fuzhou 350118, China

J.-B. Li · H.-Y. Liu
Department of Automatic Test and Control, Harbin Institute of Technology, Shenzhen, China

A. Abraham et al. (eds.), *Intelligent Data Analysis and Applications*, Advances in Intelligent Systems and Computing 370,
DOI 10.1007/978-3-319-21206-7_33

haze removal and restoring are separated into two classes: one is based on physical model of air scattering, which is oriented from physical reasons, and builds a image degradation physical model by the prior knowledge which are known or partially known, then restores haze removal image by simulating its reverse process; the other one is based on non-physical model, in other words, image enhancement method, which ignores the image degradation, and improves images contrast to realize image haze removal.

To remove haze from a single image, Tan [1] observed that the haze-free image must have higher contrast compared to the haze image and he removes the haze by maximizing the local contrast of the restored image. Fattal [2] estimates the albedo of the scene and then infers the medium transmission, under the assumption that the transmission and surface are locally uncorrelated. He [3] estimates the albedo by utilizing the dark channel prior, then corrects it by soft matting, and finally restores the image. Later He proposed the guided filter [4], and utilized it to replace soft matting, which can not only get a smooth edges, but also reduce the running time. Tarel [5] proposed a algorithm whose main advantage is its speed and its ability to handle both color images or gray level images. Histogram equalization is widely used for contrast enhancement in a variety of application due to its simple function and effectiveness. For images which contain local regions of low contrast, namely, bright or dark regions, adapted histogram equalization [6] can be used on such images for better results, but it expands the noise from small regions to global image. Contrast limited histogram equalization [7] can overcome the shortage in the AHE. The idea of the Retinex was conceived by Land [8] as a model of the lightness and color perception of human vision. Land introduced random walk [9] from one pixel to next pixel whose calculation is very complex. Later Frankle introduced the McCann [10] algorithm which is a Retinex based on multiple iteration, then McCann [11] proposed the McCann99 algorithm which utilizes the Gaussian pyramid to iteration. Later, Jobson, Rahman and Woodell proposed single-scale Retinex [12] and multiple-scale Retinex [13] which adopt center/surround Retinex to estimate brightness. Later Rahman [14] proposed the MSRCR algorithm which restores the color by utilizing color proportion factor of each waves in the original image. In this paper, we will review a variety of one-image haze removal methods based either on atmospheric scattering model or not, and for the different types of fog: dense fog, fog and mist, realize the single image haze removal with the various of algorithms. With the contrast analysis for the results, we estimate the appropriate algorithm to haze image for each kind of fog.

2 Haze Removal Algorithm

2.1 The Haze Removal Algorithm Based on Physical Model

During the light go through some mediums with scattering, there will be some light attenuation in the original direction, which will be dedicated to other directions.

In computer vision and computer graphics, the model widely used to describe the formation of a haze image is as follows [15]:

$$I(x) = t(x) J(x) + A(1 - t(x)) \tag{1}$$

where I is the observed intensity, J is the scene radiance, A is the global atmospheric light, and t is the medium transmission describing the portion of the light that is not scattered and reaches the camera. The goal of haze removal is to recover J, A, and t from I. The first term $t(x) J(x)$ on the right hand side of Equation 1 is called *direct attenuation* [1], and the second term $A(1 - t(x))$ is called *airlight* [16].

Haze removal based on one image, as it is just depending on the input I, estimating atmospheric light A and medium radio t by exploiting some assumption or knowledge prior, achieving finally the restored image J.

2.1.1 Haze Removal Using Dark Channel Prior

The dark channel prior is based on the following observation on haze-free outdoor images: in most of the non-sky patches, at least one color channel has very low intensity at some pixels. In other words, the minimum intensity in such a patch should have a very low value. Formally, for an image J, it can be defined

$$J^{dark}(x) = \min_{y \in \Omega(x)} \left(\min_{c \in \{r,g,b\}} (J^c(y)) \right) \tag{2}$$

where J^c is a color channel of J and $\Omega(x)$ is a local patch centered at x. Our observation says that except for the sky region, the intensity of J^{dark} is low and tends to be zero, if J is a haze-free outdoor image. We call J^{dark} the *dark channel* of J, and we call the above statistical observation or knowledge the *dark channel prior*.

Before we deduce the transmission $t(x)$, we need to get atmospheric light A by the dark channel prior firstly. Then we can get the haze-free image during the reverse process of atmospheric scatting model.

2.1.2 Dark Channel Prior Based on Guided Filter

He [4] simply filter the raw transmission map under the guidance of the hazy image: first apply a max filter to counteract the morphological effects of the min filter, and consider this as the filtering input of the guided filter. Using guided filter, on the one hand we can get a more smooth edges, on the other hand it has a faster running time.

We first define a general linear translation-variant filtering process, which involves a guidance image I, an filtering input image p, and an output image q. Both I and p are given beforehand according to the application, and they can be identical. The filtering output at a pixel i is expressed as a weighted average:

$$q_i = \sum_j W_{ij}(I)\, p_j \tag{3}$$

where i and j are pixel indexes. The filter kernel W_{ij} is a function of the guidance image I and independent of p. This filter is linear with respect to p.

The key assumption of the guided filter is a local linear model between the guidance I and the filtering output q. We assume that q is a linear transform of I in a window ω_k centered at the pixel k:

$$q_i = a_k I_i + b_k, \forall i \in \omega_k \tag{4}$$

This local linear model ensures that q has an edge only if I has an edge. To determine the linear coefficients (a_k, b_k). By the *linear ridge regression* model [17], the output image calculated as:

$$q_i = \frac{1}{|\omega|} \sum_{k|i \in \omega_k} \left(a_k I_i + b_k\right) \tag{5}$$

If the guidance image I and input image are identical, that is $I = p$, we can have $a_k = \sigma_k^2 / \left(\sigma_k^2 + \varepsilon\right)$,$b_k = \left(1 - a_k\right) \mu_k$. According to the different ε, with $q_i = \bar{a}_i I_i + \bar{b}_i$, we can get output image.

2.1.3 No-Black-Pixel Constraint (NBPC)

The method is proposed by Tarel [5] in 2009, its main advantage compared with other is speed, another advantage is the possibility to handle both color images or gray level images. On a gray level image, the model of the effect of the fog is established by Koschmieder:

$$I(x, y) = R(x, y)\left(1 - \frac{v(x, y)}{I_s}\right) + V(x, y) \tag{6}$$

This model is directly extended to a color image by applying the same model on each RGB component. This method can be decomposed into several steps: estimation of I_s, inference of $V(x, y)$ from $I(x, y)$, estimation of $R(x, y)$, smoothing to handle noise amplification and final tone mapping.

2.1.4 Fast Haze Removal Based Average Filter

The algorithm [18] is based on the analysis of physical-based model, and estimate airlight and atmospheric light utilizing the simple average filter.

To estimate the airlight $L(x)$, we utilize average filter to estimate $t(x)$ firstly, and set a offset for the result which is computed by average filter. for estimating the atmospheric veil, compare to dark channel prior, we apply a simply way, that is the maximum of the whole pixel RGB components in original image and average value among the maximum of dark channel.

$$A = \frac{1}{2}\left(max\left(\max_{c\in\{r,g,b\}} \left(H^c\left(x\right)\right)\right) + max\left(M_{ave}\left(x\right)\right)\right) \tag{7}$$

which not only avoids mistakes which regard the brightness gray value as global airlight directly, but also improves the speed of running time.

2.2 The Haze Removal Algorithm Based on Image Enhancement

2.2.1 The image Enhancement Based on Frankle-McCann Retinex

Retinex is the term which combines the words both Retina and Cortex. Retinex theory [19] is that as a precondition of constancy theory accorded with the human eye color. It is assumed that the foundation of establishment of the human eyes perceived color and brightness of object determined by atmospheric brightness and reflection of object surface. The main idea of preserving color constancy is that estimating brightness, extracting the influences of atmospheric brightness, finally we achieve the correct color and brightness of object.

Frankel-McCann Retinex [10] adapt one-dimensional path which is specific spiral structure to estimate brightness. In each wave of RGB space, we estimate the destination pixel brightness by make a comparison with the other pixels. If there are n pixels along the path which is between the pixel $S\left(x_1, y_1\right)$ and the pixel $D\left(x_2, y_2\right)$, whose brightness are $\left(d_1, d_2, \ldots d_n\right)$, so the bright relation between S and D can be expressed with a series of multiplication, as follows:

$$\frac{D}{S} = T\left(\frac{d_2}{d_1}\right) \times T\left(\frac{d_3}{d_2}\right) \times T\left(\frac{d_4}{d_3}\right) \times \cdots \times T\left(\frac{d_n}{d_{n-1}}\right) \tag{8}$$

where $T(\cdot)$ is threshold function $T(x) = \begin{cases} x & x \le 1 \\ 1 & x > 1 \end{cases}$.

In practice, FMR can be decomposed into for steps which are comparing, multiply, reset and averaging. There are more pixels near the predicted center, that is the reason why there is a better correlation near the center.

2.2.2 The Image Enhancement Based on McCann99

McCann99 Retinex algorithm [11] is the same as FMR algorithm in essence, and it need comparing, multiplying, reset and averaging to iterate. But McCann99 Retinex adapts image pyramid construction instead of spiral path, McCann99 Retinex restrains the length and width of input image, it need them to be formulated into $nrows \cdot 2^n \times ncols \cdot 2^n$ and $cols \geq rows$, $1 \leq cols, rows \leq 5$. This constraint arises from the fact that each level of the image pyramid differs from previous levels by a factor 2 in each dimension, the size of top is $rows \times cols$, the bottom is the original image. McCann99 algorithm estimates the reflection component by comparing each pixel to each of its 8 immediately neighboring pixels in clockwise order; then by interpolating the result of last level, make the size both the width and height same as next level; and iterating the above steps; finally, with comparing the original image to its 8 neighborhood, we can achieve the enhancement image.

2.2.3 The Image Enhancement Based on Multiple Scale Retinex with Color Restoration

Among the many improved center/surround algorithm, both single-scale Retinex (SSR) [12] and multi-scale Retinex (MSR) [13] adapt the Gaussian fuzzy about processing the brightness component. Compare to the SSR, the difference of MSR algorithm is that MSR need to provide Gaussian Fuzzy to each scale in the original image, then accumulate every scale about calculating the

$$Log\left[R(x,y)\right] = Log\left[R(x,y)\right] + Weight(i) * \left(Log\left[I_i(x,y)\right] - Log\left[L_i(x,y)\right]\right).$$

The multi-scale Retinex with color restoration combines the dynamic range compression of the small-scale Retinex and the tonal rendition of the large scale Retinex with a universally applied color restoration. MSRCR overcomes the problem that the haze-free image during the both SSR and MSR exist biasing color. In previous, more rarely, the gray-world violations can simply produce an unexpected color distortion. Jobson [14] provides good color rendition for images that contain gray-world violations.

2.2.4 Contrast Limited Adaptive Histogram Equalization (CLAHE)

Histogram equalization primary idea is that to concentrate gray histogram in the local region of original image into a whole coverage gray region. A modification of histogram equalization called the Adaptive Histogram Equalization [6] can be used on images which contain local regions of low contrast bright or dark regions for better results. It works by considering only small regions and based on their local cdf, performs contrast enhancement of those regions.

Contrast limited adaptive histogram equalization [7] is different with adaptive histogram equalization, the main difference is limited contrast. In CLAHE, the contrast of each small regions must be limited, which can overcome oversize noise about AHE. To restrict the contrast, we can clip the histogram by setting a threshold earlier before computing cdf. The value clipped from histogram, that is limited contrast, is determined from distribution of histogram, which is the size of neighbor.

2.3 Contrast Results

2.3.1 For the Mist

Figure 1 show the haze-free images of mist based on physical model method.

From Fig. 2, we can see the haze-free images of mist based on image enhancement method.

Fig. 1 Haze removal results of mist based on physical model, *left* the original images, *right* the haze-free images. From *top* to *bottom* dark channel prior, guided filter, NBPC, average filter

Fig. 2 Haze removed images of mist based on image enhancement, *left* the original images, *right* the haze-free images. From *top* to *bottom* FMR, RM99, MSRCR, CLAHE

According Figs. 1 and 2, we can find that for the mist images whose contrast are not low, and the distribution of fog is homogeneous, the algorithms based on dark channel achieve better effects while consuming more time. Other algorithms based on CLAHE or Retinex require less running time, but lead to worse haze removal ability while creating bias. It appears the algorithm based on average filter preserves the advantage of algorithms based on dark channel and decreases time cost.

2.3.2 For the Fog

Figure 3 show the haze-free images of fog based on physical model method.

From Fig. 4, we can see the haze-free images of fog based on image enhancement method.

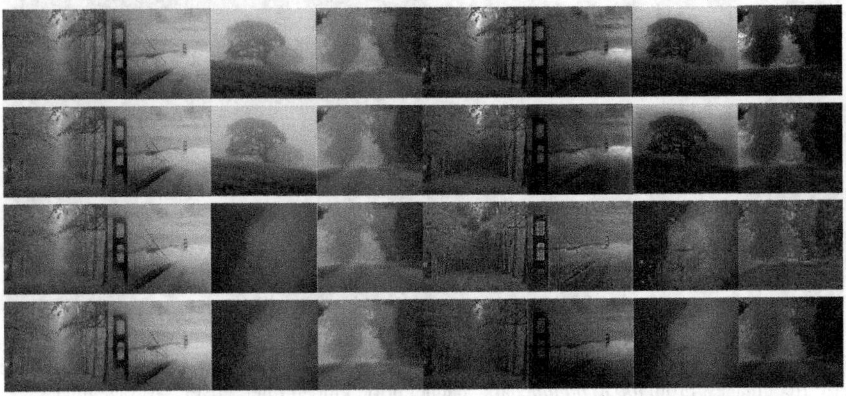

Fig. 3 Haze removal results of fog based on physical model, *left* the original images, *right* the haze-free images. From *top* to *bottom* dark channel prior, guided filter, NBPC, average filter

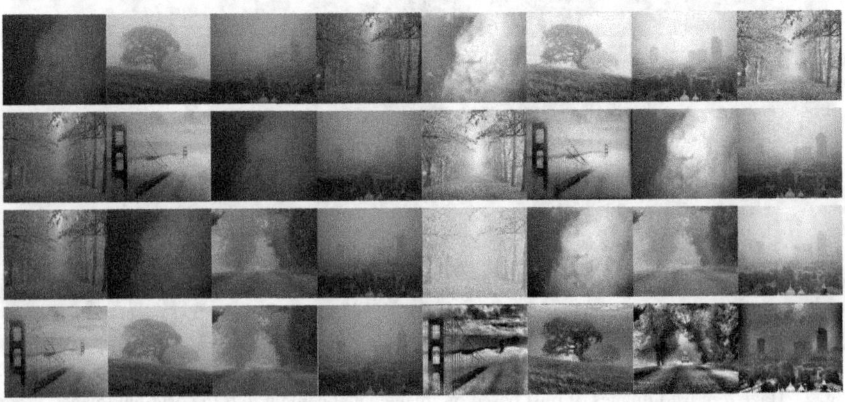

Fig. 4 Haze removal results of fog based on image enhancement, *left* the original images, *right* the haze-free images. From *top* to *bottom* FMR, RM99, MSRCR, CLAHE

Take a look at figures from Figs. 3 and 4, method based on CLAHE bring about a serious problem of blocks in images, and the haze removal techniques based on Retinex lead to image-free so bright that there is something invisible in images. Although the contrast and brightness of the fog are great obstacles, the haze removal techniques based on dark channel managed to solve the problem.

2.3.3 For the Dense Fog

Figure 5 show the haze-free images of dense fog based on physical model method.

From Fig. 6, we can see the haze-free images of dense fog based on image enhancement method.

Fig. 5 Haze removal results of dense fog based on physical model, *left* the original images, *right* the haze-free images. From *top* to *bottom* dark channel prior, guided filter, NBPC, average filter

Fig. 6 Haze removal results of dense fog based on image enhancement, *left* the original images, *right* the haze-free images. From *top* to *bottom* FMR, RM99, MSRCR, CLAHE

For the dense fog situation with which the contrast, brightness and quality of image are extremely terrible, so the haze removal techniques based on image enhancement reveal its excellent power to solve the problem compare to the ones based on physical model. From Figs. 5 and 6, we can find that the algorithms based on CLAHE and Retinex have better removal effects.

3 Conclusion

In this paper, we summed up a series of haze removal techniques and applied them with a few hazy images under several weather conditions. Firstly, The haze image under dense fog gain properties including low contrast ratio and low luminance, but the haze removal algorithms based on Retinex and CLAHE perform well on these images. Secondly, The haze removal algorithms based on dark channel present good results on the hazy images under mist. Eventually, we could use average filtering algorithm based on dark channel to deal with hazy images which have homogeneous mist and not bad contrast ratio, since it preserves the advantage of algorithms based on dark channel and decreases time cost. According to the experiment results, we could find that methods based on physical model increase the computational complexity and time cost, since the methods always use the model parameters estimated by optimal method. However, among the methods based on non-physical model, some could reach the real-time requirement, but the performance does not reach the restriction. In the practical applications of computer vision, the image haze removal algorithms retain the following problems. (1) Automatic. The methods based on multiple images or user interaction are restricted by scene, weather, imaging equipment and so on. However, the applications of intelligent transportation and video surveillance demand the algorithms focusing on single image and executing automatically. (2) Real-time. The algorithm should have the property of real-time, in order to adapt to the needs of practical application.

The image haze removal technique is a hotspot in the field of image processing and computer vision and it is widely used in the outdoor visual system. Since the robustness, aromaticity and real time of the algorithm are required in practical applications. The system should select the algorithms which perform better, and further achieve real-time processing so as to reach the requirement of practical applications.

References

1. Tan RT (2008) In: IEEE conference on computer vision and pattern recognition, CVPR 2008 (IEEE, 2008), pp 1–8
2. Fattal R (2008) In: ACM transactions on graphics (TOG), vol 27 (ACM, 2008), p 72
3. He K, Sun J, Tang X (2011) IEEE transactions on pattern analysis and machine intelligence. 33(12):2341
4. He K, Sun J, Tang X (2010) In: Computer vision-ECCV 2010 (Springer, 2010), pp 1–14

5. Tarel JP, Hautiere N (2009) In: 2009 IEEE 12th international conference on computer vision, (IEEE, 2009), pp 2201–2208
6. Pizer SM, Amburn EP, Austin JD, Cromartie R, Geselowitz A, Greer T, ter Haar Romeny B, Zimmerman JB, Zuiderveld K (1987) Computer vision, graphics, and image processing 39(3):355
7. Zuiderveld K (1994) In: Graphics gems IV. Academic Press Professional Inc, pp 474–485
8. Land EH (1986) Proceedings of the national academy of sciences 83(10):3078
9. Land EH (1983) Proceedings of the national academy of sciences of the United States of America 80(16):5163
10. Frankle JA, McCann JJ (1983) Method and apparatus for lightness imaging. US Patent 4,384,336
11. McCann J (1999) In: Color and imaging conference, vol 1999. Society for Imaging Science and Technology, pp 1–8
12. Jobson DJ, Rahman ZU, Woodell GA (1997) IEEE transactions on image processing. 6(3):451
13. Rahman ZU, Jobson DJ, Woodell GA (1996) In: Proceedings of international conference on image processing, vol 3 (IEEE, 1996), pp 1003–1006
14. Jobson DJ, Rahman ZU, Woodell GA (1997) IEEE transactions on image processing. 6(7):965
15. Narasimhan SG, Nayar SK (2002) International journal of computer vision 48(3):233
16. Koschmieder H (1925) Theorie der horizontalen Sichtweite: Kontrast und Sichtweite. Keim & Nemnich
17. Draper NR, Smith H, Pownell E (1966) Applied regression analysis, vol 3. Wiley, New York
18. Liu Q, Chen M, Zhou D (2013) In: 25th Chinese control and decision conference (CCDC), (IEEE, 2013), pp. 3780–3785
19. Land EH (1986) Vision research 26(1):7

DOA Estimation Method for Wideband Signals Based on Multiple Small Aperture Subarrays

Jiaqi Zhen and Zhifang Wang

Abstract In direction of arrival (DOA) estimation, the single small aperture subarray is hard to meet the requirement of the accuracy for multiple targets, a new method for wideband signals based on multiple small aperture subarrays is proposed to this problem. First, the signals of every frequency are focused, then a modified minimum variance estimator (MVM) is employed for DOA estimation, the method doesn't need the accurate position information among each subarray, consequently the error caused by position perturbation is avoided, and it has a preferable robustness to gain and phase intersubarray distortions, the performance has been proved by simulation results at last.

Keywords Direction of arrival · Wideband signals · Small aperture subarray · Intersubarray distortions

1 Introduction

DOA estimation is one of the important research aspects in array signal processing, it has been widely paid attention, the scholars have proposed many methods which are suitable for narrowband signals, such as harmonic analysis method [1], the maximum entropy method [2], minimum variance method [3], but they all have some boundedness as the distribution of the signal, then since the end of 1970s, the research field of spatial spectrum estimation have emerged a large number of achievements, for instance, maximum likelihood (ML) [4, 5], MUSIC [6] and ESPRIT [7], as well coherent signal method (CSM) [8] which are suitable for

J. Zhen (✉) · Z. Wang
Electronic Engineering, Heilongjiang University, Harbin, Heilongjiang, China
e-mail: zhenjiaqi2011@163.com

Z. Wang
e-mail: xiaofang_hq@126.com

A. Abraham et al. (eds.), *Intelligent Data Analysis and Applications*,
Advances in Intelligent Systems and Computing 370,
DOI 10.1007/978-3-319-21206-7_34

wideband signals, all the methods above have precise requirements on the information of array position.

Passive localization technology is a common used means in civil and military fields, small aperture array has accordingly been widely applied [9–12], but the effective aperture of single small aperture array is minor, resolution power is limited, and it can't be improved by increasing the structure size as the limitation of the environment and conditions, so it is usually impossible to search multiple targets effectively. How to use small aperture array to estimate DOA to achieve high resolution and high precision is a topic worthy of in depth study. In multiple array jointly estimation, on one hand, the distances among the subarrays usually don't satisfy the assumption of half wavelength, sometimes they are even up to a dozen wavelength, which easily leads to the position error of the array sensors, so there are usually many false peaks when conventional super resolution direction finding methods are used; on the other hand, the gain and phase offset among these subarrays are often disaccord, which will cause the calculation failure. Thus the performance of DOA estimation will reduce sharply, and they have hindered the application of these methods in the actual system.

The paper proposed a new method for wideband signals based on partly calibrated multiple small aperture subarrays, it synthesizes output information of these subarrays, then a modified minimum variance estimator is employed for DOA estimation, multiple subarrays jointly estimation extends the array aperture, improves the accuracy which is difficult to meet by the single small aperture array. The modified estimator has a preferable robustness to the position error, gain and phase offset disaccord among the subarrays.

2 Signal Model

Consider an array with K subarrays, where the kth subarray is composed of $M_k \geq 1$ sensors, the total number of sensors of all the subarrays is

$$M = \sum_{k=1}^{K} M_k \tag{1}$$

for simplicity, suppose there are three small aperture subarrays, all of them have the same number of sensors, and define $M_k = 4$, that is to say they each has four sensors which distributed on the circumference of a circle, the geometric center of the first subarray is defined as the origin of rectangular coordinate system, it is shown in Fig. 1. N far-field wideband acoustic signals arriving at the array, they are zero mean stationary random, DOAs of them are respectively $(\theta_1, \theta_2, \ldots, \theta_N)$, the noise background is Gaussian white process, and the signals and the noise are statistical independence.

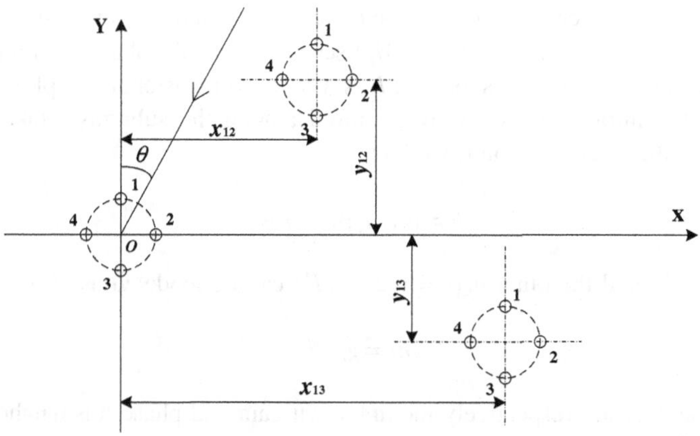

Fig. 1 The structure of array

Consider extracting a narrowband component from the wideband signal, its center frequency is f, the array manifold is expressed as

$$\boldsymbol{A} = [\boldsymbol{a}(f, \theta_1, \alpha), \boldsymbol{a}(f, \theta_2, \alpha), \ldots, \boldsymbol{a}(f, \theta_N, \alpha)] \tag{2}$$

where $\boldsymbol{a}(f, \theta, \alpha)$ is $M \times 1$ dimensional scanning vector, and α expresses the positional relation of array manifold among the subarrays, the aperture of the single small subarray is smaller and array error is easy to calibrate, so the scanning vector of the array can be expressed as

$$\boldsymbol{a}(f, \theta, \alpha) = \boldsymbol{V}\boldsymbol{h} \tag{3}$$

where

$$\boldsymbol{V} = \begin{bmatrix} V_1 & 0 & \cdots & 0 \\ 0 & V_2 & \cdots & 0 \\ \vdots & \vdots & \ddots & \vdots \\ 0 & 0 & \cdots & V_k \end{bmatrix} \tag{4}$$

is $M \times K$ dimensional scanning matrix, composing by scanning vector $\boldsymbol{V}_k (i = 1, 2, \ldots, K)$ of every small aperture subarray. If we take the first sensor of every small aperture subarray as the references, then we have

$$\boldsymbol{V}_k = \begin{bmatrix} 1 \\ \exp\{j2\pi/\lambda(\xi_{2,k}\sin\theta + \zeta_{2,k}\cos\theta)\} \\ \vdots \\ \exp\{j2\pi/\lambda(\xi_{M_k,k}\sin\theta + \zeta_{M_k,k}\cos\theta)\} \end{bmatrix} \tag{5}$$

where λ is the wavelength of the subband signal, $\xi_{m,k}$ and $\zeta_{m,k}$ are respectively the coordinates of the mth $(m = 1, 2, \ldots, M_k)$ sensor in the kth subarray with reference to the first sensor of the kth subarray, $\boldsymbol{h}$ is a $K \times 1$ dimensional complex vector, it expresses the information of array manifold among the subarrays, take the first subarray as the reference point, we have

$$\boldsymbol{h} = [h_1, h_2, \ldots, h_K]^{\mathrm{T}} \tag{6}$$

clearly, $h_1 = 1$ and the other $h_k (k = 1, 2, \ldots K)$ can be modeled as

$$h_k = g_k \mathrm{e}^{\mathrm{j}\varphi_k} \tag{7}$$

where g_k and φ_k are respectively the unknown gain and phase mismatches among the kth and the first subarray.

In practical application, as the position information among these subarrays often has some error, $\boldsymbol{h}$ can be expressed as the Hadamard product of estimated value with error

$$\boldsymbol{h} = \hat{\boldsymbol{h}} \odot \boldsymbol{e} \tag{8}$$

where $\odot$ is the Hadamard product; $\hat{\boldsymbol{h}}$ is the estimated value of $\boldsymbol{h}$, $\boldsymbol{e}$ is the vector caused by positional error, then the scanning vector can be defined as

$$\hat{\boldsymbol{a}}(f, \theta, \alpha) = \boldsymbol{V}\hat{\boldsymbol{h}} \tag{9}$$

3 MVM Estimators

Suppose there is an array consisting of M sensors, the receiving narrowband signal vector is $\boldsymbol{x}(t)$, the weighted vector of every channel is

$$\boldsymbol{w} = [\omega_1, \omega_2, \cdots, \omega_M]^{\mathrm{T}} \tag{10}$$

the output is

$$\begin{aligned} y(t) &= \boldsymbol{w}^{\mathrm{H}} \boldsymbol{x}(t) \\ &= \sum_{i=1}^{M} \omega_i^* x_i(t) \end{aligned} \tag{11}$$

average power of the whole array is

$$
\begin{aligned}
P(\boldsymbol{w}) &= \frac{1}{L}\sum_{t=1}^{L} |y(t)|^2 \\
&= \boldsymbol{w}^{\mathrm{H}} E\{\boldsymbol{x}(t)\boldsymbol{x}^{\mathrm{H}}(t)\}\boldsymbol{w} \\
&= \boldsymbol{w}^{\mathrm{H}}\boldsymbol{R}\boldsymbol{w}
\end{aligned} \tag{12}
$$

where L is the sampling times, our purpose is to guarantee to receive the actual signal we want correctly, while the signal from other directions or interferences are suppressed completely, that is

$$
\begin{cases}
\min_{\boldsymbol{w}} \boldsymbol{w}^{\mathrm{H}}\boldsymbol{R}\boldsymbol{w} \\
\boldsymbol{w}^{\mathrm{H}}\hat{\boldsymbol{a}}(\theta_d) = 1
\end{cases} \tag{13}
$$

the corresponding optimal weighted vector is

$$
\boldsymbol{w}_{\mathrm{opt}} = \frac{\boldsymbol{R}^{-1}\hat{\boldsymbol{a}}(\theta_d)}{\hat{\boldsymbol{a}}^{\mathrm{H}}(\theta_d)\boldsymbol{R}^{-1}\hat{\boldsymbol{a}}(\theta_d)} \tag{14}
$$

then curse of the spatial spectrum can be acquired by scanning the following function

$$
P(\theta) = \frac{1}{\hat{\boldsymbol{a}}^{\mathrm{H}}(\theta)\boldsymbol{R}^{-1}\hat{\boldsymbol{a}}(\theta)} \tag{15}
$$

4 Proposed Method

We can use the deriving results above, and promote them to the wideband signals, it is divided into two steps: wideband focusing and modified MVM estimator, they are discussed as follows.

4.1 Wideband Focusing

First, CSM is employed, signal subspaces of all frequencies are mapped at the same reference point, then the steady estimator is formed to calculate DOA.

The expression of the wideband signal in frequency domain is

$$
\boldsymbol{X}(f_i) = \boldsymbol{A}(f_i, \theta, \alpha)\boldsymbol{S}(f_i) + \boldsymbol{N}(f_i) \quad i = 1, 2, \ldots, J \tag{16}
$$

where $A(f_i, \theta, \alpha) = [a(f_i, \theta_1, \alpha), a(f_i, \theta_2, \alpha), \ldots, a(f_i, \theta_N, \alpha)]$ is a $M \times N$ dimensional direction matrix; $S(f_i)$ is a $N \times 1$ dimensional signal vector; $N(f_i)$ is a $M \times 1$ dimensional noise vector, J is the number of frequency point.

The covariance matrix is

$$\begin{aligned} R_{XX}(f_i) &= E[X(f_i)X^{\mathrm{H}}(f_i)] \\ &= A(f_i)R_{SS}(f_i)A^{\mathrm{H}}(f_i) + \sigma^2 I \quad i = 1, 2, \ldots, J \end{aligned} \tag{17}$$

then we can use some focusing method to acquire the average matrix $\hat{R}_Y$ by mapping the subspace of different frequencies at the reference point f_0.

$$\hat{R}_Y = \frac{1}{J}\sum_{i=1}^{J} R_{XX}(f_i) \tag{18}$$

4.2 Modified MVM Estimator

Here two errors are considered, one is amplify intersubarray distortions, it leads to the gain disaccord among the subarrays; the other is phase intersubarray distortions, it leads to the phase offset disaccord among the subarrays, we will discuss them respectively below.

4.2.1 Amplify Intersubarray Distortions

A modified MVM estimator is deduced in reference [13, 14], it can be expressed as

$$P_{BL}(\theta) = \frac{\hat{a}^{\mathrm{H}}(\theta)\hat{R}_Y^{-1}\hat{a}(\theta)}{\hat{a}^{\mathrm{H}}(\theta)\hat{R}_Y^{-2}\hat{a}(\theta)} \tag{19}$$

it can be expressed in another form

$$P_{BL}(\theta) = \frac{\hat{h}^{\mathrm{H}} B_1(\theta)\hat{h}}{\hat{h}^{\mathrm{H}} B_2(\theta)\hat{h}} \tag{20}$$

where

$$B_1(\theta) = V^{\mathrm{H}}(\theta)\hat{R}_Y^{-1}V(\theta) \tag{21}$$

$$B_2(\theta) = V^{\mathrm{H}}(\theta)\hat{R}_Y^{-2}V(\theta) \tag{22}$$

here, $\hat{a}^{\mathrm{H}}(\theta), \hat{h}, \hat{R}_Y$ and $V(\theta)$ are all used the vectors which are at the reference frequency point, thus, the estimator can be extended to wideband signals.

Here, an amplify intersubarray distortions is considered, a worst-case optimization problem is [15]

$$\max_{h} P_{BL}(\theta) \tag{23}$$

formula (23) can be deemed to be a convex quadratic optimization problem and it is expressed

$$\max_{h} \boldsymbol{h}^{\mathrm{H}}\boldsymbol{B}_1(\theta)\boldsymbol{h} \tag{24}$$

s.t.

$$\boldsymbol{h}^{\mathrm{H}}\boldsymbol{B}_2(\theta)\boldsymbol{h} = 1 \tag{25}$$

it can be solved by Lagrange function

$$f(\boldsymbol{h}) = \boldsymbol{h}^{\mathrm{H}}\boldsymbol{B}_1(\theta)\boldsymbol{h} - \mu(\boldsymbol{h}^{\mathrm{H}}\boldsymbol{B}_2(\theta)\boldsymbol{h} - 1) \tag{26}$$

by calculating its extreme value, the following equation can be acquired

$$\boldsymbol{B}_2^{-1}(\theta)\boldsymbol{B}_1(\theta)\hat{\boldsymbol{h}}(\theta) = \mu\hat{\boldsymbol{h}}(\theta) \tag{27}$$

the solution of Eq. (24) in the constraint of (25) is

$$\hat{\boldsymbol{h}}(\theta) = \eta\ell_{\max}\{\boldsymbol{B}_2^{-1}(\theta)\boldsymbol{B}_1(\theta)\} \tag{28}$$

where $\ell_{\max}\{\cdot\}$ returns the eigenvector corresponding to the maximum eigenvalue of the matrix. We can see that $\hat{\boldsymbol{h}}(\theta)$ is equal to the eigenvector of $\boldsymbol{B}_2^{-1}(\theta)\boldsymbol{B}_1(\theta)$ with η, which provided to perform by Eq. (28) reference to (24) and (25) with the constraint

$$\hat{\boldsymbol{h}}^{\mathrm{H}}(\theta)\boldsymbol{B}_2(\theta)\hat{\boldsymbol{h}}(\theta) = 1 \tag{29}$$

we can see from Eq. (27)

$$\hat{\boldsymbol{h}}(\theta)^{\mathrm{H}}\boldsymbol{B}_1(\theta)\hat{\boldsymbol{h}}(\theta) = \mu \tag{30}$$

so the spectrum of the estimator can be described as

$$P_{RBL}(\theta) = \lambda_{\max}\{\boldsymbol{B}_2^{-1}(\theta)\boldsymbol{B}_1(\theta)\} \tag{31}$$

where $\lambda_{\max}\{\cdot\}$ is the symbol returning the maximum eigenvalue of the matrix.

4.2.2 Phase Intersubarray Distortions

In many actual environment, the wave front aberration is mainly affected by random or unknown deterministic perturbations, so the amplitude distortions can be ignored, from Eq. (7), the phase distortions is described as

$$|h_k| = 1,\ k = 1, 2, \ldots, K \tag{32}$$

thus, the vector $\boldsymbol{h}$ based on worst-case optimization is

$$\min_{\boldsymbol{h}} \boldsymbol{h}^{\mathrm{H}} \boldsymbol{B}_1(\theta)\boldsymbol{h} \quad \text{s.t.} \quad |h_k| = 1,\ k = 1, 2, \ldots, K \tag{33}$$

a more accurate solution [16] based on semidefinite programming (SDP) algorithm can be used here, it is described as follows:

$$\begin{aligned} \max_{\boldsymbol{h}} \mathrm{Tr}\{\boldsymbol{B}_1(\theta)\boldsymbol{H}\} \quad \text{s.t.} \quad & \mathrm{Tr}\{\boldsymbol{B}_2(\theta)\boldsymbol{H}\} = 1, \quad [\boldsymbol{H}]_{k,k} = 1, k = 1, 2, \ldots, K \\ & \boldsymbol{H} \succ 0, \mathrm{rank}\{H\} = 1. \end{aligned} \tag{34}$$

where $\mathrm{tr}\{\cdot\}$ is the trace operator, $[\cdot]_{k,k}$ shows the kth diagonal element of the matrix, $\boldsymbol{H} \succ 0$ denotes that $\boldsymbol{H}$ is positive semidefinite.

Obviously the Eq. (34) is a non-convex problem, the non-convex rank constraint $\mathrm{rank}\{\boldsymbol{H}\} = 1$, it is necessary to change it to be convex, we employ semidefinite relaxation algorithm [17] to finish it, thus, non-convex rank constraint is dropped, it is

$$\begin{aligned} \max_{\boldsymbol{h}} \mathrm{Tr}\{\boldsymbol{B}_1(\theta)\boldsymbol{H}\} \quad \text{s.t.} \quad & \mathrm{Tr}\{\boldsymbol{B}_2(\theta)\boldsymbol{H}\} = 1, \quad [\boldsymbol{H}]_{k,k} = 1, k = 1, 2, \ldots, K \\ & \boldsymbol{H} \succ 0. \end{aligned} \tag{35}$$

in this way, the spectrum can be solved by Eq. (35), in the same way, RARE and RC can be extended to DOA estimation for wideband signals with the similar method.

5 Simulations

In the simulations, assume some wideband acoustic signals arriving at the array, it is seen from the derivation above, there is no special limitation for the correlation among signals, here assume they are all uncorrelated, the array is composed of three same circle identically oriented subarrays, each subarray has four sensors evenly distributing on the circumference, their radiuses are $\lambda/4 = 0.4$m, where λ is the wavelength corresponding to the frequency 210 Hz, bandwidth B = 50 Hz, the background noise is not related with the signal, it is a white Gaussian stationary random process with zero means and statistically independent on each sensor. We can acquire position information among the subarrays by manifold estimation and correction, here define position of the second subarray relative to the first one is $[10\lambda, 6\lambda]$, and that of the third subarray relative to the first one is $[20\lambda, -5\lambda]$, the structure is shown in Fig. 1.

The sampling frequency of simulation is taken as 600 Hz, the observation time is 16 s, it is divided into 32 periods, the whole frequency band is divided into 31 sub bands through Fast Fourier Transform (FFT), in order to reduce the spectrum leakage, Hanning window is used for FFT, overlapped data is at a rate of 20 %, the RARE, RC and MMVM methods are employed below for the performance comparison of DOA estimation.

5.1 Simulation 1 Precision of the Methods Versus SNR (Amplify Intersubarray Distortions)

Consider three far-field wideband signals arriving at the sensors from 35°, 45°, 55° with the same power, here we test the performance of these methods with amplify intersubarray distortions, phase shift ratios of the array sensors are the same, the gain ratios of three subarrays are respectively 0.9, 0.7 and 1.3, they reflect the different weightings of every subarray, snapshots in every frequency is 50, SNR varies from −5 to 15 dB, step size is 1 dB, position error is the same with simulation 1, 200 times Monte Carlo trials have run for each SNR, the average of them is regarded as the measured result for this SNR, Fig. 2 shows the RMSE of RARE, RC and MMVM methods based on Eq. (31) versus SNR.

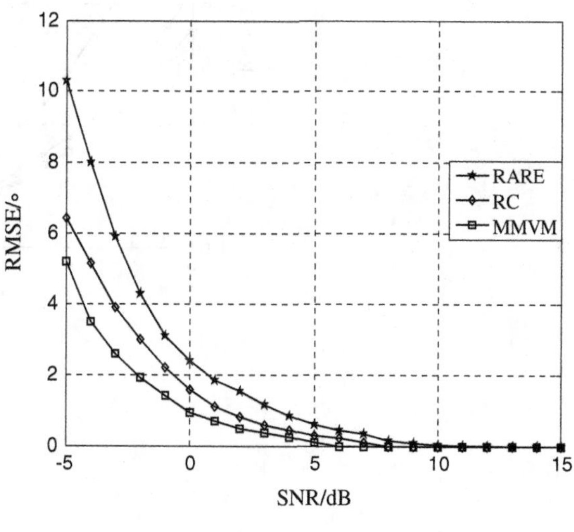

Fig. 2 RMSE of the methods versus SNR (amplify intersubarray distortions)

From Fig. 2 it is seen that RMSE of MMVM is smaller than that of the other two methods, when SNR reaches 6 dB, RMSE of MMVM is zero, but that of RC and RARE methods reach zero when SNR are respectively 8 and 10 dB.

5.2 Simulation 2 Precision of the Methods Versus SNR (Phase Intersubarray Distortions)

Consider three far-field wideband signals arriving at the sensors from 35°, 45°, 55° with the same power, here we test the performance of these methods with phase intersubarray distortions, the gain ratios of the array sensors are the same, phase shift ratios of the array sensors are 0°, 10°, 18°, it reflects the different phase offsets of every subarray, snapshots in every frequency is 50, SNR varies from −5 dB to 15 dB, step size is 1 dB, position error is the same with simulation 1, 200 times Monte Carlo trials have run for each SNR, the average of them is regarded as the measured result for this SNR, Fig. 3 shows the RMSE of RARE, RC and MMVM methods based on Eq. (35) versus SNR.

From Fig. 3 it is seen that RMSE of MMVM is smaller than that of the other two methods, when SNR reaches 9 dB, RMSE of MMVM is zero, but that of RC and RARE methods reach zero when SNR are respectively 10 and 13 dB.

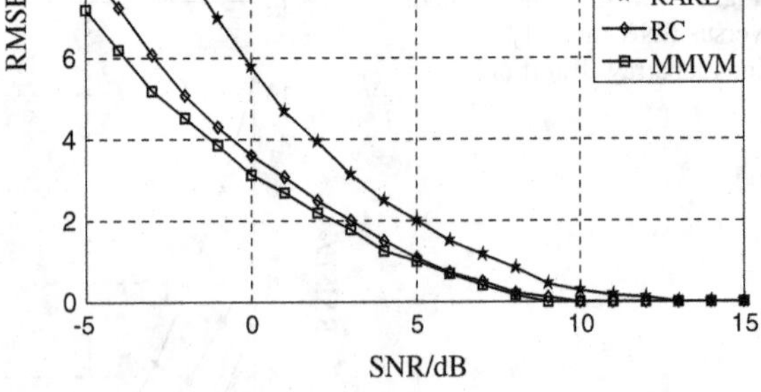

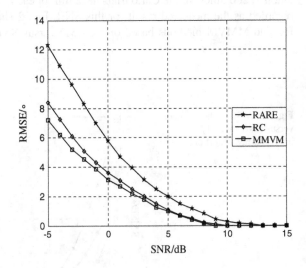

Fig. 3 RMSE of the methods versus SNR (phase intersubarray distortions)

6 Conclusions

A new DOA estimation method for wideband signal based on multiple small aperture subarrays has been proposed in the paper, it uses modified minimum variance estimator to handle the output information of multiple subarrays synthetically, eliminates the false peaks and error caused by sensor perturbation, discordance of gain and phase among subarrays effectively, and it is more preferable than RARE and RC methods in the same conditions.

Acknowledgments This work is supported by National Natural Science Foundation of China (no. 61201399), China Postdoctoral Science Foundation (no. 2012M511003), Project of Science and Technology of Heilongjiang Provincial Education Department (no. 12541638), Youth Foundation of Heilongjiang University (no. 201026), and Startup Fund for Doctor of Heilongjiang University.

References

1. Kay SM, Marple SL (1981) Spectrum analysis—a modern perspective. Proc IEEE 69 (11):1380–1419
2. Zhang Yunxia, Nan Hu, Ye Zhongfu (2013) A source enumeration method based on subspace orthogonality and bootstrap technique. Sig Process 93(4):972–976
3. Capon J (1969) High-resolution frequency-wavenumber spectrum analysis. Proc IEEE 57 (8):1408–1418
4. Wax M, Kailath T (1983) Optimum location of multiple sources by passive array. IEEE Trans Acoust Speech Signal Process 31:1210–1217
5. Ziskind I, Wax M (1988) Maximum likelihood localization of multiple sources by alternating projection. IEEE Trans Acoust Speech Signal Process 36:1553–1560
6. Schmidt RO (1986) Multiple emitter location and signal parameter estimation. IEEE Trans Antennas Propag 34(3):276–280
7. Roy R, Kailath T (1989) ESPRIT-estimation of signal parameters via rotational invariance techniques. IEEE Trans ASSP 37(7):984–995
8. Hung H, Kaveh M (1988) Focussing matrices for coherent signal-subspace processing. IEEE Trans Acoust Speech Signal Process 36(8):1272–1281
9. Schlatter NM, Grydeland T, Ivchenko N, Belyey V, Sullivan J, La Hoz C, Blixt M (2013) Radar interferometer calibration of the EISCAT Svalbard Radar and a additional receiver station. J Atmos Solar Terr Phys 105–106:287–292
10. Lopez Arteaga I, Scholte R, Nijmeijer H (2012) Improved source reconstruction in fourier-based near-field acoustic holography applied to small apertures. Mech Syst Signal Process 32:359–373
11. See C, Gershman AB (2004) Direction-of arrival estimation in partly calibrated subarray-based sensor arrays. IEEE Trans Signal Process 52(2):329–338
12. Bulgarevich Dmitry S, Watanabe Makoto, Shiwa Mitsuharu (2012) Aperture array Fabry-Perot interference filter. Opt Commun 285(24):4861–4865
13. Mavrychev EA, Ermolayev VT, Flaksman AG (2013) Robust Capon-based direction-of arrival estimators in partly calibrated sensor array. Sig Process 93(12):3459–3465
14. Lei Lei (2010) Joni Polili Lie, Alex B. Gershman, Chong Meng Samson See. Robust Adaptive Beamforming in Partly Calibrated Sparse Sensor Arrays. IEEE Trans Signal Process 58 (3):1661–1667

15. Borgiotti G, Kaplan LJ (1979) Superresolution of uncorrelated interference sources by using adaptive array technique. IEEE Trans Antennas Propag 27(6):842–845
16. Ma WK, Davidson T, Wong KM, Luo ZQ, Ching PC (2002) Quasi-ML multiuser detection using semi-definite relaxation with application to synchronous CDMA. IEEE Trans Signal Process 50(4):912–922
17. Sidiropoulos ND, Davidson TN, Luo ZQ (2006) Transmit beamforming for physical-layer multicasting. IEEE Trans Signal Process 54(6):2239–2251

Research on Heuristic Based Load Balancing Algorithms in Cloud Computing

Jengshyang Pan, Pingfei Ren and Linlin Tang

Abstract Since the proposition of the concept of cloud computing in the year 2006, cloud computing has drawn lots of attention from both industry and academic area. Several technologies such as virtualization formed the basis of cloud computing, while some other technologies acts as system improvement strategies in cloud computing. Among the used technologies, Load balancing is indispensable and extremely important in improving system performance and maintaining users' experience. In this paper, we focus on load balancing algorithms commonly used in cloud computing. Through our analysis we proposed our improved online load balancing algorithm. We use several experimental results to show its power and efficiency. We use CloudSim as a simulator to verify our thoughts.

Keywords Cloud computing · CloudSim · Load balancing algorithm · Heuristic

1 Introduction

Cloud computing is emerged as a new business model in IT Industry. Big Internet companies such as google, amazon sold its redundant computing and storage ability at a competitive price to those in need. Cloud computing paradigm is popular among both service providers and service users for its characteristics such as low cost, high scalability, on demand provisioning, and automatic management. For those service providers, cloud computing is a new profit point by making use of

---

J. Pan · P. Ren · L. Tang (✉)
Innovative Information Industry Research Center, Shenzhen Graduate School, Harbin Institute of Technology, Shenzhen 518055, China
e-mail: hittang@126.com

J. Pan
College of Information Science and Engineering, Fujian University of Technology, Fuzhou 350118, Fujian Province, China

A. Abraham et al. (eds.), *Intelligent Data Analysis and Applications*, Advances in Intelligent Systems and Computing 370,
DOI 10.1007/978-3-319-21206-7_35

hard ware resources owned. For those users, by renting virtualized resources provided by service providers, a considerable amount of investigation into IT infrastructures could be saved.

Cloud computing vendors use virtualization and multi-agent technologies to improve system's utilization, along with the famous pay per use pricing model to ensure a long term benefit, and the service providers may temporarily use other vendors' cloud computing services offered to deal with the condition where there's a sudden growth in resource requirement.

Differs from traditional supercomputer centers, data centers work in cloud computing paradigm mainly rely on cheap x86 servers hosting one or more virtual machines. Another big difference between supercomputer centers and cloud computing data centers lies in the type of tasks to be processed: traditional supercomputer centers mainly do some scientific compute, in the scenery where resources needed by submitted tasks could be known before tasks are handled, while in cloud computing sceneries data centers handle various kinds of tasks and the elastic characteristic of cloud computing determines that both of tasks' kind and size could not be known before.

Cloud computing is grow in an era where network connection becomes convenient to access. As a result, cloud computing relies on internet to offer service. Categorized by service type, most of the services could be classified into IaaS, PaaS or SaaS. Unlike traditional internet based applications, through internet, users could get virtualized storage, network, OS and a number of software.

Load balancing is extremely important in internet based applications and in grid computing scenario. We use load balancing techniques to balance the load among servers backend, where the load is occurred by systems' users' requirements. That is, in web based applications, basically the aim of the load balance is to distribute web requests among applications servers or data servers, while in grid based supercomputer scenario, we use load balancing techniques to make works submitted by end users finish as quick as possible.

Load balancing is important for cloud computing paradigm, too. But compared to grid computing and the traditional network load balancing, there exists various needs making load balancing in cloud computing a challenge work.

In load balancing, what we actually do is to find a mechanism by which tasks are distributed to process node evenly; avoiding both hot spot and hungry points exists. Thus it is a complex to make loads as balance as we expect.

In this paper, we make some research on load balancing problem in a heterogeneous cloud computing environment. We focus on load balancing's heart and soul: its algorithm. The following parts are organized in this order: In "Related works" we give some materials having done in load balancing area. In "Proposed method" we give our improved algorithm. In "Experimental Results" we use experiments done in CloudSim to prove our thoughts. Finally we give the conclusion and have a look into the future.

2 Related Work

Different needs and loads making different load balancing strategy works. Base on different kinds of load, various load balancing model and strategies are developed. Azodolmolky S, Wieder P and Yahyapour R introduced network problems exists in IaaS and cloud federation in [1]. They point out that the problem we faced by now in cloud computing network area is how to construct virtual networks. They proposed a resolution of SDN based applications. By uncouple control flows and data flows in SDN, innovated algorithms and mechanisms could be developed to meet the need of large bandwidth, elastic device arrangement, and live immigration of VMs and so on. In [2], Yeo S and Lee H H S made mathematical models about heterogeneous cloud computing environment's energy consumption and response time. Yeo S and Lee H H S analyzed electronic energy consumed and performance characteristics of cloud, finding that to achieve better performance in heterogeneous based cloud infrastructures, the worst node's response time should be no more than 3 times of the beast ones. In [3], Doyle J, Shorten R and O'Mahony D abstract network and server components in cloud into a graph. They use carbon emission, electronic consumption, together with service reach time as factors to make load balancing decisions. Doyle J, Shorten R and O'Mahony D use a linear weighted sum approach on the three factors to obtain the edges' weight value in the graph obtained and they use Voronio graphic partition to determine where to send the requests. In [4], Gulisano V, Jimenez-Peris R, Patino-Martinez M, et al. In 2012 proposed an elastic, extendable, compute engine "StreamCloud" to deal with stream data. Aiming at giving minimum resources to satisfy the given demands, they use a threshold method to offer resources while making load the balance. In [5], Zhang Y and Zhou Y proposed a new computing paradigm, called transparent computing. By making compute "transparent" to those end users, work load is distributed among backend servers and users' terminals. In [6], Xu G, Pang J and Fu X proposed a new load balancing model. By partition and switch mechanisms designed, they use a "beast-replay" evaluation standard to approach nash equilibria. The player enrolled in the game is tasks and server nodes.

Load balancing could be performed in various levels, in a centralized or decentralized form. In [7], a detailed description for load balancing methods in task level is given. In [8], a novel load balancing mechanism in VM level is given. By using methods proposed in [8], a better immigration overhead at runtime is achieved. In [9], load balancing in cloud computing is done in network level.

Load balancing could be static or dynamic. As described in [7], static load balancing methods gain information needed before tasks are executed, distribute tasks in a fixed manner. Thus static load balancing methods could not achieve a good result. Dynamic load balancing methods monitor system's state and distribute system's loads according to it. By doing this, dynamic load balancing methods could achieve better performance than static ones. But there's more overhead added to the overall system.

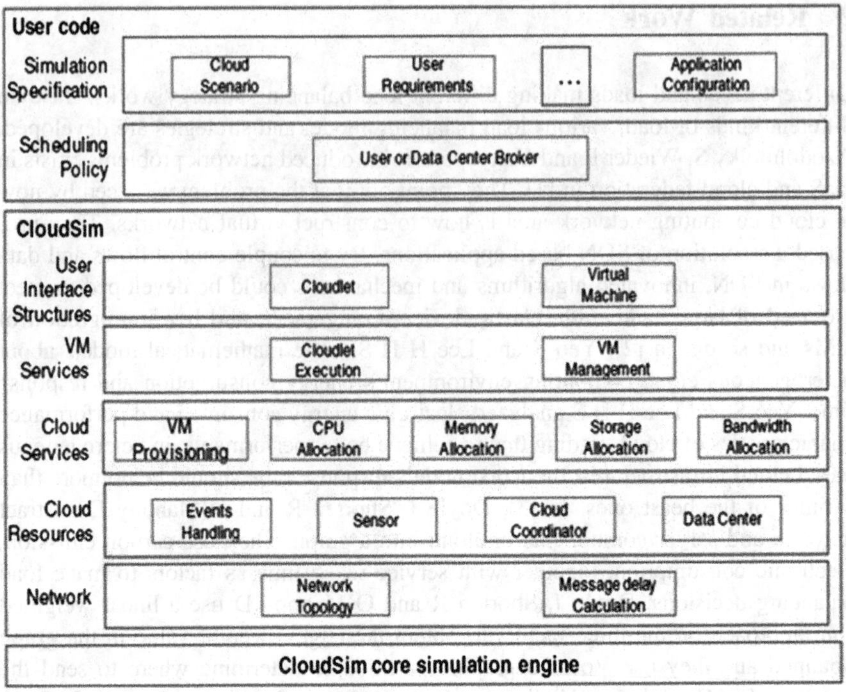

Fig. 1 CloudSim's architecture

In [10], Calheiros R N, Ranjan R, Beloglazov A, et al. offered us a simulator for test our thoughts about cloud. By using CloudSim given in [10], we can compare our proposed method to see whether it works. Its architecture is shown in Fig. 1:

3 Proposed Method

As shown in part 2, load balancing cloud be achieved in various level. But generally speaking, it could be categorized into 2 abstract levels: Job level and task level. Job is consisted of a sequence of tasks. In another word, job is a finite graph where tasks are the vertices and their relationships among are edges. In this paper we focus task level load balancing.

Before we raise the method we based and the improvement we done, we Will give our problem definition first:

1. Task number: n, Task Vector: $V_t = (T_1, T_2, \ldots T_n)^T$, here T_i is a single task with resources needed
2. Server number: m, Server Vector: $V_s = (S_1, S_2, \ldots S_m)$, here we have

$S_i = \alpha^*MIPS + \beta^*BW + \gamma^*MEMORY$ where $\alpha + \beta + \gamma = 1$, MIPS is short for million instruction per second, BW is short for bandwidth

3. Vectors implying the status of each sever $\{L_i = (LT_{1i}, LT_{2i}, \ldots LT_{ni})^T | 0 < i \leq m\}$ where L_i is the load status and $LT_{ji} \in \{0, 1\}, 0 < j \leq n$, implies whether T_j is assigned to S_i
4. Vectors implying the status of each task

$$\{J_i = (LT_{i1}, LT_{i2}, \ldots LT_{im}) | 0 < i \leq n, \sum_{x=1}^{m} LT_{ix} = 1\}$$

5. Assignment metric $A = \{L_1, L_2 \ldots L_m\} = \{J_1, J_2, \ldots J_n\}^T$
6. Value metric: $Value = \begin{pmatrix} a_{11} & \cdots & a_{1m} \\ \vdots & \ddots & \vdots \\ a_{n1} & \cdots & a_{nm} \end{pmatrix} .{*}A$ where $a_{ij} = T_i / S_j = Time_{i\_assign} + Time_{i\_proc} + Time_{i\_queue}$, here ".*" is the process of multiplying two metrics element by element, at the same position.
7. Our object is to find an A to minimal make span time. Actually we make tradeoffs among $Time_{i\_assign}$, $Time_{i\_proc}$ and $Time_{i\_queue}$ over all tasks. The difficulty lies in that if we do this in a greedy manner, we could not get a social optimal solution in the long run.

It is clearly that small assignment time may leads to big queue time and process time. In turn, small process time may obtained by a long queuing time and assignment time. Also, a small queue time cannot guarantee an ideal process time and assignment time.

Basically, it is NP-hard to get a best server to send load. Thus a suboptimal method is acceptable in most cases.

M. Mitzenmacher proposed a randomized load balancing method in [12]. He established a natural supermarket model, shown as Fig. 2, where:

1. Customers arrive as a Poisson stream of rate λ_n, $\lambda < 1$.
2. The number of servers is n

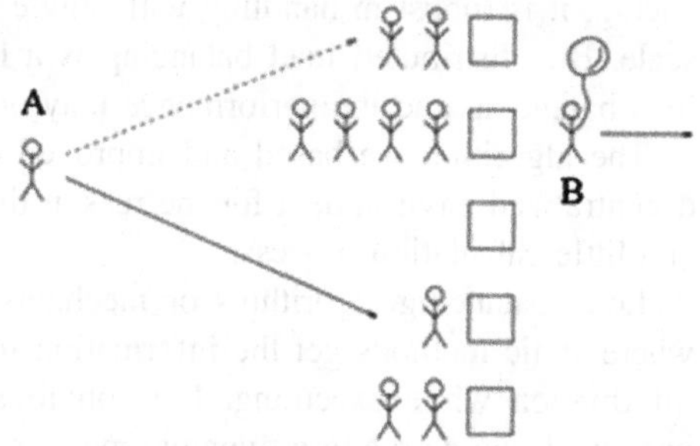

Fig. 2 Supermarket model

3. Each consumer chooses some constant d servers independently and uniformly from the n servers.
4. Customer waits for service at the one with fewest customers.
5. Service time is exponentially distributed with mean 1
6. Customers are serviced in a FIFO model.

He proved that when d = 2, there's an exponential improvement in the expected time a customer spends in system over d = 1. And when d = 3, there's a constant factor better than d = 2.

When d = 1, this method is the same with the random method. Random method is susceptible because it always could not work well in small scale and in traditional scenarios the number of backend server points is not large.

When S_i in V_s is all the same, the methods proposed upon works well, for the reason that the total time is actually determined by itself and the queue's length the consumer chosen. For the same reason, when facing about heterogeneous conditions, there exists a trap that a shorter waiting queue may have a weak server point. If the consumer chooses this queue, he may get a worse practice because of the emptiness.

Thus we modify the step in 4 to this: customer waits in the queue where its estimated time spend in total is minimal among the queues chosen. Here we use the following formula (1) to calculate the estimated execution time Task T_k may spend on processing node S_i:

$$E(Time_{ik}) = \sum_{i=1}^{m \leq i \leq k} a_{ij} \tag{1}$$

Here a_{mj} is the first task's required time in node s_j

According to the mathematical description, it is clear that this formula is used as heuristic information to make decision.

Load balancing could work in centralized or decentralized manner. On one hand, in centralized load balancing, it is easy to achieve the goals we set. But centralized load balancing has some shortcuts such as that it may become a bottleneck of the whole system. Although in a sense we can avoid this problem by adding more load balancing decision nodes, it could not deal with the situation well that the number of backend processing node is big. On the other hand, in decentralized load balancing, it is robust in handling with single point failure while elastic with system scale. But distributed load balancing is a little complex compared to centralized load balancing and the performance may not as good as centralized ones.

The algorithm we based and improved could be used in both centralized and decentralized environment for the reason that it relies on little system information and little calculation process.

Load balancing algorithms or mechanisms could be either static or dynamic, where static methods get the information before execution and dynamic ones got information while executing. It is obvious that both the proposed method and improved one works in a dynamic manner.

4 Experiments and the Results

In this part, we make some experiments based on CloudSim, we compare different commonly used methods: Random, Round-Robin, Greedy, Max-Min, Min-Min and, power-of-two (K2), improved the power of two (k2i).

Random means to assign tasks to backend servers in a fully stochastic manner. It could be considered as "a power-of-one". Max-Min and Min-Min could be performed only in a centralized way. Max-Min means assign big tasks first to the best VMs and Min-Min means assign small tasks first to the best VMs.

We did all the experiments in a centralized manner to make all the result meaningful. We make $\beta = \gamma = 0$ for the reason that the most important factor effecting speed in all kinds of tasks is CPU's ability. We did not focus on modeling various kinds of tasks, thus we make $T_i = IL$, where IL is short for instruction length.

Traditionally, stochastic methods are weak in performance for the reason that in most cases the backend servers are identical, while the scale is small.

First we make an experiment in the condition that $T_1 = T_2 = \cdots = T_n$ and $S_1 = S_2 = \cdots = S_m$. The following Fig. 3 shows its result:

In this condition, the performance of Round-Robin is the same with Max-Min and Min-Min, represented by the lowest straight line. All the stochastic methods could not perform well compared to the determined ones. Random performs worst among all the algorithms in the experiments. One reason is that it has a greater chance to assign the task to a heavy loaded backend processing point. Our improved method K2i performs beast among

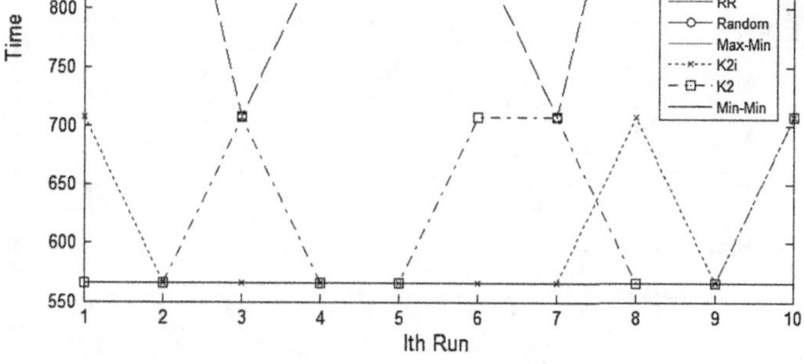

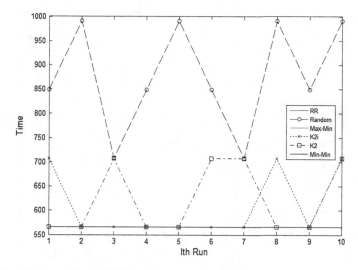

Fig. 3 Results when tasks and VM servers are same separately

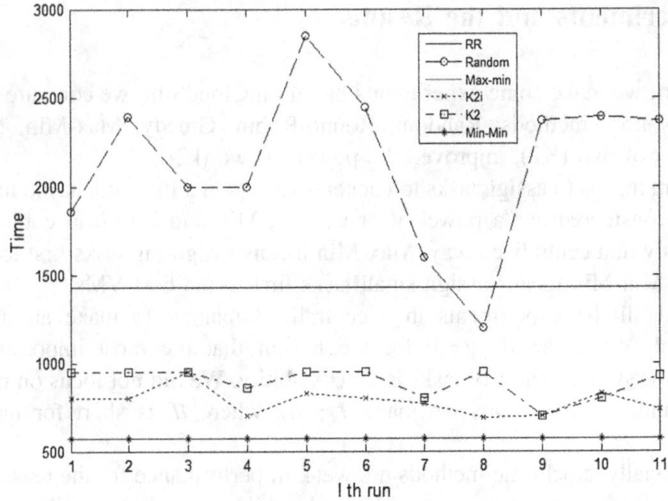

Fig. 4 Results when VM servers are descending while tasks are all the same

In the second experiment, we relax the condition where $S_i, i \leq m$ is in a descending manner while keeping tasks all the same. Results is shown in Fig. 4

In Fig. 4, Max-Min works as well as Min-Min, represented by the lowest straight line. Both K2i and K2 work better than that of the round robin, while the Random has little chance to be better than the Round Robin method. What's more, from Fig. 4 we can clearly get that K2i is more close to the best line we got in this figure.

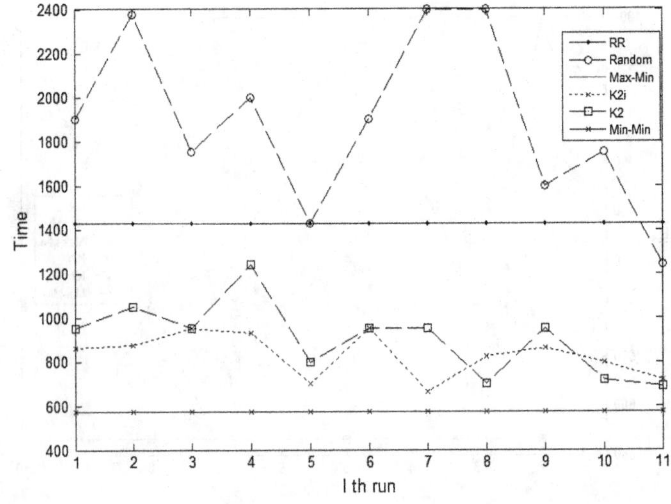

Fig. 5 Results when VM servers are not in a sequential manner

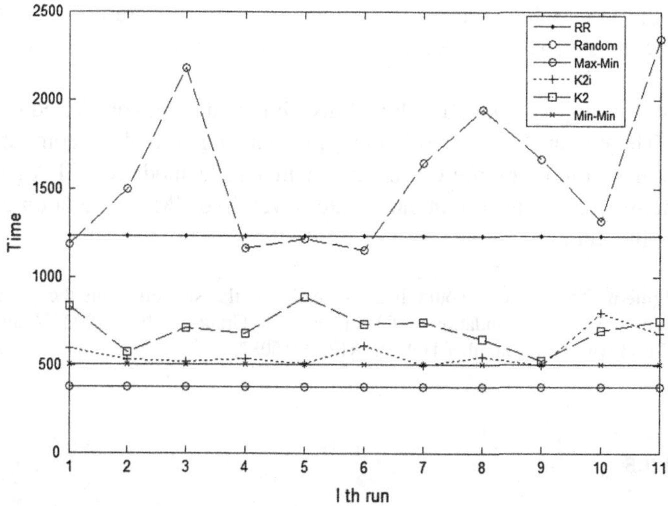

Fig. 6 Results when both VM servers and tasks are not in a sequential manner

In the third experiment, we make $S_i, i \leq m$ not in a sequential manner. In another word, we use the same configurations as in the second experiment. Following Fig. 5 illustrated the result.

Basically the result of RR, Max-Min, Min-Min remain the same with that of in Fig. 4. The line for Max-Min in hidden by that of Min-Min. The random based methods also remain the same manner compared to that in Fig. 4.

In the fourth experiment we make both tasks and VM servers are not in a sequential manner, Fig. 6 illustrated the results.

From Fig. 6 we can get that when both backend processing nodes' ability and task length vary, Min-Min method could not work better than that of the Max-Min. Min-Min strategy is the second beast strategy in Fig. 6. It is also obvious that both K2i and K2 are better than the Round Robin method. What's more, our improved algorithm performs better than that of the original method, for the reason that k2i is more close to the Min-Min method.

5 Conclusions

Load balancing techniques are indispensable for every distributed system, extremely in cloud computing scenario. The aim to make load balancing is to improving system performance while guarantee system users' performance, load balancing could be performed in various levels by a large number of methods. In this paper, we focused on the algorithm used in load balancing. Based on the theoretically proved the power-of-two-choices method, we proposed our improvement to make it more suitable in cloud computing conditions. Experiments are done using CloudSim. CloudSim is a controllable, repeated able simulator for us to test and verify our thoughts.

Cloud computing is a new developed distributed and network based computing paradigm. There's much more work about load balancing in cloud computing need to be done in the future. Future works may include tasks modeling, SLA guaranteed load balancing, load balancing in data center level, live VM immigration combined load balancing, and so on.

Acknowledgment The authors would like to thank for the support from the project NSFC (National Natural Science Foundation of China) with the Grant number 61202456 and the HIT Innovation Fund with Grant number HIT. NSRIF. 2015087.

References

1. Azodolmolky S, Wieder P, Yahyapour R (2013) Cloud computing networking: challenges and opportunities for innovations. IEEE Commun Mag 51(7):59–63
2. Yeo S, Lee HHS (2011) Using mathematical modeling in provisioning a heterogeneous cloud computing environment. Computer 44(8):55–62
3. Doyle J, Shorten R, O'Mahony D (2013) Stratus: load balancing the cloud for carbon emissions control. IEEE Trans Cloud Comput 1(1):116–128
4. Gulisano V, Jimenez-Peris R, Patino-Martinez M et al (2012) Streamcloud: An elastic and scalable data streaming system. IEEE Trans Parallel Distrib Syst 23(12):2351–2365
5. Zhang Y, Zhou Y (2013) Transparent computing: spatio-temporal extension on von neumann architecture for cloud services. Tsinghua Sci Technol 18(1):10–21
6. Xu G, Pang J, Fu X (2013) A load balancing model based on cloud partitioning for the public cloud. Tsinghua Sci Technol 18(1):34–39
7. Mathew T, Sekaran KC, Jose J (2014) Study and analysis of various task scheduling algorithms in the cloud computing environment. In: International conference on advances in computing, communications and informatics (ICACCI). IEEE, pp 658–664
8. Mour S, Srivastava P, Patel P et al (2014) Load management model for cloud computing. In: 9th international conference for internet technology and secured transactions (ICITST). IEEE, pp 178–184
9. Wang SC, Yan KQ, Liao WP et al (2010) Towards a load balancing in a three-level cloud computing network. In: 3rd IEEE international conference on computer science and information technology (ICCSIT), vol 1. IEEE, pp 108–113
10. Goyal A (2014) A study of load balancing in cloud computing using soft computing techniques. Int J Comput Appl 92(9):33–39
11. Calheiros RN, Ranjan R, Beloglazov A et al (2011) CloudSim: a toolkit for modeling and simulation of cloud computing environments and evaluation of resource provisioning algorithms. Softw Pract Experience 41(1):23–50
12. Mitzenmacher M (2001) The power of two choices in randomized load balancing. IEEE Trans Parallel Distrib Syst 12(10):1094–1104

Comparison and Test for Several Typical Cloud Computing Platforms

Xiang Xu, Fumin Zou, Quan Zhu and Xianghai Ge

Abstract Since cloud computing is proposed, increasing enterprises are investing and researching on it for the high scalability, powerful computing capability and inexpensiveness. Quite a number of cloud computing platforms were put forward. This paper analyzed several typical cloud platforms and compared the differences between these platforms. Finally, the paper deployed an experimental platform based on VMware cloud, and then test the computing performance of the platform with massive floating car data. The results show that the cloud computer platform can shorten the calculation time of mass data and improve performance significantly with high scalability.

Keywords Cloud computing · Computing performance · Virtualization · Cloud platform

1 Introduction

In recent years, new revolution was occurred in the field of information technology, the changes have brought new technical forms and business models. Cloud computing has become one of the most influential changes. As early as 1961, computer pioneer John McCarthy predicted that computing resources can be used as same as the public facilities (like water or electricity) in the future [1]. And the recent developments also proved that computing resources are becoming "ready-to-use" public facilities. Since the concept cloud computing was first put forward, it has obtained widespread attention, especially in academia.

According to the definition of cloud computing from NIST [2] (National Institute of Standards and Technology): Cloud computing is a model for enabling ubiquitous,

X. Xu (✉) · F. Zou · Q. Zhu · X. Ge
The Key Laboratory for Automotive Electronics and Electric Drive of Fujian Province, Fujian University of Technology, Fuzhou 350118, China
e-mail: xuxiang@fjut.edu.cn

A. Abraham et al. (eds.), *Intelligent Data Analysis and Applications*,
Advances in Intelligent Systems and Computing 370,
DOI 10.1007/978-3-319-21206-7_36

convenient, on-demand network access to a shared pool of configurable computing resources (e.g., networks, servers, storage, applications, and services) that can be rapidly provisioned and released with minimal management effort or service provider interaction. However, cloud computing is not a brand-new concept. It is a product of the integrative development of many traditional computing technologies and network technologies. Through continuous improvement of 'cloud' computing power, cloud computing simplify the user terminal equipment and provide with reliable, secure, powerful computing capability. Nowadays increasing enterprise-class cloud platforms were put forward; this paper analyzed several typical enterprise-class cloud platforms and made a series of tests for one of them.

2 Definition of Cloud Computing

Cloud computing is a kind of shared network interactive information service mode, and it is also an emerging business model. The foreground provides cloud service to users via the Internet transparently as a pay-per-use pattern. In the background, many cluster virtual machines form a large virtual resource pool, and encapsulate various software, hardware and data as services. Finally, the cloud computing platforms provide users with an available friendly on-demand network access [3].

In Gartner's (the world's most authoritative IT research and consulting company) report–Top 10 Strategic Technology Trends [4], the cloud computing has been nominated for six consecutive years. In its latest report, Gartner noted that the convergence of cloud and mobile computing will continue to promote the growth of centrally coordinated applications that can be delivered to any device. In fact, cloud computing has brought a lot of convenience to the public daily life. For examples, with its powerful computing capability to analyze massive case of text data, cloud computing are making diagnosis and treatment timelier and accurately [5]. Cloud navigation system recommended to user more accurate and real-time drive routes [6]. Enterprise management information system and ERP system based on cloud computing model can solve the deficiency of traditional enterprise management systems, such as in terms of storage, information integration and analysis [7]. Cloud computing can improve the precision of seismic instruments and make more accurate seismic survey by its large scale computing capabilities.

Service-oriented IT resources is an important external features of cloud computing. Cloud computing encapsulate hardware, software and data resources into services and provide to users. According to the definition of cloud computing from NIST, cloud computing services can be divided into three layers:

1. IaaS (Infrastructure as a Service), such as Amazon EC2 (elastic compute cloud), IBM smart cloud and Sun Grid. etc.
2. PaaS (Platform as a Service), such as Google App Engine and Microsoft Azure. etc.
3. SaaS (Software as a Service), such as Salesforce CRM. etc.

3 Cloud Computing Platforms

Cloud computing providers provide cloud computing services to users anytime and anywhere, the services include infrastructure, application platform and application software [8–10]. Next, the paper will analysis and compare the four kinds of typical cloud computing platforms.

3.1 Amazon Cloud

Amazon Web Services is a set of services to provide cloud computing services platform for independent software developers and vendors [11]. It allows developers access to Amazon's computing infrastructure through the program. Anyone can access the Internet can use the Web services provided by Amazon, such as storage, computing, messaging, and data sets, etc. The platform can expand or compress its computing capability elastically according to the user's requirement.

The most widely used Amazon cloud computing application is EC2. It provides scalable virtual servers by using Xen. EC2 includes the following sections:

1. AMI (Amazon Machine Image). An AMI is a special type of virtual appliance that is used to instantiate (create) a virtual machine within EC2. It serves as the basic unit of deployment for services delivered using EC2 [12]. It can be an operating system, an application server, or an application.
2. S3 (Amazon Simple Storage Service). It is widely used in storage. S3 provide four kinds of access control mechanism to ensure the security of data: IAM (Identity and Access Management policies), ACL (Access Control Lists), bucket and query string authentication.
3. Database. Amazon provides two kinds of database for applications stored on EC2. Amazon RDS or run an AMI database instance as an EC2 database instances.
4. Monitoring, automated scaling and load balancing. Amazon CloudWatch is a service which collect original data from AWS products, then transform the information to a readable, real-time data indexes. The metrics collected by Amazon CloudWatch enables Auto Scaling feature to dynamically add or remove EC2 instances [13]. Elastic Load balancing can distribute new compute tasks to EC2 instance. In this way can realize load balancing very effectively.

3.2 Google Cloud

Google cloud computing platform uses a 'top-down' design method. Starting from the upper application, based on the characteristics of particular business application to reform the infrastructure, is essentially a dedicated platform. Google cloud computing platform infrastructure includes four independent but closely linked modules:

1. Google File System. GFS is a distributed file system [14]. It is designed to provide efficient, reliable access to data using large clusters of commodity hardware. Through the joint design of server and client, GFS reached optimization of performance and availability for application support.
2. Map-Reduce programming model [15, 16]. The model is inspired by the map and reduce functions commonly used in functional programming. Mainly through two steps of Map and Reduce to deal with large data sets in parallel, distributed parallel computing cluster support, reliability and scalability are referred to the platform to deal with, thus ensuring the backstage complex parallel execution and task scheduling transparent to users and programmers.
3. Large-scale distributed database–BigTable [17]. BigTable optimizes the data read operation, and it adopted the distributed data management mode which is based on column-oriented to improve the efficiency of data read;
4. The distributed lock scheme–Chubby [18]. Chubby is a very robust coarse-grain lock. BigTable employ Chubby to save root data form pointer, that is, user can first obtain the root data form location from Chubby lock server, and then access the data.

Google App Engine (GAE) is an important product of google cloud computing. It provides PaaS services mainly, and users can run network applications on the infrastructure of GAE. App Engine application is easy to build and maintain, and it can expands elastically with the growth of traffic and data storage.

3.3 Salesforce Cloud

Salesforce cloud computing platform consists of three parts: the infrastructure and physical resource layer, PaaS middle layer and business application layer.

Multitenancy architecture, metadata drive, Web Services API and AppExchange those four technologies are the key technology of Salesforce cloud computing platform. Table 1 shows the four core technology of Salesforce cloud platform.

Table 1 Four core technologies of the Salesforce cloud computing

Technology	Characteristic
Multitenancy architecture	A multitenant application can serve the needs of multiple organizations, or tenants, by sharing a single physical instance and version of the application
Metadata drive	This model allows applications to be defined as the "Blueprint" without coding The data model, object, form and workflow are all defined by metadata
Web Services API	It provides a direct, powerful, open way to access all of the applications and data on this platform And it is independent of the development language
AppExchange directory	It contains thousands of AppExchange applications

Salesforce CRM is an online account management tools, and provides comprehensive IT support in sales, marketing, services and partner these four business areas. It also provides powerful customization and extensibility mechanism to make the user's business run better on the Salesforce platform.

3.4 VMware Cloud

VMware vCloud is a VMware cloud computing solutions with variety of techniques. It is designed for enterprises which want to access to production-level performance and reliability in the cloud internal and external. VMware provides enterprise-class hybrid cloud model, it can create a consistent policy for the entire system: cloud infrastructure and management, cloud application platform and end-user computing.

vSphere infrastructure facilities is virtualization platform, it can use unified security framework for dynamic cloud computing infrastructure environment; vCloud ensure applications can migration cross-cloud. vFabric can be used to build, operate and expand internal or external deployment of new applications. vFabric platform can provide optimal application performance, quality of service and resource utilization by using of the underlying infrastructure intelligently. VMware vFabric GemFire, as a distributed data management platform, through fusion replication, partition, data-aware type route, and continuous queries, etc. technologies, realize cloud-scale high performance data management, and can provide service with dynamic extensibility, high performance, reliability and durability similar to traditional disk-based database.

3.5 Comparison

Amazon Cloud is mainly based on IaaS cloud computing model, and it provides services for developers. Its main products are S3, EC2, SQS, RDS and so on. The famous photo sharing website SmugMug, makes full use of Amazon cloud computing services. The total size of its photos and videos which are storage on Amazon S3 and process on Amazon EC2 has reached a petabyte level. The New York Times also showed the strong ability of EC2. It processes several terabytes data by using hundreds of EC2 instance in 63 h.

Google's GAE is mainly based on IaaS cloud computing model, and it also provides services for developers. Google Analytics is the enterprise-class web analytics solution. It uses BigTable to store and retrieve data and use the Map-Reduce to analyze statistics. Thus Google Analytics can be applied to mass users and mass data environment.

Salesforce CRM uses a multi-tenancy technology to embody the SaaS idea. Many companies use CRM not only to simplify the customer relationship

Table 2 Comparison of the four cloud platforms

	Amazon	Google	Salesforce	VMware
main layer	IaaS	PaaS	SaaS	PaaS and IaaS
Type of service	Computing and storage	Web application	Application	Computing
Virtualization	Xen	Application-level virtualization	–	VMware
Open source or not	Semi-open source	Non-open source	Non-open source	Non-open source
Programming framework	AMI	Python, Java	–	–
User	Developers	Developers	End-user	Developers
products	EC2,S3	Google app engine	CRM	VFabric, Foundry
Examples of application	SmugMug	Google analytics	Business management	Suzhou industrial park

management, improve customer satisfaction, but also to create a tremendous cost-saving.

VMware cloud computing provides IaaS services by VMware vSphere and provides PaaS services by vFabric Cloud. It can provide services for developers, too. Due to the rapid expansion of information systems, Suzhou Industrial Park is having a series problem of upgrade and maintenance. By the restructuring, Suzhou industrial park takes VMware vSphere/vCenter platform as the core infrastructure and VMware View as auxiliary, forms an architecture model gradually which is suitable for application system of park government private cloud, achieves an environment friendly and efficient administrative private cloud environment (Table 2).

4 Experiments

This paper uses the cloud computing in an intelligent transportation system as an example to test the cloud computing platform. The test focuses on computing performance of the cloud transportation platform in real environment. The author chose the real data of floating car positioning system in Fuzhou as a massive amount of data. The system collected real-time data from 160000 (96000–98000 online at average) floating car in Fuzhou. The data include GPS data, vehicle state data, etc. The sampling interval is 30 s and release cycle is 5 min. Therefore, these floating car data has obvious characteristics of massive data. So the computation of whole system is very large and must ensure the result of the operation is effective and reliable.

This paper turn the hardware infrastructure to a virtualized resources pool through VMware vSphere 5.0 and build a real usable, complete cloud computing platform environment through VMware vCloud Director.

The author will compare computing performance of this cloud computing platform and traditional traffic platform.

Experiment 1: Deploy 6 fixed virtual machines in the cloud platform resource pool, run 3 servers on each GemFire data management node, and distribute 12G memory for them. The experimental data increase two million every step.

The curve of GemFire memory database computing performance with the increasing amount of data is shown in Fig. 1.

Through the performance curve, we can get the following conclusions:

The amount of data increased from 10 to 24 million, the traversal time of cloud platform only increased from 2.869 to 6.7 s. And the calculation performance curve was growth smoothly. However, the computing time of traditional transport system increased from 8.056 to 1400 s. Particularly, the time appeared a sharp increase after the amounts of data reached to 18 million even grow with 20-fold speed. Because the platform adopts the cloud computing model, cloud transportation system can avoid the performance bottleneck of disk I/O caused when the traditional database suffered large-scale data access. Computing performance has been improved significantly.

Experiment 2: Run four servers on each GemFire data management node and distribute 12G memory for them. Deploy 10 virtual machines in the cloud platform resource pool, enable the load balancing feature. The experimenter will observe the number of virtual machines actually used with the increase or decrease of mount of data.

The curve of computing performance and the number of virtual machines actually used with the change amount of data is shown in Fig. 2.

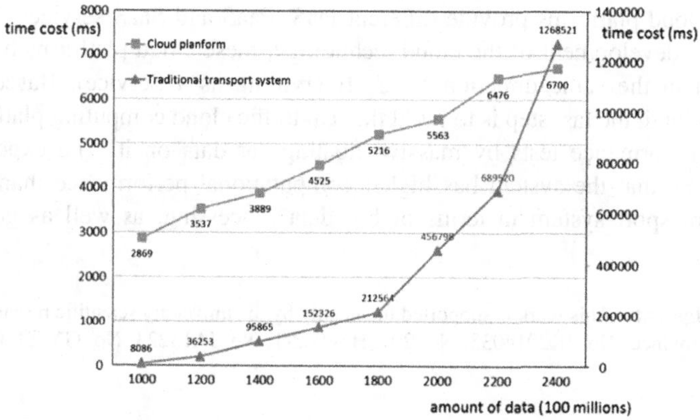

Fig. 1 Computing performance curve I

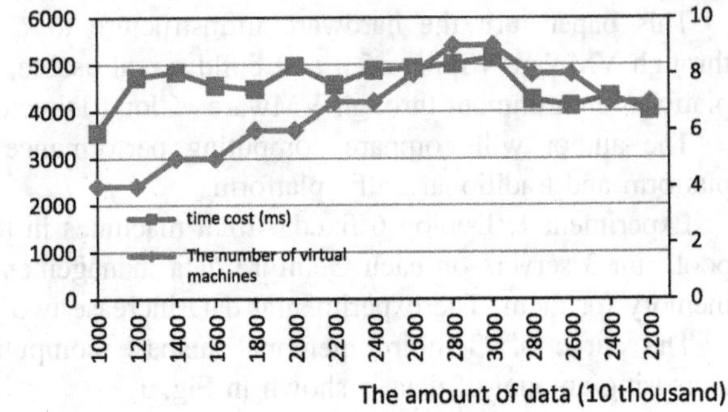

Fig. 2 Computing performance curve II

Through the curves, we can get the following conclusions:

The total data amount will be increased at 2 million, and it will begin to decline when reached to 30 million. Meanwhile it's found that the quantity of virtual machines actually used will be increased coincide with the increasing of data amount, while reducing the data amount, the quantity of virtual machines will be decreased in time consequently. This kind of elastic change ensure the calculating time of system was stable between 4000 and 5000 ms, realize a proper load balance and protect the computing performance of entire system.

5 Conclusion

The cloud makes the computing resources into ready-to-use public facilities and it simplifies the user terminal configuration. Cloud computing platform allocate all resources according to users' requirements. In this way, computing resource utilization efficiency is improved obviously. The architecture and solutions of four typical cloud computing platform are analyzed and compared in this paper, and different cloud platforms provide different IaaS, PaaS and SaaS services. With the continuous development of the cloud technology, these cloud platforms have been developed in the same direction EaaS (Everything as a Service). Based on the VMware cloud, the last step is to build the real traffic cloud computing platform and take the performance tests by massive floating car data on it. The experimental results show that the system has higher computational performance than the traditional transport system in terms of big data processing, as well as good load balancing.

Acknowledgments This work is supported financially by the university scientific research special of Fujian Province (No. JK2014033; No. 2013HZ0002-1; No. JA13223; No. GY-Z13006).

References

1. Foster I, Zhao Y, Raicu I, Lu S (2008) Cloud computing and grid computing 360-degree compared. In: Grid computing environments workshop, GCE'08, pp. 1–10, IEEE
2. Mell P, Grance T (2009) The NIST definition of cloud computing. Natl Inst Stand Technol 53 (6):50
3. Armbrust M, Fox A, Griffith R, Joseph AD, Katz R, Konwinski A, Zaharia M (2010) A view of cloud computing. Commun ACM 53(4):50–58
4. The Top 10 Strategic Technology Trends for 2015. http://www.gartner.com/technology/research/top-10-technology-trends/
5. Wei H, Yang Y, Chen H, Xu B, Li J, Jiang M, Lu A (2014) Predicting health care risk with big data drawn from clinical physiological parameters. In: Social media processing. Springer, Heidelberg, pp 88–98
6. Zhao C, Jin D, Chen G, Hao W, Song L, Zhao M (2013) Intelligent traffic service based on cloud computing. In: 4th IEEE international conference on software engineering and service science (ICSESS), 2013, pp 313–316, IEEE (2013)
7. Chen CS, Liang WY, Su HY (2015) A cloud computing platform for ERP applications. Appl Soft Comput 27:127–136
8. Qian L, Luo Z, Du Y, Guo L (2009) Cloud computing: an overview. In: Cloud computing. Springer, Berlin, pp 626–631
9. Ekanayake J, Fox G (2010) High performance parallel computing with clouds and cloud technologies. In: Cloud computing. Springer, Berlin, pp 20–38
10. Gillen A, Broussard FW, Perry R, Dowling S (2007) Optimizing infrastructure: the relationship between it labor costs and best practices for managing the windows desktop. America: Microsoft
11. Ostermann S, Iosup A, Yigitbasi N, Prodan R, Fahringer T, Epema D (2010) A performance analysis of EC2 cloud computing services for scientific computing. In: Cloud computing. Springer Berlin, pp 115–131
12. Amazon EC2. http://aws.amazon.com/ec2/
13. Cloudwatch for memory usage. https://forums.aws.amazon.com/message.jspa
14. Google (2012) Google cloud platform
15. Afrati FN, llman JD (2011) Optimizing multiway joins in a map-reduce environment. IEEE Trans Knowl Data Eng 23(9):1282–1298
16. Dean J, Ghemawat S (2008) MapReduce: simplified data processing on large clusters. Commun ACM 51(1):107–113
17. Chang F, Dean J, Ghemawat S, Hsieh WC, Wallach DA, Burrows M, Gruber RE (2008) Bigtable: a distributed storage system for structured data. ACM Trans Comput Syst (TOCS) 26(2):4
18. Burrows M (2006) The Chubby lock service for loosely-coupled distributed systems. In: Proceedings of the 7th symposium on operating systems design and implementation, pp 335–350. USENIX Association

Part V
Applications of the Big Data and the Connected Vehicles

System Architecture and Core Technologies for Transportation Internet of Things

Quan Zhu, Fumin Zou and Yanling Deng

Abstract Considering the recent application and research status for Internet of things (IoT) in transportation industry, this paper proposed the concept, system architecture and technical systems of the Transportation IoT. In this paper, several key technical problems in the development of current transportation IoT were discussed, including sensing and recognition technology, network and communications technology, cloud computing and big data processing technology and so on, which aims at exploring the future development and application of the transportation IoT further.

Keywords Internet of things · Traffic informationization · Cloud computing · Architecture · Data services

1 Introduction

After the introduction of the Computer and the Internet, IoT is known as the third wave of the world in the information industry [1], meanwhile, whose technology has become the commanding heights of the future competition in field of information industry and the key driving forces of industrial upgrading. According to the forecasts of the independent market research firm Forrester in U.S, In 2020, the ratio of IoT business with the the existing Internet business will reach to 30: l, the IoT will become the next emerging industry at level of one trillion dollars in the

Q. Zhu (✉) · F. Zou · Y. Deng
The Key Laboratory for Automotive Electronics and Electric Drive of Fujian Province, Fujian University of Technology, Fuzhou 350118, China
e-mail: loosky@fjut.edu.cn

F. Zou
e-mail: fmzou@fjut.edu.cn

Y. Deng
e-mail: dengyanling@fjut.edu.cn

A. Abraham et al. (eds.), *Intelligent Data Analysis and Applications*, Advances in Intelligent Systems and Computing 370,
DOI 10.1007/978-3-319-21206-7_37

world [2]. In recent years, with the development of supporting technologies such as RFID, sensor network, M2M and two fusion etc., the condition of developing IoT become mature gradually. The IoT has been researched and applied in various fields, and the Transportation IoT(TIoT) is exactly the concrete application that in the transportation industry.

Currently, both at home and abroad, the research and applications of IoT is still in its infancy, the starting point of research in which for different experts and scholars in the different field is far from the same. It still has some chaotic concepts of the IoT technology in positioning and characteristics, and the system architecture and technical systems are lack of the clear definition [3], which exactly makes the applications of IoT in the field of transportation faced the huge challenges [4]. As for the realistic foundation and the existing problem of key Transportation IoT technology faced, this paper make an analysis of real application demand of transportation industry, discussing the system framework and core technologies of the Transportation IoT.

2 The Concept and the Application Status of Transportation IoT

Transportation IoT can be seemed as a comprehensive architecture based on IoT, through the comprehensive utilization of all kinds of information technologies, such as cloud computing, big data processing etc., merges and processes the element information of the transportation system, to realize the data collection, transmission, processing and service of the traffic information.

The rapid development of technologies such as network communication, cloud computing and big data processing etc., provides a good opportunity to develop traffic informationization, promoting the development of applications and services for Transportation IoT at the same time. The intelligent transportation solutions, which based on Transportation IoT technology, have been achieved the preliminary application presently, such as the strategic research plan about intelligent transportation systems (ITS): 2010–2014 issued by the ministry of transportation of the U.S, Japanese VICS system and a mature ETC system.

Looking from the meetings held by the field of intelligent transportation in recent years, the application of Transportation IoT have been obtained great attention, including the 21st ITS World Congress (Detroit, 2014.9), the Second World Metropolitan Transport Development Forum(Beijing, China, 2014.7) and the fifth ITS Conference China (2014.6) in Shenzhen, China, etc. In all the meetings, Transportation IoT technology has been become the core of the debate in the aspects of reducing traffic congestion, transportation costs and other related, contains large traffic data processing, fine control, humanized service, and many other aspects.

The gradual improvement of standard architecture is the key to accelerate the research and applications of Transportation IoT, which has been attracted the great importance to several relevant departments at present including the national development and reform commission, ministry of transport and standards committee and so on. National standards committee has approved working group applied for basic standards of national IoT and six major industries IoT applied standard working group. And the standard working group in the field of Internet traffic was established in January 2012 at first, which contains 40 members, covering the local government, the competent department of industry and related businesses. It focus on the construction of application standard system of IoT in the transportation industry, covered the whole transportation, including roads, waterways, civil aviation, railways, that provides technical support for IoT application standards in transportation industry. To promote the construction of Transportation IoT standard system effectively, researching and designing a more clear architecture of Transportation IoT has become a pressing need.

3 System Requirements and the Function Model of Transportation IoT

To further study the system architecture of Transportation IoT, the first thing urge to do is understanding and analyzing the practical application requirements of Transportation IoT. From the twelfth five-year development planning for highway and waterway transportation informatization and other relevant planning files, our country put forward several concrete construction goals for transport infrastructure (transportation equipment) monitoring, traffic information resources integration, public service capability, industry management and decision support etc., proposing the construction demand of Transportation IoT at the level of national [5].

Among them, the core demand is making standard for traffic resource sharing and information exchange by carrying out the construction of Transportation IoT applications. The first thing should be done is making full use the existed information service resources of IoT, to establish the model of independent acquisition and service, integrating all of traffic elements such as people, road, vehicles, goods and the environment. And then, providing data resources for information service or enterprises, it will derive more and more applications of Transportation IoT, not only promoting the infrastructure of transportation informatization further, but also enhancing the level of traffic information service significantly, giving rise to the sustainable development of traffic information construction.

The ultimate goal of Transportation IoT, therefore, is the data services, and the core is the process of data acquisition, transmission, processing and service surrounding traffic information, whose system function model can be defined as shown in Fig. 1.

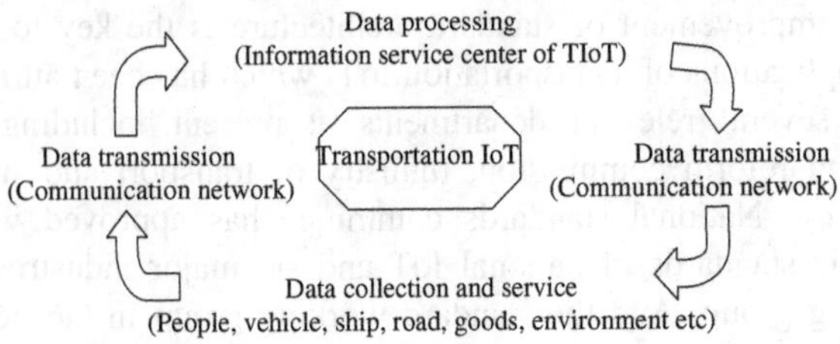

Fig. 1 System function model of transportation IoT

Above all, Transportation IoT, with the data as the core, it is a closed-loop system, by using of various information technologies comprehensively and integrating data acquisition, transmission, processing and service of various traffic elements effectively, which formed the positive feedback of data flow on the whole. Based on that positive feedback characteristics, it is predictable that the research and application of Transportation IoT technology is about to grow at a fast steady pace. Among them, many emerging information technologies will provide comprehensive technical support to the application promotion of Transportation IoT, such as overall perception technology, ubiquitous network and cloud computing etc.

4 System Framework and the Core Technologies of Transportation IoT

Based on the function model of Transportation IoT, the system framework of Transportation IoT is same as to three layer architecture of IoT recognized by the industry [6], including the sensing layer, network layer and application layer. Referring to the architecture of Transportation IoT, the core technology involved in the transportation is summarized. The overall framework is shown in Fig. 2 as follows.

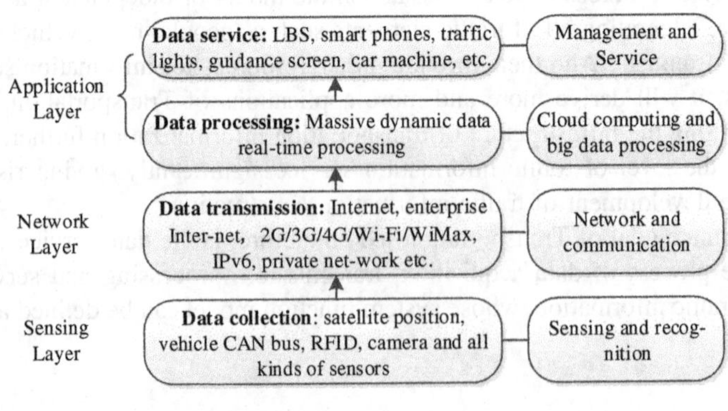

Fig. 2 System framework of transportation IoT

4.1 The Sensing Layer

In this architecture, sensing layer is mainly composed of various sensors and sensor gateway, including some perception terminal such as temperature sensor, humidity sensors, RFID tags and legibility, camera, GPS etc., the main function of which is to perceive and transport information comprehensively.

Sensing and recognition technology is the foundation of Transportation IoT, main responsible for collecting all kinds of traffic information data, including sensors, RFID, GPS, bar code.

4.1.1 Sensing Technology

By all types of sensor collaboration, sensing technology perceives and collects the basic information of traffic elements in the transportation system. At present, the hot-spot of research in this field include information collection, data transmission mechanism, the network communication protocol, network security mechanisms [11, 12]. The context network polymerization based on IP and the network transmission based on multi-homing will become the development trend in the future [13].

4.1.2 Recognition Technology

The recognition technology, basing on RFID and Qr code identification, covers many aspects, such as object recognition, location and geographical identification, that is the technical foundation to realize Transportation IoT perception comprehensively. Take RFID as an example, a blend of IT technologies including information technology, computer technology and network technology, radio frequency (RF) technology etc., the function of the RFID is to realize automatic identification of each element of the traffic by means of transmission characteristics of radio frequency signal or space coupling (inductance or electromagnetic coupling). Presently, the identification technology in the transportation industry has a wide range of applications, such as transport logistics field, the typical applications include road/water transportation administration management, port and maritime management, expressway networking toll and parking charges, multi-path recognition, vehicle management, container management, shipping management, cargo management, yard management and so on. With the development of label technology and the expanding scale of application, the security of traffic industry is becoming the focus of everyone's attention [14, 15].

4.2 The Network Layer

The network layer, made up of all sorts of private network, Internet, mobile communication network, network management system, etc., its main function is ensuring the ubiquity of two-way transmission between collecting data and service data.

Network, the infrastructure platform supported by information transfer and services of Transportation IoT, it covers a lot content in multiple levels, such as the Internet, 3G/4G mobile communication networks, wi-fi/WiMAX, etc. Communication network in our country has strong foundation. The development progress of the next generation key technology IPv6 is synchronized with the world. With independent intellectual property rights, the third generation mobile communication standard, TD-SCDMA, has been widely accepted and approved by the country and the world. On January 20, 2012, the TD-LTE our country put forward became one of the 4G standards officially, that has network deployment on a large scale in the domestic currently.

The Multi-Radio gateway, developed in the reference [16], support 802.11n, IPv6 and operation of three embedded WiFi module at the same time. As shown by experimental results, the Multi-Radio gateway can meet the demand of traffic network transmission of IoT completely. However, represented by sensor network, peripheral network faces with the access problems to backbone network after large-scale application, and need to be fully coordinated with the backbone network, that will be faced with new challenges including the need to study the cable, wireless and mobile Ad hoc network technology, autonomous computing and networking technology, etc. [13].

4.3 The Application Layer

The application layer, interface between IoT and the user including people, organizations, and other systems, it combines with industry demand, achieving some specific applications for a variety of service data, and ultimately, realizing the intelligent traffic.

4.3.1 Cloud Computing Platform

As a supporting platform for the application layer, cloud computing platform, mainly is responsible for the information collection and integration, and providing flexible processing capacity according to the different needs of the business. Based that, the cloud computing platform carry out the analysis and process for all sorts of real-time/offline data, to generate the service data of each business system.

Look from the function and nature of the Transportation IoT, the key of it lies in organic integration of massive traffic information data, contains data acquisition, transmission, processing and service process. Controlling the implementation of intelligent traffic elements, the application services can be offered, such as telemetric service provider (TSP), vehicle on-board application services, realizing the development goal “intelligent traffic”. Among them, what the important and difficult is the analysis and processing of the huge amounts of data, and as for that, the elastic capacity of the cloud computing exactly provides the necessary technical support to it [7, 8].

4.3.2 Cloud Computing and Big Data Processing Technology

Cloud computing, as an emerging computing model and business model, the essence of it is not a new concept, which is derived from cluster computing, utility computing, grid computing and service computing technology, the main key technologies include data storage technology, data management, programming model and task scheduling model, security and privacy protection etc. Seen from the system architecture of Transportation IoT, cloud computing technology provides important support platform for Transportation IoT, and the processing capacity for smart transportation service is one the most dependent on.

Considering the requirement of transportation industry such as large amount of information, high requirement of the information real-time processing, high availability and high stability of data sharing, it is need to set up a unified data processing platform by cloud computing technology to realize sharing and collaboration of the data information, meanwhile, realize dynamic and real-time processing of the massive multi-source traffic information data by use of data mining technology [7, 19]. In addition, by virtualization technology, the cloud computing platform integrate many hardware resources such as servers, storage, network and others, optimize the proportion of system resource allocation, that is to provide flexible and extensible processing platform, realizing the flexibility of application deployment, at same time, improving resource utilization and reducing the total energy consumption as well as operational costs [20].

Among them, one of major challenge of large-scale development of Transportation IoT application is the analysis and processing technology for huge amounts of traffic information data. Currently, to solve the problem of flexibility requirements of mass data processing, what's widely used that distributed processing development framework, similar to Map Reduce, its research interests surrounding Map Reduce include the expansion of application domain, performance improvement and increasing ability of the real-time data stream processing, etc. [21–23].

As for the efficiency problem of the large data processing, the traditional disk I/O becomes the bottleneck of processing performance optimization. Through the operation of data in memory, memory database avoid the I/O operations, to gain the processing performance better than the traditional disk database. In order to improve the processing performance of application system of Transportation IoT further, the distributed data processing technology is able to execute data computing tasks concurrently by mean of scattering the massive data on the multiple server nodes of cloud platform. Based on this, the distributed memory database is very suitable for real-time processing of massive traffic data [24], that is expected to solve the bottleneck problem of Transportation IoT in calculating massive information data [25].

4.3.3 Management and Service Technology

(1) System Management
With the enlargement of transportation network scale for the IoT, the diversification of business bearing, improvement of service quality requirements and increasing factors influencing the normal operation of network, management and support technology is the key to ensure the Transportation IoT be operational, manageable and controlled, which include the operational security system, the standard safeguard system, network management and security, etc.

(2) Application Service
With the rapid development of wireless communication technology and intelligent mobile terminal, location-based services (Location-based Service, LBS) has been widely applied in the field of transportation, logistics, which can provide personalized services according to the position information of moving objects [27].
However, not limited in the transportation sector, the LBS cover almost all aspects of life. There is no doubt that LBS is the huge gold mine in the era of mobile Internet and all the manufacturers go all out. As one of the core technology providing on-demand service, the LBS, supported by the GIS platform, it is capable of offering a value-added service more intelligent to users. LBS include two meanings: the first is to determine the geographic location of mobile device or user, the second is to provide all kinds of information service related to the location. However, under the background of cloud computing, LBS is facing many challenges, including current/future location index of mass moving objects, privacy protection etc. [28, 29].

5 Conclusion

The Internet of things is the mainstream in the development of information technology. The research and development of Transportation IoT is both opportunity and challenge. Based on Internet of things and its development status in the field of traffic, the paper summarized the concept, system framework and core supporting technologies of Transportation IoT by deep research and analysis of the actual application requirements. Considering the problem urged to solve for the development of the Internet of things, this paper gives a comprehensive analysis and summary from the three layers of Transportation IoT system, including sensing and identification technology, network and communication technology, cloud computing and big data processing technology, management and service, etc. The popularity of the Transportation IoT still have a long process, from the opposite side, its development space is enormous. In the process of popularization, focusing on core technology needed to solve, the Transportation IoT can fuse and process information of traffic factors more reasonable and comprehensive, providing better and better services for the transportation industry.

Acknowledgments This work is supported financially by the university scientific research special of Fujian Province (No. JK2014033; No. 2013HZ0002-1; No. 2014H0008; No. 2014-G-83).

References

1. Chen HM, Cui L, Xie KB (2013) A comparative study on architectures and implementation methodologies of internet of things. J Chinese J Comput 36(1):168–188
2. Xu LD, He W, Li S (2014) Internet of things in industries: a survey. IEEE Trans Ind Inf 10 (4):2233–2243
3. Liu C, Li Y, Ji S (2012) Application of the internet of things technology in explosive dangerous goods transport. In: 2nd International conference on civil engineering, architecture and building materials, CEABM 2012. Trans Tech Publications, Yantai, China, pp 1725–1728
4. Wang Y, Cao K (2014) A proactive complex event processing method for large-scale transportation internet of things. J Int Distrib Sens Netw 2014:1–8
5. Ministry of Transport Planning Division of People's Republic of China. In: Highway and waterway transportation information, "Twelfth Five-Year Plan". EB/OL.2014-5-18. http://www.moc.gov.cn/zhuzhan/xiazaizhongxin/ziliaoxiazai/guihuatongji/201105/P020110516379012274071.doc
6. Qian Z, Wang Y (2012) IoT technology and application. J Acta Electronica Sinica 5:026
7. Tsai C, Lai C, Chiang M et al (2014) Data mining for internet of things: a survey. IEEE Commun Surv Tutor 16(1):77–97
8. Velte T, Velte A, Elsenpeter R (2009) Cloud computing, a practical approach. M. McGraw-Hill, Inc., New York
9. Liu VY, Yu Z (2013) Wireless Sensor Networks for Internet of things: a systematic review and classification. Inf Technol J 12(16):3581–3585
10. Hu Y, Sun Y, Yin B (2012) Information sensing and interaction technology in internet of things. Chin Comput 35(6):1147–1163
11. Sharma A, Chaki R, Bhattacharya U (2011) Applications of wireless sensor network in intelligent traffic system: a review. In: 2011 3rd International conference on electronics computer technology (ICECT). IEEE, Kanyakumari, pp 53–57
12. Soumyasri SM, Ballal R (2013) A review: preserving privacy in wireless sensor networks. In: National conference on challenges in research and technology in the coming decades (CRT 2013). IET, Ujire, pp 1–7
13. Qian ZH, Wang YJ (2013) Internet of things-oriented wireless sensor networks review. J Electron Inf Technol 35(1):215–227
14. Kai Z, Lina G (2013) A survey on the internet of things security. In: 2013 9th International conference on computational intelligence and security (CIS). IEEE, Leshan, pp 663–667
15. Chukwunonyerem J, Aibinu AM, Onwuka EN (2010) Review on security of wireless body area sensor network. In: 2014 11th International conference on electronics, computer and computation (ICECCO). IEEE, Abuja, pp 1–10
16. Zhu Q, Jiang XH, Zou FM (2013) Research on the Trunk Line's transmission performance of multi-hop WMN Based on 802.11n .M. In: Business, economics, financial sciences, and management. Springer, Berlin, pp 631–636
17. Behl A, Behl K (2012) An analysis of cloud computing security issues. In: 2012 World congress on information and communication technologies (WICT). IEEE, Trivandrum, pp 109–114
18. Bouayad A, Blilat A, El Houda Mejhed N et al (2012) Cloud computing: security challenges. In: 2012 Colloquium in information science and technology (CIST). IEEE, Fez, pp 26–31
19. Xiaomin Z, Wei S, Liming L (2014) An Implementation approach to store GIS spatial data on NoSQL database. In: 2014 22nd international conference on geoinformatics (GeoInformatics). IEEE, Kaohsiung, pp 1–5

20. Deng W, Liu FM, Jin H et al (2013) Leveraging renewable energy in cloud computing datacenters: state of the art and future research. Chin J Comput 36(3):582–598
21. Qin XP, Wang HJ, Li FR et al (2013) New landscape of data management technologies. Ruanjian Xuebao/J Softw 24(2):175–197
22. Shen DR, Yu G, Wang XT et al (2013) Survey on NoSQL for management of big data. J Softw 24(8):1786–1803
23. Jing H, Haihong E, Guan L et al (2011) Survey on NoSQL database. In: 2011 6th international conference on pervasive computing and applications (ICPCA). IEEE, Port Elizabeth, pp 363–366
24. Iwazume M, Iwase T, Tanaka K et al (2014) Big Data in memory: benchimarking in memory database using the distributed key-value store for machine to machine communication. In: 2014 15th IEEE/ACIS International conference on software engineering, artificial intelligence, networking and parallel/distributed computing (SNPD). IEEE, Sanya, China, pp 1–7
25. Xu X, Zou F, Liao L et al (2013) Experimental analysis for calculation performance of mass data based on GemFire. J. Comput Appl 33(1)
26. Da Silva CA, Ferreira AS, De Geus PL (2012) A methodology for management of cloud computing using security criteria. In: IEEE Latin America Conference on cloud computing and communications (LATINCLOUD), 2012. IEEE, Porto Alegre, pp 49–54
27. Zhou AY (2011) Location-based services: architecture and progress. Chin J Comput 34 (7):1155–1171
28. Abbas F, Hussain R, Jun S et al (2013) Privacy preserving cloud-based computing platform (PPCCP) for using location based services. In: 2013 IEEE/ACM 6th International conference on utility and cloud computing (UCC). IEEE, Dresden, pp 60–66
29. Hao Z, Wen YG, Yu N et al (2013) Privacy-preserving computation for location-based information survey via mobile cloud computing. In: 2013 IEEE/CIC International conference on communications in China (ICCC). IEEE, Xi'an, pp 100–105

A Power Consumption Balancing Algorithm Based on Evolved Bat Algorithm for Wireless Sensor Network

Jeng-Shyang Pan, Lingping Kong, Pei-Wei Tsai, Snasel Vaclav and Jiun-Huei Ho

Abstract It is known that sensor nodes equipped with limited battery power in wireless sensor network, the sensor network will paralyze if sensor node run out of its energy. Therefore energy awareness is an essential consideration for wireless sensor network. Recent advances in wireless sensor networks have led to many new protocols specifically for saving sensor node's energy. A new scheme for predetermining the optimized routing path is proposed based on the evolved bat algorithm (EBA) in this paper. This is the first leading precedent that the EBA is employed to provide the routing scheme for the WSN. A simulation is given and the results obtained by the EBA are compared with the AODV, the LD method based on ACO. The simulation results indicate that our proposed method shows good performance on average power consumption.

Keywords Wireless sensor network · Evolved bat algorithm · Power consumption

J.-S. Pan · P.-W. Tsai (✉)
College of Information Science and Engineering,
Fujian University of Technology, Fuzhou 350118, Fujian Province, China
e-mail: peri.tsai@gmail.com

J.-S. Pan
e-mail: jengshyangpan@gmail.com

J.-S. Pan · L. Kong
Innovative Information Industry Research Center, Shenzhen Graduate School,
Harbin Institute of Technology, Shenzhen 518055, China
e-mail: konglingping2007@163.com

S. Vaclav
Department of Computer Science, VSB Technical University of Ostrava,
70833 Ostrava, Czech Republic
e-mail: vaclav.snasel@vsb.cz

J.-H. Ho
Department of Computer Science and Information Engineering,
Cheng Shiu University, 83347 Kaohsiung, Taiwan
e-mail: jhho@csu.edu.tw

A. Abraham et al. (eds.), *Intelligent Data Analysis and Applications*,
Advances in Intelligent Systems and Computing 370,
DOI 10.1007/978-3-319-21206-7_38

1 Introduction

Wireless sensor network (WSN) is an active research area in computer science and telecommunications because it is extensively used in a variety of domains, such as the environmental observation, the air pollution monitoring, the natural disaster prevention, the healthcare, and etc. The WSN is composed of a few to several hundreds or even thousands sensor nodes. The nodes cooperatively pass the collected data through the network to the sink node, which is regarded as data collection and processing control center. Although the nodes deployed in the adverse environment are very cheap, the energy supply of the node is limited. In many proposed methods, for their wireless sensor networks, nodes are stochastic deployed by the airplane, the location and density are unpredictable. But these researches have one common difficult problem, the nearer to the sink node, the more power consumption is produced. Since all the nodes need to pass the data to sink node with the help of delay node. Apparently the chance is much higher for the inlayer nodes (close to sink node) than outer layer nodes (far away from sink node). In this situation, the inlayer nodes are easily dead before the outer layer nodes. Actually when the inlayer nodes are dead because of overload, the wireless sensor networks are paralyzed. There are no delay nodes for transmitting data to sink. Therefore some modifications for the wireless sensor network to prolong the lifetime of inlayer nodes, prolong the lifetime of networks are needed. On the other hand, not only the network connectivity plays the critical role in the wireless sensor networks, but the coverage of wireless sensor network is a vital factor for evaluating the pros and cons of the network. Image that an adverse environment area deployed a lot of sensors to detect natural disasters, you would not tolerate that there are some locations without any monitoring in the wireless sensor network, the good wireless sensor network need to meet the users' requirement that anywhere of the environment region can be detected, and the delay time from the place where the event is sensed to the data collection and processing control center should be within an acceptable range. That is to say, it's important for the nodes to find a convenient and efficient routing. This paper proposed a new routing algorithm based on evolved bat algorithm (EBA) [1, 2] aim at saving nodes energy, and balancing coverage problem and connectivity problem by modifying network nodes deployment.

The rest of the paper is composed as follows: the related works is reviewed in Sect. 2; our proposed method is explained in detail in Sect. 3; a simulation result is given and is compared with the AODV, the LD methods with ACO in Sect. 4; and finally, the conclusion is made in Sect. 5.

2 Related Works

2.1 Minimum-Number-Node Theory

Wireless sensor network is a promising technology that can improve the way we communicate and the daily life. A typical WSN is composed of a sink and a lot of sensors which are powered by battery. It's the easily and popular way that scatting a lot of redundant sensors are just to keep high connectivity status with full sensing coverage. However it is very wasteful to deploy too many redundant sensors. Thus the issue on the minimum requirement of the sensor node number has been discussed in many existing literature [3, 4], in 2006, Liu et.al present the definition of coverage intensity for a specific point and network coverage intensity. At the same time, sensors are independently and uniformly deployed in the environment. Hence the expectation of coverage intensity for a specific point is equal to the network coverage intensity. In 2015, Kong [5] used the minimum sensor node requirement number theory to present a kind of node deployment model, which satisfied network high connectivity and coverage intensity. This paper uses the same network model as [5].

2.2 Evolved Bat Algorithm

In the field of swarm intelligence, many optimization algorithm are proposed in recent years. Bat inspired algorithm (BA) is a metaheuristic optimization algorithm developed by XS Yang in 2010 [1]. The bat algorithm is based on the echolocation behavior of microbats with varying pulse rates of emission and loudness. Inspired by the original BA, an Evolved Bat Algorithm (EBA) with better efficiency and higher accuracy on finding the near best solution is proposed by Tai et al. [2] in 2012. The process of EBA can be depicted in 4 steps:

Step 1. Initialization: Randomly spread the bats into the solution space.

Step 2. Move the bats by Eqs. (1–2). Generate a random number. If it is greater than the pulse emission rate, move the bat by the random walk process, which is defined by Eq. (3). In EBA, the medium for spreading the sonar wave defined to be the air. It results in the distance between a virtual bat to a target coordinate can be calculated by Eq. (1).

$$D = 170 \times \Delta T(m/s) = 0.17 \times \Delta T(km/s) \tag{1}$$

where D denotes the distance and ΔT means the time difference between sendingsound wave and receiving the echo. The value of ΔT is assigned to be a random value in the range of [–1, 1]. The negative part of ΔT comes from the moving direction in the coordinate. ΔT is given with a negative value when the transmission direction of the sound wave is opposite to the axis of the coordinate. The movement of the bat in EBA is defined by Eq. (2).

$$x_i^t = x_i^{t-1} + D \quad (2)$$

where x_i denotes the location of the *i*th bat in the solution space and t indicates current iteration. Moreover, if a virtual bat moves into the random walk process, its location will be changed by Eq. (3).

$$x_i^{t_R} = \beta * (x_{best} - x_i^t) \quad (3)$$

where $x_i^{t_R}$ indicates the new location of the bat after the random walk process, β is a random number in the range of [0, 1], x_{best} denotes the coordinate of the near best solution found so far, and x_i^t is the present location of the virtual bat.

Step 3. Evaluate the fitness of the bats and update the global near best solution.

Step 4. Check the termination condition to decide whether go back to step 2 or terminate the program and output the near best result.

3 Applying the EBA into Wireless Sensor Network

In the wireless sensor network, we need to build the routing path for every sensor node. The routing path is different in the number of relay nodes for each concentric circle node. We use the same sensor deployment strategies Kong [5]. 2688 nodes deployed in two-dimensional field with radius 350 circles. The circle is divided into seven concentric circles and each of concentric circles contains the same number of sensor nodes, 384 nodes. The outer layer is defined as sectors on the 7th concentric circles and the inlayer is defined as sectors on the 1th concentric circles. The transmission range is set 50. So the node located in the fifth concentric circles needs four relay nodes to transmit package to sink node. Here we utilize node *tar* from concentric circle five as an example to apply in EBA algorithm.

Fitness Function. The fitness function (also called the object function or the evaluation function) plays the principal role in the whole process. We use the Eq. (4) as fitness function to evaluating bat.

Path amount x_1. The sum of any connected path. Connect the source node to any node in one bat to the sink node where each of nodes comes from the different concentric circles. It is not allowed to transmit package between the same concentric circles.

Total Power Consumption x_2. The total power consumption of all connected paths exists in one bat.

Relay amount x_3. If node *a* could transmits package to node b where node b is the next relay step for node a. Relay number plus one.

$Path_{ratio} = x_1 / 3^l$ the ratio of path number to idealize path number in one cat. *l* is the position circles minus one. The ratio of And idealize path number is 3^l.

$$Average\,power{:}\,Ave_{power} = x_2 / x_1$$

$Realy\,value{:}\,R_{value} = x_3 / 3^{l-1}$ the x_3 is the relay amount, idealize relay amount is 3^{l-1}.

$$F_x = \alpha_1 \times x_1 \;/\; 3^l + \alpha_2 \times (1 - x_2 \;/\; x_1) + \alpha_3 \times \left(1 - x_3 \;/\; 3^{l-1}\right) \tag{4}$$

Here $\alpha_1, \alpha_2, \alpha_3$ is numbers in the range of [0–1] where the sum of the four numbers is one. F_x is the fitness function used to evaluating the performance of a bat. Using fitness function to evaluate all the bats, then store the near best bat x_{best}.

Step 1. Initialization: Randomly initialize 16 bats, each bat is represented as:

$$bat_i = \{(x_{11}, x_{12}, x_{13}), (y_{21}, y_{22}, y_{23}), \ldots, (n_{l1}, n_{l2}, n_{l3})\} \tag{5}$$

where *i* is the identifier of bats and *l* is the identifier of the concentric circles. Each routing path composes of four sensor nodes for node tar, this four nodes come from differently concentric circles, and the order of path is the exactly the relay node order for node tar to transmit package to sink node. All path nodes are constrained in the same quadrant.

Step 2. Move the bats by Eqs. (1–2). Generate a random number. If it is greater than the pulse emission rate, move the bat by the random walk process, which is defined by Eq. (3).

Step 3. Evaluate the fitness of the bats using Eq. (4) and update the global near best solution.

Step 4. Check the termination condition to decide whether go back to step 2 or terminate the program and output the near best result.

4 Experimental Result

The experimental result of our proposed routing method base on EBA algorithm is given in this section. As mentioned above, the experiment environment includes 2688 nodes deployed in a two-dimensional field with radius 350 circles. The circle is divided into seven concentric circles and each of concentric circles contains the same number of sensor nodes, i.e., 384 nodes. The experimental result of the total power consumption produced by our proposed method is compared with LDACO [6], and AODV [7]. The results are given in Table 1.

Table 1 The results of power consumption

Nodes methods	All	One	Two	Six	Seven
Our method	4.8855×10^5	1.5620×10^5	8.6828×10^4	3.0869×10^4	2.9806×10^4
AODV	8.9432×10^5	2.3334×10^5	1.9600×10^5	5.8560×10^4	2.8286×10^4
LD_ACO	8.8571×10^5	2.2583×10^5	1.9073×10^5	6.2966×10^4	2.5069×10^4

In our simulation, we only count the routing power consumption. The routing path is from one node that senses an event to the sink. In the first column, "all" stands for the power consumption caused by all sensor nodes deployed in the environment, "one" means the power consumption caused by all sensor nodes located in the most inner layer, "two" is the power consumption caused by the sensor nodes located in the 2nd layer, and "seven" is the power consumption caused by the sensor nodes located in the most outer layer. The experimental result indicates that our proposed method with EBA reduces the total power consumption to almost half than other two methods. In addition, EBA shares parts of the payload from the inner layers to the outer layers. In other words, the power consumption is balanced between different layers.

5 Conclusions

In this paper, we present a power consumption balancing method by EBA to reduce the total power consumption in the whole WSN. The experimental result is compared with the AODV and the LDACO method. The experimental result indicates that our method significantly reduces the power consumption to almost only half than other methods. Moreover, the payload of the sensor nodes are shared by finding the balanced transmitting path via EBA.

Acknowledgments This work is supported by Department of Computer Science, VSB Technical University of Ostrava, College of Information Science and Engineering in Fujian University of Technology and Innovative Information Industry Research Center in Shenzhen Graduate School.

References

1. Yang XS (2010) a new metaheuristic bat-inspired algorithm. In: Gonzalez JR et al (eds) Nature inspired cooperative strategies for optimization (NISCO 2010). Studies in computational intelligence. Springer Berlin, vol 284, pp 65–74
2. Tsai PW, Pan JS, Liao BY, Tsai MJ, Istanda V (2012) Bat algorithm inspired algorithm for solving numerical optimization problems. Appl Mech Mater 148–149:134–137

3. Lin JW, Chen YT (2008) Improving the coverage of randomized scheduling in wireless sensor networks. IEEE Trans Wireless Commun 7(12):4807–4812
4. Liu C, Wu K, Xiao Y, Sun B (2006) Random coverage with guaranteed connectivity: joint scheduling for wireless sensor networks. IEEE Trans Parallel Distrib Syst 17(6):562–575
5. Kong LP, Pan JS, Tsai PW, Vaclav S, Ho JH (2015) A balanced power consumption algorithm based on enhanced parallel cat swarm optimization for wireless sensor network. Int J Distrib Sensor Netw 729680:10
6. Perkins CE, Royer EM (1990) Ad-hoc on-demand distance vector routing. In: Proceedings of the 2nd IEEE workshop on mobile computing systems and applications (WMCSA '99), pp 90–100
7. Ho JH, Shih HC, Liao BY, Chu SC (2012) A ladder diffusion algorithm using ant colony optimization for wireless sensor networks. Inf Sci 192:204–212

Research and Simulation on the MB Parallel Interleaver Cognitive Ultra Wideband System

Bing Zhao, Zhifang Wang and Jiaqi Zhen

Abstract In order to improve the transmission rate and spectrum utilization of cognitive ultra wideband system, this paper combining with the band-limited and orthogonal characteristics of finite prolate spheroidal wave functions, proposes a parallel multiband orthogonal transmission of cognitive ultra wideband system based on the multiband pulse system, uses a pair of orthogonal pulses to replace the single subband pulse, the information is modulated to orthogonal pulses, and introduces interleaver in the modulation front, further improvs the system reliability of information transmission. On this basis, the system simulation model is built, to analyze the system BER performance on the Gauss white noise background and the interference of pulse correlation. The simulations reveal that orthogonal pulses pair can greatly improve the information transmission rate, meanwhile does not affect system BER in low SNR, the value is similar to the Gaussian white noise channel theory value.

Keywords Cognitive ultra wideband · PSWF · Multiband · Parallel transmission · Interleaver

1 Introduction

Recently, with the low power consumption, high transmission rate, high multipath resolution and other advantages, ultra wideband (UWB) technology has been used in the short distance wireless communication widely [1]. Because the traditional single band UWB signal occupies a wide spectrum and lacks information of RF environment in the communication process, lead to the flexibility of spectrum utilization is weak, which restricts the further improves of transmission performance. Therefore, cognitive radio (CR) technology is introduced into

B. Zhao (✉) · Z. Wang · J. Zhen
Electronic Engineering, Heilongjiang University, Harbin, Heilongjiang, China
e-mail: zb0624@163.com

A. Abraham et al. (eds.), *Intelligent Data Analysis and Applications*,
Advances in Intelligent Systems and Computing 370,
DOI 10.1007/978-3-319-21206-7_39

the UWB system, cognitive ultra wideband (CUWB) wireless communication system is designed [2, 3]. According to the information of CR spectrum sensing, multiband parallel CUWB divides the available spectrum into multiple subbands, constructs the adaptive spectrum emission mask, and makes the data parallel transmission, which can increase the system capacity dynamically and improve the spectrum utilization flexibility.

Academic circles and industrial circles pay more attention to CUWB system. In 2004, Stkphane Paquelet [4] proposed a idea of multiband pulse system and proved that this method can improve the system transmission rate effectively. Martin Mittelbach [5] corrected the system model and proved that the transmission rate can be increased to Gbit/s. Aamish Hasan [6] and H.-U. Dehner [7] optimized the model from multiple users and interference suppression standpoint. But the above methods use band-pass filtering way to divide band, which makes complex of the filter design and realization. Literature [8] proposed a multiband adaptive pulse design, but still used serial transmission, did not improve the transmission rate. Literature [9] proposed multiband parallel transmission design, but did not fully consider the phase orthogonal properties between subband pulses.

In view of the above problems, this paper combines with the orthogonal character of PSWF pulse, generats a pair of orthogonal pulses in each subband, one pulse transmits the original information, another pulse transmits the data after interleaving. Receiver uses the dual mask, demodulates two orthogonal signals in each sub at same time, one path demodulation data is deinterleaved, then is accumulated with another path data, orthogonal pulses replace the repetition code of traditional system, improves the information transmission rate and obtains good signal diversity characteristics. Besides, the interleaver can improve the system capability of anti burst interference, make a plurality of successive bits damaged in the channel are dispersed in time, so they are regarded as random error, reduce the system bit error rate.

2 Multiband Parallel Interleaver CUWB System Model

Multiband parallel interleaver CUWB system is proposed on the basis of IR-UWB, according to the spectrum sensing results, the spectrum of FCC regulations is divided into multiple subbands flexibly, the original information and the interleaved information are modulated into orthogonal pulse on different subbands and parallel transmitted. It not only can improve the system transmission rate, use spectrum flexibly, obtain diversity characteristics, and can avoid the frequency band of narrowband interference, so as to improve the receiver performance. This paper uses orthogonal pulses pair to replace single pulse in each subband, and deletes the low efficiency repetitive code module. The principle diagram is shown in Fig. 1.

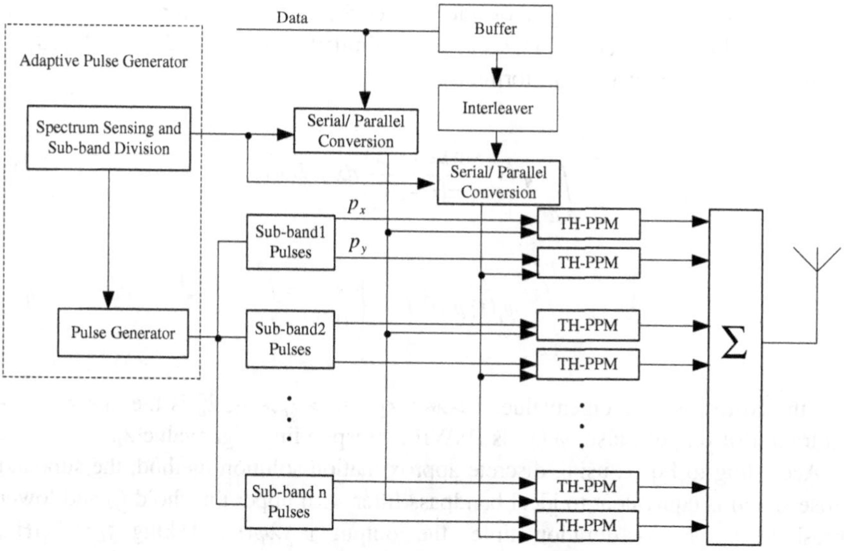

Fig. 1 Principle Diagram of MB parallel Interleaver CUWB system

According to results of sensing module, the subband division module has the flexibility to allocate each subband. The role of pulse generator is to produce a series of subband pulse, each subband has a pair of orthogonal pulse p_x and p_y. Information data goes through the serial/parallel conversion module into multiple, each way uses the pulse p_x of different subband to parallel transmit respectively. At the same time, the information data after interleaving operation uses pulse p_y, two orthogonal UWB signals are superimposed in each subband. In transmitter, all subband pulses are weighted sum in the time domain and composed a line to send out through a single antenna. In receiver, signal through a set of band-pass filter is decomposed into subbands, then is coherent demodulated using a pair of orthogonal masks mask_x and mask_y in each subband respectively, finally, accumulate the two groups of demodulated data corresponding different masks in other subbands, get two path complete data, one path data corresponding mask_y is deinterleaved and merged with another path.

3 Subband Pulse Generator

The late 1950s, bell labs D. Slepian and H.O. Pollack first proposed the prolate spheroidal wave function (PSWF), that is calculate the expression in time domain UWB pulse according to FCC emission mask. Multiband CUWB system requires a pulse signal both band-limited and time-limited to satisfy the high-speed transmission and low inter-symbol interference.PSWF is the complete orthogonal basis in band limited space $[-\Omega, \Omega]$ and time limited space $[-T/2, \quad T/2]$ [10]. It satisfies the following integral formulas:

$$\int_{-T/2}^{T/2} \varphi(x) \frac{\sin \Omega(t-x)}{\pi(t-x)} dx = \lambda\varphi(t) \tag{1}$$

$$\int_{-T/2}^{T/2} \varphi_i(t)\varphi_j(t)dt = \begin{cases} \lambda & i=j \\ 0 & i \neq j \end{cases} \tag{2}$$

λ is the corresponding eigenvalue, $1 > \lambda_0 > \lambda_1 > \cdots > \lambda_i > \cdots$, λ_n is the energy concentration of output pulse, $\varphi_i(x)$ is PSWF corresponding eigenvalue λ_i.

According to Eq. 1, using discrete approximation solution method, the subband pulse signal is equivalent to ideal bandpass filter with upper threshold f_H and lower threshold f_L, T_m is duration time, the output is $\lambda\varphi(t)$. Taking $f_L = 4$ GHz, $f_H = 5$ GHz, pulse duration $T_m = 2ns$, the pulses power spectrum density and frequency domain phase of $\varphi_0(x)$ and $\varphi_1(x)$ are shown in Fig. 2.

In the frequency domain, the odd order and even order PSWF have similar power spectrum density, but the phase difference is $\pi/2$, they are a pair of orthogonal pulses. Select $\varphi_0(x)$ and $\varphi_1(x)$ to be ith subband pulses $p_{x,k}(t)$ and $p_{y,\ k}(t)$.

Figure 3 is time domain waveform of ith subband, although using different transmission code, two signals within the same subband still overlap in the time domain, but because two pulses phase are orthogonal, so the receiver can correctly receive data using orthogonal mask.

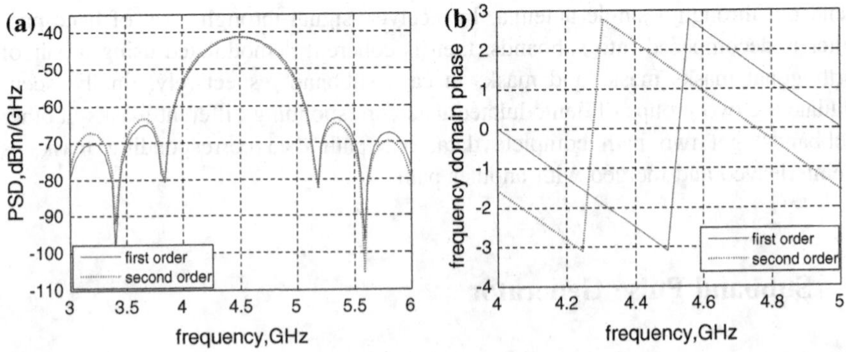

Fig. 2 PSD of Adaptive Pulse **a** power spectrum density, **b** frequency domain phase

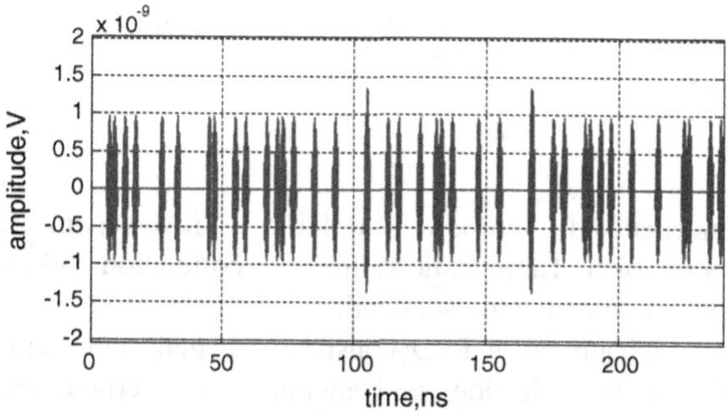

Fig. 3 Time Domain Waveform

4 Interleaver Operation

The interleaving operation can disrupt the relationship between the information bits, the groups of burst errors in transmission channel are converted random errors, to improve the system reliability. This paper uses interweave encoder, the output information of channel coding module is divided into n code blocks, each block is composed of m data, form a matrix of $m \times n$, named interleaver matrix. Data interleaving operation is completed base on date input matrix in written-row and output in read-column, data can also be read from matrix according to other sequence, the ultimate aim is to disrupt the order of input data. If j is the location of input data in the original sequence, $f(j)$ indicates the location of interleaver matrix output sequence, then the interleaver output can be represented by the following formula:

$$f(j) = [(j-1) \bmod n]gm + \lfloor (j-1)/n \rfloor + 1 \tag{3}$$

$\lfloor g \rfloor$ represents a rounding function, mod represents remainder function.

The advantages of interleaving are increasing the ability of anti burst error, but system does not increase new supervision code, so the coding efficiency does not be reduced. In theory, the numbers of code group increase, the ability of anti burst error is stronger, but the requirement of decoder buffer is bigger, and the decoding delay is also corresponding increase.

5 Performance Analysis

5.1 Pulse Correlation Property

Theoretically, different orders PSWF within each sub-band are mutually orthogonal, so cross-correlation value of any two pulses is zero, and because the PSWF energy is concentrated in band $[-\Omega, \Omega]$, the pulses in different subbands are

Table 1 Correlation between orthogonal pulses in same subband

Correlation calculation	p_x	p_y
p_x	1	−3.5491e-012
p_y	−3.5491e-012	0.9981

independent, there is no interference. However, in practice, PSWF are not strictly orthogonal, Table 1 shows the auto-correlation and cross-correlation of orthogonal pulses pair in same subband.

Subband pulse PSD consists of a main-lobe and some side-lobes, side-lobes diffuse to both sides and attenuate in a certain interval, thus cause the energy leakage. Meanwhile, for this reason, there is interference between adjacent sub-bands, controlling the time-bandwidth product parameter C can reduce energy leakage, in the condition of certain subband width, with the increase of C, the main-lobe energy increases, but the pulse time domain duration rises.

5.2 Information Transmission Rate

Traditional single band UWB system and most MB UWB system use the packet coder of repetition code (N_s, 1) in sender, in order to introduce redundancy, but it may reduce the information transmission rate. In this paper, packet coder is replaced by a pair of orthogonal pulses. Transmission code module adopts integer values sequence C, the elements are integers, and $0 \leq C_j \leq N_h - 1$, period of C is N_p, transmission code can be used for multiple access code. System uses binary orthogonal TH-PPM modulation, in the condition of certain pulse duration T_m, assuming single band system transmission rate is R_1, ordinary multiband parallel system transmission rate is R_2, transmission rate of multiband parallel system using orthogonal pulses is R_3, then:

$$R_1 = \frac{1}{N_s N_h T_m} \tag{4}$$

$$R_2 = \frac{M}{N_s N_h T_m} \tag{5}$$

$$R_3 = \frac{M}{N_h T_m} \tag{6}$$

M is the number of sub bands.

Formulas (4)–(6) shows that in the same conditions, R_3 is N_s times of R_2, $M \times N_s$ times of R_2, information transmission rate is improved greatly. FCC agreed to allocate 7500 MHz of spectrum for unlicensed use of UWB devices for communication applications in the 3.1–10.6 GHz frequency band, the instantaneous bandwidth minimum allowed by the FCC ruling is 500 MHz, so maximum value of

M is 15. In practical application, the narrower band width, the longer pulse duration, energy leakage is more serious, this will lead to interference between adjacent subband be larger.

5.3 System Bit Error Rate

According to Fig. 1, system emission signal is the sum of a series of orthogonal pulses pairs, it can be expressed as follows:

$$s(t)=\sum_{i=1}^{M}s_i(t)=\sum_{i=1}^{M}\sum_{j=-\infty}^{\infty}\sqrt{E_p^{(i)}}p_{x,i}(t-jT_s-\theta_{x,j}^{(i)})+\sqrt{E_p^{(i)}}p_{y,i}(t-jT_s-\theta_{y,j}^{(i)}) \quad (7)$$

where $s_i(t)$ is the emission signal of ith subband; E_p is pulse carrying energy; $p_i(t)$ is pulse waveform; θ_j is time random displacement.

$$\theta_j=C_jT_c+a_j\varepsilon=\eta_j+a_j\varepsilon \quad (8)$$

where η_j is random TH jitter; $a_j\varepsilon$ is PPM modulation displacement

In Gauss white noise channel, the receiver can receive M subbands signals and noise, the sum $r(t)$ can be expressed as:

$$r(t)=\sum_{i=1}^{M}\sum_{j=-\infty}^{\infty}\left[\sqrt{\alpha^{(i)}E_p^{(i)}}p_{x,i}(t-jT_s-\theta_{x,j}^{(i)}-\tau_x^{(i)})+\sqrt{\alpha^{(i)}E_p^{(i)}}p_{y,i}(t-jT_s-\theta_{y,j}^{(i)}-\tau_y^{(i)})\right]+n(t) \quad (9)$$

Assuming receiver and sender are synchronous,namely $\tau=0$, and $p_i(t)$ is the energy normalized pulse signal. The BER of multiband parallel interleaver CUWB system is:

$$P_{r_b}=\frac{1}{2}erfc\left(\sqrt{\frac{1}{2}\left(\left(\frac{2E_b^{(1)}}{N_0}\right)^{-1}+\left(\frac{1}{2\sum_{i=1}^{M}\sum_{\substack{j=1\\ i\neq j}}^{M}R_{p_ip_j}^2(0)}\right)^{-1}+\left(\frac{1}{\sum_{i=1}^{M}R_{p_{x,i}p_{y,i}}^2(0)}\right)^{-1}\right)^{-1}}\right) \quad (10)$$

Formula 10 shows that system BER is determined by thermal noise, interference between subbands and interference between orthogonal pulses, the second part and third part of BER are influenced by subband number M and pulse cross correlation function $R_{pp}^2(t)$, these parameters smaller, influence between bands less.

Fig. 4 System Bit Error Rate

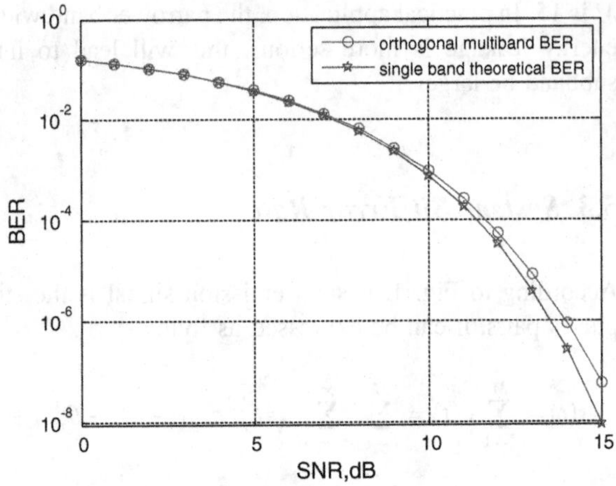

Figure 4 illustrates the BER of multiband parallel interleaver CUWB system, simulation condition is 7 subbands, TH-PPM modulation. In low SNR, thermal noise is the most limiting factor for system, when SNR exceeds 8 dB, the difference in BER curves is primarily due to pulse cross correlation function, system is influenced by multiband and orthogonal pulses.

6 Conclusion

Subband width of multiband modulation can be adjusted according to the results of spectrum sensing module, this can use UWB spectrum more efficiently, improve the flexibility of spectrum utilization. Because the RF signal spectrum is composed of a series of discrete subbands, therefore, the mutual interference between different subbands is very small. Orthogonal pulses in same subband provide diversity characteristic, instead of the repetition code in traditional UWB system, enhance the system transmission rate greatly. Meanwhile, the interleaver can enhance the ability of anti burst interference, improve the system reliability. Due to the various subbands are independent of each other, when narrowband system or narrowband interference occurs within the working frequency band, system can remise subbands according to the actual situation, achieve coexistence with existing wireless systems.

Acknowledgments This work is supported by Heilongjiang Provincial Education Department Science and Technology Research Project (NO.12531492), and National Natural Science Foundation of China (No.61201399). Many thanks to the anonymous reviewers, whose insightful comments made this a better paper.

References

1. Yang L, giannakis GB (2004) Ultra-wideband communications: an idea whose time has come. IEEE Signal Process Mag 21(6):26–54
2. Chen XB, Chen H, Chen Y (2011) Spectrum sensing for cognitive ultra-wideband based on fractal dimensions. In: Fourth international workshop on chaos-fractals theories and applications. IEEE Press, Hangzhou, pp 363–367
3. Mitola J, Maguire GQ (1999) Cognitive radio: making software radios more personal. Pers Commun 6:13–18
4. Batra A, Balakrishnan J, Aiello GR, Foerster JR, Dabak A (2004) Design of a multiband OFDM system for realistic UWB channel environments. IEEE Trans Microw Theory Tech 52 (9):2123–2138
5. Mittelbach M, Moorfeld R, Finger A (2006) Performance of a multiband impulse radio UWB architecture. In: 3rd International conference on mobile technology, applications and systems. IEEE Press, Bangkok, Thailand, p 17
6. Hasan A, Anwar A, Mahmood H (2011) On the performance of multiuser multiband DS-UWB system for IEEE 802.15. 3a channel with hybrid PIC rake receiver. In: 2011 International conference on computer networks and information technology. IEEE Press, Abbottabad, pp 65–69
7. Dehner HU, Moorfeld R, Jakel H, Burgkhardt D, Finger A, Jondral FK (2009) Multi-band impulse radio–an alternative physical layer for high data rate UWB communication. Frequenz. 63(9):200–204
8. Li B, Zhou Z, Zou WX (2009) A novel spectrum adaptive UWB Pulse: application in cognitive radio. In: 2009 IEEE 70th vehicular technology conference fall. IEEE Press, Anchorage, pp 1–5
9. Chen LL, Yu X, Dou Z (2014) Research on the performance of multiband impulse radio UWB communication system based on PSWF. J Harbin Eng Univ 35(4):499–503
10. Slepian D (1978) Prolate spheroidal wave functions, fourier analysis, and uncertainty. Bell Syst Tech J 40:43–64

Part VI
Machine Learning Algorithms for Big Data

A Novel Uninorm-Based Evolving Fuzzy Neural Networks

Rong Hu, Ye Sha and Hai Yan Yang

Abstract Due to the constantly increasing rate in size and temporal availability of data, learning from data stream is a contemporary and demanding issue. This work presents a structure and introduces a learning approach to train an uninorm-based evolving fuzzy neural networks (UEFNN). The fuzzy rules can be stitched up or expelled by of statistical contributions of the fuzzy rules. The learning and modeling performances of the proposed UEFNN are validated using a benchmark problem. Simulation result and comparisons with state-of-art evolving neuro-fuzzy methods and demonstrate that our new method can compete and in some cases even outperform these approach in terms of RMSE and complexity.

Keywords Uninorm-based · Evolving fuzzy neural networks · Fuzzy rules

1 Introduction

Fuzzy logic system proposed by Zadeh [1] is intelligible using fuzzy linguistic rule and can realize approximate reasoning to deal with imprecision and uncertainty in a decision-making process. Unfortunately some traditional approaches in designing the fuzzy system are over dependent on expert knowledge [2] and necessitate tedious annual interventions usually. Noticeably the designers have to spend a laborious time to examine all input-output relationships of a complex system to

R. Hu (✉) · H.Y. Yang
College of Information Science and Engineering, Fujian University of Technology, 350108 Fuzhou, China
e-mail: hurong@fjut.edu.cn

Y. Sha
College of Ecological Environment and Urban Construction, Fujian University of Technology, 350108 Fuzhou, China
e-mail: YeSha@fjut.edu.cn

A. Abraham et al. (eds.), *Intelligent Data Analysis and Applications*, Advances in Intelligent Systems and Computing 370,
DOI 10.1007/978-3-319-21206-7_40

elicit a representative rule base. So it constrains its practicability in evolving dynamic and time critical environments.

This issue has led to the development of neuro-fuzzy systems (NFSs) [3] which is a powerful hybrid modeling approach that assimilates the learning ability, robustness, and parallelism of neural networks with the advantage of the fuzzy logic systems like linguistic and approximate reasoning.

The generalization of t-norm and s-norms called uninorm was studied in [4]. Uninorms were used to develop fuzzy neurons model and networks in [5] recently. Aiming at flexibility these developments adopt uninorms at some level of the neuron model as a uninorm can have its identity element anywhere in the unit interval.

Recently, coping with nonstationary environments has drawn intensive research works to deal with large volume of high dimensional data in an online manner. Evloving NFS (ENFS) based on the concept of incremental learning [6] accordingly opens a new unchartered territory. Evolving models update their structures and parameters in an online fashion, learning continuously from a stream of data. It can be used to improve model performance as new data is input through gradual model construction, inducing model adaptation and refinement without catastrophic forgetting while keeping current model useful [7]. Some successful examples of evolving systems have been reported in [8–11].

This paper begins considering a neuron model based on uninorms called unineuron. Here, the unineurom uses uninorms at both, local synaptic processing level and global aggregation level. We study a feedforward network structure with two parts named a novel uninorm based evolving neural network (NUENN). The first part is a fuzzy inference system, while the second part acts as an aggregation neural network. The first part has fuzzy membership functions in the input layer neurons and a subsequent layer of unineurons. The second part uses sigmoidal neurons to perform aggregation of the first part outputs. NUENN is likewise capable of starting its rehearsal process from scratch with an empty rule base. The fuzzy rules can be henceforth extracted and removed during the training process based on a novelty of a new incoming training pattern and the contribution of an individual fuzzy rule to the system output. A novel aspect concerns the rule pruning methodology, which employs an extended rule significance (ERS) concept, supplying the blueprint of rule contributions. Two fuzzy sets, which are similar to each other, are grouped and are turn blended to be one single fuzzy set exploiting a kerned-based metric [12] in conjunction with a transparent explanatory module (rule).

The rules are evolved dynamically based on a potential of data point being learned by the use of datum significance (DS) criterion. The initialization of new rules is furthermore consummated in a new way assuring completeness of the rule base as well as fuzzy partitions, thus leading to an adequate coverage of the input space. Whenever no new rule is evolved, the focal points and radii of the fuzzy rules are adjusted by means of extended self-organizing map (ESOM) theory to up date neighboring rules with a higher intensity than rule lying farer away.

Another important facet of NUENN is the use of a recursive version of extreme learning to adjust the weights.

The remainder of this paper is organized as follows: Sect. 2 presents the structure of the model. Section 3 elaborates the incremental learning policy of NUENN, which involves a rule base management. Section 4 explores the empirical studies and discussions on benchmark problem. With this, the new evolving method will be compared against various other state-of-art evolving neuro-fuzzy approaches in terms of predictive quality. Section 5 concludes this paper.

2 Uninorm-Based Neural Network

The structure of the uninetwork adopted in this paper is similar to the one studied in [13]. It has a feedforward topology with two major parts: a fuzzy inference system and an aggregation neural network, assembled into three layers. The fuzzy inference system is composed by the first two layers. The input layers consists of neurons whose activation function are membership functions of fuzzy sets that granulate the input space to form fuzzy portions. We suppose that C_i is the centroid of the *i*th fuzzy rule. Gaussian membership functions centered at C_i with dispersion (radius) σ are adopted. The membership degree of input vector x is computed using (1). For the membership degree of input x_i there are L fuzzy sets $A_i^{li}, l_i = 1, \ldots, L$. L corresponds to the number of clusters or, equivalently, the number of fuzzy rules that composes the model

$$a_{li} = e^{-(x_i - c_{li})^2/2\sigma^2} \tag{1}$$

Here $l = 1, \ldots L$, $i = 1, \ldots, n$ and c_{li} is the *i*th coordinate of the *l*th cluster center. The radius σ is the spread of the set.

The second (hidden) layer contains U_U unineuron to aggregate the outputs of the input neurons weighted by synaptic weights w_{li}:

$$z_l = U_{i=1}^{n}(a_{li} u w_{li}) \tag{2}$$

where z_l, l = 1, … L, is the output of the *l*th unineuron and w_{li} are the synaptic weights. The third layer is a neural network layer with sigmoidal activation functions *f*(.) whose outputs are

$$\hat{y}_j = f\left(\sum_{l=1}^{L} r_{jl} z_l\right) \tag{3}$$

where m is the dimension of the output space, j = 1,… m, and r_{jl} are the output weights connecting the *j*th output with the *l*th U_u neuron. Figure 1 depicts the UEFNN structure.

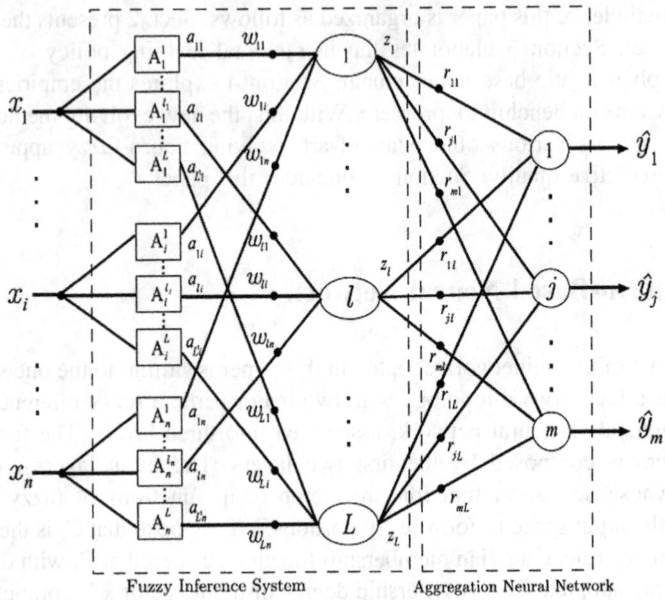

Fig. 1 UEFNN structure

3 The Incremental Learning Policy of NUENN

This section elaborates the incremental learning policy of NUENN.

3.1 Initialization of New Fuzzy Rules Based on ε-Completeness Criterion

A selection of a new antecedent fuzzy set constitutes an indispensable constituent in the automatic proliferation of the fuzzy rules [19, 20]. In NUENN learning platform, the allocation of the new premise parameters is enforced according to the profound concept of the *ε-completeness* criterion deliberated by [14]. In this viewpoint, for any inputs in the operating range, there exists at least a fuzzy rule so that a match degree is no less than ε for every individual input injected. Whenever the new fuzzy rule is appended as a reaction to troublesome training datum supplied, NUENN shall plug in this input datum as focal point or center of the new rule. The new rule is setting the width of the Gaussian function as follows.

Remark: In fuzzy application, ε is usually selected as 0.5 [15].

$$c_{i+1} = x^n \tag{4}$$

$$\sigma_{i+1} = \frac{\max(|c_i - c_{i-1}|, |c_i - c_{i+1}|)}{\sqrt{\ln(\frac{1}{\epsilon})}} \tag{5}$$

An ideal state to augment a fuzzy rule base is when a datum fed evokes a high system error. In NUENN the potential of the injected datum is designated by the statistical of a datum when the number of observations approaches to infinity. This algorithmic backbone is called the DS criterion proposed in [16] as follows:

$$D_n = |en| \frac{\det(\sum_{r+1})^l}{\sum_{i=1}^{r+1} \det(\sum_i)^l} \tag{6}$$

$\sum_i$ is a dispersion or covariance matrix $\sum_i \in R^{l \times l}$ of ith rule, whose elements are the spreads of the Gaussians in each direction (dimension) σ_{ki}. If an observation complies with (6), the new datum suffixes to be hired as an extraneous rule. As $g < D_n$, the DS method appraise this datum high descriptive power and generalization potential. In the other contrary if $g > D_n$, the current fuzzy rules have confirmed its completeness in seizing the available training stimuli.

3.2 Rule Updating with ESOM

Self-organizing map (SOM) theory is originated by Kohonen [17] The crux of the SOM method is appealing to progressively adjust the focal points (centers) of the Gaussian functions thus adapting to the changing patterns in the possible of this approach is that a spatial proximity in determining the winner is obtained via the Euclidean distance. That is, the distance is only calculated based on points without considering the interdependence between the input variables and the size of the fuzzy region. To correct this shortcoming, ESOM method the resorts in overcoming these drawbacks with the compatibility measure of the firing strength and the use of sequential maximum-likelihood estimation proposed by [18] in adapting the zone of influence of the cluster. Hence, the center C_i and the size of fuzzy region σ of ith rule are updated by

$$\begin{aligned} C_i^n &= C_i^{n-1} + \beta^n R_v^n h_v^n (X^n - C_i^{n-1}) \\ \sigma_i^n &= (\frac{N_i}{N_i+1})\sigma_i^{n-1} + (\frac{1}{N_i+1})(X - C_i)^T (X - C_i) \\ h_i^n &= \exp(-(c_v^n - c_i^n)^T \sigma_i^{-1} (c_v^n - c_i^n) \end{aligned} \tag{6}$$

where β^n is the learning rate, R_v^n is the firing strength of wining rule, and h_i^n is the neighborhood function, which estimates the similarities of two neighboring rules.

$$\beta^n = 0.1 * \exp(-n) \tag{7}$$

3.3 Pruning of Inconsequential Rules and Merger of Similar Fuzzy Sets

This section aims to explore the rule pruning and rule merging adornments in NUENN. The rule base simplification technology of the so-called generalized growing and pruning radial basis function [16] is used in this paper. The statistical contribution estimation of the fuzzy rule is doable as follows:

$$E_{\inf}(i) = |\delta_i| \frac{\det(\Sigma_i)^l}{\sum_{i=1}^{r} \det(\Sigma_i)^l} \tag{8}$$

where $\delta_i = \sum_{k=1}^{l+1} \omega_{1i} + + \cdots + \omega_{l+i} x_l$.

If the contribution vector contains the fuzzy rule contributions less than equal k_{err}, then these fuzzy rules are classed as outdated fuzzy rules. They should be dispossessed to mitigate the rule base complexity thereby eradicating the vulnerabilities of over complex network structures gained. Although fuzzy sets are well scattered originally, they are still susceptible to be significantly overlapping to each other. If the membership functions are very similar to each other, they can be fused into one new membership function to obviate fuzzy rule redundancies. Similarity between two Gaussian membership functions A and B can be found by benefiting a kernel-based metric method comparing the centers and widths of two fuzzy sets in one formula as follows:

$$\begin{aligned} c_{new} &= (\max(U) + \min(U))/2 \\ \sigma_{new} &= (\max(U) - \min(U))/2 \end{aligned} \tag{9}$$

where $U = \{c_A \pm \sigma_A, c_B \pm \sigma_B,\}$, the underlying construct is to reduce the approximate merging of two Gaussian kernels to the exact merging of two of their a-cut for a specific value of a. Here, we choose $a = e^{1/2} \approx 0.6$ which the membership degree is of the inflection points $c \pm \sigma$ of a Gaussian kernel. This also guarantees ε-completeness of the fuzzy partition as cuts the outer contours of the two sets at membership value 0.6, thus arrive at a large coverage span than the original sets.

3.4 Output Layer Weights Update

As outlined in Sect. 1, NUENN plugs in ERS methods and ESOM in deriving and updating the fuzzy consequent parameters. Suppose it arrive at U_U neurons and next is uses the recursive least squares (RLS) algorithm with forgetting factor λ to update the output layer weights $\mathbf{r}^t = [r^t_{11} \cdots r^t_{jl} \cdots r^t_{mL^t}]$:

$$\begin{aligned} P^t &= Q^{t-1} z^t \{\lambda + z^t Q^{t-1} (z^t)^T\}^{-1} \\ Q^t &= (I^t_L - P^t z^t) \lambda^{-1} Q^{t-1} \\ r^t &= r^{t-1} + (p^t)^T (f^{-1}(y^t) - z^t (r^{t-1})^T) \end{aligned} \tag{10}$$

where $f^{-1}(y^t) = \log(y^t) - \log(1 - y^t)$. Initialization of Q is commonly $I_{L^t}\omega$, $\omega = 1000$, where I_{L^t} is the identity matrix.

4 Experiment

In this section we summarize the computational experiments considering classic benchmarks in time series of forecasting. Simulations were performed using a 1.8 GHz four-core personal computer and MATLAB. The root mean squared error (RMSE) at step t is computed using

$$RMSE = \sqrt{\frac{1}{t} \sum_{j=1}^{t} (y^j - \hat{y}^j)^2} \tag{11}$$

4.1 Mackey-Glass Time Series

The Mackey-Glass time series is a well known benchmark to evaluate forecasting algorithms. Data is generated using

$$\frac{dx}{dt} = \frac{Ax^{t-\tau}}{1 + (x^{t-\tau})^C} - Bx^t, \quad A, B, C > 0 \tag{12}$$

Semi-periodic or chaotic behavior depends on the parameter values chosen. Several studies suggest: A = 0.2, B = 0.1, C = 10 and τ = 17, with a time step of 0.1 for integration, 3200 were used for training and testing simultaneously. The goal is to forecast the value of x^t 85 steps ahead, that is

Table 1 Forecasting modeling performances for Mackey-glass

eTs	RMSE	Num of rules
UEFNN	0.0443	4
FBeM	0.0985	14
EfuNN	0.0770	124
ANFIS	0.0673	26
DENFIS	0.0730	6

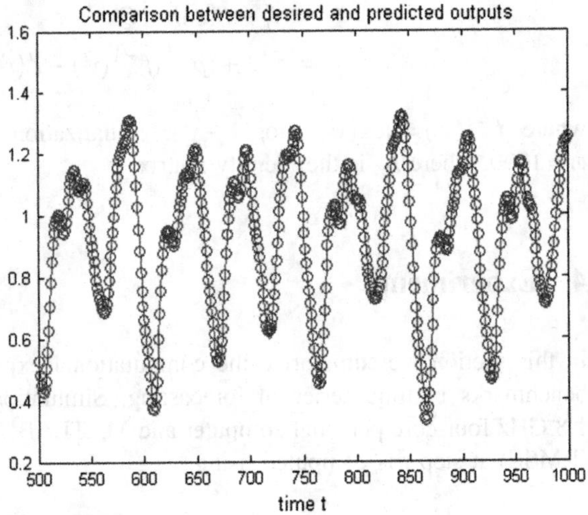

Fig. 2 Mackey-Glass time series

$$x^{t+85} = p(x^t, x^{t-6}, x^{t-12}, x^{t-18}) \quad (13)$$

Results comparing the lowest and average RMSE values with alternative approaches reported in the literature are summarized in Table 1.

For this case the UEFNN RMSE was better than the others. Training took an average of 6.42 s for the entire dataset, an average of 3.32 ms per sample.

Figure 2 depicts the results of UEFNN and UEFNN for the first 500 data samples. Figure 3 shows the evaluation of fuzzy rule.

The methods addressed in this section utilize parameters to tune the final number of rules achieved. Here the goal was to demonstrate that the UEFNN does provide competitive results with considerable lower training time and effort.

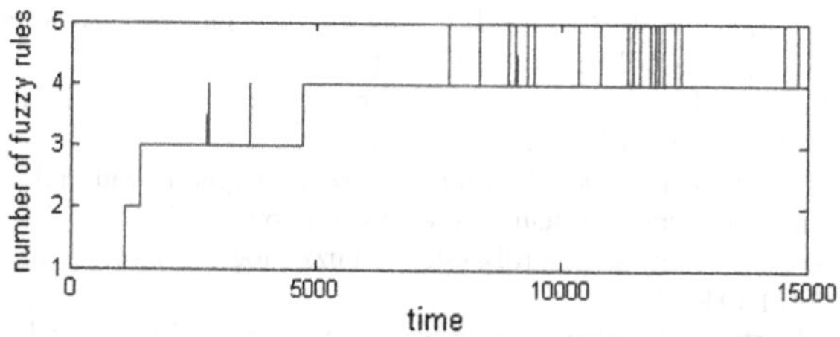

Fig. 3 Fuzzy rule evaluation in one of trials

5 Conclusion

This paper has introduced learning approaches for hybrid fuzzy neural networks based on uninorms. Computational results show that the fuzzy neural network is competitive with alternative methods. The comparisons against numerous state-of-the-art works have been enforced, which astonishingly infers UEFNN as an overwhelming breakthrough to the evolving neuro-fuzzy field marginalizing its counterpart's predictive accuracy and structural complexity facets. As future work, the application of UEFNN in dealing with the classification case is the subject of our further investigation.

Acknowledgments The authors acknowledge the support of Scientific Research Starting Funds of Fujian University of Technology under Grants GY-Z13103, and Scientific research project in fujian province education department under JA13223/GY-Z13082.

References

1. Zadeh LA (1994) Soft computing and fuzzy logic. IEEE Softw 11(6):48–56
2. Whye LT, Chai Q (2010) eFSM-A novel online neural fuzzy semantic memory model. IEEE Trans Neural Netw 21(1):136–157
3. Lin C-T, Lee CSG (1996) Neural fuzzy systems: a neuro-fuzzy synergismto intelligent systems. Prentice-Hall, Upper Saddle River
4. Yager RR, Rybalov A (1996) Uninorm aggregation operators. Fuzzy Sets Syst 80(1):111–120
5. Lemos A, Caminhas W, Gomide F (2010) New uninorm-based neuron model and fuzzy neural networks. In: Annual meeting of the North American fuzzy information processing society, pp 1–6
6. Lughofer E (2011) Evolving fuzzy Systems—methodologies, advanced concepts and applications. Springer, New York
7. Kasabov N (1998) ECOS: evolving connectionist systems and the ECO learning paradigm. In: ICONIP'98, Kitakyushu, Japan, pp 123–128
8. Kasabov N (2001) Evolving fuzzy neural networks for supervised/unsupervised online knowledge-based learning. IEEE Trans Syst Man Cybern Part B: Cybern 31(6):902–918
9. Kasabov N, Song Q (2002) Denfis: dynamic evolving neural-fuzzy in fuzzy system and its application for time-series prediction. IEEE Trans Fuzzy Syst 10(2):144–154
10. Kasabov N (2007) Evolving connectionist systems: the knowledge engineering approach evolving connectionist systems. Springer, London

11. Leite D, Gomide F, Ballini R, Costa P Jr (2011) Fuzzy granular evolving modeling for time series prediction. In: FUZZ-IEEE, pp 2794–2801
12. Lughofer E, Bouchot J-L, Shaker A (2011) On-line elimination of local redundancies in evolving fuzzy systems. Evol Syst 2(3):380–387
13. Hell M, Costa P, Gomide F (2008) Hybrid neuro fuzzy computing with null neurons. In: IEEE international joint conference on neural networks, pp 3653–3659
14. Lee CC (1990) Fuzzy logic in control systems: Fuzzy logic controller. II. IEEE Trans Syst Man Cybern 20(2):404–436
15. Chiu S (1994) Fuzzy model identification based on cluster estimation. J Intell Fuzzy Syst 2 (3):267–278
16. Rong HJ, Sundararajan N, Huang GB, Saratchandran P (2006) Sequential adaptive fuzzy inference system (SAFIS) for nonlinear system identification and time series prediction. Fuzzy Sets Syst 157(9):1260–1275
17. Kohonen T (1982) Self-organized formation of topologically correct feature maps. Biol Cybern 43(1):59–69
18. Vigdor B, Lerner B (2007) The Bayesian ARTMAP. IEEE Trans Neural Netw 18 (6):1628–1644
19. Rong HJ, Han S, Bai JM et al (2014) Improved adaptive control for wing rock via fuzzy neural network with randomly assigned fuzzy membership function parameters. Aerosp Sci Technol 39:614–627
20. Reiner P, Wilamowski BM (2015) Efficient incremental construction of RBF networks using quasi-gradient method[J]. Neurocomputing 150:349–356

Quaternion Fisher Discriminant Analysis for Bimodal Multi-feature Fusion

Meng Chen, Xiao Meng and Zhifang Wang

Abstract Aiming at the accuracy and security of pattern recognition system, this paper proposes a quaternion based multi-modal recognition algorithm that is more accurate and safe than unimodal, and fuses more features than most existing methods. Our algorithm fuses four features that involve two linear features and two non-linear features of two kinds of modalities. We fuse features into quaternion and the process of recognition is dealt in quaternion field. The equal error rate (EER) and DET curves given by the experiment we did on Yale face database and PolyU palm print database show that the quaternion based algorithm we proposed improves the recognition rate observably.

Keywords Quaternion space · PCA · KPCA · Multi-feature fusion · Fisher

1 Introduction

Identity authentication is a common issue that people face in daily life. For example internet bank, e-business, public safety, etc. With the develop of technology, recognition based on human biological features has been accepted by public because of the convenience, uniqueness and security. These features involved face, palm print, iris, fingerprint, handwriting, etc. However, people never satisfied with conditions as they are.

To improve the recognition rate and security, people aimed at authenticate multiple features at the same time. It not only inherited the advantages of single

M. Chen (✉) · X. Meng · Z. Wang
Electronic Engineering, Heilongjiang University, Harbin, Heilongjiang, China
e-mail: 181363682@qq.com

X. Meng
e-mail: 354911741@qq.com

Z. Wang
e-mail: xiaofang_hq@126.com

A. Abraham et al. (eds.), *Intelligent Data Analysis and Applications*,
Advances in Intelligent Systems and Computing 370,
DOI 10.1007/978-3-319-21206-7_41

feature but also has enhancing in many aspects. The recognition rate of multi-features will be higher than single feature. For example, if the recognition rate of face is 90 % and the handwriting is 85 %. It must be higher than 90 % after fused. Considering the security, on the one hand, people must provide all the features to complete the recognition process. So attackers also must steal all of them, it is obviously more difficult. On the other hand, attackers cannot restore the human biological information without fusion method though they have the feature.

It is the starting stage of multi-modal identity verification from 1995 to 2000. The academia realized the advantage of multi-modal identity verification, did experiments to explore and a series of fruition is achieved [1–3]. Bigun and Duc [1, 4] came up with the concept of multi-modal in 1997, they fused voice and face by supervised learning and Bayesian theory to verify identity, and there is a high accuracy with their method. Verlinde [5, 6] build the decision vector parallelly with vocal print and visual feature output by different classifier, classify further with the KNN algorithm and use the method of weighted distance and vector quantization. After 2000, multi-modal technology entered a fast-developing period and had a great development in theory and application. More and more experts and institutions joined in the related research and proposed many new fusion system. For example fingerprint/face [7], palm print/hand vein [8], etc.

However, these methods mentioned above can just fuse two single feature modalities or one modality with two kinds of feature. That's not enough. In this paper, we will propose a new fusion algorithm based on quaternion that can fuse two modalities with four different features. And our algorithm involved the linear feature and the non-linear feature of one modality at the same time. The recognition rate and safety will be further improved with this method.

The content of the thesis can be summarized as follows: Firstly, introduce the method of feature extraction. Then, give the related concept of quaternion and quaternion matrix, and explain the solving method of the orthogonal eigenvectors in quaternion matrix. On this basis, expand the PCA and the Fisher to quaternion field, and fuse these four features with our algorithm. Finally, give the result of experiment and conclude.

2 Propose Algorithm

See Fig. 1.

2.1 Multi-feature Extraction

Our algorithm fuses different biological features at the feature layer, so features must be extract at the first. A successful recognition algorithm highly depends on the extract and choice of biological feature. It aims at extracting obviously

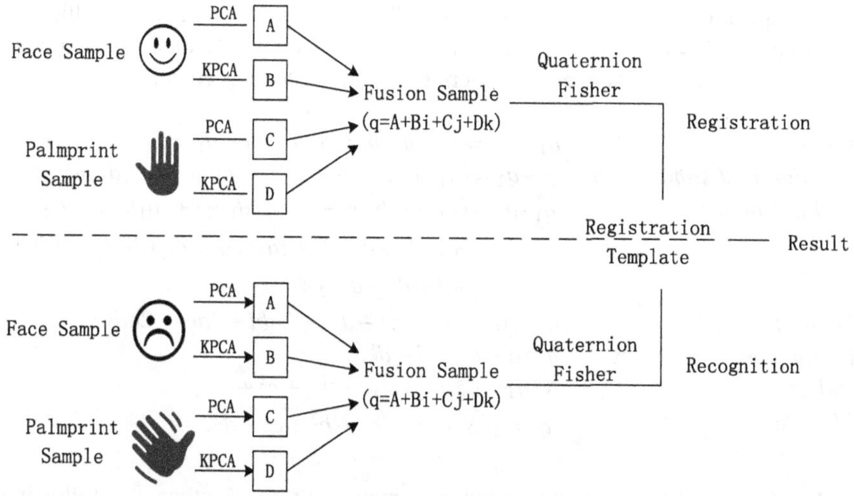

Fig. 1 The process of the proposed algorithm

differentiable features and reducing dimension of pattern space at the same time to improve the running speed.

As you know, the Principal Component Analysis (PCA) [10] and the Kernel Principal Component Analysis (KPCA) [11] both are widely used in the research of identity authentication. PCA reconstitutes the modal sample to find the principal projection vector. It gives linear features. KPCA introduces the kernel function on the base of PCA and projects the input space to high-dimensional feature space by the nonlinearity projection. Then run the process of PCA in the high-dimensional feature space. So it has strong non-linear solution ability and gives non-linear features. The Fisher Discriminant Analysis (Fisher) [9] is also very famous, it is different with PCA as a supervised project method. Fisher builds the discriminant vector to classify the sample and different groups of data are easier to differentiate with this method. In this paper, we respectively extract the face feature with PCA and KPCA and then extract the palm print feature with PCA and KPCA. In this way, we get four different features of two kinds of modalities that can be fused.

2.2 Quaternion Based Multi-modal Fusion

2.2.1 Quaternion and Quaternion Matrix

If $a, b, c, d \in R$ and i, j, k meet $i^2 = j^2 = k^2 = -1$, $ij = -ji = k, jk = -kj = i$, $ki = -ik = j$, then $q = a + bi + cj + dk$ is named quaternion. Quaternion q is consist of the real part and imaginary part, respectively are a and $bi + cj + dk$. We can also rewrite quaternion q as $q=(a+ib)+(c+di)j$ according to imaginary multiplication rule. Let $q_1=a_1+b_1i+c_1j+d_1k,\ q_2=a_2+b_2i+c_2j+d_2k,$ then we can define the basal operation principles of quaternion by the following equation.

Equality: $q_1=q_2 \Leftrightarrow a_1=a_2, b_1=b_2, c_1=c_2, d_1=d_2.$

Addition and subtraction: $q_1 \pm q_2=(a_1 \pm a_2)+(b_1 \pm b_2)i+(c_1 \pm c_2)j+(d_1 \pm d_2)k.$

Multiplication:

$$q_1 \cdot q_2=(a_1a_2-b_1b_2-c_1c_2-d_1d_2)+(a_1b_1+b_1a_2+c_1d_2-d_1c_2)i+(a_1c_2+a_2c_1+b_2d_1-d_2b_1)j+(a_1d_2+d_1a_2+b_1c_2-c_1b_2)k.$$

Square: $q^2=(a^2-b^2-c^2-d^2)+2abi+2acj+2adk.$

Conjugate: $\bar{q}=a-bi-cj-dk.$

Rules: $N(q)=q\bar{q}=\bar{q}q=a^2+b^2+c^2+d^2 \geq 0.$

Modulus: $|q|=\sqrt{N(q)}=\sqrt{a^2+b^2+c^2+d^2}.$

Let Q says the set of quaternion and matrix $A \in Q^{n \times n}$, then the following concepts and propositions are used in the algorithm.

(1) If $A^H=A,$ then name A self-conjugate quaternion matrix. The set of n order self-conjugate quaternion matrix write $SC_n(Q)$;

(2) If exist $\lambda \in Q$ and $0 \neq a \in Q^{n \times 1}$ make $Aa=a\lambda$(or $Aa=\lambda a$), then call λ the right (or left) eigenvalue of A and a is the eigenvector belong to right (or left) eigenvalue of A. If λ is both the right and left eigenvalue of $A,$ then name λ the eigenvalue of A;

(3) If $U \in Q^{n \times n}$ and $UU^H=U^HU=I,$ then name U extended unitary matrix. It means $U^H=U^{-1}$ if U is an extended unitary matrix;

(4) We mentioned that quaternion matrix can expressed as the sum of two plural matrices in the first paragraph. That is $A=A_1+A_2j,$ then name plural matrix $A^\sigma=\begin{pmatrix} A_1 & -A_2 \\ \overline{A_2} & \overline{A_1} \end{pmatrix}$ the induced matrix of quaternion matrix A;

(5) If A is self-conjugate quaternion matrix, then name A^σ Hermite matrix;

(6) Eigenvalues of quaternion matrix A are the same as eigenvalues of its induced matrix A^σ;

(7) If A^σ is the induced matrix of quaternion matrix A, and λ is eigenvalue of A^σ, the corresponding eigenvector is $\begin{pmatrix} \alpha_1 \\ \alpha_2 \end{pmatrix}$, then the eigenvector of the matrix A eigenvalue λ is $\alpha_1+\overline{\alpha_2}j$;

The trigonometric expression of $q=a+bi+cj+dk$ can be written as the following form.

$$q = |q|(\cos\theta + I sin\theta)$$
$$I = \frac{1}{\sqrt{b^2+c^2+d^2}}(bi + cj + dk)$$
$$\theta = arctan\frac{\sqrt{b^2+c^2+d^2}}{a}$$

Then the formula of quaternion n-th root expresses as

$$\sqrt[n]{q} = \sqrt[n]{|q|}\left(\cos\frac{\theta+2k\pi}{n} + I\sin\frac{\theta+2k\pi}{n}\right), k = 0, 1, 2, \ldots, n$$

If Q is a quaternion vector, the Euclidean norm of Q expresses as

$$\|Q\|_2 = (Q, Q)^{\frac{1}{2}} = \left(\sum_{i-1}^{n} q_i^2\right)^{\frac{1}{2}}$$

2.2.2 Orthogonal Eigenvectors

It's easy to verify the associative law and the commutative law for quaternions addition, and the associative law for quaternion multiplication. But the commutative law does not apply to quaternion multiplication. That makes quaternion matrix calculation much more complex than real or plural matrix. It's necessary to figure out the orthogonal eigenvectors of quaternion matrix in order to expand the PCA and Fisher to quaternion. In theory, the proposition 7 in the above section shows a way to figure out the eigenvectors of quaternion matrix. However, in the practice programme process, though the eigenvectors of the induced matrix can be figured out and build the eigenvectors of the quaternion matrix, it's uncertain that these eigenvectors of quaternion matrix are mutually orthogonal. In view of this, we reference a reasonable way proposed by LANG Fang-Nian [12] to solve this problem.

A random $n \times n$ self-conjugate quaternion matrix A (the train sample scatter matrix in this multi-modal fusion recognition application is a self-conjugate quaternion matrix), its induced matrix is A^{σ}. And $A^{\sigma}X = \lambda X$ is the feature equation of A^{σ}, that is $(A^{\sigma} - \lambda I)X = 0$. Let $\lambda_i \in R$ be the eigenvalues and the corresponding eigenvectors of A^{σ} are $X_i, i = 1, 2, \ldots, n$. Build the following expressions with different λ_i, λ_j.

$$I - (A^{\sigma} - \lambda_i I)^{+}(A^{\sigma} - \lambda_i I)$$
$$I - (A^{\sigma} - \lambda_j I)^{+}(A^{\sigma} - \lambda_j I)$$

If $\begin{pmatrix}\alpha_1\\ \alpha_2\end{pmatrix}$ and $\begin{pmatrix}\beta_1\\ \beta_2\end{pmatrix}$ respectively are the first column vector of the above two expressions, and α_1, α_2, β_1, β_2 have the same dimensions. Then λ_i, λ_j are the eigenvalues of quaternion matrix A, and $\alpha_1 + \overline{\alpha_2}j$, $\beta_1 + \overline{\beta_2}j$ are the corresponding eigenvectors.

2.2.3 Expand to Quaternion Field

As you know, quaternion q is consist of four parts which are one real part a and three imaginary parts $bi + cj + dk$. Let these features extracted before build the four parts of quaternions. For example, we name the face feature extracted with PCA is A, and similarly there are B, C and D. Let A be the real part of quaternion and B be the first imaginary part, C be the second and D be the last. In this way, We can get an quaternion matrix consist of these four different features. Training and testing sample set of fused of multi-feature can be built in this method. Then we expand Fisher to quaternion field and deal with the sample set. It's worth mentioning that different features may have different dimensions, so we add zeros to make sure that they are in the same dimension.

We reduce the dimension of the sample set before apply Fisher to it, so it is necessary to expand the general PCA to quaternion field. If $\{X\|x_i, i = 1, 2, \ldots, m\}$ is the multi-feature fusion sample set and $x_i = (\alpha_{i1}\alpha_{i2}\cdots\alpha_{in}), \alpha_{ik} \in Q$. Then the generating matrix should be

$$\Sigma = \frac{1}{M}\sum_{i=1}^{M}(x_i - \mu)(x_i - \mu)^H$$

x_i is the ith sample vector, μ is the average vector of sample set and M is the total of sample. Figure out the orthogonal eigenvector set of the generating matrix, and choose same eigenvectors which correspond these bigger eigenvalues as principal vectors, then the PCA can be applied to quaternion with general definition in the field of real number.

For Fisher in quaternion field, if the number of class is c, the between-class scatter matrix and within-class scatter matrix are respectively expressed as

$$S_b = \sum_{i=1}^{c} P(\omega_i)(m_i - m_0)(m_i - m_0)^H$$

$$= \sum_{i=1}^{c} n_i(m_i - m_0)(m_i - m_0)^H$$

$$S_w = \sum_{i=1}^{c} P(\omega_i)D\left\{\frac{(X - m_i)(X - m_i)^H}{\omega_i}\right\}$$

$$= \sum_{i=1}^{c}\sum_{x \in X_i}(x - m_i)(x - m_i)^H$$

$P(\omega_i)$ is the prior probability, n_i is samples belong to class i and m_i is the mean value of the sample of class i.

$$m_0 = E\{X\} = \sum_{i=1}^{c} P(\omega_i)m_i = \frac{1}{n}\sum_{i=1}^{c} n_i m_i$$

m_0 is the mean value of total samples. And the follow quaternion Fisher is similar to general Fisher in the field of real number. It's worth noting that the generating matrix structured with above method is a self-conjugate quaternion matrix. With these above definitions, we can achieve our algorithm in quaternion field.

3 Experiment Results and Analysis

Aiming at the superiority and feasibility of the algorithm, we performed an experiment. Yale face database is build by Yale university center for computational vision and control, it involves 15 volunteers, 165 pictures with different illumination, expression and posture. PolyU palm print database is build by The Hong Kong Polytechnic University and involves 100 volunteers with 6 pictures each. In the experiment, modal images are provided by these two databases. We ran the algorithm on these two databases to verify multi-feature recognition.

False match rate (FMR) and False non-match rate (FNMR) are two major parameters to evaluate the performance of the recognition algorithm. EER unify FMR and FNMR to one parameter to measure the overall performance of algorithm. In the same coordinate, FNMR and FMR have a intersection that means they have the same value. The value of this print is EER. FMR and FNMR both should be smaller in same threshold value for a better algorithm. The DET curve is similar to EER, FMR is the x-axis and FNMR is the y-axis. Better algorithm has lower DET curve.

Table 1 gives the data of EER of four single biological features recognition and the fused feature with the algorithm we proposed. From the table, we can clearly see that the lowest EER of these single features is Face with PCA and it's 13.3 %. And the highest is Palmprint with KPCA, it even be 19.7 %. However, it has a huge change after fuse. The EER of fused feature with the algorithm we proposed is just 6.6 % and it is only half of the lowest single feature. This is an acceptable improvement. Figure 2 is the DET curves of these algorithms. It also shows the same result that the algorithm we proposed is feasible and accurate.

Table 1 EER of two-modal (four features) fusion

Feature	EER (%)
Face with PCA	13.3
Face with KPCA	15.8
Palmprint with PCA	17.7
Palmprint with KPCA	19.7
Proposed algorithm	6.6

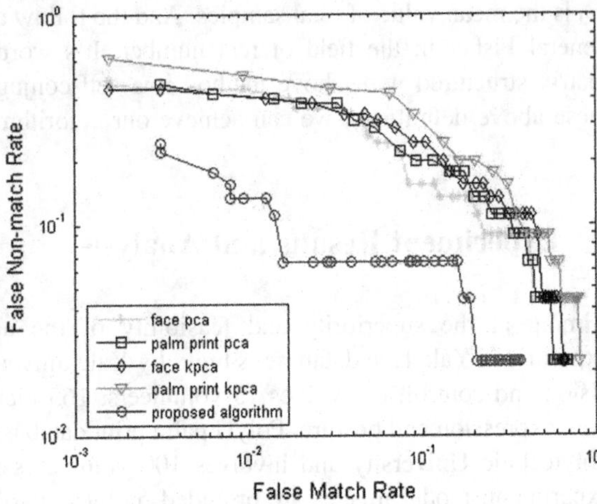

Fig. 2 DET of two-modal (four features) fusion

4 Conclusions

Using the knowledge of quaternion and quaternion matrix, the general PCA and Fisher in real field has been generalized to quaternion field. Based on that, we proposed a new multi-feature fusion algorithm. We fused different features into quaternion, and extracted the feature of the quaternion matrix. Compare with these existing algorithms, we enhance the features from two to four. And the recognition rate have greatly improved than single feature. Our experiments show that the algorithm we proposed is feasible and accurate. However, there are still some deficiencies. For example the modality, the experiment just involved two kinds of modalities though we fused four features in the experiment. We will find more human biological features that can be used in our algorithm.

Acknowledgments This work is supported by National Natural Science Foundation of China (no. 61201399), China Postdoctoral Science Foundation (no. 2012M511003), Project of Science and Technology of Heilongjiang Provincial Education Department (no. 12521418), Youth Foundation of Heilongjiang University (no. 201026), and Startup Fund for Doctor of Heilongjiang University.

References

1. Bigun E, Bigun J, Duc B et al (1997) Expert conciliation for multimodal person authentication systems using bayesian statistics. In: Proceeding of 1st international conference on audio-and video-based biometric person authentication. Springer, Crans-Montana, Switzerland, pp 291–300
2. Hong L, Jain AK, Pankanti S (1999) Can multibiometric improve performance? In: Proceedings of IEEE workshop on automatic identification advanced technologies. IEEE, Morristown, NJ, USA, pp 59–64

3. Hong L, Jain AK (1998) Integrating faces and fingerprints for personal identification. In: Proceedings of 3rd Asian conference on computer vision. IEEE, Hong Kong, China, pp 1295–1307
4. Duc B, Bigiin ES, Bigiin CJ et al (1997) Fusion of audio and video information for multimodal person authentication. Pattern Recognit Lett 18(9):835–843
5. Verlinde P, Maitre G (1997) Decision fusion in a multimodal identity verification system using a multilinear classifier. Techical report, Idiap Research Institute, Martigny, Switzerland
6. Verlinde P, Chollet G (1998) Combining vocal and visual cues in an identity verification system using K-n based classifiers. In: Proceedings of IEEE workshop on multimedia signal processing. IEEE, Redondo Beach, CA, USA, pp 59–64
7. Bouchaffra D, Amira A (2008) Structural hidden Markov models for biometrics-fusion of face and fingerprint. Pattern Recognit 41(3):852–867
8. Ma H (2014) Research of recognition method based on the feature layer fusion of palmprint and hand vein. Tianjin University of Technology
9. Yang J, Yang J, Frangi A (2003) Combined Fisherfaces framework. Image Vis Comput 21 (12):1037–1044
10. Jolliffe IT (1986) Principal component analysis. Springer, New York
11. Scholkopf B, Smola AJ, Muller KR (1998) Nonlinear component analysis as a kernel eigenvalue problem. Nerual Comput 10(5):1299–1319
12. Lang FN et al (2008) Obtain method of quaternion matrix orthogonal eigenvector set and its application in color face recognition. Acta Automatica Sinica 34(2):121–129

Algorithm SESP of Wireless Sensor Network Node

Xuhong Huang

Abstract In order to extend lifespan of the Wireless Sensor Node, the research propose a self-adaptive algorithm (SESP) based on energy assessment of WSN nodes with identical hardware model and same parameters. The algorithm SESP is able to choose the cluster head without sending message to learn the surplus energy of other nodes, thus reducing in-cluster communication and simplifying the calculation of energy assessment, and eventually reducing energy consumption of WSN nodes and extending their lifespan.

Keywords WSN · Energy assessment of wireless sensor node · Self-adaptive algorithm

1 Introduction

The node energy of Wireless Sensor Network (WSN) is finite. And WSN is out of use with the exhaustion of the node energy. Many studies have shown that the wireless transmission consumes the largest proportion of the whole node energy [1, 2], with the consumption of transmitting a bit of datum 800 times more than that of executing a command [1]. Therefore, minimizing the transmission energy consumption is the key to reduce the node energy consumption and prolong the life of WSN. Also important for the extension of the lifespan is to balance node energy consumption and make full use of the energy of each node. To tackle the issue, a self-adaptive algorithm SESP is based on the assessment of the residual energy of nodes.

X. Huang (✉)
School of Information Science and Engineering, Fujian University of Technology, Fuzhou, China
e-mail: Huangxuh1@163.com

A. Abraham et al. (eds.), *Intelligent Data Analysis and Applications*,
Advances in Intelligent Systems and Computing 370,
DOI 10.1007/978-3-319-21206-7_42

2 The Concumption of Node Energy and Energy-Saving Measures

The major energy consumers of WSN node are sensor modulating circuit of data collection module, micro controller and memory of data processing module and the RF circuit of data transmission module.

There is little room left to tap the sensor modulating circuit for lessening the energy consumption as it consumes quite a little. The key then lies in reducing the energy dissipation of micro-controller and memory of data processing module and the RF circuit since the energy consumed in the process of data transmission results mainly from signal processing and offsetting transmission path attenuation.

The power dissipation of microprocessor consists of dynamic power dissipation and static one [4], with the former as the main contributor to energy saving. According to Vasanth and Michael [5] item 5 of the references, the dynamic power of microprocessor is closely related to supply power voltage, physical capacitance and physical capacitance, which is formulated as follow:

$$P_D \propto \alpha C V^2 f \tag{1.1}$$

In the formula, PD stands for dynamic power, V is supply power voltage, C physical capacitance, f physical capacitance and α active factor.

Therefore, reduction of dynamic power dissipation can be realized by lowering the clock frequency and supply power voltage. Vasanth and Michael [5] has demonstrated that lowering both the supply power voltage and clock frequency can result in the reduction of dynamic power dissipation. And changing the working conditions of the microprocessor from 200 MHz and 1.5 V to 150 MHz and 1.2 V could reduce up to 52 % of the dissipation.

The management of dynamic power dissipation could also be realized by adopting Dynamic Voltage Supply (DVS) technology. DVS will dynamically adjust the working voltage and frequency of the microprocessor with the change of the node's working load, thus reduce the unnecessary power output in the comparatively free moments.

The RF circuit, among the components of the node, consumes the largest proportion of the energy. In accordance with the technological requirements of WSN node, RF circuit usually is made from popularly utilitized parts of low energy consumption, low price and with small size.

In a WSN with nodes of N number, apart from reducing the power dissipation of the hardware of each node, energy saving can also be achieved by reducing the workload of nodes through the energy management of in-between nodes communication.

Suppose the set of nodes is U, and $U = \{u_i = (x_i, y_i), i = 1, 2, \ldots, n\}$, with (x_i, y_i) indicating the location of a node. With the completion of the distribution of a WSN of general purpose, its nodes locations are fixed.

The model for adjusting energy is proposed by item 2 of the references. Given that the needed energy for transmitting a unit of data from node i to node j is E_{ijs}, the energy for signal processing E_{proc}, and the energy for offsetting transmission path attenuation E_{was}, then

$$E_{ijs} = E_{proc} + E_{was} \tag{1.2}$$

E_{proc} here is relevant to the energy for encoding, modulation and wave filtering, and E_{was}, to the distance between node i and node j. And if the distance is designated as d_{ij}, then

$$E_{was} \propto dij \tag{1.3}$$

Node j also needs energy to receive information. Suppose the energy for node j to receive a unit of data is E_{jr}. E_{jr} then is relevant to the energy consumed by decoding, demodulation and wave filtering.

As the nodes locations are fixated when completing the distribution of a WSN of general purpose, the value of d_{ij} becomes fixed. And when the node hardware of WSN is determined, E_{proc} and E_{jr} remain almost unchanged too. Accordingly, reducing the volume of transmitted data is the key for saving energy.

The major tasks done by WSN are collecting data from sensors and transmitting them to base station. And because its transmitting power is rather low, a WSN node has to communicate directly with its neighboring nodes. To ensure that those comparatively distant nodes could transmit data to base station, the nodes could be distributed in the form of a multiple hops self-organizing network. In the self-organizing network, clustering networking algorithm is capable of effectively reducing the volume of data transmission [6]. Every cluster elects a node as its own head which then collects data transmitted from other nodes and merges them. However, the cluster head would consume more energy than other nodes. In order to balance the energy consumption and prolong the life of the network, every node in a cluster should be chosen to serve as the head in turn periodically. As such, assessing the node energy consumption and electing a suitable one as the head would be of great help to extend the lifespan the network.

3 Energy Assessment of Neighboring Nodes

Within any of the clusters of WSN, if the neighboring nodes' residual energy could be learnt, the node with much residual energy could be chosen as new cluster head so as to balance energy consumption and maximize the lifespan of each cluster.

Different nodes in a WSN typically are made from the same model of hardware or parts. Therefore, the energy E_{proc} needed by different nodes to transmit a unit of data is approximately equivalent. So is the energy E_{jr} needed for receiving a unit of data. With the completion of the distribution of WSN of general purpose, d_{ij} is fixated, and E_{was}^{ij} (from node i to node j) and E_{was}^{ji} (from node j to node ij) are approximately identical.

Given that node t is the old cluster head, node i is one of the other nodes in the same cluster, E_{tis} represents the needed energy for transmitting a unit of data from node t to node i, E_{proc} the energy needed for signal processing, E_{was}^{si} the energy offsetting the transmission path attenuation, according to (1.2), then

$$E_{tis} = E_{proc} + E_{was}^{si} \tag{2.1}$$

Let E_{its} stand for the energy needed for transmitting a unit of data from node i to node t, and E_{was}^{it} for the energy offsetting the transmission path attenuation, then

$$E_{its} = E_{proc} + E_{was}^{it} \tag{2.2}$$

Because

$$E_{was}^{ti} = E_{was}^{it} \tag{2.3}$$

Substitute formula (2.3) for its counterpart in formula (2.2), and then compare the resulted with (2.1), then

$$E_{tis} = E_{its}$$

That is, the energy demanded for transmitting a unit of data from node i to node t is equivalent to that of from node t to node i.

So is the energy E_{tr} needed for node t to receive a unit of data and E_{ir} for node i to receive a unit of data.

If in a time cycle node i receives m units of data and dispatches n units of data, then the total energy consumption of node i $e_i(k)$ is

$$e_i(k) = mE_{tr} + nE_{tis} \tag{2.4}$$

In other words, so long as the data volume transmitted and received by node i with its neighboring nodes in a time cycle can be obtained, the total energy consumption of node i in the time cycle can be calculated and thus its residual energy can be figured out.

With new cluster head being chosen, the old heads will transmit with minimal transmitting power within them the messages from the new head. And the new head, with its own confirmation of the identity, will broadcast among its neighboring heads and make them know its identity message with rather big transmitting power.

4 SESP Self-adaptive in-Cluster Route Algorithm

SESP assumes that every node has different scales of transmission power. The lowest scale is used for in-cluster transmission, with a transmission radius of R_1. All nodes of the cluster are located within the circular area with a radius of $R_1/2$, which means all cluster nodes can communicate directly with each other. The higher scales of transmission power are used for inter-cluster transmission, with a transmission radius of R_2.

If the network lifespan is divided into several life cycles based on a fixed time step length, and the length of a life cycle is l, then the time step length is $1/l$. And the whole network life cycle L is

$$L = Nl \tag{3.1}$$

Before the end of every cycle, the current cluster head assesses each node's residual energy with the above-mentioned SESP, and chooses the next head for head shift.

Apart from the head, other nodes within the cycle (suppose it is the k-th one) communicate only with the head, thus forming a head-centered star-network structure as indicated in Fig. 1.

The data volume received is n and the data volume transmitted i is m. At the cluster head, the data volume m and n of node i can be calculate. At the end of the k-th cycle, the data volume n received by the head are respectively equivalent to that transmitted by node i and the data volume m transmitted by the head are respectively equivalent to that received by node i. Substituting m and n for their

Fig. 1 Network structure

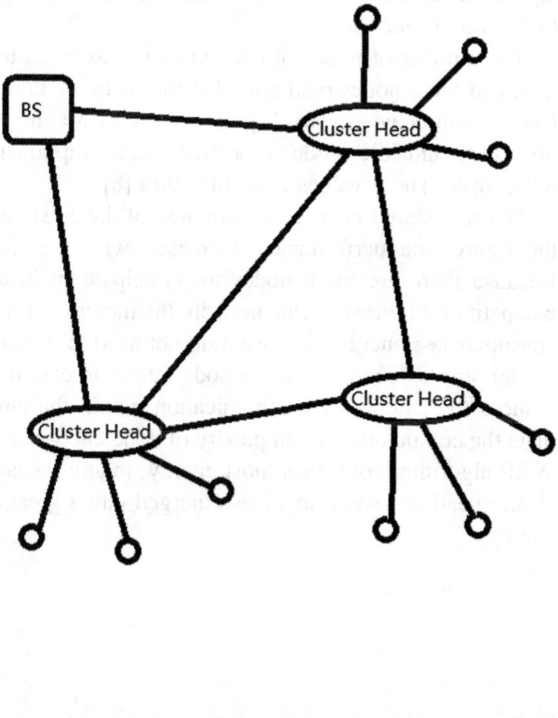

counterparts in the formula (2.4), the energy consumption $e_i(k)$ in the cycle of node i can be obtained. Combined with the energy consumed in the previous cycles, the total energy consumption at present of node i $E_i(k)$ is

$$E_i(k) = e_i(k) + \sum_{r=1}^{k-1} e_i(r) \tag{3.2}$$

Let the initial energy of each node be E_i^0, and the present residual energy of node i be $E_i^r(k)$, then

$$E_i^r(k) = E_i^0 - E_i(k) \tag{3.3}$$

Choose the node with maximal residual energy $E_i^r(k)$ as the next cluster head.

Every node saves a message of its energy consumption in the cluster. Meanwhile, the head builds a list of total energy consumption, recording each in-cluster node's $\sum_{r=1}^{k-1} e_i(r)$, the head's own message included, and the locations of the nodes. In the head shift, the list will be transferred to the new head. At the end of next cycle, the new head can use the list to assess the energy consumption of the other nodes and choose the subsequent head. Such a mode will be repeated till the end of the life of the network.

5 Simulation

Simulation of the SESP algorithm is compared with WAF [7] algorithm. WAF algorithm calculate small CDS (connected dominating set, CDS) and connect them to form a cluster.

The number of nodes in the simulation were set to 50, 100, 150, 200, 250 and 200, and these nodes randomly distributed in the area of 200 by 200 m; each node has the initial energy of 1 j; R1 is set to 30 m. Nodes in sending, receiving, dormancy and idle mode respectively consumption power of 24, 14.4, 14.4 and 0.015 mW. These values from literature [8].

Figure 2 shows energy consumption of the clustering process. Can be seen from the figure, the performance decreased with the increase of the node density. Because there are more nodes to participate in local information exchange and competition of wireless channel with the increase of nodes density. SESP algorithm consumes less energy, because it do not need to initiate communication to learn the cluster surplus energy of other node in the process of choosing cluster head. So It reduces the amount of communication within the cluster, also simplifies the estimate the computational complexity of node energy and prolong the life of network. WAF algorithm consumes most energy, mainly because the sequence selection of cluster head and some small sets merged into a great one need to consume a lot of energy.

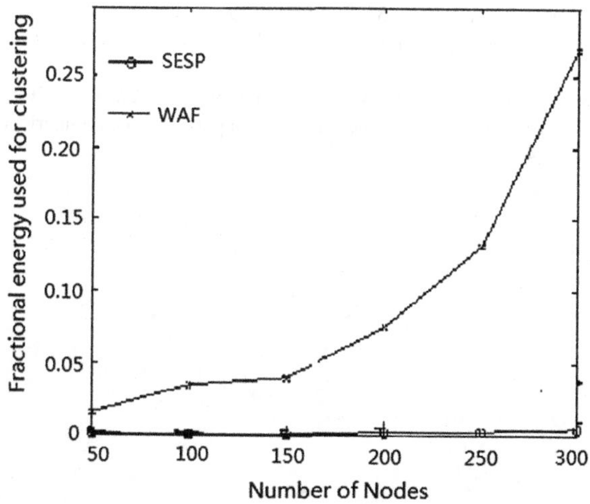

Fig. 2 Energy consumption of the clustering process

6 Conclusion

The self-adaptive algorithm SESP i.e. based on energy assessment in a WSN network in that there are a lot of node with same parameters. By the algorithm, no communication is needed to obtain the residual energy of other nodes when choosing cluster head, thus reducing both the in-cluster communication volume and the complexity of energy assessment calculation. Simulation of the SESP algorithm compared with others illustrate that SESP would save the node energy and effectively extend the lifespan of the network.

Acknowledgments This research is partially supported by Natural Science Foundation of Fujian Province Of China (2008J0178).

References

1. Madden S, Franklin MJ, Hellerstein JM, HongW (2002) Tag: a tiny aggregation service for ad-hoc sensor networks. In: Proceedings of OSDI
2. Heinzelman W, Chandrakasan A, Balakrishnan H (2002) An application-specific protocol architecture for wireless sensor networks. IEEE Trans Wireless Commun 1(4):660–670
3. Amit S, Anantha C (2001) Dynamic power managementin wireless sensor networks. Sel Test Comput 18(2):62–74
4. Ju Z, Hua Y (2009) Wireless sensor network node hardware modular design. J Univ Shanghai Sci Technol 6:20–22
5. Vasanth V, Michael F (2005) Power reduction techniques for micropro cessor system. ACM Comput Surv 37(3):195–237
6. Amis AD, Prakash R (2000) Load balancing clusters in wireless Ad Hoc networks. In: Proceedings of ASSET, March 2000, pp 25–32

7. Wan PJ, Alzoubi KM, Frieder O (2004) Distributed construction of connected dominating sets in wireless Ad Hoc networks. In: ACM/Kluwer mobile networks and applications, no 2, pp 141–149
8. Basagni S, Mastrogiovanni M, Panconesi A, Petrioli C (2006) Localized protocols for Ad Hoc clustering and backbone formation: a performance comparison. IEEE Trans Parallel Distrib Syst 17(4):292–306

Part VII
Intelligent Data Analysis and Processing

Automated Enzyme Function Classification Based on Pairwise Sequence Alignment Technique

Mahir M. Sharif, Alaa Tharwat, Aboul Ella Hassanien and Hesham A. Hefeny

Abstract Enzymes are important in our life due to its importance in the most biological processes. Thus, classification of the enzyme's function is vital to save efforts and time in the labs. In this paper, we propose an approach based on sequence alignment to compute the similarity between any two sequences. In the proposed approach, two different sequence alignment methods are used, namely, local and global sequence alignment. There are different score matrices such as BLOSUM and PAM are used in the local and global alignment to calculate the similarity between the unknown sequence and each sequence of the training sequences. The results which obtained were acceptable to some extent compared to previous studies that have surveyed.

Keywords Enzyme · Classification · Prediction · Global alignment · Local alignment · BLOSUM · PAM

1 Introduction

Recent advancements in sequencing technologies have seen an exponential growth in protein sequences. One of the modern studies is to identify the function of enzymes. In labs, it is a time-consuming experiment to decipher the function of the sequences.

M.M. Sharif · H.A. Hefeny (✉)
Institute of Statistical Studies and Researches (ISSR), Cairo University, Giza, Egypt
e-mail: Aboitcairo@gmail.com

A. Tharwat
Faculty of Engineering, Suez Canal University, Ismaïlia, Egypt

A.E. Hassanien
Faculty of Computers & Information, Cairo University, Cairo, Egypt

A.E. Hassanien
Faculty of Computers and Information, Beni Suef University, Beni Suef, Egypt

A.E. Hassanien
Scientific Research Group in Egypt (SRGE), Cairo, Egypt

A. Abraham et al. (eds.), *Intelligent Data Analysis and Applications*,
Advances in Intelligent Systems and Computing 370,
DOI 10.1007/978-3-319-21206-7_43

Moreover, it is necessary to predict whether the protein is an enzyme or non-enzyme. The classification process of enzymes is organized based on its biochemical reactions by the Enzyme Commission of the International Union of Biochemistry and Molecular Biology (NC-IUBMB) [1].

According to main Enzyme Commission EC (is a numerical classification scheme for enzymes, based on the chemical reactions) enzymes are classified into six main classes such as (I) oxidoreductase, catalysing oxidation-reduction reactions; (II) transferase, transferring a chemical group from one substrate (the donor) to another (the acceptor); (III) hydrolase, These enzymes catalyse the hydrolytic cleavage of bonds; (IV) lyase, catalysing the nonhydrolytic and cleave C–C, C–O, C–N and other bonds by means other than hydrolysis or oxidation; (V) isomerase, catalysing geometrical or structural changes within one molecule; (VI) ligase, catalysing the joining together of two molecules coupled with hydrolysis of a pyrophosphate bond in ATP or a similar triphosphate. The recommended name often takes the form A-B ligase [2].

For prediction of enzyme function, there are many approaches are widely used. The first method was to predict enzyme functions based on sequence similarity [3]. Faria et al. [3], developed a machine learning methodology called Peptide Program (PPs) that achieved a high accuracy when used small dataset. Des Jardins et al. [4], proposed a novel approach to predict the function of a protein from its amino acid sequences, they achieved performance better than famous algorithms, e.g. Basic Local Alignment Sequence Tool (BLAST) and Support Vector Machine (SVM) in case using a small dataset. Many other studies were based on extracting features from sequences. Mohammed and Guda proposed a model [5], reviewed a group of studies, which used a number of computational to predict and classify enzymes based on the extracted features, e.g. molecular weight, Polar, etc. Moreover, many studies predicted the function of enzymes based on combining two different methods. Omer et al. present a new method for extracting features from motif content and protein composition for protein sequence classification [6]. Bum et al. proposed new PNPRD features representing global and/or local differences in sequences, based on positively and/or negatively charged residues, to assist for predicting protein function, and useful feature subset for predicting the function of various proteins [7]. Moreover, Chien et al. reported that, some specific features represent the physiochemical properties of protein complex subunits [8].

In this paper, we proposed an approach to predict enzyme function class based on different sequence alignment techniques. We compare between the local and global sequence alignment technique to select the best technique in predicting the enzyme function class. Moreover, to improve the classification accuracy, different score matrices such as BLOSUM and PAM are used. The computer program which was used for the implementation of this approach has been written and developed by the authors (using MATLAB 7.11 (R2010b)) and without the need for any other ready software or online tools, which adds more facility to use different assessment methods and develop our approach to increase the accuracy.

The rest of this paper is organized as follows. Section 2 describes the theoretical background of the sequence alignment technique. Moreover, the local and global sequence alignment techniques are explained; Sect. 3 introduces the proposed approach; Sect. 4 presents experimental scenarios, results, and discussion; finally, Sect. 5 summarizes the conclusions and future work.

2 Preliminaries

2.1 Enzyme Prediction and Classification

There are three famous approaches widely used for enzyme prediction and classification based on protein's structure, protein's features, and sequence alignment. Many studies were classified the enzymes based on structure (e.g. the structure information of protein more conserved from sequence information, protein functions governed by its structure, and sometimes sequence similarity between two proteins exists). But, this method not strong enough to produce an unambiguous alignment). Whoever, recent approaches used the simple sequence, structure properties and many features that extracted from the enzymes (i.e. predicted secondary structures, physiochemical properties, sequence motifs, and highly conserved regions) [3]. The third approach is based on sequence similarity measure. This approach is the most common way of functional prediction, and the majority of sequences in the protein databases is annotated using just sequence comparison [9].

2.2 Sequence Alignment Technique

Sequence alignment is a process of arranging two (Pairwise Alignment) or more (Multiple Alignment) sequences of characters to define the region of similarity and calculate the similarity score. There are many problems and challenges that face the process of calculating similarity scores such as, (i) the sequences usually are in different length and different gaps, (ii) There may be, the matching (similar) region between two sequences is a relatively small region with respect to the length of each sequence [10, 11].

Consider X and Y is a pair of two different protein or DNA sequences, where $X \equiv x_1x_2 \ldots x_m$, $Y \equiv y_1y_2 \ldots y_n$, where x_i and y_i are letters chosen from the alphabet. To calculating the distance or similarity between X and Y, the two sequences are aligned first using sequence alignment technique such as local or global sequence alignment techniques.

In each sequence alignment process, gaps are an arbitrary number of null characters or spaces (represented by dashes) may be placed and it is called indel. Each indel represents an insertion of a character into one sequence or deletion of a character from the other one as shown in Fig. 1.

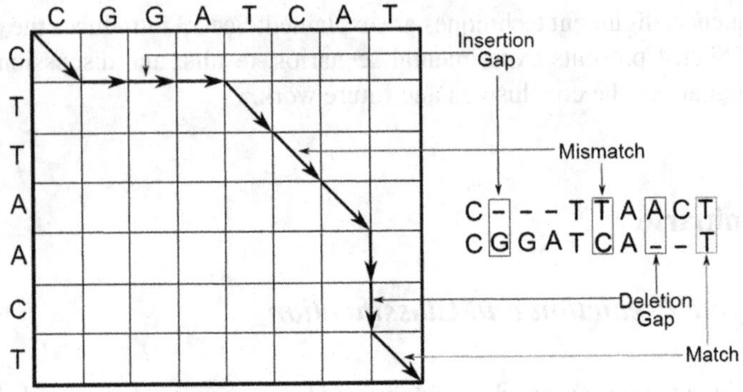

Fig. 1 Pairwise sequence alignment process, indicates match, mismatch, insertion, and deletion gap cases

There are many techniques to calculate the score of alignment gaps with letters (Gap score). The constant gap penalty is considered one of the simplest method. In this technique, a gap with any size, receives the constant negative penalty $-g$. In a Linear gap penalty, the penalty depends on the length of the gap. So, $-g$ in this technique represents a penalty per unit length of the gap. Gap scores are independent of letters. However, gap penalty value is subtracted from the total alignment score to calculate the final score [10].

After aligning the pair of sequences, we need to measure the score of alignment or similarity. Calculating the scores or similarity of alignment depends on the values of matching scores or mismatch penalty scores. In other words, it depends on the scoring or substitution matrix that have the values of matching scores and mismatching penalty scores [11].

Scoring matrix or substitution matrix is a set of values representing the likelihood of one residue being substituted by another. Since, the proteins composed of 20 amino acid, the scoring matrices for proteins are 20×20 matrix. Two well-known scoring matrices for proteins are PAM and BLOSUM [10].

In PAM (Point Accepted Mutation or Percent Accepted Mutation), a substitution of one amino acid by another that has been fixed by natural selection. In PAM1 (i.e. original PAM score matrix) 1 %, of the amino acids in a sequence are expected to accept mutation, so 1 % sequence divergence. PAM score matrix applies when the divergence is low. The higher suffix number, the better it is in dealing with distant sequence alignment [12]. Increasing the suffix number of PAM scoring matrix will increase the score of alignment between two different sequences, which reflects that, when the suffix number increased the aligned distant sequences can be identified better. The (ith,jth) cell in a PAM matrix denotes the probability that amino-acid i will be replaced by amino-acid j [12].

In BLOSUM (BLOck SUbstitutions Matrices), the blocks are ungapped multiple sequence alignment corresponding to the most conserved regions of the sequences involved. Default BLOSUM matrix is BLOSUM62 (used in BLAST algorithms), which represents the sequences used to create the BLOSUM62 matrix have approximately 62 % identity. Hanikoff and Hanikoff tested the performance of BLOSUM and PAM and found that BLOSUM matrices performed better than PAM matrices. In contrast, with PAM matrices, the higher the suffix number in BLOSUM matrix, the better it is in dealing with closer sequences [12].

The alignment score is the sum of the scores for aligning pairs of letters (alignment of two letters) and gap scores (alignment of a gap with a letter). For example, if the matching score is five, mismatch penalty is −3, and gap or space penalty is −2, so the alignment score of the two sequences in Fig. 1 equal to $= 5-2-2-2+5-3+5-2-2+5 = 7$. In the next section, global and local sequence alignment techniques which considered the most popular sequence alignment techniques are discussed.

Global Sequence Alignment: In Global Alignment technique, all letters and null in each sequence must be aligned from one end to the other as shown in Eq. (1), where $s(a,b)$ represents score for aligning letters a and b, g represents gap score, x_i is the partial sequence consisting of *ith* letter of $X \equiv x_1x_2\ldots x_m$, y_i is the partial sequence consisting of *jth* letter of $Y \equiv y_1y_2\ldots y_n$, and $SIM(i,j)$ represents the similarity of x_i and y_j. In the first line of Eq. (1), x_i and y_i are aligned. But, in the second line x is aligned with a null, while in the third line y is aligned with a null [13].

$$SIM(i,j) = max\begin{cases} SIM(i-1,j-1) + s(x_i + y_i)) \\ SIM(i-1,j) + g \\ SIM(i,j-1) + g \end{cases} \tag{1}$$

Example 1 Given two different sequences X and Y, where $X = TGCCGTG$ and $Y = CTGTCGCTGCACG$. The matching score is equal to four, mismatch penalty equals −2, and gap penalty is also −2.

For the pair of sequences with length m and length n, the size scoring matrix is needed for their alignment must be $(m+1) \times (n+1)$ as shown in Table 1.

The first step in any pairwise sequence alignment technique is to fill the entire matrix as shown in Table 1. In the global alignment algorithm, the scoring matrix step starts by assigning the two sequences' letters and gap penalties on the left and top headers of the table as shown in Table 1. To fill all cells of the matrix, first assigns the gap penalty in the top row and the left column in a cumulative way. Then, assigns the maximum of all three cells to the new cell as in Eq. (1).

After finishing from assigning values to all cells as shown in Table 1, the next step in pairwise sequence alignment is to trace back according to the values that are calculated before in the scoring matrix. As shown in Table 1 we note that, the bold fonts show the trace-back of Example 1. The final step is to make alignment by tracing back methods as shown in Fig. 1. As shown in the figure we note that, going straight right or down means, insertion or deletion gap, respectively. While, moving in diagonal represents match or mismatch cases. The steps and directions we follow it to move from one cell to the other (forward) are the same directions we trace it back to align the two sequences. Finally, the final alignment is shown in Fig. 2.

Table 1 Score matrix of the global sequence alignment techniques using the two sequences in Example 1

	Gap	T	G	C	C	G	T	G
Gap	**0**	−2	−4	−6	−8	−10	−12	−14
C	−2	−2	−4	0	−2	−4	−6	−8
T	−4	**2**	0	−2	−2	−4	0	−2
G	−6	0	**6**	4	2	2	0	4
T	−8	−2	4	**4**	2	0	6	4
C	−10	−4	2	8	**8**	6	4	4
G	−12	−6	0	6	6	**12**	10	8
C	−14	−8	−2	4	10	**10**	10	8
T	−16	−10	−4	2	8	8	**14**	12
G	−18	−12	−6	0	6	12	12	**18**
C	−20	−14	−8	−2	4	10	10	**16**
A	−22	−16	−10	−4	2	8	8	**14**
C	−24	−18	−12	−6	0	6	6	**12**
G	−26	−20	−14	−8	−2	4	4	**10**

Global Alignment	Local Alignment
X=[C T G T C G C T G C A C G]	X=[C T G T C G C T G C A C G - -]
Y=[- T G C C G - T G - - - -]	Y=[- - - - - - - T G C - C G T G]

Fig. 2 Example of local and global sequence alignment techniques for pair of sequences in Example 1 and Example 2

Local Sequence Alignment: Smith-Waterman proposed the first algorithm that used local alignment to measure the similarity between different sequences as in Eq. (2). Global alignment works better if the sequences are similar in length and characters while the local alignment is useful when the sequences are not similar in length or characters. However, local alignment finds the most similar regions in any two sequences being aligned as shown in Fig. 2 [12]. In Fig. 2 we note that, the global alignment technique aligns the whole length of the two sequences, while the local alignment technique aligns the most similar region in the two sequences. So, the local alignment technique isolates regions in the sequences, hence it is easy to detect repeats while the global alignment is suitable for overall similarity detection.

$$SIM(i,j) = max \begin{cases} SIM(i-1,j-1) + s(x_i + y_i) \\ SIM(i-1,j) + g \\ SIM(i,j-1) + g \\ 0 \end{cases} \quad (2)$$

Table 2 Score matrix of the local sequence alignment techniques using the two sequences in Example 2

	Gap	T	G	C	C	G	T	G
Gap	0	0	0	0	0	0	0	0
C	0	0	0	1	1	0	0	0
T	1	0	0	0	0	0	1	0
G	0	0	1	0	0	1	0	1
T	0	1	0	0	0	0	2	1
C	0	0	0	1	1	0	1	1
G	0	0	1	0	0	2	1	2
C	**0**	0	0	2	1	1	1	1
T	0	**1**	0	1	1	0	2	1
G	0	0	**2**	1	0	2	1	3
C	0	0	1	**3**	2	1	1	2
A	0	0	0	**2**	2	1	0	1
C	0	0	0	1	**3**	2	1	0
G	0	0	1	0	2	**4**	3	2

Example 2 Given two different sequences X and Y with different lengths m and n respectively, where $X = [TGCCGTG]$ and $Y = [CTGTCGCTGCACG]$ as in Example 1. The matching score is four, mismatch penalty equals −2, and gap penalty is also equal to −2. The size of the scoring matrix is needed for their alignment must be $(m + 1) \times (n + 1)$.

The first step in any local sequence alignment algorithm is to initialize first row and first column to be 0 as shown in Table 2. To calculate the value of that cell as in Eq. (2), we assign the maximum of all three cells or zero to the new cell. After filling all cells, trace back as in the global alignment technique. Finally, the final alignment is as shown in Fig. 2.

3 The Proposed Enzyme Function Classification Approach

The first step in the proposed approach is to collect labelled enzyme sequences with different functions or categories as shown in Fig. 3. In the proposed approach, we have two techniques to compute the pairwise alignment between any two sequences, namely, global and local alignment techniques. Thus, the similarity between the unknown sequence and each sequence in the labelled (i.e. training) sequences are computed using the global and local sequence alignment techniques. The similarity or score reflects the distance between the unknown sequence and all the labelled sequences. The class function of the sequence that has the maximum score

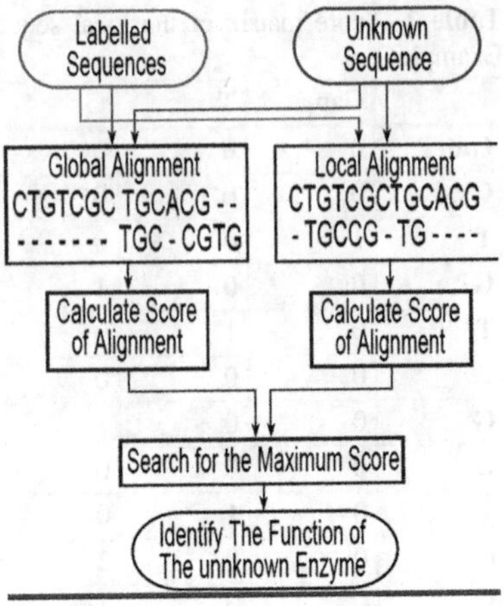

Fig. 3 Block diagram of the proposed approach

(i.e. maximum similarity) is assigned to the unknown sequence. The details of the proposed approach are summarized in Algorithm (1).

Algorithm 1 Enzyme function classification algorithm

1: **Input:** Given a set of labelled sequences with their functions as follows, $S_i, i = 1, 2, \ldots, N$, where N represents the total number of labelled sequences.
2: Given an unknown sequences T.
3: Align the unknown sequence T with all labelled sequences S.
4: Compute the score matrix $M_i, i = 1, 2, \ldots, N$ of all pairs of alignment, where M_i represents the matching score between the unknown sequence T and the ith labelled sequence (S_i).
5: Search for the maximum score M_{max} which represents the nearest sequence from the unknown sequence and assigns the label (function) of M_{max} to the unknown sequence T.

4 Experimental Results and Discussion

There are two scenario's which designed to test the proposed approach. In all classes we used one against all approach or, in other words, for each unknown sequence we calculate the sequence alignment between an unknown sequence and all other sequences in the dataset and select the maximum similarity. To save time and memory storage, we used a subset of the whole dataset which represents approximately a half ($\frac{3470}{6923} = 50.12\,\%$) of the total number of all sequences as shown in Table 4.

Table 3 Distribution of all sequences across different classes in SWISS-PROT enzyme database

Class name	Number of sequences	Percentage
Oxidoreductase	1079	15.6
Transferase	2338	33.8
Hydrolase	2536	36.6
Lyase	258	3.7
Isomerase	171	2.5
Ligase	541	7.8
Total	**6923**	**100**

Table 4 Accuracy and CPU time of the local and global sequence alignment techniques using different score matrices (i.e. the first and second experimental scenarios)

Score matrix	Local Seq. alignment		Global Seq. alignment	
	Acc. (in %)	CPU Time (in S)	Acc. (in %)	CPU Time (in S)
BLOSUM62	93.3	0.0093	88.2	0.0092
BLOSUM100	92.9	0.0061	87.4	0.0036
BLOSUM30	92.9	0.0069	89.9	0.0014
PAM10	90.5	0.004	85.4	0.0014
PAM100	91.2	0.0043	86.3	0.0014

4.1 Dataset

The dataset, which was used in all experiments were obtained from SWISS-PROT enzyme database. It consists of sequences of enzymes from [14], that have 6923 instances and the details of each class are shown in Table 3. The distribution of sequences across different classes was not equivalent as shown in Table 3.

4.2 Experimental Scenarios

The first experiment scenario is conducted to compare between the global and local alignment techniques using the default BLOSUM score matrix (BLOSUM62). The summary of this scenario is shown in Fig. 4)and Table 4.

The second experiment scenario is conducted to understand the effect of changing score matrix of the global alignment and local sequence alignment techniques. The summary of this scenario is shown in Table 4.

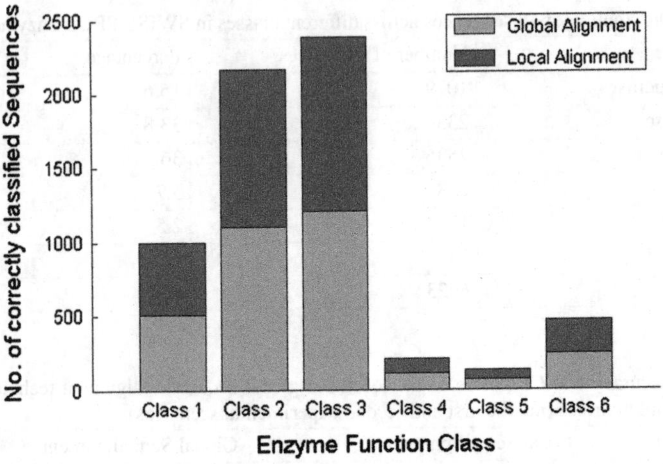

Fig. 4 Accuracy of all enzyme function classes using local versus global sequence alignment techniques

4.3 Discussion

Figure 4 and Table 4 show the accuracy (i.e. the percentage of the total number of predictions that were correct) of local alignment technique achieved a good relatively accuracy ranged from 91.2 to 93.3 %, while the global sequence alignment technique achieved accuracy ranged from 85.4 to 89.9 %, which reflects that, the local alignment technique is better to align two sequences and calculate an accurate similarity score than global sequence alignment technique. On the other hand, as shown in Table 4 the global alignment technique needs CPU time less than the local alignment technique. Moreover, it is apparent from Fig. 4 that the first three classes achieved accuracy better than the other three classes because the number of sequences of the first three classes represents about 85.9 % of the total number of sequences (2980/3470). Furthermore, the third class (Hydrolases) achieved the best accuracy among all classes while the fifth class (Isomerases) achieved the worst one. Table 4 illustrates the local sequence alignment technique achieved accuracy better than global sequence alignment technique using different score matrices. The best accuracy of local alignment achieved using BLOSUM62 score matrix while the best accuracy of global alignment processes achieved when using BLOSUM30 score matrix.

Table 4 and Fig. 4 show the accuracy of the proposed approach is better than Jardins et al. who have developed a model achieved 74 % in first level and 68 % in second level of Enzyme Commission (EC) [4], and Kumar et al. [15] achieved accuracy near to 80.37 %. Finally, our proposed approach is fully self-implemented without any needs to online tools.

5 Conclusions and Future Work

In this paper, we have proposed, developed, and implemented an approach to identify the class function of an unknown sequence using sequence alignment technique. In the proposed approach, two different sequence alignment techniques are used, namely, local and global sequence alignment techniques. Through the application of global alignment and local alignment to compute the pairwise alignment for any sequence and assign it to the suitable functional class when it gets the maximum score of alignment. The approach which we developed using MATLAB achieved a reasonable classification accuracy reached 89.9, 93.3 % when we apply global alignment technique using BLOSUM30, and local alignment technique using BLOSUM62, respectively. The results which achieved by the proposed approach are very promising and opens prospects for future development. In the future work, we will use different information fusion methods to combine the scores of the local and global techniques to increase the accuracy of our approach.

References

1. Enzyme-Nomenclature, N.C.o.t.I.U.o.B. (1992) Molecular biology (NC-IUBMB). Academic Press, N.h. ISBN 0-12-227164-5
2. Lu L, Qian Z, Cai YD, Li Y (2007) Ecs: an automatic enzyme classifier based on functional domain composition. J Comput Biol Chem 31(3):226–232
3. Faria D, Ferreira AE, Falcao AO (2009) Enzyme classification with peptide programs: a comparative study. BMC Bioinform 10(1):1–9
4. Des Jardins M, Karp PD, Krummenacker M, Lee TJ, Ouzounis CA (1997) Prediction of enzyme classification from protein sequence without the use of sequence similarity. Proc Int Conf Intell Syst Mol Biol 5:92–99
5. Mohammed A, Guda C (2011) Computational approaches for automated classification of enzyme sequences. J Proteomics Bioinform 4:147–152
6. Sarac OS, Gursoy-Yuzugullu O, Cetin-Atalay R, Atalay V (2008) Subsequencebased feature map for protein function classification. Comput Biol Chem 32(2):122–130
7. Lee BJ, Shin MS, Oh YJ, Oh HS, Ryu KH (2009) Identification of protein functions using a machine-learning approach based on sequence-derived properties. Proteome Sci 7(1):27
8. Huang CH, Chou SY, Ng KL (2013) Improving protein complex classification accuracy using amino acid composition profile. Comput Biol Med 43(9):1196–1204
9. Syed U, Yona G (2009) Enzyme function prediction with interpretable models. In: Computational systems biology. Springer, pp 373–420
10. Zou Q, Shan X, Jiang Y (2012) A novel center star multiple sequence alignment algorithm based on affine gap penalty and k-band. J Phys Procedia 33:322–327
11. Blazewicz J, Frohmberg W, Kierzynka M, Wojciechowski P (2013) G?msaa gpu?based, fast and accurate algorithm for multiple sequence alignment. J Parallel Distrib Comput 73(1): 32–41
12. Xiong J (2006) Essential bioinformatics. Volume 1. Cambridge University Press
13. Huang X (1994) On global sequence alignment. J Comput Appl Biosci CABIOS 10(3): 227–235

14. Bairoch A (2000) The enzyme database in 2000. J Nucleic Acids Res 28(1): 304-305. http://www.expasy.ch/enzyme/(2000)). doi:10.1093/nar/28.1.304
15. Naik PK, Mishra VS, Gupta M, Jaiswal K (2007) Prediction of enzymes and non-enzymes from protein sequences based on sequence derived features and pssm matrix using artificial neural network. J Bioinform 2(3):107–112

Predicting Personality Traits and Social Context Based on Mining the Smartphones SMS Data

Fatma Yakoub, Moustafa Zein, Khaled Yasser, Ammar Adl and Aboul Ella Hassanien

Abstract Reality Mining is one of the first efforts that have been exerted to utilize smartphone's data; to analyze human behavior. The smartphone data are used to identify human behavior and discover more attributes about smartphone users, such as their personality traits and their relationship status. Text messages and SMS logs are two of the main data resources from the smartphones. In this paper, The proposed system define the user personality by observing behavioral characteristics derived from smartphone logs and the language used in text messages. Hence, The supervised machine learning methods (K-nearest nighbor (KNN), support vector machine, and Naive Bayes) and text mining techniques are used in studying the textual matter messages. From this study, The correlation between text messages and predicate users personality traits is broken down. The results provided an overview on how text messages and smartphone logs represent the user behavior; as they chew over the user personality traits with accuracy up to 70 %.

Keywords Personality traits · Reality mining · Behavioral patterns · Behavioral analysis · Supervised machine learning · Short message service (SMS)

1 Introduction

Every time the smartphone user makes a call, sends a Short Message Service (SMS), or uses a social network, a few bits of information left behind. This information describes some of the users behavior, and relationship with friends and family.

Scientific Research Group in Egypt (SRGE), http://www.egyptscience.net.

F. Yakoub (✉) · M. Zein
Faculty of Computers & Information, Cairo University, Cairo, Egypt
e-mail: Aboitcairo@gmail.com

K. Yasser · A.E. Hassanien
Faculty of Engineering, Cairo University, Cairo, Egypt

A. Adl · A.E. Hassanien
Faculty of Computers and Information, Beni Suef University, Beni Suef, Egypt

A. Abraham et al. (eds.), *Intelligent Data Analysis and Applications*,
Advances in Intelligent Systems and Computing 370,
DOI 10.1007/978-3-319-21206-7_44

By analyzing this information, exploring the capabilities of the smartphone's data enables the researchers to investigate human interaction with Smartphones. In this paper, there are three main concepts that are followed, as the following: (1) *Reality mining:* which studies the human interactions based on the usage of wireless devices such as smartphones, which leads to provide a more accurate explanation of what people do, where they go and with whom they communicate, rather than the subjective sources such as a person's own account. The rapid, universal growth of smartphone usage [1, 2] makes us need to study the social science through smartphone telephony. Smartphones provide a new vision for investigating smartphone phone usage [2, 3], (2) *Smartphones call and text message (SMS) logs* [4]: The Reality mining concepts using these logs are applied, which can provide information for every smartphone user: (a) The message writer mood: the user used some words to express a situation that would reflect a behavior; to it will be used as a behavioral indicator and (b) According to the research in this field, different individuals have different ways of expressing through the written word with different situations, different moods, and personality traits [5]. The SMS is chosen, because of sending SMS is less expensive, as most of the smartphone operators have offered prepaid SMS packages at very low cost. SMS is viewed as a trusted service for sharing confidential information between people [12], and (3) *Machine learning techniques:* After defining the links between personality and behavioral patterns, which are derived from contextual data. The machine-learning techniques are applied to convert these behavioral patterns into personality types of the users and to understand the various expressions that may be detected in the context of smartphone applications [6].

The introduced important topic has many studies; this topic identified users' personality traits and their relationship status with recipients using their smartphone data, and predicting user interests which user talks about. The prediction of personality traits using a small dataset with a single type of smartphone data (mobility data or text messages data) is unclear resource. When focussing on mobility data, personality traits, the effect of gender differences, and ignore the user's attitude to solve the problem, it is not enough to predict the human behavioral patterns.

There is a one idea about extract human personality traits and their inner personal inclinations using smartphone SMS text mining with the message logs [8]. Although few studies study the effect of text mining on identifying humans' relationship status. These works focus on the smartphone social networks apps. We want to focus on the smartphone content analytics. The text message will represent the case study of smartphone data in this study. The users of smartphones cannot know the nature of their relationships changing through the time and the relation between the way of dealing with the people and user's behavior. Studying the text massages logs only is not enough to predict the peoples' interests and their inner personal inclinations. Using the SMS logs and text massages content to predicate user personality with life facts and relations analysis is not implemented.

At that point, a great deal of questions that this survey demands. Some of these questions will be done. The questions are: what is the effect of smartphone logs on the user personality? How can useing the SMS logs and text massages content to predicate user personality? How can useing the text massages content to determine the user relationship status with his contacts? How can the of use the text massages content to determine user interests and his inner personal inclinations?

The main objective of the paper is to introduce an affective solution in the prediction and identification of human behavioral patterns from the text messaging content and logs. The introduced model will show that the use of message logs is not enough to extract the personality traits and user's relationship status. On the other hand, the use of the SMS content and logs together may be used in predicting personality traits based on the Big-Five personality traits.

The rest of this paper is structured as follows. Section 2 describes most relevant works related to this study. Section 3 describes the methodologies that used in this study start with the data set and the proposed solution of the problem statement. Moreover, the results will be discussed in the Sect. 4 and by comparing the results with the previous results in the related studies. Finally, the conclusion and future work will be represented in Sect. 5.

2 Related Work

Many studies have been introduced in the research of personality and human behavior. In addition, There are studies that had introduced the solutions in extracting human behavioral patterns from text messaging identification and analysis. Thither is a study used the logs of calls, SMS, Bluetooth scans and application usage from smartphone phones to investigate between behavioral characteristics derived from rich smartphone data and self-reported personality traits.

One of the prevoius studies extracts various features such as Call Logs, SMS Logs, and applications Logs and Machine learning techniques to classify user personality. The study experimental results on 83 users showed that could be classified users based on the Big-Five personality traits (Extraversion, Agreeableness, Conscientiousness, Emotional Stability and Openness to Experience) up to 75.9 % accuracy [7]. Another study used logs such as phone calls; text messages sent and received, from 69 participants? smartphones to provide the evidence that personality that are reliably predicted from standard smartphone logs.

Another prevoius study introduces five features categories: Basic phone use, Active user behaviors, Location, Regularity, and Diversity. The experimental results of this study showed that able to predict users' personality with a mean accuracy across traits of 42 % better than random, Moreover the experimental results of the study are reaching up to 61 % accuracy on a three-class problem [7]. The purpose of prevoius study was to undertake some analyses of how the language used in text messaging varies as a function of personality traits and the interpersonal context. This study uses 224 students last 20 text messages and with indication of time, date, and location of massages. Define linguistic alterations feature as a function of both personality traits and relationship status [8]. Some Authors seek to provide a platform for collecting text messages from social media (WhatsApp), and classifying them into different personality categories and measure happiness [5].

A prevoius study provided the evidence that daily stress that will be reliably recognized based on smartphone activities (call log, SMS logs, and Bluetooth interactions). In this study, 25 calls and SMS features and 9 proximity features had been derived. These features, grouped into four broad categories, characterize the general phone usage, diversity, active behavior and regularity. The study experimental results on 69 users showed that able to predict individual daily stress up to 72.39 % accuracy for 2-class classification problem [9]. Moreover, another introduced an article in studying the problem of predicting personality with features based on social behavior [10].

There is a study introduced an approach extracts meta-attributes from texts and does not work directly with the content of the messages. The system does not work directly with the Tweets but instead uses information extracted from the Tweets. The information extracted from the Tweets divided into grammar and social behavior. The system was pplied to predict the personality of Tweets taken from three datasets available in the literature, and resulted in an approximately 83 % accurate prediction, with some of the personality traits presenting better individual classification rates than others [11].

3 Methodology

In this study, the tartegt is extracting human personality traits and their inner personal inclinations using smartphone SMS text and logs. A personality traits prediction model is built to extract the personality traits. The model mainly depends on SMS text and some methodologies used in text mining as below:

3.1 Dataset and Data Pre-Processing

The original corpus SMS was collected by an honors year undergraduate project student [13, 14]. The dataset does not contain contributors send messages only, but also contains contributors' background information as (age, gender, city, and country), their text habits (input method, number of sent daily, and years of using SMS), and information about their phones (brand, smartphone or not). This information provided beside sent SMS content. SMS logs are extracted such sender phone number, receiver phone number, time, and date. The authors save the privacy of the sender and receiver phone numbers. The authors adopt Data Encryption Standard (DES) to create a one-way enciphering of the phone numbers, which replaces the originals in the corpus. Moreover, to ensure SMS body privacy, they adopt a stricter standard in dealing with sensitive data such as email addresses, URLs, and IP addresses, and others in the messages. Such information is captured using regular expressions and replaced by the corresponding semantic placeholders, For example, any detected email address will be replaced by the code (EMAIL). There is a sample of the original data set in Table 1.

Table 1 A sample of original dataset, which is used in this study, where *L* is language and *Fre* is the frequency and its more than 50 SMS daily

Serial number	Time	Text	L	Age	Experience	Fre	Gender	Input method	Native speaker
79780a9dbe83fd1e 5dd2bd2543e7da2a	2010.10.24 11:59	Lets go	En	21–25	3 to 5 years	50	Female	Multi-tap	Yes
79780a9dbe83fd1e 2dd2bd2543e7da2a	2010.10.28 11:53	Studying?	En	21–25	3 to 5 years	50	Unknown	Multi-tap	Yes
79780a9dbe83fd1e 3dd2bd2543e7da2a	2010.10.26 22:21	Vch photo	En	21–25	3 to 5 years	50	Male	Multi-tap	Yes

Table 2 The used data set after pre-processing step

Sender number	Receiver number	Date	Time	Text	Language	Frequency	Native speaker
001	1000	2010.10.24	00:55	Monday I am free	En	More than 50 SMS daily	Yes
	1001	2010.10.28	00:53	We r here	En	More than 50 SMS daily	Yes
	1002	2010.10.26	00:52	Pls call me	En	More than 50 SMS daily	Yes
	1003	2010.10.25	01:32	Her is go	En	More than 50 SMS daily	Yes

The sample of data includes Serial number for every sender, time of message sending, SMS text, Sender language, sender age, the experience of sender in using a smartphone, number of messages that sender send them through a day, sender sex, input message, native language. Original dataset contains different 45062 SMS collected from 170 contributors [13, 14], here the system is used the date of contributors that their native language is English language and covered different countries. Just use 22532 SMS transferred between 90 contributors and 789 receivers', range of contributors' age between 16 to 50 years. useing only text SMS, time as SMS logs, their daily text frequency, and phone numbers of contributors and receivers to solve our problem.

The data set needs some changes that have been made. These changes do not effect on the data set content. Due to the big length of the cipher receiver and sender phone numbers, Each unique cipher number is replaced with small unique random numbers. Then text messages that send are grouped using unique sender with user's receivers. The Time column in original dataset is divided which contain time and date of SMS into two columns, one of them for a date and the second for time as in Table 2. The effective columns is used from Table 1 in Table 2, which related to this study. Further, using this data to extract feature that uses as indication of user text behavior, which reflect user's personality, and user's relationship status with his destinations. Moreover, about 80 % from dataset is selected as a training dataset and the rest ones used as test data set.

3.2 Text Features Extraction

After preprocessing step, high-level features are extracted from the SMS dataset. These features will represent contributor's text behavior. The text behavior features extracted information from raw SMS text. To extract the features, applying the

Table 3 The extracted expressive words and their related category

Category	Samples words
EMOTIONS	
Positive effect	Cheerful, enjoy, fun
Anxiety	Afraid, fear, phobic
Sadness	Depression, dissatisfied, lonely
Affection	Affectionate, marriage, sweetheart
Aggression	Angry, harsh, sarcasm
Expressive behavior	Art, dance, sing
Glory	Admirable, hero, royal
SECONDARY PROCESS	
Abstraction	Know, may, thought
Social behavior	Say, tell, call
Instrumental behavior	Make, find, work
Restraint	Must, stop, bind
Order	Simple, measure, array
Temporal references	When, now, then
Moral imperative	Should, right, virtue

content analysis on contributors SMS. The English Regressive Imagery Dictionary (RID) is used from WordStat6.1, which is a content analysis and text-mining software [18], it is used to extract and analyze information from large amounts of documents. WordStat 6.1 simply analyzed word-by-word, each word being compared against the RID dictionary to extract common feature from the dataset. Table 3 contains a sample of RID dictionary categories, which contain 29 categories of primary process cognition, seven categories of secondary process cognition, and seven categories of emotions [3, 4].

Since the RID was not developed for use with text messages. Hence, it does not contain categories that capture the various linguistic alterations that can occur in text messages. Because of this reason, some categories are developed to it like personal pronoun's (first and second person pronoun) singular and plural, impersonal pronoun's, emotions, and positive words. Our text features are generated by WordStat 6.1, which counts the frequency of SMSs that contain dictionary categories, the Fig. 1, represents a sample of frequency analysis of contributor's SMS and dictionary categories. The frequencies represent the total number of contributor SMS that contain words belong to the category. Also calculating user total SMS, average SMS length, total number of SMS's contacts, average of daily texting, and average send SMS at night and morning as additional features to text behavior. sample data is visualized for 3 contributors each one represents with different color and their SMS content frequency analysis in this figure.

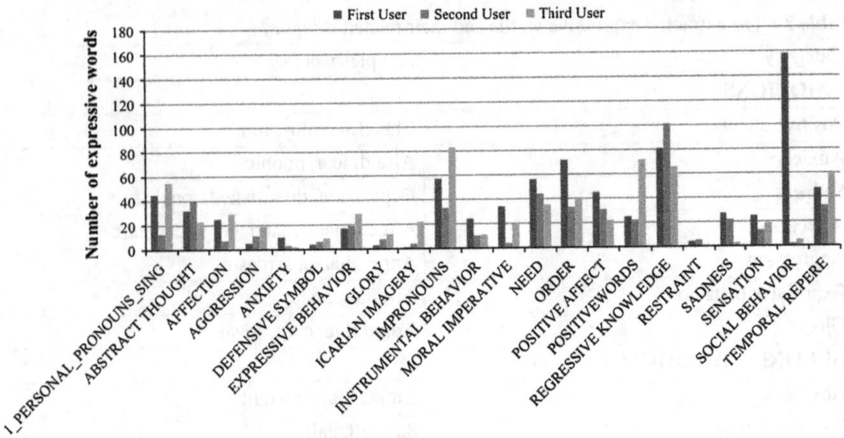

Fig. 1 A sample of text features data for three users and their SMS text analysis based on RID dictionary categories

3.3 Personality Traits Classification

The user personality traits are classified based on the previous features, which extracted from the SMS text content and logs. the need is to clarify what a trait is; it is an internal characteristic that corresponds to an extreme position on a behavioral dimension. Personality traits are basic tendencies that remain stable across the life span. Based on the Big-Five model, the following five factors, Extraversion (outgoing, amicable, and assertive), Neuroticism (anxious, insecure, and sensitive), Agreeableness (cooperative, helpful, and nurturing), Conscientiousness (responsible, organized, and persevering), and Openness to Experience have been shown to account for most individual personality differences [6].

From previous studies some of personality traits have high correlated relation between them and text content categories [8, 15, 16], from those studies Extraversion trait was associated with the more frequent use of person pronouns, positive emotion words and words associated with social processes, and talkativeness [6, 8], neuroticism was associated with few frequent use of positive emotion words [8], more frequently to use of anxiety words and Agreeableness was correlated with positive words [8], was mostly associated with low frequently use of anxiety words. The extracted features cover some categories, which describe three personality traits (Extraversion, Neuroticism, and Agreeableness). Therefore, these categories are used as output of our model.

Randomly about 80 % is selected from dataset as a training dataset and the rest ones used as test data set. Because of the relationship between personality traits and numerous behavioral can be non-linear; the used techniques are Naive Bayes classifier, K-nearest neighbor (KNN) with k = 2 and k = 3. From Our experiments, In case of k = 3 the KNN algorithm the model gets better accuracy with Euclidean distance

Table 4 Classification results with percentage of accuracy

	KNN accuracy	NB accuracy	SVM accuracy
Extraversion	100.0 %	50.0 %	50.0 %
Agreeableness	33.33 %	100.0 %	33.33 %
Neuroticism	66.67 %	33.33 %	66.67 %

metric. Another techniques is used that is the multiclass Support Vector Machine (SVM) with quadratic and RBF kernel functions. The quadratic kernel function gets better accuracy. All this classifier was implemented through MATLAB Bioinformatics Toolbox [17] to build predicting models.

Within the accuracy that is in Table 4, the results show that the users can be classified significantly above chance levels for three traits; the traits for which accuracy was highest are Extraversion and Neuroticism. This may be happening because the extracted features to text behavior of the SMSs are not enough to describe to this personality trait, or our labeling criteria may lead to this. Further, Adding the features to describe this trait more accurate like other traits and enhance labeling criteria to be automatic and more accurate.

4 Results and Discussion

All the time smartphone users try to find expressive and abbriviated words to wirte them in SMS. theses words represent their emotions and what they feel. In this study, we tried to prove emotions refer to the users personality. The results in Table 4 indicate that personality traits could be predicted through text features extracted from original SMS content. The accuracy of the model is calculated with the average accuracy of used classifiers. Our study presents accuracy up to 70 % better than other personality study [7], which used standard mobile phone logs such as phone calls, text messages sent and received to predict personality from it, this study get accuracy up to 61 %. Since personality provides some guidance to identify new users? preferences, it might be an effective approach to build an automatic human personality model. After the implementation, steps that are shown above, we can deduce a personality traits prediction model (PTPM) based on SMS text.

PTPM is a sequential design process where progress is a classification through the phases of Initiation, Normalization, Analysis, Feature extraction, Classification, and Behavior detection. In the first phase, get the user SMS text and logs as input to the model. The data may as undesired elements and fake data. In the second phase, raw data enter into normalization that aims into preprocessing the raw data, in this phase; the data will be extracted and will be as a specified input to the third phase. In the third phase, some data elements will analyze to be a meta-data for feature extraction phase. In the fourth data, the features will be extracted from data using RID dictionary. In the fifth phase, the user features is entered into prediction technique. The prediction technique classifies user personality as Extraversion, Agreeableness and Neuroticism based on his text features and using machine-learning algorithms. The target output from this model is extracting behavioral patterns that represent the user personality traits.

5 Conclusion and Future Work

The forecast of personality traits for cell phone clients is an intriguing issue in the reality mining pattern. The study concentrated on anticipating the identity attributes by utilizing SMS content. Also examining the SMS content from extensive genuine SMS information sets and anticipated an essayist's identity qualities. To the best of our insight, this is the first endeavor at gathering a SMS essayist's identity through programmed substance investigation. we constructed a model to concentrate behavioral examples that speak to the client identity characteristics with five stages (Initiation, Normalization, Analysis, Feature extraction, Classification, and Behavior detection). The model is shown that it is possible to reliably interfere personality without self-reported personality. The plan is to use the MMS as another input with SMS to predict the personality traits for smartphone users. In addition to, use the concepts of image processing and pattern recognition in identifying the MMS purpose and advantages of detecting the personality traits. Moreover, the text messaging influenced by the user's location and his relationships with his contacts. Therefore, using the location and relations of the sender as features beside the previous features and the MMS features.

References

1. Amicha Y, Wainapel G (2004) "On the Internet No One Knows I'm an Introvert": Extroversion, Neuroticism, and Internet Interaction. Cyber Psychol Behav 5(2):125–128
2. Chittaranjan G, Idiap Res I (2011) Who's who with big-five: analyzing and classifying personality traits with smartphones. In: 15th IEEE annual international symposium on wearable computers (ISWC), pp 29–36
3. Philip E (2011) Top 11 technologies of the decade. IEEE Spectr 48(1):27–63
4. Min J, Wiese J (2013) Mining smartphone data to classify life-facets of social relationships. In: Proceedings of the 2013 ACM conference on computer supported cooperative work, pp 285–294
5. Saez Y, Navarro C (2014) A system for personality and happiness detection. Int J Interact Multimedia Artif Intell 2(5):7–15
6. Chittaranjan G, Blom J (2011) Mining large-scale smartphone data for personality studies. Pers Ubiquit Comput 17(3):433–450 (Springer)
7. de Montjoye Yves-Alexandre, Quoidbach Jordi, Robic Florent, Pentland Alex (Sandy) (2013) Predicting personality using novel mobile phone-based metrics. In: Greenberg Ariel M, Kennedy William G, Bos Nathan D (eds) SBP 2013, vol 7812., LNCSSpringer, Heidelberg, pp 48–55

8. Holtgraves T (2011) Text messaging, personality, and the social context. J Res Pers 45(1):92–99
9. Bogomolov A, Lepri B (2014) Pervasive stress recognition for sustainable living. In: IEEE international conference on pervasive computing and communications workshops (PERCOM Workshops), vol 45, issue (1), pp 345–350
10. Adal? S, Golbeck J (2014) Predicting personality with social behavior: a comparative study. Social Network Analysis and Mining, Springer
11. Lima A, Castro D (2014) A multi-label, semi-supervised classification approach applied to personality prediction in social media. Neural Networks 122–130
12. Ahmed I, Ali R (2014) Semi-supervised learning using frequent itemset and ensemble learning for SMS classification. Expert Syst Appl 42(3):277–285
13. Collecting SMS Messages for a Public Research Corpus. http://wing.comp.nus.edu.sg:8080/SMSCorpus/listMsg.do?category=collect_timeŹcriteria=2011/1ŹpageNo=202
14. Chen T, Kan M (2012) Creating a live, public short message service corpus: the NUS SMS corpus. Lang Res Eval 47(2):299–335 (Springer)
15. Golnoosh F, Susana Z (2013) Recognising personality traits using facebook status updates. Comput Pers 14–18
16. arkoni T (2010) Personality in 100,000 words: a large-scale analysis of personality and word use among bloggers. J Res Pers 44(3), 263–373
17. MATLAB Bioinformatics Toolbox. http://www.mathworks.com/products/bioinfo
18. WordStat 6 - Provalis Research. www.provalisresearch.com/Documents/WordStat6.pdf

Machine Learning-Based Measurement System for Spinal Cord Injuries Rehabilitation Length of Stay

Rehab Mahmoud, Nashwa El-Bendary, Hoda M.O. Mokhtar, Aboul Ella Hassanien and Hala A. Shaheen

Abstract Disabilities, specially Spinal Cord Injuries (SCI), affect people behaviors, their response, and the participation in daily activities. People with SCI need long care, cost, and time to improve their heath status. So, the rehabilitation of people with SCI on different period of times is required. In this paper, we proposed an automated system to estimate the rehabilitation length of stay of patients with SCI. The proposed system is divided into three phases; (1) pre-processing phase, (2) classification phase, and (3) rehabilitation length of stay measurement phase. The proposed system is automating International Classification of Functioning, Disability and Health classification (ICF) coding process, monitoring progress in patient status, and measuring the rehabilitation time based on support vector machines algorithm. The proposed system used linear and radial basis (RBF) kernel functions of support vector machines (SVMs) classification algorithm to classify data. The accuracy obtained was full match on training and testing data for linear kernel function and 93.3 % match for RBF kernel function.

Keywords Rehabilitation · Length of stay (LOS) · Spinal cord injuries · International classification of functioning · Disability and health classification (ICF) · Support vector machines (SVMs)

R. Mahmoud (✉)
Faculty of Computers and Information, Fayoum University, Fayoum, Egypt
e-mail: Aboitcairo@gmail.com

N. El-Bendary
Arab Academy for Science, Technology, and Maritime Transport, Cairo, Egypt

H.M.O. Mokhtar · A.E. Hassanien
Faculty of Computers and Information, Cairo University, Cairo, Egypt

H.A. Shaheen
Faculty of Medicine, Fayoum University, Fayoum, Egypt

A.E. Hassanien
Faculty of Computers and Information, Beni Suef University, Beni Suef, Egypt

R. Mahmoud · N. El-Bendary · A.E. Hassanien
Scientific Research Group in Egypt (SRGE), Cairo, Egypt

A. Abraham et al. (eds.), *Intelligent Data Analysis and Applications*,
Advances in Intelligent Systems and Computing 370,
DOI 10.1007/978-3-319-21206-7_45

1 Introduction

According to the International Health Organization (WHO), statistics of people with spinal cord injuries (SCI) in 2013 are between 250,000 and 500,000 persons. SCI may be caused due to preventable causes, such as road traffic crashes, falls, or violence as well as sports. Only 5–15 % of people with spinal cord injuries in low and middle income countries can access the devices they need [1–3].

In 2001, the International Health Organization (WHO) developed and International Classification of Functioning, Disability and Health classification (ICF) to provide a unified and standard language and framework for the description of health and health-related states. ICF aims to improve integration of health information and ensure the quality of the collection of accurate health data. ICF framework describes the health status of individuals via four perspectives: (1) Body function, (2) Body structure, (3) activities and participations, and (4) Environmental factors [4, 5]. Also, ICF established a common language for describing health and health-related comparison of data across countries, and provide a systematic coding scheme for health information systems. The ICF is based on three levels of functioning body functions and structures, activities, and participation with parallel three levels of disability impairments, activity limitations and participation restrictions, in addition to contextual factors that cover environmental factors, and personal factors. Disability denotes the negative aspects of the interaction between an individual with a health condition and individual's contextual factor [4, 6]. The main components of ICF framework are divided into chapters that determine subcategories of each component expressed by numerical values. In the ICF code, the prefix denotes categories followed by the chapter number. For example, body function component with chapter *one* refers to mental functions, chapter *two* refers to sensor functions and pain, chapter *three* refers to voice and speech functions, and so on. Moreover, for each chapter, there are multiple levels, which are expressed by numerical values as well, in order to present more details for individual's case. For example, considering chapter *two* with sensor functions and pain, it can include additional levels of details such as seeing and related functions, hearing and vestibular functions, and so on [2]. After determining the ICF code, the degree of injuries, which is called *ICF qualifier*, is needed to be determined as well. The three components of functioning and disability are used with the same qualifier. The qualifier may equals to 0, which means that there is no impairment, equals to 1 that means mild impairment, equals to 2 that means moderate impairment, equals to 3 that means severe impairment, or equals to 4 that means complete impairment [5, 7].

Support vector machines (SVMs) algorithm [8, 9] is a classification technique used to classify supervised data via performing classification by finding the hyper-plane that maximizes the margin between the two classes. The strength of SVM is the linearly separate data. An ideal SVM analysis should produce a hyper-plane that completely separates the vectors (cases) into two non-overlapping classes. However, perfect separation may not be possible, or it may result in a model with so many cases that the model does not classify correctly. In this situation SVM finds the hyper-plane that maximizes the margin and minimizes the mis-classifications. SVM separates data using mathematical functions called kernel functions such as linear, polynomial, radial basis function (RBF) and sigmoid.

In this paper we propose an automated system to record patient status on different periods of time using ICF code, monitor the progress in patient status using Functional independent measure (FIM), and measure the rehabilitation length of stay for the patient with spinal cord injuries based on Machine learning (ML) support vector machines (SVM) [8, 9] classification algorithm. The proposed system is divided into three phases *(1) pre-processing phase* that converts the entire description of patient status into its equivalent ICF code and store the code in database on different period of times, *(2) classification phase* that uses ICF classifier to classify patients into tree groups (truma, non-truma, and neuro) using linear and RBF support vector machines (SVM) kernel functions, and *(3) rehabilitation length of stay measurement phase* that measures FIM for patients and predicts the length of stay (LOS) based on patient group, FIM vlaue, and their ages.

The remaining of this paper is organized as follows. Section 2 discusses related research work. Section 3 explains preliminaries of the support vector machines (SVMs) classification algorithm, functional independent measure, in addition to benchmarks of rehabilitation length of stay for spinal cord injuries. Section 4 depicts experimental results. Section 5 presents conclusions and discusses future work.

2 Related Work

This section reviews a number of current researches related to spinal cored injuries and ICF coding based rehabilitation systems.

In [7], authors developed a methodology for recording spinal cord neural activity in order to detect and classify external sensory information. They proposed dual-stage classification scheme using principal component analysis (PCA) and k-means algorithms to classify patients into eight classes based on the type of injuries. The mean accuracy they obtained is 96 %.
In [10], authors aimed to reduce the overall length of stay (LOS) of patients with spinal cord injuries by improving the patients outcomes. They developed decision-making tools to guide the discussion in meetings of teams. Also, they developed a revised patient census tool that was implemented to facilitate the communication between team members and collection of data. It has been found that the use of benchmarking and decision support tools improved rehabilitation efficiency, and increased the standardization in LOS determination.

In [1], authors developed ICF Code selection tools based on domiciliary mental health care reports. The developed search tool is based on tracking back in ICF conceptual tree using a directory tool. Also, a full-text search tool for ICF code selection used natural language processing (NLP) technology for displaying codes of each sentence in mental health care reports. The study in that research performed ICF coding of 3199 sentences in 375 reports on 80 patients. Data was divided as 943 sentences of 65 patients for training and 212 sentences of 15 patients for testing. Authors found that 90.1 % of sentences were assigned the correct ICF codes.

Also, in [11], authors enhanced an existing natural language processing (NLP) system to encode rehabilitation discharge summaries based on the ICF framework. A formal evaluation has been conducted where six subjects have used the ICF coding to encode functional status information in 25 discharge summaries. Comparative analysis has been conducted among the coding performed by expert coders, non-expert coders, and the NLP proposed system. Expert coders have obtained better performance than both the non-expert coders and the NLP system. Also, the three coding groups achieved better performance when assigning the first qualifier than the second one.

Moreover, authors in [12] examined the use of natural language processing (NLP) for automated functional status information coding systems. An existing NLP system that was used to encode clinical information has been utilized in that research. Authors created and modified coding table in pre-processing program, where different senses in the functional status information (FSI) domain such as transfers and bladder be recognized as related terms, and post-processing program, which identifies all ICF main codes and related qualifiers, and performs final code assignments, in system's lexicon. Authors developed automated selection approach for ICF codes and performed 549 entries to the lexicon and 181 entries to the coding table. They found many challenges that faced the development of medical language processing software for the FSI as well as the similarity between phrases and ideas, which required a complex medical inference and difficult to formalize.

On the other hand, authors in [6] made 12-Item Short Form Health Survey (SF-12) that's been coded using the ICF coding in order to assess the feasibility of using a generic health measure to create coded functional status indicators. They used the functional indicators to get the characterization of a stroke population, and compare it against health-related quality-of-life summary measures. The SF-12 measures two scores, the first is for physical health and the other one is for mental health. Experimental results showed that persons with stroke scored, on average, approximately 10 points lower than controls on physical and mental health. The ICF coding provided enhanced information about specific functional limitations for persons with stroke. Also, the format of the ICF is more compatible with the structure of administrative health databases and enriches these databases. A limitation of the SF-12 is that it does not capture the full parts of functional limitations that are considered as a consequence of many conditions or accidents.

The automated system proposed in this paper aims to record patient status on different periods of time using ICF codes, monitor the progress in patient status using functional independent measure (FIM), and measure the rehabilitation length of stay for the patient with spinal cord injuries based on Machine learning (ML) support vector machines (SVMs) classification algorithm. The proposed system uses ICF classifier to classify patients into three groups (truma, non-truma, and neuro) using linear and RBF support vector machines (SVMs) kernel functions. Then, it measures FIM for patients and predicts the LOS based on patient group, FIM value, and their ages.

3 Preliminaries

This section provides a brief explanation of support vector machines algorithm, and functional independent measure in addition to benchmarks of rehabilitation length of stay for spinal cord injuries.

3.1 Support Vector Machines (SVMs)

The Support Vector Machines algorithm is founded in 1992 by Boser, Guyon, and Vapnik, and defined as a group of supervised learning used for classification and regression analysis of the high dimensional dataset. SVM solve binary problems and aims to maximize the margin between two classes by using kernel function. There are four types of kernel functions:(1) linear function, (2) polynomial function, (3) radial basis function (RBF), and (4) sigmoid function [8, 9].

Given a set of n input vectors x_i and outputs $y_i \in \{-1, +1\}$, one tries to find a weight vector w and offset b defining a hyperplane that maximally separates the examples. This can be formalized as the maximize problem in Eq. (1).

$$maximize \sum_{i=1}^{n} \alpha_i - \frac{1}{2} \sum_{i,j=1}^{n} \alpha_i \alpha_j y_i y_j . K(x_i, x_j) \tag{1}$$

$$Subject\,to : \sum_{i=1}^{n} \alpha_i y_i, 0 \leq \alpha_i \leq C$$

where the coefficients α_j are non-negative. The x_i with $\alpha_j > 0$ are called support vectors. C is a parameter used to trade off the training accuracy and the model complexity so that a superior generalization capability can be achieved. K is a kernel function transforms the data into a higher dimensional feature space to make it possible to perform the linear separation.

Linear function is a method in which no kernel is used. Eq. (2) is defined as a decision boundary which divided the space into two sets.

$$\Theta^T X = \sum \Theta^n_{i=0} X^n_{i=0} \tag{2}$$

where θ is a wight vector, X is a set of data contains on n cases on patterns, and x_i is a pattren in X.

Table 1 Eighteen items/dimensions of Functional Independence Measure (FIM) scale [13]

1- Eating	2- Grooming
3- Bathing	4- Upper body dressing
5- Lower	6- Body Dressing Toileting
7- Bladder management	8- Bowel management
9- Bed to chair transfer	10- Toilet transfer
11- Shower transfers	12- Locomotion (wheelchair level)
13- Stair	14- Cognitive comprehension
15- Expression	16- Social interaction
17- Problem solving	18- Memory

3.2 *Functional Independent Measure and Benchmarks of the Rehabilitation Time of SCI*

Functional independent measure is a standard score developed by rehabilitation center to evaluate the progress in patient status and the rehabilitation outcomes. FIM consists of eighteen element, as shown in Table 1. These elements are scored from 1 to 7 based on level of dependency. Score 7 means full Independence and score 1 means total dependency [10, 13].

The FIM is measured by Eq. (3) [14].

$$FIM = \sum_{i=1}^{18} items. \tag{3}$$

The benchmarks of rehabilitation length of stay for spinal cord injuries are developed in 2013 by (Burns et al. 2013). They classified patients into three groups according to type of injuries (traumatic, non-traumatic, and neuro), also they used FIM to determine the required rehabilitation length of stay. They used 12 item of FIM motor tasks (excludes shower transfer). Table 2 shows the benchmarks which they found.

4 Proposed Automated System for SCI Rehabilitation LOS

In this section we will discuss model and phases of the proposed system. The system model consists of three phases: pre-processing phase, classification phase, and rehabilitation length of stay for spinal cord injuries, as shown in Fig. 1.

Table 2 Benchmarks of rehabilitation length of stay for spinal cord injuries based on category (Trumatic, Non-trumatiic, and Neuro), FIM Admision motor, and age [10]

Category	Admission motor FIM	Age	Predicted LOS
Neuro	12–32	N/A	56 days
	33–55	N/A	41 days
	56–74	N/A	30 days
	75–84	N/A	20 days
Traumatic	12–16	N/A	125 days
	17–41	≥31 years	83 days
	17–41	≤31 years	92 days
	42–84	N/A	44 days
Non-traumatic	12–28	N/A	76 days
	29–54	≥51	46 days
	29–54	≤51	63 days
	55–72	N/A	29 days
	73–84	N/A	23 days

Fig. 1 Proposed system model

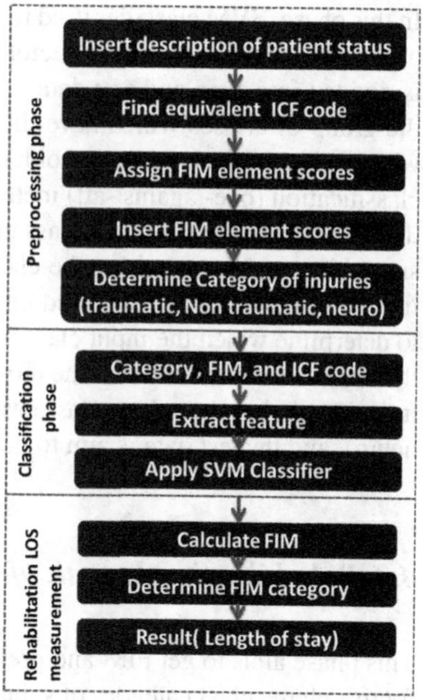

4.1 Preprocessing Phase

This phase aims to preparing data that required as a input in the other phases. The first step aims to convert the whole description of patient status into equivalent ICF code. The input of this step the description of patient status and the target output is the equivalent ICF code, also this step is required to repeated on different period of times to monitoring the progressive in patient with spinal cord injuries. The whole details of this phase discussed and proposed in (Mahmoud et al. 2014) [15]. The second step of this phase is assigning and entering FIM scores, we manually assigned the FIM scores from 0 to 7 based on expert report of patient status, then we used these scores as a input to get FIM of patient and stored in database. The third step of this phase is determining category of spinal cord injuries. We checked and read the report of patient status, and classified the patient into three groups or category based on type of injuries (traumatic, non traumatic, and neuro).

4.2 Classification Phase

In this phase, SVM classifier used to classify input matrix using train model, the aim of this phase is classifying the vector based on feature of each class. The input matrix is divided into train and test data using svmtrain method and the desired output is the group of classes which have the same feature. The SVM is a binary approach which solve the (one-to-one) problems. In the proposed system we used multi-class classification (one- against-all) method to divide the matrix into three classes.using the SVM classifier the multi-class classification decompose the whole classes into small binary classes and get the classification. So based on the SVM classifier the input matrix is randomly divided into train and test data. The output of this phase is to determine which the input class belong to previous defined classes by measuring the distances matrix and find the most fitting class. According to the input matrix the train data divided into three class based of type of injuries (traumatic, non-traumatic, neuro), and the test data is aim to get the most fitting class for the input matrix.

4.3 Rehabilitation Length of Stay Measurement Phase

This phase aims to get FIM and predicted rehabilitation length of stay. The input of for this phase, class category of spinal cord injuries, age, and 12 items of FIM admission motor (excludes shower transfer). By using Eq. (3) we get the FIM admission motor as summation of 12 items. The score of FIM admission motor and class category (traumatic, non traumatic, and neuro) is used to determine the predicated LOS as shown in Table 2. So the target output of this phase is rehabilitation length of stay. We could used the FIM efficiency after the first rehabilitation the patient obtained.

5 Experimental Results and Discussion

5.1 Dataset

The dataset that used in the proposed system divided into four categories; namely *ICF Core data set for spinal cord injuries, seventy case-studies for real patients with spinal cord injuries, benchmark cases of FIM, benchmark for rehabilitation Length of stay*. The ICF Core data set for spinal cord Injuries was developed with a cooperative effort between the ICF Research Branch, the Classification, Assessment and Terminology team and the Disability and Rehabilitation team at the World Health Organization (WHO). It was started in 2005 and first version is finished in 2007 [15, 16]. Case-studies for real patients with spinal cord injuries were gathered by the Swiss Paraplegic Research for the aim to contribute to optimal functioning, social integration, and health and quality of life for people with SCI through clinical and community-oriented research [15, 17]. The collected case-studies cover rehabilitation of spinal cord injury patients of different age, gender, and with different level of injuries.Benchmarks of FIM were used to evaluate the efficiency of the progress in SCI patient's status [2, 13]. Finally the benchmark for rehabilitation Length of stay, which developed by (Burns et al. 2013) to measure the rehabilitation length of stay based of type of injuries (traumatic, non-traumatic, neuro), FIM admission motor, and age.

5.2 Results and Discussions

The proposed system is preformed on actual data of seventy cases for patients with spinal cord injuries. The system using linear and RBF kernel function to classify the patients. To measuring the accuracy of kernel functions we used the accuracy in Eq. (4), and we obtain full match on training and testing data for linear kernel function and 0.933 for two function respectively as shown in Fig. 2. Table 3 shows results of FIM and LOS for the tested 28 cases.

$$Accuracy = \frac{Number\ of\ correctly\ classified\ cases}{Total\ number\ of\ testing\ cases} * 100 \quad (4)$$

Figure 3 shows the relationship between FIM measure admission motors and Rehabilitation LOS for the three categories Trauma, non trauma, and Nuero. Figure 3 shows that the LOS value is decreasing when the FIM is increasing for the three category. Also, Fig. 3 shows that trauma cases have the higher values of predicated LOS than in all neuro cases. Trauma cases have the higher values of predicated LOS on different FIM values than non-trauma in all cases expected expected when FIM equals 42. When the FIM equals 42 in trauma the predicted LOS still fixed on 44 day ignoring the increasing of FIM value. Otherwise in non- trauma the predicted LOS for FIM value 42 are 46 days or 63 days for ≥51, or ≤51 age respectively.

Fig. 2 Accuracy of linear and RBF kernel functions

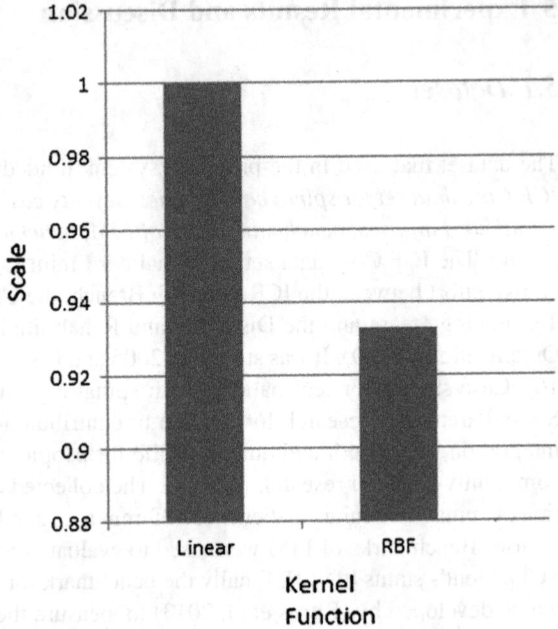

Table 3 Eighteen items/dimensions of Functional Independence Measure (FIM) scale

Case no.	Category	Age	FIM	LOS	Case no.	Category	Age	FIM	LOS
1	Traumatic	79	15	125	**14**	Traumatic	22	34	92
2	Traumatic	79	45	44	**15**	Neuro	23	29	56
3	Traumatic	26	12	125	**16**	Traumatic	41	15	125
4	Traumatic	26	31	92	**17**	Traumatic	41	33	83
5	Neuro	23	19	56	**18**	Non-traumatic	15	42	63
6	Traumatic	26	36	92	**19**	Traumatic	41	35	83
7	Traumatic	35	19	83	**20**	Traumatic	36	12	125
8	Non-traumatic	15	39	63	**21**	Traumatic	36	17	83
9	Traumatic	35	29	83	**22**	Neuro	23	33	41
10	Traumatic	35	33	83	**23**	Traumatic	36	19	83
11	Neuro	23	34	41	**24**	Traumatic	17	28	92
12	Traumatic	22	14	92	**25**	Non-traumatic	15	56	29
13	Traumatic	22	27	92	**26**	Traumatic	17	34	92
14	Traumatic	17	36	92	**28**	Traumatic	17	37	92

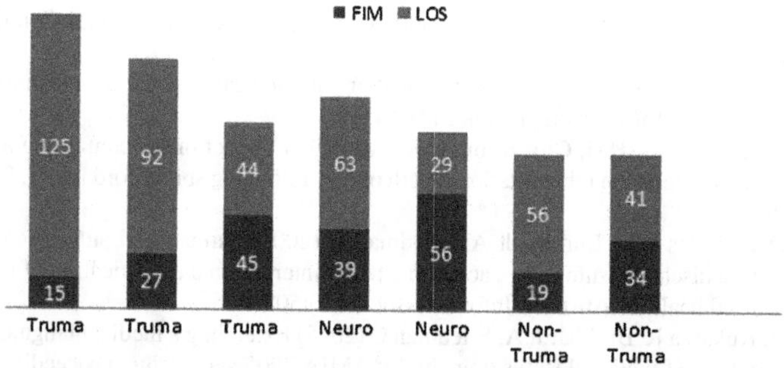

Fig. 3 Relationship between functional independent measure admission motors and Rehabilitation length of stay for spinal cord injuries

6 Conclusions and Future Work

In this articles we propose a system that measure the rehabilitation length of stay for patient with spinal cord injuries based on functional independent measure admission motors, category (traumatic, non traumatic, and neuro), and age. The system classified data using support vector machines linear and RBF kernel functions to classify based of three categories, then the proposed system calculated the FIM admission motors, and finally the output of rehabilitation LOS is measured for patient with spinal cord injuries. The accuracy for linear is totally match for training and test data, and 93.3 % match for RBF kernel function. Our future work aims to apply collaborative filtering recommender system, and the similarity methods on patient with spinal cord injuries to recommend the rehabilitation methods.

References

1. Joan L (2014) Healdly, "world health organization releases internship perspective on spinal cord injuries." J Vent Assist Living 28(1):1–2
2. World Health Orgnization (WHO) (2013) Spinal cord injury. http://www.who.int/mediacentre/factsheets/fs384/en/. Accessed 20 Feb 2015
3. Center, National, others (2012) Spinal cord injury facts and figures at a glance. J Spinal Cord Med 36:1–2
4. World Health Organization (WHO) (2001) ICF: international classification of functioning, disability and health (ICF manual). World Health Organization, Geneva
5. Boldt C (2013) The international classification of functioning, disability and health (ICF) in nursing: persons with spinal-cord injury as an example. J Adv Nurs 46(2):162–170
6. Mayo NE, Poissant L, Ahmed S, Finch L, Higgins J, Salbach NM, Soicher J, Jaglal S (2004) Incorporating the international classification of functioning, disability, and health (ICF) into an electronic health record to create indicators of function: proof of concept using the SF-12. J Am Med Inform Assoc 11(6):514–522
7. Mayo N, Lowell T, James TW, Gary E, Matt S (2006) Detection and classification of sensory information from acute spinal cord recordings. IEEE Trans Biomed Eng 53(8):1715–1719

8. Wu Q, Zhou D-X (2006) Analysis of support vector machine classification. J Comput Anal & Appl 8(2):99–119
9. Vanitha L, Venmathi AR (2011) Classification of medical images using support vector machine. Int Conf Inf Network Technol IACSIT 4
10. Burns A, Yee J, Flett HM, Guy K, ournoyea M (2012) Impact of benchmarking and clinical decision making tools on rehabilitation length of stay following spinal cord injury. Spinal Cord 51(2):165–169
11. Kukafka R, Bales ME, Burkhardt A, Friedman C (2006) Human and automated coding of rehabilitation discharge summaries according to the international classification of functioning, disability, and health. J Am Med Inform Assoc 13(5):508–515
12. Bales M, Kukafka R, Burkhardt A, Friedman C (2005) Extending a medical language processing system to the functional status domain. In: AMIA 2005 symposium proceedings, p 888
13. Rehabilitation Institute of Chicago, Center for Rehabilitation Outcomes Research and Northwestern University Feinberg School of Medicine Department of Medical Social Sciences Informatics group (2013) "Rehab measures—functional independence measure, the rehabilitation measures database. http://www.rehabmeasures.org/Lists/RehabMeasures/PrintView.aspx?ID=889. Accessed 25 Feb 2015
14. Meyer M, Britt E, McHale H, Teasell R (2012) Length of stay benchmarks for inpatient rehabilitation after stroke. J Disabil Rehabil 34(13):1077–1081
15. Mahmoud R, Elbendary N, Mokhtar H, Hassanien A (2014) ICF based automation system for spinal cord injuries rehabilitation ICCES 2014:192–197 (Cairo, Egypt)
16. Icf-research-branch.org (2007) ICF core set projects, ICF RESEARCH BRANCH - Development of ICF Core Sets for Spinal Cord Injury (SCI). http://www.icf-research-branch.org/icf-core-sets-projects-sp-1641024398/neurological-conditions/development-of-icf-core-sets-for-spinal-cord-injury-sci. Accessed 20 Feb 2015
17. Jaimie F, Cieza A, Hoogland-Eriks I, Rauch A, Stucki G, Wong V (2009) ICF case studies implementation of the international classification of functioning, disability and health (ICF) in rehabilitation practice. http://www.icf-casestudies.org/. Accessed: 15 Feb 2015

Region-based Image Fusion Approach of Panchromatic and Multi-spectral Images

Reham Gharbia, Ali Hassan El Baz, Aboul Ella Hassanien and Vaclav Snasel

Abstract In this paper, a region-based image fusion approach were proposed based on the stationary wavelet transform (SWT) in conjunction with marker-controlled watershed segmentation technique. The SWT is redundant, linear and shift invariant and these properties allow SWT to be realized exploiting a recursive algorithm and gives a better approximation than the DWT. The performance of the fusion approach is illustrated via experimental results obtained with a broad series of images and the experimental results used the MODIS multi-spectral bands and Spot panchromatic band to validate the proposed image fusion technique. Moreover, the visual presentation and different evaluation criteria including the standard deviation, the entropy information, the correlation coefficient, the root mean square error, the peak signal to noise ratio and the structural similarity index was used to evaluate the obtained results. The proposed approach achieves superior results compared with the existing work.

Keywords Remote sensing · Multi-sensor image fusion · Region based image fusion · Stationary wavelet transform(SWT) · Marker-controlled watershed segmentation technique · Panchromatic image · Multi-spectral images

---

Scientific Research Group in Egypt (SRGE) http://www.egyptscience.net.

R. Gharbia (✉)
Nuclear Materials Authority, Cairo, Egypt
e-mail: Aboitcairo@gmail.com

A.H. El Baz
Faculty of Science, Damietta University, Damietta, Egypt

A. E. Hassanien
Faculty of Computers & Information, Cairo University, Cairo, Egypt

A. E. Hassanien
Faculty of Computers and Information, Beni Suef University, Beni Suef, Egypt

V. Snasel
Faculty of Electrical Engineering and Computer Science, Department of Computer Science and IT4Innovations, VŠB-Technical University of Ostrava, Cairo, Czech Republic

A. Abraham et al. (eds.), *Intelligent Data Analysis and Applications*,
Advances in Intelligent Systems and Computing 370,
DOI 10.1007/978-3-319-21206-7_46

1 Introduction

The most earth observation satellite systems such as Landsat 7, SPOT, IKONOS and QuickBird provides sensors only with high spatial resolution panchromatic (Pan) or multi-spectral (MS) bands simultaneously. Many remote sensing applications necessity high spatial resolution and multi-spectral resolution in the same time, until, it is very hard to produce a sensor with both high spatial and spectral characteristics; due to the limitation of the technologies. Multi-sensor image fusion has received significant attention in many applications such as remote sensing image fusion is a procedure to integration of disparate and complementary data to increment the information content in the output image and then increasing the ability to the interpretation. This pays to more accurate data and utility for interpretation.

The image fusion would take place at three different levels [1]; The pixel level, the feature level and the decision level. The pixel level image fusion is the simplest level and it interested in the information associated with each pixel and the fused image is resultant from the corresponding pixel values of source images. The pixel level is performed in the spatial domain or the transform domain [2]. Some multi-resolution is presented in the image fusion pixel level, like the pyramid and the wavelet. The feature level fusion, source images are segmented into regions, and it considered on the feature or object. The features like pixel intensities, edges of texture, are utilized for fusion. It overcomes some potential drawbacks of the pixel based approaches [3]. The decision level fusion is a high level fusion which is based on statistics, fuzzy logic, prediction and heuristics, etc. [4]. The decision level fusion depends on integrating the information at a higher level of abstraction, achieves the results from multiple algorithms by applying decision rules to produce a final fused image and reinforce common interpretation. The decision level fusion which deals with symbolical representation of images is very complicated [5]. The objects and the region details are more advantageous than the individual pixels [6]. In the region based image fusion approach, the input images are divided into several regions and the important features and objects of each region are extracted. These features and objects are used to decide the input image from which a particular region is to be selected in the fused impersonation. The region image approach is more flexible in adapting to intelligent fusion rules than pixel based approach [7]. The quality of any region based image fusion technique relies on how well the regions are extracted from the source images [8]. In this paper, a region image fusion approach were introduced using the SWT and the watershed transform. This approach is based on segmenting two source images into regions of interest, by using a suitable segmentation technique. This process is followed by fusion of images.

The rest of this paper is organized as follows, Sect. 2 presents an overview of the main techniques used in this paper including the stationary wavelet transform and marker-controlled watershed segmentation. Section 3 explains the phases of the proposed region based image fusion technique and its detailed steps. Experimental scenarios are introduced in Sect. 4. Finally, the conclusions are illustrated in Sect. 5.

2 Preliminaries

2.1 The Stationary Wavelet Transform

The wavelet transform is used extensively in image processing and provides a good resolution in both the time and frequency domains. It provides a multi-resolution decomposition of an image on a biorthogonal basis [9]. The discrete wavelet transform plays an important role in image fusion. It is most widely used in multi resolution transform for image fusion. Because it minimizes the structural distortions compared with the various others transforms. The Wavelet decompositions are very suitable to detect edges,that is essentially important in remote sensing where the images have low contrast or the features in the images are not distinct. However, the wavelet transform suffers from lack of shift invariance, poor directional selectivity and absent from phase. These disadvantages are eliminated by Stationary Wavelet transform (SWT). The most important advantages of SWT transform is translation invariant, this is done without downsampling process. The SWT is introduced in 1996, and it is similar to discrete wavelet transform. The SWT transform is described as the following:

Let's $f(x)$ is a function projected at every step j on the subset U_j where $(\ldots\subset U_2 \subset U_1 \subset U_0)$. The projection can be gained by the scalar product of $C_{i,j}$ of $f(x)$ with the scaling function $\phi(X)$ which is dilated and translated.

$$C_{i,j} = < f(x), \phi(x) > \tag{1}$$

$$\phi_{i,j}(x) = 2^{-j}\phi(2^{-j}x - k) \tag{2}$$

where $\phi(x)$ is the scaling function, it is a low-pass filter and $C_{i,j}$ is identified by a discrete approximation at the resolution 2^j.

For SWT transform, in state of downsampling like DWT, an upsampling process is done before performing filter convolution at all scales. The distance between samples increasing by a factor of two from scale j to the next,The coefficients $h(l)$ form low-pass filter and the coefficients $g(l)$ high-pass filter where $C_{j+1,k}$ is give by the following equation:

$$C_{j+1,k} = \sum h(l)C_{j+1,k} \tag{3}$$

and the discrete wavelet coefficients is defined as follows:

$$Z_{j+1,k} = \sum g(l)C_{j+1,k} \tag{4}$$

2.2 Marker-Controlled Watershed Segmentation

The watershed transform is a powerful mathematical morphological approach and it is looked as region based segmentation technique. The watershed transform considers an image of two dimensional as a topographic relief. The pixel value regards as its elevation. The watershed lines split the image into catchment basins, so that each basin is connected with one minimum in the image. The watershed transform is usually applied to the gradient function of an image. The gradient determines transitions between regions, for it has high values on the borders between objects and minimum values in the homogeneous regions. In this way, if the peak lines in the gradient image coincide with the edges of image objects, watershed transform splits the image into meaningful regions [10]. The watershed transform is implemented on gray scale image to solve a variety of image segmentation problems. The watershed ridge lines in an image is considered as a surface where light pixels are high value, and dark pixels are low value. To overcome the over segmentation problem, marker-controlled watershed is introduced by [11]. The major idea is to perform watershed segmentation around user specific markers rather than the local maxima in the input image.

3 The Proposed Region-Based Image Fusion Approach

In this section, a novel of the region based image fusion approach using the SWT and marker-controlled watershed segmentation is submitted. This approach is implemented on remote sensing data. The proposed region based image fusion technique framework contains several processes described in the following subsections.

3.1 Image Registration and Resampling

The image registration is the alignment and overlaying for two or more images of different times for the same scene, by different sensors or from different Points of view. We used the ground control point technique to register the MS image to the PAN [12]. Since the input images have different spatial resolution so the input images have different size. To perform the image fusion process; the input images must have the same size. Prior any image fusion technique, a resampling of the MS images should made. The resampling operator has an influence upon the final result. Bilinear resampling approach were applied and it offers a good compromise between the required computer time and the accuracy of the result.

3.2 Histogram Matching and SWT Decomposition

The histogram matching was implemented on the panchromatic image to ensure the mean and standard deviation value of the panchromatic image and multi-spectral images are within the same range. The SWT decomposes the two source images;

the MS (3bands) image and the Pan (1 band) image at every level resulting in three details subbands and one approximation subband (HL, LH, HH and LL bands). LH (horizontal information in high frequency), HL (vertical information in high frequency), HH (diagonal information in high frequency) and the approximation imag, symbolized by LL, which consist of the low frequency information. The SWT transform decomposed the LL subband recursively to obtain the level 2.

The coefficients of approximation and detail images (sub-band images) are substantial features, which are helpful for image fusion. The decomposed images are more suitable for segmentation techniques to extract the major features in the region based fusion. The features derived from the SWT transform images are used to segment source images precisely by virtue of multi-scale approach.

3.3 Segmentation Using Marker Controlled Watershed Segmentation

The major function of image segmentation is to gain a set of segmentation regions $K(k^1, k^2, k^3..k^n)$, k^n represents the segmentation regions at level n. This representation will control the other procedures of the region based image fusion. The accuracy of the segmentation technique plays a vital role the fusion process. Segmentation process is given in Algorithm (1) (Fig. 1).

Algorithm 1 Segmentation Algorithm

1: Determine a segmentation function, which is the image whose dark regions are considered as the objects which are used to be segmented. Use the Gradient Magnitude as the segmentation function.

$$\begin{aligned} g(x, y) &= grad(f(x, y)) \\ &= \sqrt{(f(x, y) - f(x-1, y))^2 + (f(x, y) - f(x, y-1))^2} \end{aligned} \tag{5}$$

2: Determine the foreground markers. that are connected blobs of pixels within every of the objects. The morphological techniques called opening by reconstruction and closing by reconstruction to clean up the image are used.

3: Determine background markers. Actually, These are pixels that are not part of any object. The background pixels are in black, which are too close to the edges of the objects; Which are wanted to segment. The background will be thin. This can be done by computing the watershed transform and then looking for the watershed ridge lines.

4: Improve the segmentation function so that it has minima at the foreground and background marker positions. Modify an image so that it has regional minima only in certain desired locations and modify the gradient magnitude image so that its only regional minima occur at pixels that at foreground and background marker.

$$g(x, y) = max\big(grad(f(x, y)), g^{\theta}\big) \tag{6}$$

Where g^{θ} means threshold value.

5: Compute the watershed transform of the improvement segmentation function.

6: Finally, the region matrix label is obtained for every band of the input images.

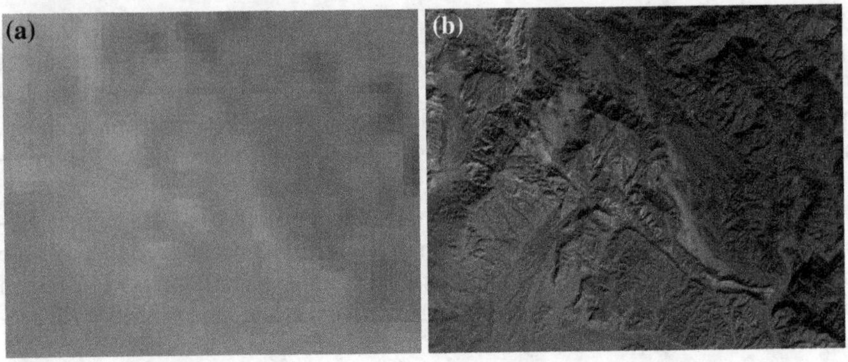

Fig. 1 The input MODIS MS and Spot Pan images **a** the MODIS MS image, **b** the Spot Pan image

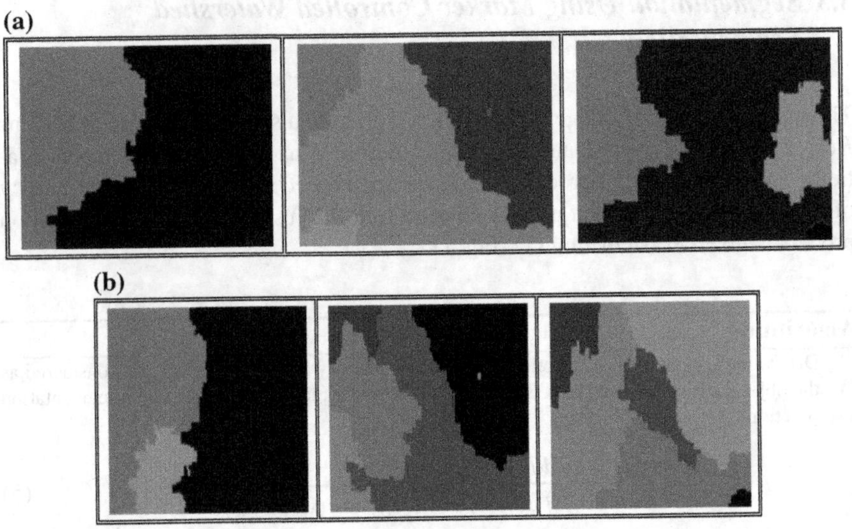

Fig. 2 Region segmentation of the approximate MS image at two levels of SWT decomposition **a** MODIS MS image 3 bands at level1, **b** MODIS MS image 3 bands at level2

The segmentation map of MS image (3 bands) at two levels of SWT transform decompositions in Fig. 2. The segmentation map of Pan image(1 band) at two levels of SWT transform decompositions in Fig. 3. This means that is at each level k, the segmented images R^k which consists of the different regions R at this level.

3.4 Fusion Rules

The feature selection of the image fusion rules are used to build the detail images decision map and the region activity table is produced based on the regions assigned

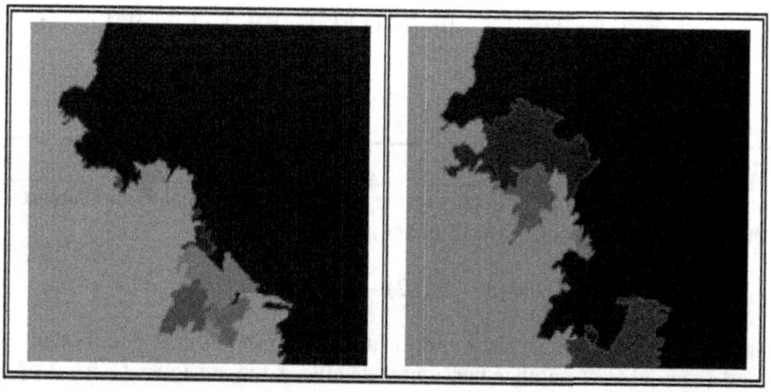

Fig. 3 Region segmentation of the approximate pan image at two levels of SWT decomposition **a** Spot Pan image at level1 and level2

to the coarse approximation image using the watershed transform. The region and feature fusion rules are then applied to the corresponding activity table to produce the fusion decision map that will determine how the multiscale representations will be involved to build the fused wavelet coefficient map.

Decision map This part is the core of the combination technique. The decision map is a matrix of the same size as the source images that is used to fuse. it could be considered as a guide that tells the fusion technique from which one of the input images each region will be taken from. Its output governs the actual combination of the coefficients of the SWT decompositions of the various images. For each level k, orintation p, and position (i,j) the decision process assigns a value $d^{(k)}(i,j|p)$.

Activity measures To determine all the regions in the source images MS and Pan, the two region representations are over laid onto each other to create a joint region map R_{Fn} at each level of decomposition. The region activity procedure is give in Algorithm (2).

Algorithm 2 Region activity algorithm

1: The regions determined in the multi-scale joint region maps are assigned a label,

$$R = R_n^k \quad (7)$$

2: Calculate the area of the regions. This is given by the total number of pixels within the boundary of the region.

3: Overlay the boundaries of the joint region map onto the source images.
To be able to exploit region based fusion it is vital to segment the image into regions.

4: The activity measure at each of the regions for image A, known as the feature activity information are calculate using:

$$A_A(X) = \frac{1}{S_i} \sum_{s_i} a_A(i,j) \quad (8)$$

However, the fused the detail coefficients of the SWT at level 2 ia given in Algorithm (3).

Algorithm 3 Region activity algorithm

1: the absolute values of horizontal details of the first image(Pan image) and subtract the second part of the second image(band 1 at MS image) from first.

$$D = (abs(H1L2) - abs(H2L2)) >= 0 \tag{9}$$

2: And then for fused horizontal part multiplicative D and horizontal detail of first image(Pan image) and then subtract another horizontal details of second image(1 band of MS image) multiplied by logical not of D from first.
3: Find D for vertical and diagonal parts and obtain the fused vertical and details of image.
4: Repeat for fusion at first level.
5: Finally, Fused image is obtained by taking inverse of the SWT.

4 Experimental Results and Discussion

The MODIS bands 1, 4 and 3 as multi-spectral image and Spot panchromatic as pan images are used in the experiments and the following are the description of the tested images and their characteristics.

4.1 Data Sets: Panchromatic and Multi-spectral Images

Remote sensing optical images are acquired over the earth surface in visible, near infrared and short-wave infrared bands. the wavelengths range of the optical bands is from 0.30 to 15.00 mm. The optical images are classified into four categories, Panchromatic (Pan), multi-spectral (MS), super-spectral and hyper-spectral.

A panchromatic (Pan) sensor acquires a single band detector which is sensitive to radiation within a wide spectral range covering visible as well as IR wavelengths. It is a gray image. The Pan image always has greater resolutions than the MS images from the same satellite. It is due to the much more energy per unit area gathered by a Pan sensor due to its wider bandwidth. SPOT images provides a high resolution optical imaging Earth observation satellite. The panchromatic band is 10 m as spatial resolution. The sensor of a multi-spectral image is acquired from a set of band sensors (less than 10 bands). Compared to Pan image, the recorded radiation of a multispectral image is within a narrow range of wavelength for each band. The multispectral images have a low resolution and are color images. This natural color image acquired on August 2011 and it is the combination of bands 1, 4 and 3, shows South Western Sinai, Egypt. Figure 1 shows the input MODIS MS and Spot Pan

Fig. 4 The fused image

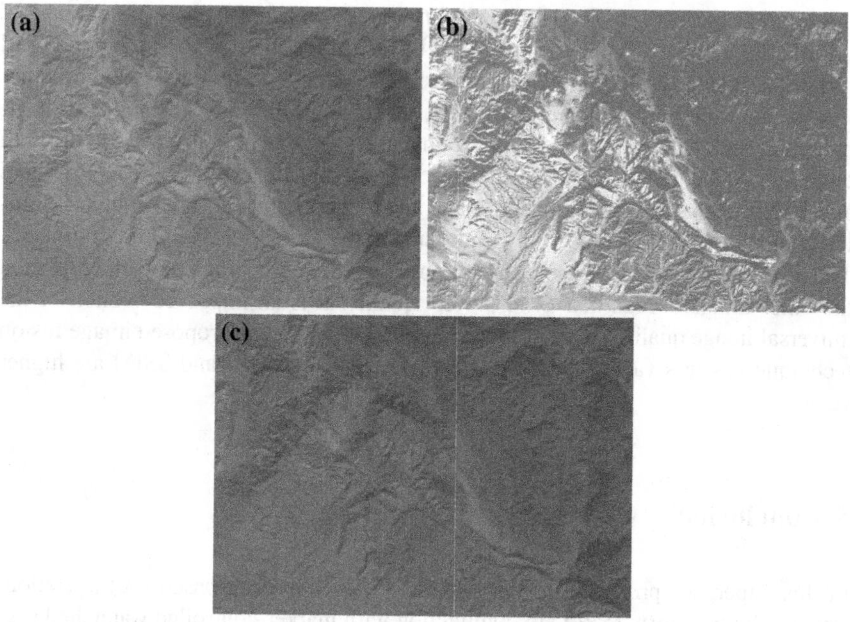

Fig. 5 The image fusion results of the proposed algorithm against different techniques on pixel level a WT image. b PCA image. c Proposed technique fused image

images. Its a natural color image acquired on August 2011 and it is the combination of bands 1, 4 and 3, that shows South Western Sinai, Egypt (Figs. 4 and 5).

Table 1 shows a a comparative analysis of the proposed approach against published approaches and the proposed approach results exposures the best performance more than the other technique. The proposed technique has the smallest value of SD. SD value reflects the deviation from the mean of the image .SD increasing leads to increase the dispersion. Correlation Coefficient (CC) measures how the Convergence between the input and output images (CC) is widely used for comparing image [15]. CC expresses the spectral information contained in the fused image depending on the

Table 1 Comparative analysis results of the wavelet transform and PCA pixel image fusion and the proposed region based image fusion

Image	SD	EI	CC	RMSE	PSNR	SSIM
WT	27.0241	6.4741	0.899	29.03	39.0999	0.6622
PCA	36.8083	6.3922	0.7856	36.0846	7.60182	0.4977
The proposed	23.8017	6.5191	0.9043	20.5178	35.0095	0.7681

original MS image and the ideal value is 1 [13]. The proposed technique has highest value of CC; this means; the proposed technique fused image is more closely to the original MS image. The entropy information (EI) of an image is a measure of information content. The highest value in EI means the information in the proposed technique image fused is more than any other technique. The PSNR is commonly used as measure of quality reconstruction of image. In this case; the signal is original image and the noise is the error introduced. High value of PSNR infer to the high quality of image. It depends on MSE and corresponding distortion metrics [16]. MSE and PSNR are used for measuring image spatial quality [14]. The Structural Similarity index (SSIM) is a method for measuring the similarity between two images. The SSIM index is a quality measure of one of the images being compared provided the other image is regarded as of perfect quality. It is an improved version of the universal image quality index proposed before [15, 16]. The proposed image fusion technique has less value of RMSE. On contrary; The PSNR and SSIM are higher value.

5 Conclusions

In this paper, we proposed a region-based fusion approach based on the stationary wavelet transform (SWT) in conjunction with marker-controlled watershed segmentation technique. The performance of the fusion approach were illustrated via experimental results obtained with a broad series of images and the experimental results used the MODIS multi-spectral bands and Spot panchromatic band to validate the proposed technique. The proposed approach provides improved subjective and objectives results compared to the previously reported pixel and region-based fusion methods.

Acknowledgments This work was supported by the IT4Innovations Centre of Excellence project (CZ.1.05/1.1.00/02.0070), funded by the European Regional Development Fund and the national budget of the Czech Republic via the Research and Development for Innovations Operational Programme and by Project SP2015/146 "Parallel processing of Big data 2" of the Student Grant System, VŠB - Technical University of Ostrava.

References

1. Pohl C, Van Genderen JL (1998) Review article multisensor image fusion in remote sensing: concepts, methods and applications. Int J Remote Sens 19(5):823–854
2. Jiang Y, Wang M (2014) Image fusion with morphological component analysis. Inf Fusion 18:107–118
3. Liling G, Yingjie Z (2008) Image fusion based on segmentation and iterative strategies. In: 3rd IEEE international conference on innovative computing information and control, (ICICIC'08), pp 390–390
4. Hima Bindu C, Veera Swamy K (2014) Medical image fusion using content based automatic segmentation. In: Recent advances and innovations in engineering (ICRAIE), pp 1–5
5. Mishra HOS, Bhatnagar S, Shukla A, Tiwari A (2014) Medical image fusion based on wavelet transform. Int J Sci Eng Res 5(2):772–778
6. Nirmala DE, Paul ABS, Vaidehi V (2013) Improving independent component analysis using support vector machines for multimodal image fusion. J Comput Sci 9(9):1117–1132
7. Wan T, Canagarajah N, Achim A (2009) Segmentation-driven image fusion based on alpha-stable modeling of wavelet coefficients. IEEE Trans Multimedia 11(4):624–633
8. Zaveri T, Zaveri M, Shah V, Patel N (2009) A novel region based multifocus image fusion method. IEEE Int Conf Digital Image Proc 50–54
9. Chaithanya JK, Ramashri T (2014) Area level fusion of multi-focused images using Feature Level Image Fusion (FLIF) Algorithm. pp 285–295
10. Tarabalka Y, Chanussot J, Benediktsson JA (2010) Segmentation and classification of hyper-spectral images using watershed transformation. Pattern Recognit 43(7):2367–2379
11. Meyer F, Beucher S (1990) Morphological segmentation. J Vis Commun Image Represent 1(1):21–46
12. Gharbia R, Ahmed SA, ella Hassanien A (2015) Remote sensing image registration based on particle swarm optimization and mutual information. In: Information systems design and intelligent applications, pp 399–408
13. Yang X, Huang F, Li G (2009) Urban remote image fusion using fuzzy rules. Int Conf Mach Learn Cybern 1:101–109
14. Antonino F, Edmondo T (2011) Applications. In: Antonino F, Edmondo T (eds) Hybrid random fields, vol 15. ISRLSpringer, Heidelberg, pp 121–150
15. Rana A, Arora S (2013) Comparative analysis of medical image fusion. Int J Comput Appl 73(9):10–13
16. Wang Z, Bovik AC, Sheikh HR, Simoncelli EP (2004) Image quality assessment: from error visibility to structural similarity. IEEE Trans Image Proc 13(4):600–612

Biomarker-Based Water Pollution Assessment System Using Case-Based Reasoning

Asmaa Hashem Sweidan, Nashwa El-Bendary, Osman Mohammed Hegazy and Aboul Ella Hassanien

Abstract This paper presents Case-Based Reasoning (CBR) system to asses water pollution based on fish liver histopathology as biomarker. The proposed approach utilizes fish liver microscopic images in order to asses water pollution based on knowledge stored in the case-based database and stores likelihood description of the previous solutions in order to make the knowledge stored more flexible. The proposed case-based reasoning system consists of 5 phases; namely case representation (pre-processing and feature extraction), retrieve, reuse/adapt, revise, and retain phases. After applying pre-processing and feature extraction algorithms on the input images, similarity between the input and case base database is being calculated in order to retrieve similarity. Experimental results show that the performance of CBR systems increases according to the number of retrieved cases in each scenario against each strategy. The proposed system achieved 95.9 % accuracy for all water quality degrees.

Keywords Case-based reasoning · Similarity · Fish liver · Water pollution

A.H. Sweidan (✉)
Faculty of Computers and Information, Fayoum University, Fayoum, Egypt
e-mail: Aboitcairo@gmail.com

N. El-Bendary
Arab Academy for Science, Technology and Maritime Transport, Cairo, Egypt

O.M. Hegazy · A.E. Hassanian
Faculty of Computers and Information, Cairo University, Cairo, Egypt

A.E. Hassanien
Faculty of Computers and Information, Beni Suef University, Beni Suef, Egypt

A.H. Sweidan · A.E. Hassanien
Scientific Research Group in Egypt (SRGE), Cairo, Egypt

A. Abraham et al. (eds.), *Intelligent Data Analysis and Applications*,
Advances in Intelligent Systems and Computing 370,
DOI 10.1007/978-3-319-21206-7_47

1 Introduction

Water pollution significantly affects people's ability to obtain and use water, as well as for maintaining the standards of healthy life [1]. Water quality refers to the chemical, physical and biological properties of water [2]. It is a measure of the status of water relative to the requirements of one or more biotic and/or any human need or purpose [3]. It is most commonly used with referring to a set of standards that can assess compliance. The most common standards used to assess water quality related to the health of ecosystems [4].

The aquatic environment with its water quality is considered the major factor controlling the case of health and disease in both wild and farmed fish [5]. In Egypt, effluents discharged from the various factories directly to the river Nile and the wide use of agricultural drainage water and waste municipal water in fish culture disturb the water quality and increase the water heavy metals content. To undertake chemical analysis is an ongoing complex and expensive, and also provides limited data regarding chemical compounds, which ignores the effect of the excluded data in the analysis process [6, 7].

Biomarkers, also known as Bioindicators, are used increasingly in order to solve the problems that previously mentioned. Biomarkers are also considered as witnesses of the water conditions. There was a growing demand for comfort and sensitive tools for the detection of biomarkers in the aquatic environment [8].

There exist quite a few systems in the literature that monitor the environment, in general, or sea/fresh water, in specific, and alert the user about possible dangers or water suitability. This section reviews current approaches tackling the problem of biomarkers based water quality monitoring. In [9], authors proposed an approach based on the use of a hybrid case-based reasoning system for monitoring water quality based on chemical parameters and biomarker "algae population". The proposed approach, which depends on Machine Learning (ML) techniques, can be employed to support environmental control, by monitoring water quality based on chemical parameters and algae population. The proposed system used the pattern generation algorithm extracts a case from the original case base, returns the N most similar cases from case retrieval. Finally, the generated datasets are used to train the committee of ML algorithms. The multilayer perceptron (MLP), support vector machines (SVM) and M5 algorithms are trained individually using the adaptation pattern dataset generated.

In [10], authors proposed CBR tool for the identification of pollutant transport models for the reaches of Somes River, flows through Romania and Hungary. The proposed approach requires building a case base, development of the CBR tool, compare the input case with the stored case to identify the most similar using K-Nearest Neighbor Algorithm, retrieve the related information for input case then added to the case base, and further used in studies related to the modeling of pollutant transport in other rivers.

In [11], authors proposed an approach based on CBR fuzzy inference systems (FIS) to assess water quality. The proposed approach aims to water quality monitoring

recruitment of extracting knowledge adaptation and automatic adjustment of status based on the induction of adaptive algorithms models ML. The dataset used in that paper includes physical and chemical data as well as the concentration of planktons in the water.
Moreover, in [12], authors proposed an hybrid CBR system for monitoring water quality based on the sensors values (chemical and physical parameters and algae population). The proposed CBR system system employed the MLP, SVM and M5 technique as an estimator for case adaptation, trained with adaptation instances generated using K most similar case. The obtained results in the experiments show that the use of hybrid technology to improve the ability to predict and CBR (SV M - *) models produced the best accuracy.

In this paper, we used histopathology of fish as biomarker for water quality in order to provide sort of early detection of biological changes due to exposure to chemical pollutants, which may result in long-term physiological disturbances. Also, this paper aims to use the existing knowledge stored in case base in order to assess the water pollutant via employing the case-based reasoning (CBR) approach.

The rest of this article is organized as follows. Section 2 describes the fundamentals of case-based reasoning approach. Section 3 describes the different phases of the proposed case-based reasoning system. Section 4 introduces the tested fish liver microscopic images dataset and discusses the obtained experimental results. Finally, Sect. 5 presents conclusions and discusses future work.

2 Preliminaries

2.1 Case-Based Reasoning System

Case-based reasoning (CBR) uses the similar case information available as historical precedence for proposing solutions to current problems. The most important aspects of the existing cases are first stored and indexed. New problem situations are then presented and similar, existing cases are identified from the knowledge base. Finally, the previous problem solutions are adapted and the revised solutions are proposed for the current situation. This happens according to the processes that occur when CBR is applied that is, retrieve, reuse, revise, and retain [13–16]. Figure 1 introduces the main tasks of these processes.

2.2 Distance-based similarity measure

In order to perform similarity measurement, we used distance-based similarity measure. In this paper we selected distances from different families that explained as follows:

- **Euclidean distance**: This type of distance metric is a member of *Lp* Minkowski distance metric family and is based on the value of *p*, which belongs to *Lp* distance metric family. *Eclid* stated that the distance between points *p* and *q* is the length of the line segment connecting them, as shown in Eq. (1) [17, 18].

$$d_{Euc}(x,y) = \sqrt{x+y} \tag{1}$$

- **City block distance**: The city block distance between two points, *x* and *y*, with *k* dimensions is calculated as shown in Eq. (2):

$$d_{City} = \sum_{j=1}^{j=n} |x_j - y_j| \tag{2}$$

- **Canberra distance**: It is a member of *L*1 Minkowski distance metric family where all distance metrics as shown in Eq. (3) [17]

$$d_{Can}(x,y) = \sum_{j=1}^{j=n} \frac{|S_{j1} - S_{j2}|}{(S_{j1} - S_{j2})} \tag{3}$$

- **Squared chord**: This distance cannot be used for feature space with negative values as shown in Eq. (4) [17]

$$d_{Sq}(x,y) = \sum_{j=1}^{j=n} (\sqrt{x_j} - \sqrt{y_j})^2 \tag{4}$$

- **Squared Chi-square**: It is calculated by Eq. (5)

$$d_{Chi}(x,y) = \frac{1}{Sum_j}\left(\frac{x_j}{size_{x_j}} - \frac{y_j}{size_{y_j}}\right) \tag{5}$$

3 The Proposed Case-Based Reasoning System

In this article, case-based reasoning system has been proposed for assessing water pollution based on fish liver microscopic images. As Tilipia fish is pollution resistant species, they are ideal to be used as biomarker for water pollution. The datasets used for experiments were constructed based on real sample images for fish liver, in different histopathlogical stages, exposed to copper and water pH. The collected datasets contain colored JPEG images as 200 images and 45 images that were used as training and testing datasets, respectively. Training dataset is representing the different histopathlogical change and water quality degrees. Features have to be extracted from

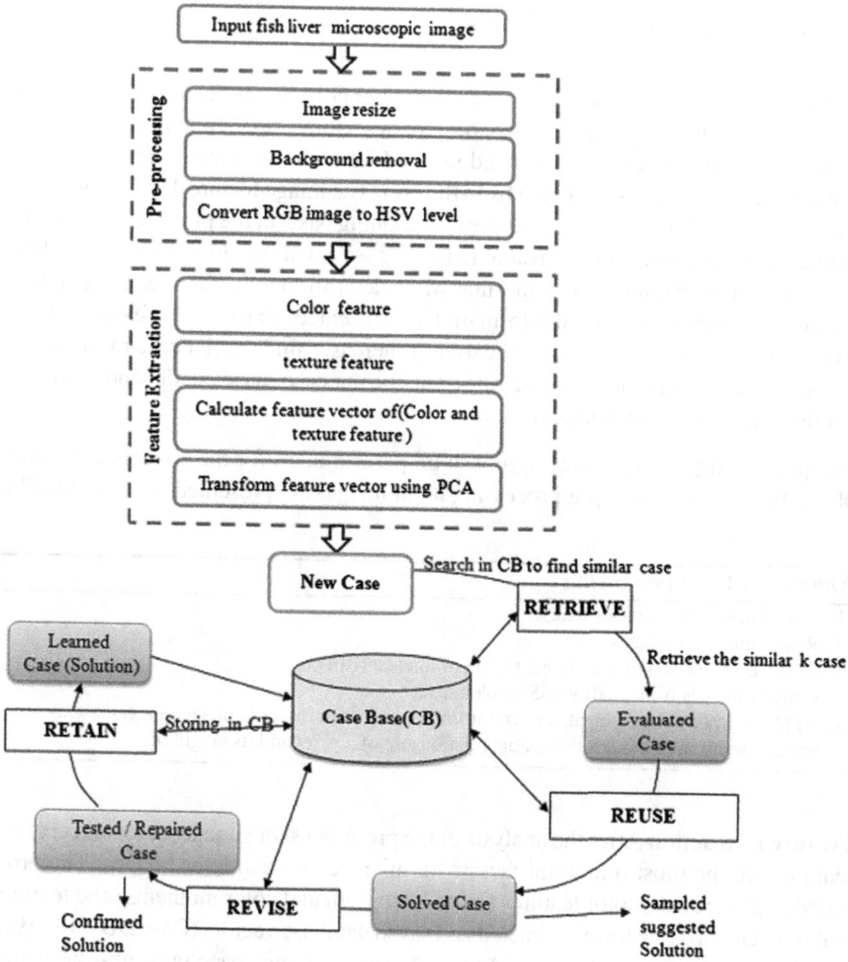

Fig. 1 Architecture of the proposed system

the dataset images by using digital image processing techniques used as case representation for proposed system. The proposed system utilizes texture and color feature extraction methods and case-based reasoning for assessing water pollution. The proposed case-based reasoning system consists of 5 phases; namely case representation (pre-processing phase, feature extraction phase), retrieve phase, reuse/adapt phase, revise phase and retain phase as depicted in Fig. 1.

3.1 Case Representation Phase

The first requirement of the CBR system for water pollution is deciding what to store in a case, finding an appropriate structure for describing case contents, and deciding how the case should be organized and indexed for effective retrieval and reuse. It is the start time to get a new problem from the user. The image feature database is part of the proposed system, it acts as a source of reasoning basis in the proposed system. By standardizing the information available for all cases, it allows for effective searching and retrieval algorithms. It is important when determining the features to include in a case, that all of the essential information be presented. Extracts the features of fish liver image which is uploaded by the user. It then uses find similar case in case base. It can be used as an index for case retrieval. During case retrieval, depending on the similar image index case base database.

Pre-processing The proposed approach prepares images for the features extraction phase, the major steps in pre-processing input images are presented in algorithm (1).

Algorithm 1 Pre-processing

1: Input microscopic images dataset
2: Resize the input images
3: Remove image background to get region of interest (RoI)
4: Convert images from RGB to HSV color space
5: Apply contrast enhancement, so that the contrast of a microscopic image in a given gray level texture descriptors models the spatial relationship of a pixel and its neighbors

Feature Extraction After the analysis of the pre-processing image, color and texture features are the most important factors on microscopic fish liver images. The proposed approach uses color features (color histogram and color moments) and texture features (Gabor Transform) extracted and saved database vectors (Case DB) [18, 19]. Principal component analysis (PCA) transforms the input space into sub-spaces for dimensionality reduction. Then, it transformed the input space into sub-spaces for dimensionality reduction and calculated 1D 8*2*2 HSV histogram, 8 level for hue and 2 level for each of saturation and value. In addition, nine color moments, three for each channel (H, S and V channels), are calculated. Texture representation is created using Gabor filters on each image(eg.Gabor filters are a group of wavelets. For a given image) with 4 scales and 6 orientations and an array of magnitudes is obtained. The mean (μ_{mn}) and standard deviation (ρ_{mn}) are calculated as the feature components. Scales and orientations are used in common implementation and the feature vector are calculated by Eq. (6) [20].

$$f = (\mu_{00}, \rho_{00}, \mu_{01}, \rho_{01}, \dots, \mu_{35}, \rho_{35}) \tag{6}$$

Then, a feature vector will be formed as a combination of HSV histogram, the nine color moments, and Gabor texture.

3.2 Retrieve Phase

Case structure and index are important factors that affect the efficiency of case retrieval. After features are known from the input then an initial inspect is done on the case base for similarity matching. When establishing the case base, firstly, we summarize and extract the character attributes of fish liver. Then, we classify case base as water pollution degrees, then index definite by water pollution degrees after extract attributes of fish liver image. Each retrieved problem has two parts: the problem, the problem solution. If one of the retrieved problems is similar to the new case, the obtained solution for that problem in the case base will be revealed; thus, the end of the process. Otherwise, the Reuse stage will begin to work.
Different matching algorithm can be used to estimate the similarity between the feature values of a current case and the previous case. In the proposed system, to perform similarity measurement; we used Distance based similarity measure. In this paper we are selecting distance measures (Euclidean distance, City Block Distance,Canberra distance, Squared Chord and Squared Chi-square) [17]. This method locates the case that is most similar to the input case. The interest rate of a case is estimated based on the presence or absence of certain features. These features identify which case is supposed to be retrieved.

3.3 Reuse/Adaptation Phase

Retrieved cases are reviewed and adjusted to fit the new case. The proposed system assesses the feature attributes and selects a most similar case from past cases in the case base database is obtained by a retrieval phase. Then checks the solution of the retrieved similar case and reuses the solution to inform the input case. After being shown the solutions, the system checks the case against the cases in the database to see if a similar case just there or whether to save up the new case.Distance based similarity metrics are trained using the adaptation dataset whereas the outputs are the corresponding to water quality degree equivalent to each image in the testing dataset then the solutions also are saved in the database. These solutions are called when an input a new similar problem; otherwise, referring to the required modifications in the component of the retrieved solution. Finally, these modifications apply to the current item in order to obtain the solution for the new problem (New Solution). Therefore; the solutions are stored in the database for references to future access.

3.4 Revise Phase

Whether the case generated in the previous stage includes a solution. If an acceptable solution is found in this initial generation, it will be reported and the Retain stage will begin to work. This phase is used only if case reuse is not happening. In this case, the system has to adapt to a given situation to the list. This adaptation is used to solve a new problem.

3.5 *Retrain Phase*

To improve the CBR method under an incremental learning process, the case base should gradually become more complete. If the introduced problem is a new one, after the solution is found. This solution is saved to the case base and updating the case base with the new case for new problem solving takes place.

4 Experimental Analysis and Discussion

Nile Tilipia "Oreochromis niloticus" is pollution resistant species ideal for usage as biomarker of water pollution. The datasets used for experiments were constructed based on real sample microscopic images for fish liver in different histopathlogical change stages exposed to copper and water pH. Fish images were collected from Abbassa farm, Abo-Hammad, Sharkia Governote, Egypt. Some samples of both training and testing datasets are shown in Fig. 2 [21].

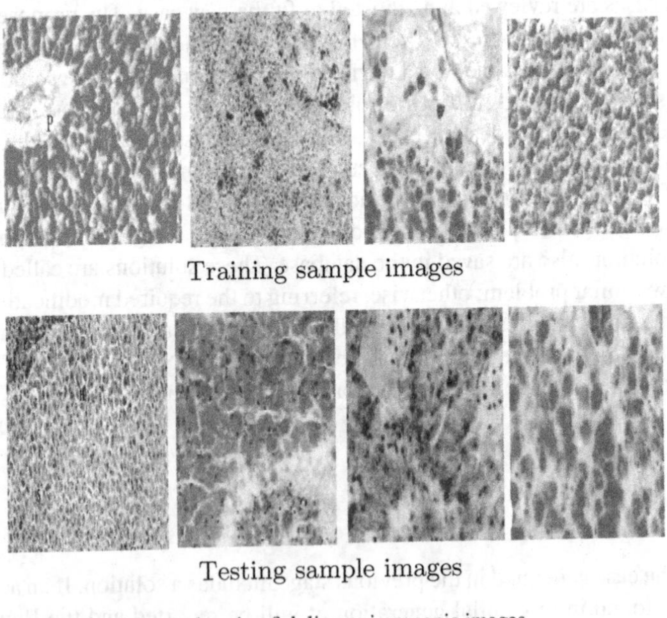

Fig. 2 Examples of training and testing fish liver microscopic images

For the experiments carried out, three different scenarios for the generation of retrieval solution and test 10-fold-cross-validation for the proposed system were investigated: **Scenario I:** the five most similar case is retrieved for a given input case, **Scenario II:** the ten most similar cases are retrieved for a given input case, and **Scenario III:** the fifteen most similar cases are retrieved for a given case. Figure 3 depicts experimental results that show CBR accuracy obtained via applying each distance measures function. The results of the proposed CBR system were evaluated against human expert assessment for measuring obtained accuracy. As shown in Fig. 3, with CBR systems, Squared Chord distance measures function being used in scenario III, the proposed system achieved 95.9 % accuracy for all output (water quality degrees). The results show that the performance of CBR systems increases according to the number of retrieved cases.
The accuracy is computed using the following equation:

$$Accuracy = \frac{Number\ of\ correctly\ retrieved\ images}{Total\ number\ of\ testing\ images} * 100 \qquad (7)$$

Many points of research assessing water pollution based on using fish liver microscopic images as biomarker but this research using experimental laboratory. However, none of them used not computer-based system on the experimented dataset(s). So, to the best of our knowledge, this article is the first research work aims at highlighting the most appropriate CBR system, for classifying water quality degree using fish liver microscopic images as biomarker for assessing water pollution.

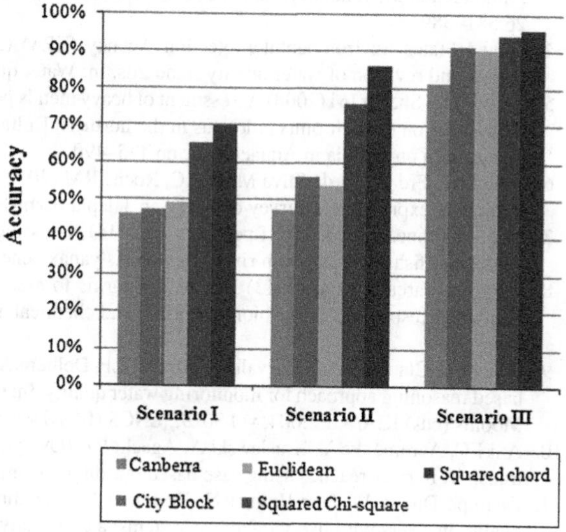

Fig. 3 Accuracy of CBR system using distance measure functions

5 Conclusions and Future Work

In this article, the proposed CBR system utilized texture and color feature extraction methods and case-based reasoning for assessing water pollution based on fish liver microscopic images. The proposed case-based reasoning system consists of 5 phases; namely case representation phase (pre-processing, feature extraction), retrieve phase, reuse/adapt phase, revise phase and retain phase. After pre-processing and feature extraction phase, similarity between the input and case base database has been calculated and retrieve similarity case otherwise requiring adapt the solution. Based on the obtained results the performance of CBR systems increases according to the number of retrieved cases in each scenario each strategy .The proposed system achieved 95.9 % accuracy for all water quality degrees. For future work, additional similarity functions and other classifiers are planned to be considered for experimental analysis.

References

1. Tantawi PI, OShaughnessy NJ, Gad KAR, Ragheb MAS (2009) Green consciousness of consumers in a developing country: a study of egyptian consumers. J Contemp Manage Res 5(1): 29–50
2. Diersing N (2009) Water quality: frequently asked questions. PDA. NOAA. http://floridakeys.noaa.gov/pdfs/wqfaq.pdf, pp 08-24
3. Johnson DL, Ambrose SH, Bassett TJ, Bowen ML, Crummey DE, Isaacson JS, Johnson DN, Lamb P, Saul M, Winter-Nelson AE (1997) Meanings of environmental terms. J Environ Qual 26:581–589
4. United States Environmental Protection Agency (EPA) (2006) Washington procedures for review and revision of water quality standards. In: Water quality handbook
5. Saeed, SM, Shaker IM (2008) Assessment of heavy metals pollution in water and sediments and their effect on Oreochromis Niloticus in the northern Delta lakes, Egypt. In: 8th international symposium on Tilapia in Aquaculture, pp 475–490
6. Viana AP, Frdou FL, da Silva Montes C, Rocha RM (2012) Heavy metal pollution and metallothionein expression: a survey on egyptian Tilapia Farmsl. Springer, vol 35, no 3, pp 395
7. Trujillo-Jimnez P, Sedeo JE, Lpez-Lpez E (2014) Assessing the health condition profile in the freshwater fish in Champoton river, Mexico Astyanax aeneus. J Environ Biol 1:137–145
8. Quesada-Garca A et al (2013) Use of fish farms to assess river contamination: combining biomarker responses, active biomonitoring, and chemical analysis. J Aquat Toxicol 140:439–448
9. Policastro Claudio A, de Carvalho André CPLF, Delbem Alexandre CB (2004) A hybrid case based reasoning approach for monitoring water quality. In: Orchard Bob, Yang Chunsheng, Ali Moonis (eds) IEA/AIE 2004, vol 3029., LNCS (LNAI)Springer, Heidelberg, pp 492–501
10. Ani EC, Avramenko Y, Kraslawski A, Agachi P (2009) Sl, "Selection of models for pollutant transport in river reaches using case based reasoning. Comput Aided Chem Eng 27:537–542
11. Ocampo-Duque W, Ferr-Huguet N, Domingo JL, Schuhmacher M (2006) Assessing water quality in rivers with fuzzy inference systems: a case study. Environ Int 32:733–740
12. Policastro CA, Carvalho ACP (2007) Applying case based based reasoning to sensor fusion. IEEE ISSNIP 419–424
13. Richter MM, Aamodt A (2005) Case-based reasoning foundations. Knowl Eng Rev 20(3): 203–207

14. Richter MM, Weber RO (2013) Basic CBR elements. In: Book case-based reasoning. Springer, Berlin, pp 17–40
15. Kevin D (2001) Ashley, "Case-based reasoning for interpretation of data from non-destructive testing". J Eng Appl Artif Intell 14(4):401–417
16. Lopez De Mantaras R, McSherry D, Bridge D, Leake D, Smyth B, Craw S, Faltings B, Maher ML, Cox MT, Forbus K (2005) Retrieval, reuse, revision and retention in case-based reasoning. J Knowl Eng Rev 20(3):215–240
17. Choi S, Cha S, Tappert C (2010) A survey of binary similarity and distance measures. J Syst Cybern 8(1):43–48
18. Ashok Kumar D, Esther J (2011) Comparative study on CBIR based by color histogram gabor and wavelet transform. Int J Comput Appl 17(2):37–44
19. Shahbahrami A, Borodin D, Juurlink B (2008) Comparison between color and texture features for image retrieval. In: Proceedings 19th annual workshop on circuits, systems and signal processing, Veldhoven, The Netherlands
20. Kaur Amandeep, Gupta Savita (2012) Texture classification based on gabor wavelets. Int J Res Comput Sci 2(4):39–44
21. Mohamed A-K, El-sayed N, El-Shershaby A-F, Hassan Zaghloul KH (2006) Taxicological and histopathological studies on Nile tilapia Oreochromis niloticus exposed to copper individually and in mixture with zinc at different pH value, Ph.D. thesis Cairo univeristy

Erratum to: Personalized Source Selection Process: a Social Profile Based Adaptation Technique

Zakaria Saoud and Samir Kechid

Erratum to: Chapter 'PERSONALIZED Source Selection Process: A Social Profile Adaptation Technique' in: A. Abraham et al. (eds.), *Intelligent Data Analysis and Applications*, Advances in Intelligent Systems and Computing 370, DOI 10.1007/978-3-319-21206-7_18

The original version of this chapter was inadvertently published with an incorrect title. The correct title is shown below:

Personalized Source Selection Process: a Social Profile Based Adaptation Technique.

The online version of the original chapter can be found under
DOI 10.1007/978-3-319-21206-7_18

Z. Saoud (✉) · S. Kechid
Computer Sciences Department University of Sciences and Technologies Houari Boumediene, BP 32 EL ALIA Bab Ezzouar, 16111 Algiers, Algeria
e-mail: zakaria.saoud@live.fr

S. Kechid
e-mail: skechid@usthb.dz

A. Abraham et al. (eds.), *Intelligent Data Analysis and Applications*,
Advances in Intelligent Systems and Computing 370,
DOI 10.1007/978-3-319-21206-7_48

Author Index

A. Abraham et al. (eds.), *Intelligent Data Analysis and Applications*,
Advances in Intelligent Systems and Computing 370,
DOI 10.1007/978-3-319-21206-7

Printed in the United States
By Bookmasters